THE DYNAMICAL BEHAVIOUR OF OUR PLANETARY SYSTEM

THE DYNAMICAL BEHAVIOUR OF OUR PLANETARY SYSTEM

Proceedings of the Fourth Alexander von Humboldt Colloquium on Celestial Mechanics

Edited by

R. DVORAK

Institut für Astronomie, Universitätssternwarte, Vienna, Austria

and

J. HENRARD

Département de Mathématique, FNDP, Namur, Belgium

Partly reprinted from Celestial Mechanics and Dynamical Astronomy
Volume 65, Nos. 1–2, 1996/7

SPRINGER SCIENCE+BUSINESS MEDIA, B.V.

A C.I.P. Catalogue record for this book is available from the Library of Congress.

ISBN 978-94-010-6320-3 ISBN 978-94-011-5510-6 (eBook)
DOI 10.1007/978-94-011-5510-6

Printed on acid-free paper

TABLE OF CONTENTS

Proceedings of the
Fourth Alexander von Humboldt Colloquium on Celestial Mechanics

THE DYNAMICAL BEHAVIOUR OF OUR PLANETARY SYSTEM

Theoretical Dynamics

Numerical Methods of Analysis

Planets and Satellites

Asteroids

Miscellaneous

Participants in the Fourth Alexander von Humboldt Colloquium on Celestial Mechanics

PREFACE

It is now a well established tradition that every four years, at the end of winter, a group of "celestial mechanicians" from all over the world gather at the "Alpengasthof Peter Rosegger" in the Styrian Alps (Ramsau, Austria).

This time the colloquium was held from March 17 to March 23, 1996 and was devoted to the Dynamical Behaviour of our Planetary System. The papers covered a large range of questions of current interest: theoretical questions (resonances, universal properties, non integrability, transport, . . .) and questions about numerical tools (symplectic maps, indicators of chaos, . . .) were particularly well represented; the never ending problem of the sculpting of the asteroid belt was also quite popular.

You will find in the following pages a *pot-pourri* of what we listen to; you will miss of course the diversity of accents with which the tunes were delivered: from China, from Japan, from Brazil, from the United-States of America and from all over Europe, East and West.

Let us not forget that the comet *1996 B2 (Hyakutake)* came to visit us; many an evening was spent on the deck of the Alpengasthof contemplating this celestial visitor who liked to play hide-and-seek behind the spruce trees.

To be able to organize this *Fourth Alexander von Humbolt Colloquium*, we have to thank primarily the generosity of the Austrian Ministry of Science. Also the Steiermärkischen Wissenschafts- und Landesfonds, the Österreichische Forschungsgemeinschaft and especially the **Bank - Austria** made it possible to invite participants from abroad. Latter supported our meeting also in form of "Tagungsunterlagen"; coffee during the whole meeting time was donated by Hornig-Kaffee Graz.

Many thanks are due to the Chairwoman A. Lemaître and Chairmen: G. Contopoulos, R. Dvorak, H. Eichhorn, C. Froeschlé, S. Ferraz-Mello, P.K. Seidelmann, V. Szebehely, Yi-Sui Sun . They acted as Editors of the papers presented during "their" session, arranging for competent and fast refereeing so that the papers could be reviewed and, when necessary, corrected.

Our model hosts Barbara and Fritz Walcher contributed a lot to the success of the meeting, as they did their best providing pleasant and friendly surroundings. We all are looking forward to the next millennium when we plan to meet again at the Alpengasthof Rosegger.

R. Dvorak J. Henrard

SCATTERING BY RESONANCES

ANATOLY NEISHTADT
Space Research Institute,
Profsoyuznaya 84/32, Moscow 117810, Russia

Abstract. Small perturbations imposed on an integrable system produce an evolution of the motion. In the course of this evolution the system can pass through a state of resonance. Quasi-random phenomena of capture into the resonance and scattering on the resonance are discussed in this paper for the case of a two-frequency system. Major topics are calculation of probability of capture, description of the motion of the captured phase points, calculation of a probabilistic distribution of the scattering amplitude.

Key words: Resonance – adiabatic invariant – capture into resonance – scattering

1. Introduction

Consider a system which differs from an integrable one by a small perturbation. The motion in such a system can be approximately represented as a slowly evolving unperturbed one. For the description of this evolution the classical averaging method prescribes to average the differential equations of the perturbed motion over the phases of the unperturbed oscillations. In the course of the evolution the frequencies of motion are changing slowly and at some time they become resonant, i.e. linearly rationally dependent. Due to the influence of resonances the actual motion can be considerably different from the one predicted by the averaging method.

The two basic phenomena that are associated with the effect of a single resonance are capture into the resonance and scattering by the resonance.

Capture into a resonance can be described as follows. First the system evolves as it is predicted by the averaging method. At a certain time the system approaches a resonance. After that the system evolves in such a way that the resonant relation between frequencies is approximately kept. As a result, after a time $\sim 1/\varepsilon$ the state of the system is completely different from the one predicted by the averaging method. Here ε is a small parameter characterizing the magnitude of the perturbation. The set of initial conditions for trajectories with capture has a small measure $O(\sqrt{\varepsilon})$ (we consider capture at time $t \in (0, 1/\varepsilon)$). Initial conditions for trajectories with capture and trajectories without capture are mixed if ε is small. Thus, it is reasonable to consider the capture as a random event and to calculate a probability of this event. Capture into a resonance was first discussed in [(10; 17)] in connection with problems of celestial mechanics.

Scattering by a resonance can be described as follows. Far from the resonance the system oscillates with an amplitude $\sim \varepsilon$ near the state predicted by the averaging method. During a passage through a narrow neighborhood of the resonant state, the parameters undergo a jump $\sim \sqrt{\varepsilon}$. The amplitude of this jump is very sensitive to the variation of initial conditions if ε is small. Thus, it is reasonable to consider this jump as a random scattering by the resonance. Scattering by a resonance was

Celestial Mechanics and Dynamical Astronomy **65**: 1–20, 1997.

first discussed in [(8)] for a Hamiltonian system depending on a slowly varying parameter and in [(4)] in a general context.

The word "scattering" in the title of the paper is used in a broad sense which includes the changes in the system state due both to capture into resonance and scattering by a resonance in the restricted sense that was described above.

Major topics of this review paper are:

-calculation of probability of a capture into resonance,

-description of motion after capture into resonance,

- calculation of the statistical distribution of the amplitude of scattering by a resonance.

For Hamiltonian systems with slowly varying parameters these topics are considered in [(25; 26)].

2. Perturbed Systems with Rotating Phases

The differential equations of motion of a system which differs by a small perturbation from an integrable one can often be written in the following standard form (see, e.g., [(1; 2)]):

$$\dot{I} = \varepsilon f(I,\varphi,\varepsilon), \quad \dot{\varphi} = \omega(I) + \varepsilon g(I,\varphi,\varepsilon). \tag{1}$$

Here ε is a small parameter characterizing the amplitude of the perturbation. If $\varepsilon = 0$, the system is unperturbed. Variables $I \in D \subset R^n$ are called slow variables. Variables φ modd $2\pi \in T^m = R^m/(2\pi Z)^m$ are called fast variables or phases, ω is a vector of (unperturbed) frequencies, the functions f, g are 2π-periodic in the components of φ. System (1) is called a system with rotating phases.

The averaging method gives a recipe for the approximate description of the slow variables I evolution in a time interval of order $1/\varepsilon$. According to this method, one should replace the system (1) with the averaged system

$$\dot{J} = \varepsilon F(J), \; F(J) = \frac{1}{(2\pi)^m} \oint_{T^m} f(J,\varphi,0)\, d\varphi. \tag{2}$$

We denote by $J(t)$ the solution of the averaged system (2) with initial condition $I_0 \in D_0 \subset D$ at $t = 0$ and by $(I(t), \varphi(t))$ the solution of system with rotating phases (1) with initial condition $(I_0, \varphi_0) \in D_0 \times T^m$. So, $I(0) = J(0) = I_0$. We suppose that the solution $J(t)$ is defined and kept at a positive distance from the boundary of the domain D for $0 \le t \le 1/\varepsilon$.

If the system possesses only one frequency $(m = 1)$ and this frequency ω does not vanish, then the following estimate is valid: $|I(t) - J(t)| = O(\varepsilon)$, $0 \le t \le 1/\varepsilon$ (see, e.g. [(1; 7)]).

In general multi-frequency case ($m \ge 2$, $\frac{\partial \omega}{\partial I} \ne 0$) a point I is called a resonant one, if $(k, \omega(I)) = 0$ for some integer vector $k \in Z^m$; here $(\cdot,\cdot)$ is the standard scalar product in R^m. The set of resonant points for a given k is a surface in the space of the slow variables, which is called a resonant surface. When a phase point of (1) approaches a resonant surface, it can be captured into the resonance and undergo a scattering as it is described in the introduction.

Henceforth we will consider two-frequency systems: $m = 2, \varphi = (\varphi_1, \varphi_2), \omega = (\omega_1, \omega_2)$. We will assume that $\omega, f|_{\varepsilon=0}, g|_{\varepsilon=0} \in C^4$, and $f, g \in C^2$.

In the two-frequency case, resonances are represented, in the frequency plane, by straight lines with rational slopes passing through the origin of the coordinates. Such a simple structure of the resonant set simplifies the problem essentially. This first has been used in [(9)] for a Hamiltonian system depending on a slowly varying parameter and in [(4)] for general system in the form (1). Under some additional assumptions (see (3), (5)below) it is possible to investigate the effect of each resonance separately and combine the results afterwards.

A two-frequency system is considered in [(4; 2)] under the assumption that the rate at which the frequency ratio changes along the trajectories of the system (1) is different from zero:

$$|(\omega_2 \frac{\partial \omega_1}{\partial I} - \omega_1 \frac{\partial \omega_2}{\partial I}) f(I, \varphi, 0)| > c_1^{-1}, \tag{3}$$

$c_1 = \text{const} > 0$. This condition prevents capture into resonances. It is shown that

$$|I(t) - J(t)| = O(\sqrt{\varepsilon}),\ 0 \leq t \leq 1/\varepsilon. \tag{4}$$

Henceforth system (1) is considered under the assumption that the rate at which the frequency ratio changes along the trajectories of the averaged system (2) is different from zero [(4)]:

$$\begin{aligned} |\mathcal{L}| &> c_1^{-1},\ c_1 = \text{const} > 0, \\ \mathcal{L} &= (\omega_2 \frac{\partial \omega_1}{\partial I} - \omega_1 \frac{\partial \omega_2}{\partial I}) F(I). \end{aligned} \tag{5}$$

Under this assumption and some additional "general position" assumptions it is shown [(19)] that

$$|I(t) - J(t)| = O(\sqrt{\varepsilon} \log \varepsilon),\ 0 \leq t \leq 1/\varepsilon, \tag{6}$$

for all initial conditions except of a set of a measure $O(\sqrt{\varepsilon})$ (see, also, [(15)]).

Henceforth we will use assumption (5) also with some additional "general position" assumptions that are formulated in Subsection 3.3.

3. Reduction Near a Resonant Surface

Near the resonant surface of a given resonance system, (1) can be transformed into a standard "perturbed pendulum-like system" form. This transformation has been used in a number of works (see [(5; 8; 13; 16; 17; 18; 19)]).

Let $k = (k_1, k_2)$ be a vector with integer co-prime components. The value $|k| = |k_1| + |k_2|$ is called the order of the k- resonance. We assume that $|k| \sim 1$.

By definition, the resonant surface of the k-resonance is the surface $k_1\omega_1(I) + k_2\omega_2(I) = 0$ in the I-space. A reduction to a pendulum-like form is carried out in a $O(\sqrt{\varepsilon})$-neighborhood of the resonant surface. This neighborhood is called the resonant zone.

3.1. Partial Averaging

There exist co-prime integers l_1, l_2 such that $k_1l_2 - k_2l_1 = 1$. Let us make the transformation of variables $(\varphi_1, \varphi_2) \mapsto (\gamma, \psi)$ using formulas

$$\gamma = k_1\varphi_1 + k_2\varphi_2,\ \psi = l_1\varphi_1 + l_2\varphi_2\ . \tag{7}$$

The new phase $\gamma = k_1\varphi_1 + k_2\varphi_2$ is called the resonant phase. In a resonant zone γ changes slowly and ψ changes rapidly: $\dot\gamma \sim \sqrt{\varepsilon}$, $\dot\psi \sim 1$. The first step of our reduction is the averaging of the system over ψ.

PROPOSITION 3.1. *In an $O(\sqrt{\varepsilon})$-neighborhood of the resonant surface there exists a 2π-periodic in (γ, ψ), $O(\varepsilon)$-close to identity transformation of variables*

$$(I, \gamma, \psi) \mapsto (\bar I, \bar\gamma, \bar\psi)$$

such that in the new variables the system has the form (bars over the new variables are omitted)

$$\begin{aligned} \dot I &= \varepsilon \bar f(I,\gamma) + O(\varepsilon^2), \\ \dot\gamma &= (k, \omega(I)) + \varepsilon(k, \bar g(I,\gamma)) + O(\varepsilon^2),\ \ \dot\psi = (l, \omega(I)) + O(\varepsilon). \end{aligned} \tag{8}$$

Here $\bar f, \bar g$ are the average of $f|_{\varepsilon=0}$, $g|_{\varepsilon=0}$ over ψ. Symbols $O(\cdot)$ admit one differentiation without changes in the smallness order.

The transformation of variables in Proposition 3.1 is a standard transformation of variables for the partial averaging of a system (see, for example, [(5)]).

3.2. An Expansion in a Neighborhood of a Resonant Surface

The resonant surface is defined by the equation $(k, \omega(I)) = 0$. Condition (5) implies that $\partial(k,\omega)/\partial I \neq 0$ on this surface.

Let $\sigma \in R^{n-1}$ be a coordinate on the resonant surface. A point I in an $O(\sqrt{\varepsilon})$-neighborhood of the resonant surface is specified by coordinates $\rho = (k, \omega(I))/\sqrt{\varepsilon} = O(1)$ and σ; here σ is defined locally via projection of I onto the resonant surface parallel to a given direction transversal to the resonant surface.

We introduce also the slow time $\theta = \sqrt{\varepsilon}t$ and use a prime to denote differentiation with respect to θ. We then get the system

$$\begin{aligned} \gamma' &= \rho + \sqrt{\varepsilon}\alpha_1(\gamma,\sigma) + O(\varepsilon), \\ \rho' &= Q(\gamma,\sigma) + \sqrt{\varepsilon}\rho\alpha_2(\gamma,\sigma) + O(\varepsilon), \\ \sigma' &= \sqrt{\varepsilon}\alpha_3(\gamma,\sigma) + O(\varepsilon). \end{aligned} \tag{9}$$

Here Q, α_i are some functions 2π-periodic in γ. Function $Q(\gamma, \sigma)$ can be represented in the form

$$\begin{aligned} Q(\gamma,\sigma) &= L(\sigma) - \partial V(\gamma,\sigma)/\partial\gamma, \\ L(\sigma) &= (k_1/\omega_2)\mathcal{L}(I) \neq 0. \end{aligned} \tag{10}$$

In the last expression I is a point on the resonant surface and L is the mean value of Q with respect to γ, function $\mathcal{L}$ is defined in (5).

For an approximate description of the motion near the resonant surface during the time interval $0 \le t \le 1/\varepsilon$ (or, $0 \le \theta \le 1/\sqrt{\varepsilon}$) we shall omit in (9) terms $O(\varepsilon)$. We get

$$\begin{aligned} \gamma' &= \rho + \sqrt{\varepsilon}\alpha_1(\gamma,\sigma), \\ \rho' &= L(\sigma) - \partial V(\gamma,\sigma)/\partial\gamma + \sqrt{\varepsilon}\rho\alpha_2(\gamma,\sigma), \\ \sigma' &= \sqrt{\varepsilon}\alpha_3(\gamma,\sigma). \end{aligned} \tag{11}$$

3.3. Pendulum-like Systems

System (11) still contains a small parameter and can be treated by perturbation theory. Putting in (11) $\varepsilon = 0$ we get for ρ, γ a Hamiltonian system with one degree of freedom; this system contains σ as a parameter. Hamiltonian of this system is

$$E(\rho,\gamma,\sigma) = \frac{1}{2}\rho^2 + V(\gamma,\sigma) - L(\sigma)\gamma + b(\sigma), \ \sigma = \text{const} \tag{12}$$

Here b is an arbitrary function of σ. The equation of motion can be written in the form

$$\gamma'' = -\partial V(\gamma,\sigma)/\partial\gamma + L(\sigma). \tag{13}$$

This system is called *a pendulum-like system* as it resembles a pendulum under the action of a constant torque (for a pendulum $V \sim -\cos\gamma$).

There are two basic types of phase portraits of the equation (13): with an oscillation region and without such a region.

If

$$\min_\gamma \frac{\partial V}{\partial \gamma} < L < \max_\gamma \frac{\partial V}{\partial \gamma}, \tag{14}$$

then there are stable and unstable stationary points on the phase portrait. The separatrices of the unstable stationary points enclose the oscillation regions. For simplicity we assume that there are no unstable stationary points inside the oscillation regions. The phase portrait is shown in Fig. 1,a (for $L < 0$ and under assumption that there is just one oscillation region on the phase cylinder $(\rho, \gamma \bmod 2\pi)$).

If the opposite to (14) inequality is satisfied, then there are no oscillation region. The phase portrait is shown in Fig. 1,b.

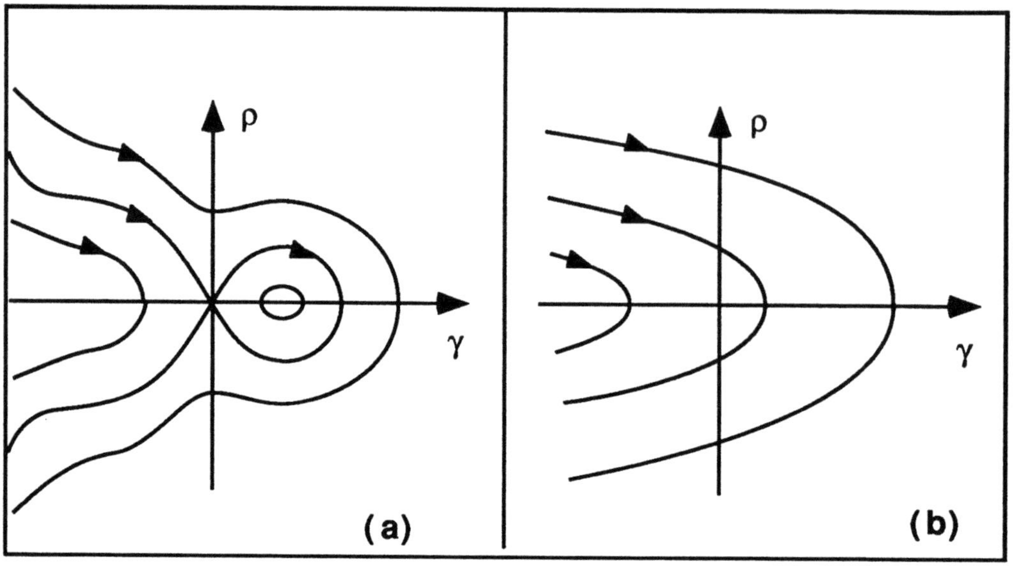

Fig. 1. Phase portraits of the pendulum-like system.

PROPOSITION 3.2. *The system (13) has an oscillation region only for a finite (i.e. not depending on ε) number of resonant vectors $k = (k_1, k_2)$.*

Indeed,the condition $f|_{\varepsilon=0} \in C^4$ implies estimate $\partial V/\partial\gamma = (k_1/\omega_2)O((|k_1| + |k_2|)^{-4})$. The assumption (5) and formulas (10) imply that L is bounded from below by the value $c_1^{-1}|k_1/\omega_2|$. Hence, condition (14) can be satisfied for a finite number of vectors k only.

Let l be a separatrix passing through a saddle point C in Fig. 1,a and let in expression (12) $b(\sigma)$ be chosen in such a way that $E = 0$ at the point C and therefore on l. We introduce

$$\Theta(\sigma) = -\oint_l (\frac{\partial E}{\partial\gamma}\alpha_1 + \frac{\partial E}{\partial\rho}\rho\alpha_2 + \frac{\partial E}{\partial\sigma}\alpha_3)d\theta \tag{15}$$

Integral in (15) is improper (motion along a separatrix takes an infinite time), but it converges.

Remark The meaning of the integral Θ (15) is as follows: the value $-\sqrt{\varepsilon}\Theta$ is the leading term in the energy E change for the system (11) along a segment of a trajectory close to the separatrix in Fig. 1, a.

Henceforth we assume that equations (11), (13) meet the following "general position" conditions.

1. The unstable stationary points are nondegenerate.
2. The separatrices do not connect different saddles.

3. For every separatrix the value of the corresponding integral (15) is different from zero: $\Theta(\sigma) \neq 0$.

In Fig. 1 the resonant surface is represented by the coordinate line $\rho = 0$. A resonant zone is a strip $|\rho| <$ const. A motion in the oscillation regions in Fig 1,a corresponds to a motion of phase points that are captured into the resonance. A motion in the supplementary region in Fig. 1,a (the rotation region) corresponds to a motion of phase points that cross a resonant zone without capture.

Now take into account terms $O(\sqrt{\varepsilon})$ in equations (11). Under the action of these terms the phase point may cross a separatrix in Fig. 1,a and change the regime of motion from a rotation to an oscillation or from an oscillation to a rotation. This means a capture into the resonance or an escape from the resonance respectively.

4. The Effect of a Single Resonance

4.1. A Model Problem with a Single Resonance

In this section the effect of a single resonance is described for a model problem in which only a single resonance exists. This means that the functions f, g in (1) depend on the phases only through $\gamma = k_1\varphi_1 + k_2\varphi_2$. The results of this Section are extended in part to the general problem in Section 5.

Denote $\hat{M}_0 = (\hat{I}_0, \hat{\varphi}_0)$. Let U^κ be the ball of radius κ with center in $\hat{M}_0$. Henceforth we consider trajectories with initial conditions at $t = 0$ in U^κ. We introduce also the slow time $\tau = \varepsilon t$.

Consider a trajectory of the averaged system with an initial condition $\hat{I}_0$ at $\tau = 0$. Let this trajectory cross the resonant surface only once at the moment $\tau = \hat{\tau}_* < 1$ at the point $\hat{\sigma}_*$. Let us suppose that there are oscillation regions on the phase portrait of the pendulum-like system (13) for the point $\sigma = \hat{\sigma}_*$. Henceforth it is assumed for simplicity that there is just one oscillation region on the phase cylinder $(\rho, \gamma \bmod 2\pi)$. If there are several such regions, then the results of this Section are valid separately for each of them.

4.2. Probability of Capture into Resonance

Initial conditions for trajectories with capture into resonance and without capture are mixed if ε is small. Variation $\sim \varepsilon$ in initial value of I can change the motion considerably. It is reasonable to consider a capture as a random event and to calculate the probability of this event. This approach was suggested in [(14)] for problems with separatrix crossings. We use a definition of a probability analogous to that in [(3)] (see also [(10; 20; 23)]). By $U_c^{\kappa,\varepsilon}$ denote a subset of U^κ that is formed by the initial conditions of trajectories with capture into the resonance.

There is some ambiguity in the definition of $U_c^{\kappa,\varepsilon}$, because we still do not have a precise definition of capture. Further results are valid under any natural definition. To be definite, adopt the following agreement: if a phase point spends time $\geq 1/\varepsilon^{3/4}$

in the oscillation region of a resonance, then this point is captured into this region (it is important that $1/\sqrt{\varepsilon} \ll 1/\varepsilon^{3/4} \ll 1/\varepsilon$).

DEFINITION 4.1. *The value*

$$Pr(\hat{M}_0) = \sqrt{\varepsilon} \lim_{\kappa \to 0} \lim_{\varepsilon \to 0} \frac{mesU_c^{\kappa,\varepsilon}/\sqrt{\varepsilon}}{mesU^\kappa} \tag{16}$$

is called the probability density at the point $\hat{M}_0$ for capture into the oscillation region of the resonance (or, briefly, the probability of the capture into resonance for the point $\hat{M}_0$).

Here mes is the standard phase volume in the space (I, φ).

Thus, the probability of capture for the point $\hat{M}_0$ is an asymptotic expression for the part of the phase volume that is occupied in a small neighborhood of $\hat{M}_0$ by the initial points for trajectories with capture.

PROPOSITION 4.1. *Limit (16) exists and can be calculated by the following formula:*

$$Pr(\hat{M}_0) = \begin{cases} \sqrt{\varepsilon}\Theta(\hat{\sigma}_*)/(2\pi|L(\hat{\sigma}_*)|) & \text{if } \Theta(\hat{\sigma}_*) > 0 \\ 0 & \text{if } \Theta(\hat{\sigma}_*) < 0 \end{cases} \tag{17}$$

The meaning of formula (17) can be explained as follows (cf. [(10; 14; 23)]). Points from U^κ approach the resonant surface at $\tau \approx \hat{\tau}_*$. The volume distortion during the time interval $0 \le \tau \le \hat{\tau}_*$ is approximately the same for all subsets of U^κ because the motion is approximately described by the averaged system. A motion near the resonant surface can be analyzed in terms of the approximate system (11). Consider a phase point which crosses the ray $C\rho$ in Fig. 1,a at some (slow) instant $\tau_* \in [\hat{\tau}, \hat{\tau} + \Delta\tau]$ and has at this moment the value of the Hamiltonian $E = h_*$. (We recall that at the saddle point C $E = 0$.)

If $c_3\varepsilon^{3/4} < h_* < \sqrt{\varepsilon}[\Theta(\hat{\sigma}_*) - c_2\Delta\tau - c_3\varepsilon^{1/4}]$, then the phase point will cross the separatrix in Fig.1a and will be captured into the resonance. Here $c_{2,3}$ are positive constants. (We use here the property that the value of E changes approximately by $-\sqrt{\varepsilon}\Theta(\hat{\sigma}_*)$ along a segment of the perturbed trajectory near the separatrix). If $\sqrt{\varepsilon}[\Theta(\hat{\sigma}_*) + c_2\Delta\tau + c_3\varepsilon^{1/4}] < h_* < 2\pi|L(\hat{\sigma}_*)| - c_2\Delta\tau - c_3\sqrt{\varepsilon}$, then the phase point will cross the resonant surface without capture. If $h_* > 2\pi|L(\hat{\sigma}_*)| + c_2\Delta\tau + c_3\sqrt{\varepsilon}$, then the phase point will arrive at the ray $C\rho$ again (on the phase cylinder $(\rho, \gamma \bmod 2\pi)$). In a Hamiltonian system the phase flux through any curve segment is equal to the difference between values of the Hamiltonian at the endpoints of this segment. Now consider on the cylinder $(\rho, \gamma \bmod 2\pi)$ the phase volume captured into the resonance during the time interval $[\hat{\tau}, \hat{\tau} + \Delta\tau]$, $\Delta\tau > \sqrt{\varepsilon}|\log\varepsilon|$. We easily see that this phase volume is equal to

$$\sqrt{\varepsilon}\Theta(\hat{\sigma}_*)\Delta\tau + O(\sqrt{\varepsilon}\Delta\tau^2) + O(\varepsilon^{3/4}\Delta\tau). \tag{18}$$

Analogously, the phase volume that passes through the resonance without being captured is equal to

$$2\pi|L(\hat{\sigma}_*)|\Delta\tau + O(\Delta\tau^2) + O(\sqrt{\varepsilon}\Delta\tau). \tag{19}$$

By Proposition 4.1, the probability of capture is asymptotically equivalent to the ratio of the values (18) and (19) (if $\Theta(\hat{\sigma}_*) > 0$).

4.3. A Description of a Motion with a Capture into a Resonance. An In-out Function

The motion of phase points captured into resonance is approximately described by system (11). Let us make in this system the transformation of variables $(\gamma, \rho, \sigma) \mapsto (\chi, h, \sigma)$, where h is the value of the Hamiltonian E and χ is the "angle"-variable (from pair "action-angle") for pendulum-like system (13) in the oscillation region. In the new variables we get the system with one rotating phase χ and a small parameter $\sqrt{\varepsilon}$. In accordance with the averaging method for the approximate description of the motion one should average this system over χ. We obtain the so called *inner averaged system* of the form

$$h' = \sqrt{\varepsilon}\beta_1(h, \sigma), \ \sigma' = \sqrt{\varepsilon}\beta_2(h, \sigma). \tag{20}$$

Let at some (slow) instant $\tau = \tau_l$ a phase point be inside the oscillation region and have $h = h_l < 0$, $\sigma = \sigma_l$. Consider the solution $(\bar{h}(\tau), \bar{\sigma}(\tau))$ of the inner averaged system (20) with the initial condition (h_l, σ_l) at the moment $\tau = \tau_l$. In the averaging method approximation the time for capture into the oscillation region $\tau_* < \tau_l$ and the time for escape from the oscillation region $\tau_{**} > \tau_l$ are determined as follows: if $\tau = \tau_*$ or $\tau = \tau_{**}$, then $\bar{h}(\tau) = 0$; if $\tau_* < \tau < \tau_{**}$, then $\bar{h}(\tau) < 0$ (Fig. 2). The values $\sigma_* = \bar{\sigma}(\tau_*)$ and $\sigma_{**} = \bar{\sigma}(\tau_{**})$ determine points of capture and escape by or from the resonant surface. The map $\sigma_* \mapsto \sigma_{**}$ is called the in-out function (see example of calculation of the in-out function and capture probability in [(27)]).

Consider a phase point with an initial condition $(I_0, \varphi_0) \in U^\kappa$. Assume that its trajectory has a segment with capture into resonance. The motion of this phase point can be described approximately by gluing together three motions, (I), (II), (III), Fig.3.

The motion (I) is the motion during the (slow) time interval $0 \le \tau \le \tau_*$ along the trajectory of the averaged system with initial condition I_0 at $\tau = 0$. Here τ_* is the time when the trajectory of the motion (I) hits the resonant surface. Let σ_* be the endpoint of the motion (I). We assume that $\tau_* < 1$. Motion (II) is the motion during the time interval $\tau_* \le \tau \le \tau_{**}$ along the trajectory of the inner averaged system (20) with initial condition $\sigma = \sigma_*$, $h = 0$ at $\tau = \tau_*$. Here τ_{**} is the time of escape from the resonance for the inner averaged system provided that the capture occurs at the moment τ_* at the point σ_*. We assume that $\tau_{**} < 1$. Let σ_{**} be the endpoint of the motion (II). Motion (III) is the motion during the time interval

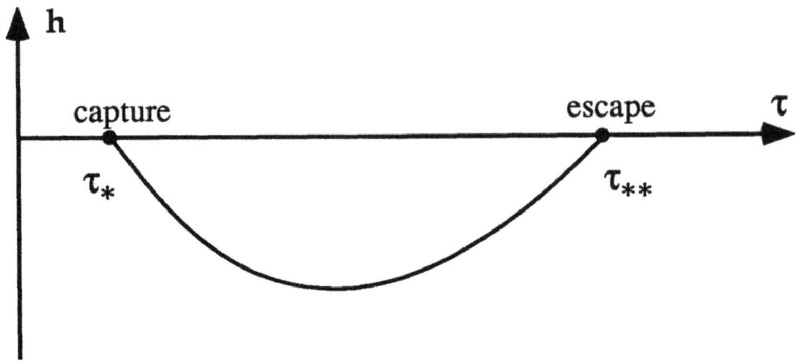

Fig. 2. Evolution of the "pendulum" energy in the inner averaged system.

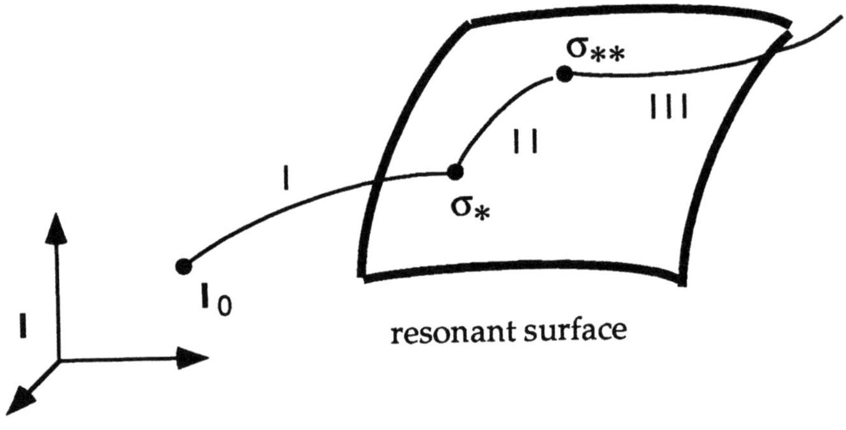

Fig. 3. The glued averaging method approximation.

$\tau_{**} \le \tau \le 1$ along the trajectory of the averaged system with initial condition σ_{**} at $\tau = \tau_{**}$. We call a description of the motion by gluing (I), (II), (III) a glued averaging method approximation. The accuracy of this approximation is described by the following Theorem.

Let p be any prescribed positive number.

THEOREM 4.1. *The set U^κ may be represented in the form $U^\kappa = U_a \cup U_g \cup v$ as follows.*

1. *For initial conditions in U_a the averaging method describes the behavior of I with an accuracy $O(\sqrt{\varepsilon}\log\varepsilon)$ for $0 \le \varepsilon t \le 1$.*
2. *For initial conditions in U_g the glued averaging method approximation describes the behavior of I with an accuracy $O(\sqrt{\varepsilon}\log\varepsilon)$ for $0 \le \varepsilon t \le 1$. For*

$\varepsilon t \in [\tau_*', \tau_{**}'] \subset [0,1]$, *where* $\tau'_{*,**} = \tau_{*,**} + O(\sqrt{\varepsilon}\log\varepsilon)$, *the phase point belongs to the oscillation region of the resonance and the behavior of* E, σ *is described by the solution of the inner averaged system with an accuracy* $O(\sqrt{\varepsilon}\log\varepsilon)$.

3. *mes* $U_g = O(\sqrt{\varepsilon})$.
4. *mes* $v = O(\varepsilon^p)$.

The proof is based on estimates [(19; 20; 21)].

In Subsection 4.2 the set $U_c^{\kappa,\varepsilon}$ was introduced as the set of initial conditions for trajectories with a capture. Obviously, $U_g \subset U_c^{\kappa,\varepsilon} \subset U_g \cup v$. Thus, we can replace $U_c^{\kappa,\varepsilon}$ by U_g in definition (16) of capture probability, and formula (17) remains valid.

4.4. Scattering by a Resonance

Consider the motion of a phase point $(I(t), \varphi(t))$ that passes through the resonance without capture. Far from the resonant surface $I(t)$ oscillates with an amplitude $O(\varepsilon)$ near some solution of the averaged system. Before the passage through resonance this is a solution $J_-(t)$, $J_-(0) = I(0)$. After passage through the resonance this is a solution $J_+(t)$, $J_+(1/\varepsilon) = I(1/\varepsilon)$. Therefore, the result of the passage through a narrow neighborhood of the resonant surface (actually, through a resonant zone) can be interpreted as a jump from one solution of the averaged system to another one. We give here an asymptotic formula for this jump.

Let the motion start at the moment $t = 0$ at the point $(I_0, \varphi_0) \in U^\kappa$. The trajectory of the averaged system with such initial condition crosses the resonant surface at the time τ_* and at the point $\sigma = \sigma_*$ (or, $I = I_*$). Denote by γ_* the value of the angle γ at the point of intersection of the exact trajectory with the resonant surface. Denote by h_* the value of the Hamiltonian E for $\rho = 0$, $\gamma = \gamma_*$ and $\sigma = \sigma_*$. We can assume without loss of generality that $b(\sigma) = 0$ in (12). Thus, $h_* = V(\gamma_*, \sigma_*) - L(\sigma_*)\gamma_*$. Let $h_{c,*}$ be the closest from the left to h_* value of E at its saddle point. By definition, put

$$\Delta J = J_+(\tau_*/\varepsilon) - J_-(\tau_*/\varepsilon).$$

PROPOSITION 4.2. *For all initial points in* U^κ *except for a set of a measure* $O(\sqrt{\varepsilon})$ *the following formula is valid*

$$\Delta J = \Delta J_0 + O(\varepsilon/(h_* - h_{c,*})) + O(\varepsilon/(h_* - h_{c,*} - 2\pi|L(\sigma_*)|)), \qquad (21)$$

$$\Delta J_0 = -s\sqrt{\varepsilon}\int_{s\cdot\infty}^{\gamma_*} \frac{2\tilde{f}(I_*, \gamma)d\gamma}{\sqrt{2(h_* - V(\gamma, \sigma_*) + L(\sigma_*)\gamma)}},$$

where $\tilde{f}(I,\gamma) = \bar{f}(I,\gamma) - F(I)$ *is the pure periodic part of* $\bar{f}$ *(see (8)), and* $s = \operatorname{sign}(L(\sigma_*))$.

Indeed, in some neighborhood of the point I_* the vector field of the averaged system can be rectified: it is possible to make a transformation of variables $I \mapsto$

$(x_1, x_2, \dots, x_{n-1}, \eta)$ such that in the new variables the averaged system has the form

$$\dot{x}_i = 0,\ i = 1, 2 \dots, n-1,\ \dot{\eta} = \varepsilon. \tag{22}$$

We introduce $x = (x_1, x_2, \dots, x_{n-1}, \eta - \tau)$. Then the exact perturbed system in the new variables has the form

$$\dot{x} = \varepsilon \tilde{X}(x, \gamma, \tau),\ \tilde{X} = (\partial I/\partial x)^{-1} \tilde{f}(I, \gamma). \tag{23}$$

Let $x(t),\ y_\pm(t)$ be solutions $I(t),\ J_\pm(t)$ written in the x-variables. Denote $x_* = y_-(\tau_*/\varepsilon)$, $\Delta y = y_+(\tau_*/\varepsilon) - y_-(\tau_*/\varepsilon)$. Let $t_\pm$ be two instants at distances $\sim 1/\varepsilon$ from τ_*/ε, $0 < t_- < \tau_*/\varepsilon < t_+ < 1$. We shall assume that $\varepsilon t_\pm$ are close enough to τ_* in order to $y_\pm(t_\pm), x(t_\pm)$ have been well-defined. Then $y_\pm(\tau_*/\varepsilon) = y_\pm(t_\pm) = x(t_\pm) + O(\varepsilon)$. Now we have

$$\Delta J = (\frac{\partial I}{\partial x})_{I_*} \Delta y + O(|\Delta y^2|) = (\frac{\partial I}{\partial x})_{I_*} (x(t_+) - x(t_-)) + O(\Delta y^2) + O(\varepsilon) \tag{24}$$

Using (23), we get

$$x(t_+) - x(t_-) = \varepsilon \int_{t_-}^{t_+} \tilde{X}(x, \gamma, \tau) dt \tag{25}$$

The integral in (25) can be divided into two parts: one part corresponds to the motion far from the resonant surface, another part corresponds to the motion near the resonant surface. The first part is $O(\varepsilon)$. In the second part we can in the main approximation replace x by x_* and calculate the integral along the trajectory $E = h_*$ of the system with the Hamiltonian E and $\sigma = \sigma_*$. Thus we have (21).

Remark If the pendulum-like system (13) does not have equilibria, then the error term in (21) is just $O(\varepsilon)$.

The main term in formula (21) was obtained (in a slightly different form) in [(8)] for the case of a Hamiltonian system depending on a slowly varying parameter.

The solution $J_+(t)$ of the averaged system crosses the resonant surface at the time $t = \tau_*^+/\varepsilon$ at the point I_*^+, where

$$\tau_*^+ - \tau_* = -(k, \partial\omega/\partial I \Delta J)/L(\sigma_*) + O(|\Delta J^2|), \tag{26}$$

$$I_*^+ - I_* = F(I_*)(\tau_*^+ - \tau_*) + O(|\Delta J^2|). \tag{27}$$

Formulas (26) allow to interpret the scattering by the resonance as a jump at the moment τ_* from the point I_* to another point I_*^+ on the resonant surface combined with a jump in time to a moment τ_*^+.

Value γ_* in (21) is very sensitive to changes of initial conditions. A change in I_0 of order ε produces change in γ_* of order 1. Therefore, it is reasonable to consider γ_* as a random value. Then ΔJ should be considered as a random value as well. Define a probability distribution of ΔJ. Denote $\xi = \text{Frac}(h_*/2\pi|L(\sigma_*)|)$. Let $U^{\kappa,\varepsilon}_{(\alpha,\beta)}$ be the set of the initial points in U^κ such that $\xi \in (\alpha, \beta) \subset (0, 1)$.

DEFINITION 4.2. *The value*

$$Pr(\xi \in (\alpha,\beta)) = \lim_{\kappa\to 0}\lim_{\varepsilon\to 0} \frac{mesU^{\kappa,\varepsilon}_{(\alpha,\beta)}}{mesU^{\kappa}} \tag{28}$$

is called the probability of the event $\xi \in (\alpha,\beta)$.

PROPOSITION 4.3. $Pr(\xi \in (\alpha,\beta)) = \beta - \alpha$.

Therefore, the probability of the event $\xi \in (\alpha,\beta)$ is an asymptotic expression for the part of the phase volume that is occupied in U^{κ} by the initial conditions for trajectories with $\xi \in (\alpha,\beta)$. Proposition 4.3 states that the distribution of ξ is uniform on the interval $(0,1)$.

The proof of Proposition 4.3 is based on the calculation of the phase fluxes through the intervals $\alpha \le \xi \le \beta$ and $0 \le \xi \le 1$. The ratio of this fluxes is equal to $\beta - \alpha$ in the main approximation. On the other hand, definition (28) implies that this ratio coincides in the main approximation with the probability of the event $\xi \in (\alpha,\beta)$.

The distribution of ΔJ is known now (it is obtained through formula (21) from the uniform distribution of ξ). The mean value of ΔJ (and any smooth function of ΔJ) with respect to initial conditions coincides in the main approximation with the mean value of ΔJ (respectively, function of ΔJ) with respect to ξ.

5. Description of Motion over Time Intervals $\sim 1/\varepsilon$

Consider a system with two rotating phases (1). For any vector $k \in Z^2$ with co-prime components, the pendulum-like system is defined that corresponds to the resonance $k_1\omega_1 + k_2\omega_2 = 0$ (Subsection 3.3). Only for a finite number of resonances there exist oscillation regions on the phase portrait of the pendulum-like system (Proposition 3.2). These resonances are called *strong* ones, all other resonances are called *weak* ones. The capture into a weak resonance is impossible.

A motion without capture into resonance is approximately described by the averaged system. A motion with one capture into a resonance is approximately described by the glued averaging method approximation (Subsection 4.3). For an approximate description of a motion with several consequent captures into resonances and escapes from resonances we shall glue consequently to each other the trajectories of the glued averaging method approximations of Subsection 4.3. The approximate solutions obtained in such a way are called glued averaging method approximations too. For a given initial point I_0 there is only a finite number of glued averaging method approximations. Indeed, the number of strong resonances is finite; for any strong resonance there is only a finite number of oscillation regions on the phase portrait of the pendulum-like system.

We shall consider motions with initial conditions in a ball U^{κ} in the I-space. Assume that the projection of U^{κ} onto I-space does not intersect the resonant surfaces of the strong resonances. Assume that for initial conditions in U^κ and for $0 \le \tau \le 1$ the trajectory of the averaged system and the trajectories of all glued averaging method approximations are well-defined.

Let p be any prescribed positive number.

THEOREM 5.1. *The set U^κ may be represented in the form $U^\kappa = U_a \cup U_g \cup v$ as follows.*

1. *For initial conditions in U_a the averaging method describes the behavior of I with an accuracy $O(\sqrt{\varepsilon}\log\varepsilon)$ for $0 \le \varepsilon t \le 1$.*
2. *For initial conditions in U_g one of the glued averaging method approximations describes the behavior of I with an accuracy $O(\sqrt{\varepsilon}\log\varepsilon)$ for $0 \le \varepsilon t \le 1$.*
3. *mes $U_g = O(\sqrt{\varepsilon})$.*
4. *mes $v = O(\varepsilon^p)$.*

The proof is based on estimates [(19; 20; 21)].

It looks like a measure of initial conditions for trajectories with two captures into resonances is $O(\varepsilon)$, if $n \ge 2$ (this estimate is based on the hypothesis that consequent captures can be treated as independent random events). If this estimate is valid, then in Theorem 5.1 for $p = 1$ we can in statement 2) restrict ourselves to glued averaging method approximations with one capture into resonance only. Furthermore, U_g can be represented as the union of a finite number of sets $U_{g,j}$ that contain initial conditions for different single captures into resonances. The measure of every set $U_{g,j}$ is estimated by the probability of the corresponding capture into resonance in accordance with Subsection 4.2.

For initial conditions in U_a the deviation of the exact solution from the averaged system solution over the time $1/\varepsilon$ in the main approximation is the sum of the deviations produced by different resonances. In this sum should be included only the resonances whose resonant surfaces are crossed by the trajectory of the averaged system for $0 \le \varepsilon t \le 1$. For any single resonance this deviation can be computed in the main approximation by formula (21) (if an averaged system trajectory starts or finishes at the resonant surface of some resonance, then the contribution of this resonance is one-half of the magnitude in formula (21)). Thus, in the main approximation the deviation of the exact solution from the averaged system solution over the time $1/\varepsilon$ is represented as a sum of a random values series (this series converges). An important statistic characteristics of this sum are its mean value and variance. The mean value of the sum is equal to the sum of the mean values of the terms. It looks like the variance of our sum is equal in the main approximation to the sum of the variances of the terms, if $n \ge 2$. The reason for this hypothesis is as follows. If two terms correspond to resonant surfaces at distance ~ 1 to each other, then these terms seem to be statistically independent in the main approximation (see the discussion below). If two terms correspond to resonant surfaces that are close to each other, then at least one of the resonances that create these surfaces has a high order and the contribution of this resonance into the variance is small.

The reason for the hypothesis about statistical independence of results of successive passages through resonances is as follows. Consider two passages of a phase point through the resonances and the corresponding random values ξ_i, $i = 1,2$ (see Subsection 4.4). It looks like values ξ_1 and ξ_2 may be considered as statistically independent in the limit as $\varepsilon \to 0$. Indeed, a small variation in ξ_1 by a value $\Delta\xi_1$ produces an additional jump in $I \sim \sqrt{\varepsilon}\Delta\xi_1$ and then, if $n \geq 2$, an additional variation in the phase $\sim (1/\sqrt{\varepsilon})\Delta\xi_1$ as a result of the motion during the time $\sim 1/\varepsilon$ between passages through the resonances. This gives a variation in ξ_2 by $\Delta\xi_2 \sim (1/\sqrt{\varepsilon})\Delta\xi_1 \gg \Delta\xi_1$ and statistical independence of ξ_1 and ξ_2 in the limit as $\varepsilon \to 0$ (yet there is no accurate proof here). If $\xi_{1,2}$ are statistically independent, then jumps in the slow variables due to passages through the resonances are statistically independent as well.

6. On the Evolution under the Combined Action of Hamiltonian and non-Hamiltonian Perturbations

For many problems in celestial mechanics an important case is when the perturbation in (1) consists in two parts, a Hamiltonian and a non-Hamiltonian one, and the non-Hamiltonian part is much smaller, than the Hamiltonian one. In other words, system (1) has the form [(17)]

$$\begin{aligned} \dot{I} &= -\varepsilon\frac{\partial H_1(I,\varphi,\varepsilon)}{\partial\varphi} + \delta f(I,\varphi,\varepsilon,\delta), \\ \dot{\varphi} &= \omega(I) + \varepsilon\frac{\partial H_1(I,\varphi,\varepsilon)}{\partial I} + \delta g(I,\varphi,\varepsilon,\delta)\,. \end{aligned} \tag{29}$$

Here ε, δ are small parameters, $\delta \ll \varepsilon$. We shall assume that $\delta = \varepsilon^r$, $r \geq 2$.

The averaged system is defined by the following formula:

$$\dot{J} = \delta F(J), \; F(J) = \frac{1}{(2\pi)^m}\oint_{T^m} f(J,\varphi,0,0)\,d\varphi, \tag{30}$$

because the Hamiltonian part averages to 0. We suppose that r.h.s. of (29) is of smoothness C^∞. We consider the two-frequency case ($m = 2$) under the assumption that the following two conditions are satisfied:

$$|\mathcal{L}| > c_1^{-1}, \; |\det A| > c_2^{-1},$$

where

$$\begin{aligned} \mathcal{L} &= (\omega_2\frac{\partial\omega_1}{\partial I} - \omega_1\frac{\partial\omega_2}{\partial I})F(I), \\ A &= \begin{pmatrix} \frac{\partial\omega}{\partial I} & \omega \\ \omega^T & 0 \end{pmatrix}, \; c_{1,2} = \text{const} > 0. \end{aligned}$$

We restrict ourselves to the model problem containing only one resonance (cf. Section 4). In this case the r.h.s. of (29) depends on phases only through a resonant phase $\gamma = k_1\varphi_1 + k_2\varphi_2$. The analysis is based on the fact that the effects of Hamiltonian and non-Hamiltonian parts of the perturbation in (29) are quite different, and in all asymptotic expansions it is enough to keep the main term of the Hamiltonian part and the main term of the non-Hamiltonian part. The results here are analogous to those of Section 4. We use the notations of Section 4.

First make a canonical transformation of the variables $(I_1, I_2, \varphi_1, \varphi_2) \mapsto (R, \sigma, \gamma, \psi)$ using formulas

$$\begin{aligned} \gamma &= k_1\varphi_1 + k_2\varphi_2,\ \psi = l_1\varphi_1 + l_2\varphi_2, \\ R &= l_2 I_1 - l_1 I_2,\ \sigma = -k_2 I_1 + k_1 I_2. \end{aligned} \tag{31}$$

where $k_1 l_2 - k_2 l_1 = 1$. The equations of motion in the new variables do not contain ψ in the right hand side. Equations for γ, R, σ have the form

$$\begin{aligned} \dot\gamma &= (k,\omega) + \varepsilon\frac{\partial H_1}{\partial R} + \delta u_1, \\ \dot R &= -\varepsilon\frac{\partial H_1}{\partial \gamma} + \delta u_2,\ \dot\sigma = \delta u_3\ . \end{aligned} \tag{32}$$

Here $u_i = u_i(R, \sigma, \gamma, \varepsilon, \delta)$.

The resonant surface can be locally represented in the form $R = a(\sigma)$. Introduce $P = (R - a(\sigma))/\sqrt{\varepsilon}$, $\theta = \sqrt{\varepsilon}t$. By expanding the equations of motion near the resonant surface and keeping the main terms in the Hamiltonian part and in the non-Hamiltonian part, we get the system

$$\begin{aligned} \gamma' &= \frac{\partial E}{\partial P} + \hat\delta\sqrt{\varepsilon}\alpha_1(\gamma,\sigma), \\ P' &= -\frac{\partial E}{\partial \gamma} + \hat\delta L(\sigma) + \hat\delta\sqrt{\varepsilon}P\alpha_2(\gamma,\sigma), \\ \sigma' &= \hat\delta\sqrt{\varepsilon}\alpha_3(\gamma,\sigma), \end{aligned} \tag{33}$$

where

$$\begin{aligned} E &= E(P,\gamma,\sigma) = \frac{1}{2}g(\sigma)P^2 + V(\gamma,\sigma) + b(\sigma), \\ g &= (k_1k_2/\omega_1\omega_2)\det A,\ \ L = -(\omega_2/k_1)(\mathcal{L}/\det A), \\ V &= H_1|_{\varepsilon=0},\ \ \hat\delta = \delta/\varepsilon,\ \ (') = d/d\theta. \end{aligned} \tag{34}$$

(Values $\omega_1, \omega_2, A, \mathcal{L}, H_1$ are calculated at a point of the resonant surface).

Putting in (33) $\hat\delta = 0$ we obtain a pendulum-like system which is Hamiltonian system for γ, P with Hamiltonian E and $\sigma =$ const. Its phase portrait is shown in Fig. 4 (under assumption that there is just one oscillation region on the phase cylinder $(\rho, \gamma \bmod 2\pi)$). For simplicity we assume that there are no unstable stationary points inside the oscillation regions.

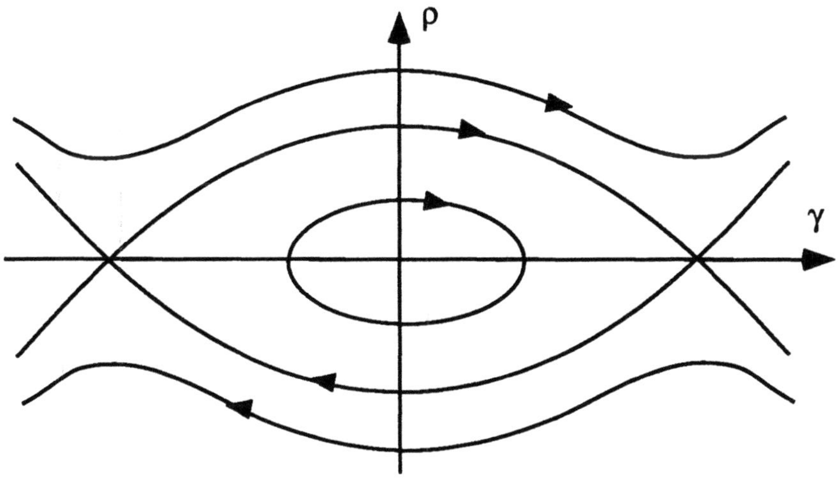

Fig. 4. Phase portrait of the pendulum-like system.

Let l be the separatrix in Fig. 4 and let $b(\sigma)$ in expression (34) be chosen in such a way that $E = 0$ at the saddle point and, therefore, on l. Without loss of generality we may assume that $b(\sigma) = 0$. We introduce

$$\Theta(\sigma) = -\oint_l \left(\frac{\partial E}{\partial \gamma}\alpha_1 + \frac{\partial E}{\partial P}P\alpha_2 + \frac{\partial E}{\partial \sigma}\alpha_3\right)d\theta. \tag{35}$$

We assume that the following "general position" conditions are satisfied.

1. The unstable stationary points are nondegenerate.
2. $\Theta(\sigma) \neq 0$.

Let $\tau = \delta t$ be the slow time. Let the solution of the averaged system with some initial condition I_0 crosses the resonant surface at the slow time instant τ_* at the point with $\sigma = \sigma_*$.

The probability of capture into the resonance is defined in accordance with formula (16). It turns out that this probability can be calculated by means of formula (17).

The motion of a phase point captured into the resonance is approximately described by the inner averaged system. This system is obtained by averaging the equations (33) over oscillations of the pendulum-like system analogously to Subsection 4.3. This allows to define the in-out function (see Subsection 4.3) as well. Motion of a phase point with segment of capture is approximately described by the glued averaging method approximation analogously to Subsection 4.3.

The accuracy of the averaging method approximation is described by the following Theorem (see notations in Subsection 4.3).

THEOREM 6.1. *The set U^κ may be represented in the form $U^\kappa = U_a \cup U_c$ as follows.*

1. *For initial conditions in U_a the averaging method describes the behavior of I with an accuracy $O(\sqrt{\varepsilon})$ for $0 \le \delta t \le 1$.*
2. *mes* $U_c = O(\sqrt{\varepsilon})$.

The proof is based on estimates [(19; 20; 21)].

The jump of the slow variables on the resonant surface can be defined in analogy with Subsection 4.4. The components of this jump, ΔR and $\Delta\sigma$, can be calculated by the following formulas: for all initial conditions except of a set of a measure $O(\sqrt{\varepsilon})$ we have:

$$\begin{aligned} \Delta R &= \Delta R_0 + O(\varepsilon \log \varepsilon),\ \Delta\sigma = \Delta\sigma_0 + O(\varepsilon \log \varepsilon), \\ \Delta R_0 &= \sqrt{\varepsilon}[\mathrm{sign}(L_*)\frac{S_*}{2\pi} - \mathrm{sign}(g_*)\frac{da(\sigma_*)/d\sigma}{\pi|L_*|}\int_0^{2\pi} \tilde{u}_3^* \sqrt{-2V_*/g_*}d\gamma], \\ \Delta\sigma_0 &= -\sqrt{\varepsilon}\mathrm{sign}(g_*)\frac{1}{\pi|L_*|}\int_0^{2\pi} \tilde{u}_3^* \sqrt{-2V_*/g_*}d\gamma. \end{aligned} \tag{36}$$

Here $L_* = L(\sigma_*),\ g_* = g(\sigma_*),\ V_* = V(\gamma,\sigma_*),\ \tilde{u}_3^*(\gamma)$ is the pure periodic part of $u_3(a(\sigma_*),\sigma_*,\gamma,0,0)$, and S_* is the area of oscillation region in Fig. 4 for $\sigma = \sigma_*$.

Formulas (36) show that in the main approximation the jump in the slow variables does not depend on the initial phase. The quasi-random dependence on initial phase is contained in the small residue terms in (36).

To derive formulas (36) one should write

$$\begin{aligned} \Delta R &= \int_{t_-}^{t_+} \left(-\varepsilon\frac{\partial H_1}{\partial\gamma} + \delta\tilde{u}_2\right) dt + O(\varepsilon \log \varepsilon), \\ \Delta\sigma &= \int_{t_-}^{t_+} \delta\tilde{u}_3 dt + O(\varepsilon \log \varepsilon). \end{aligned} \tag{37}$$

Here $\tilde{u}_{2,3}$ are the pure periodic parts of $u_{2,3}$; $t_\pm$ are two instants at distances $\sim 1/\delta$ from $\tau_*/\delta,\ 0 < t_- < \tau_*/\delta < t_+ < 1$. Asymptotics of the integrals in formulas (37) are given by the formulas for ΔR_0 and $\Delta\sigma_0$ (36).

The results for the model problem with one resonance allow to formulate the following hypothesis concerning general problem with many resonances: under some "general position " conditions the averaged system (30) describes the evolution of the slow variables in the exact system (29) over the time interval $0 \le t \le 1/\delta$ with an accuracy $O(\sqrt{\varepsilon})$ for a majority of initial conditions; the exceptional set, for which this estimate is not valid, has a measure $O(\sqrt{\varepsilon})$.

A system of the form (29) describes, in particular, the tidal evolution in the planar restricted 3-body problem with a moderate eccentricity of the massless body. In this case the masses of the heavy bodies are 1 and ε, a tide is characterized by a parameter δ. A tidal evolution of the heavy bodies orbits lead to resonant effects in the motion of the massless body. Form (29) is also appropriate for a description of the motion of the massless body in the planar restricted 3-body problem under the action of a small accelerating or drag force. However, in many cases of celestial mechanics problems with small Hamiltonian and smaller non-Hamiltonian perturbations, the equations of motion have a more complicated structure than (29). This is due to the proper degeneration of the n-body problem and to the special interest in motions with small eccentricities and inclinations (cf. [(6; 11; 12; 20; 28)]).

Acknowledgements

This work was made possible in part by Grants No. 97-01-00350 of Russian Fundamental Research Foundation, No. 93-339-ext of International Association for the promotion of cooperation with scientists from the Independent States of the Former Soviet Union (INTAS) and No. RM1-184 of the U.S. Civilian Research & Development Foundation for the Independent States of the Former Soviet Union (CRDF). The author thanks V.V.Sidorenko for useful discussion.

References

[1] Arnold V.I. (1978) *Mathematical methods of classical mechanics*. Springer, New York

[2] Arnold V.I. (1983) *Geometrical methods in the theory of ordinary differential equations*. Springer, New York

[3] Arnold V.I. (1963) Small denominators and problems of stability of motion in classical and celestial mechanics, *Usp. Mat. Nauk*, **18**, pp. 91-192 (Russian) [English transl.: (1963) *Russ. Math. Surv.*, **18**, pp. 85-192]

[4] Arnold V.I. (1965) Conditions of the applicability and estimates of the error of averaging method for systems which pass through states of resonance in the course of their evolution, *Dokl. Akad. Nauk SSSR*, **161**, pp. 9-12 (Russian) [English transl.: (1965) *Soviet Math. Dokl.*, **6**, pp. 331-334]

[5] Arnold V.I., Kozlov V.V, Neishtadt A.I. (1988) *Mathematical aspects of classical and celestial mechanics. Encyclopaedia of mathematical sciences, vol. 3*. Springer, Berlin

[6] Beaugé C., Ferraz-Mello S. (1994) Capture in exterior mean-motion resonances due to Poynting-Robertson drag, *Icarus*, **110**, pp. 239-260

[7] Bogolyubov N.N., Mitropol'skii Yu.A. (1961) *Asymptotic methods in the theory of nonlinear oscillations*. Gordon and Breach, New York

[8] Chirikov B.V. (1959) Passage of nonlinear oscillatory system through resonance, *Dokl. Akad. Nauk SSSR*, **125**, pp. 1015-1018 (Russian) [English transl.: (1959) *Soviet Phys. Dokl.*, **4**, pp. 390-394]

[9] Dirac P.A.M. (1925) The adiabatic invariance of the quantum integrals. *Proc. Roy. Soc.*, **A107**, pp.725-734

[10] Goldreich P., Peale S. (1966) Spin-orbit coupling in the Solar System, *Astron. J.*, **71**, pp. 425-438

[11] Greenberg R. (1978) Evolution of satellite resonances by tidal dissipation, *Astron. J.*, **78**, pp. 338-346

[12] Henrard J., Lemaitre A. (1983) A second fundamental model for resonance, *Celest. Mech. Dynam. Astron.*, **27**, pp. 3-22

[13] Kevorkian J.(1974) On a model for reentry roll resonance, *SIAM J. Appl.Math.*, **26**, pp. 638-669

[14] Lifshits I.M., Slutskin A.A., Nabutovskii V.M. (1961) On phenomenon of scattering of charged quasi-particles at singular points in p-space, *Dokl. Akad. Nauk SSSR*, **137**, pp. 553-556 (Russian)

[15] Lochak P., Meunier C. (1988) *Multiphase averaging for classical systems*. Springer, New York

[16] Melnikov V.K.(1963) On the stability of the center for time-periodic perturbations, *Tr. Mosk. Mt. O.-va*, **12**, pp. 3-52 (Russian)[English transl.: (1965) *Trans. Mosc. Math. Soc. 1963*, pp. 1-56

[17] Moltchanov A.M. (1968) The resonant structure of the Solar System, *Icarus*, **8**, pp. 203-215

[18] Morozov A.D. (1976) A complete qualitative investigation of Duffing's equation, *Differ. Uravn.*, **12**, pp. 241-255 (Russian) [English transl.: (1976) Differ. Equations, **12**, pp. 164-174]
[19] Neishtadt A.I. (1975) Passage through resonance in the two-frequency problem, *Dokl. Akad. Nauk SSSR*, **221**, pp. 301-374 (Russian) [English transl.: (1975) *Soviet Phys. Dokl.*, **20**, pp. 189-191]
[20] Neishtadt A.I. (1975) Passage through a separatrix in a resonance problem with slowly varying parameter, *Prikl. Mat. Mekh.*, **39**, pp. 621-632 (Russian) [English transl.: (1975) *J.Appl.Math.Mech.*, **39**, pp. 594-605]
[21] Neishtadt A.I. (1975) *Some resonant problems in nonlinear systems*. Ph.D.Thesis. Moscow State Univ, Moscow (Russian)
[22] Neishtadt A.I. (1990) Averaging, capture into resonances and chaos in nonlinear systems. In: Campbell D. (ed.)*Chaos/Xaoc*. AIP, New York, pp. 261-275.
[23] Neishtadt A.I. (1991) Probability phenomena due to separatrix crossing, *Chaos*, **1**, pp. 42-48
[24] Neishtadt A.I. (1993) Probability phenomena in perturbed systems, *Selecta Matematica Sovetica*, **12**, pp. 195-209
[25] Neishtadt A.I. (1993) On destruction of adiabatic invariants in multi-frequency systems, In: Perelló C., Simó C., Solà-Morales J. (Eds.) *Equadiff 91, International conference on differential equations*, Vol. 1, pp.195-207.
[26] Neishtadt A.I. (1996) On adiabatic invariance in two-frequency systems. In: A.Delshams, R.de la Llave, C., Simó (Eds.) *Hamiltonian systems with 3 or more degrees of freedom. Proceedings of NATO ASI*, in press
[27] Sidorenko V.V. (1994) Capture and escape from the resonance in the dynamics of the rigid body in viscous medium., *Journal of Nonlinear Science*, **4**, pp. 35-57
[28] Sinclair A.T. (1972, 1974) On the origin of commensurabilities among the satellites of Saturn, I, II, *Monthly Notices Roy. Astron. Soc.*, **160**, pp. 169-187, **166**, pp. 165-179.

NON-INTEGRABILITY IN HAMILTONIAN MECHANICS

SIMOS ICHTIAROGLOU
Department of Physics
University of Thessaloniki
54006, Greece

Abstract. We review theorems for proving non-integrability of Hamiltonian dynamical systems, which are based on properties of the variational equations in real or complex time or on the destruction of the resonant tori of an integrable system under a perturbation.

Key words: Hamiltonian systems, non-integrability

1. Introduction

A Hamiltonian system of n degrees of freedom is *integrable* if it possesses a set of n independent integrals in involution which are analytic in open domains of the phase space (e.g. Arnold, 1978, p. 271). Then the system can be integrated by quadratures. If such a complete set of integrals does not exist, the system is termed *non-integrable* and, in this case, chaotic motion is expected in general to occur.

Most of the theorems for proving non-integrability of Hamiltonian systems fall mainly in three groups, each one exploiting different properties of integrable or non-integrable behaviour:
(a) Properties of the variational equations around periodic solutions, in the real or complex time domain.
(b) Destruction of the resonant tori of an integrable system under a perturbation.
(c) Existence of chaotic motion in a perturbed integrable system.

A feature common in the above groups, is that the main ideas in all three of them are already present in Poincaré's work (1892-1899). The methods involved in group (c), generally known as *homoclinic Melnikov theory*, have found their way in text-books (e.g. Guckenheimer and Holmes, 1983; Wiggins, 1990) and will not be presented in this review.

2. Isolated Periodic Orbits and Non-Existence of Integrals in Involution

Consider the n degree-of-freedom Hamiltonian system

$$\dot{\eta} = \Omega \nabla H \tag{1}$$

where η are a set of canonical coordinates and

$$\Omega = \begin{pmatrix} 0 & I \\ -I & 0 \end{pmatrix}$$

is the standard symplectic structure. Let $\eta_0(t) = \eta_0(t+T)$ be a T-periodic solution of (1) and ξ be the first variation around this solution. Then ξ is a solution of the linear system of *variational equations* (VE)

$$\dot{\xi} = A(t)\xi \tag{2}$$

Celestial Mechanics and Dynamical Astronomy **65**: 21–31, 1997.

where $A = \Omega\nabla^2 H$ is a T-periodic matrix, evaluated on the above periodic solution.

Consider the fundamental solution $\Delta(t)$ of VE, such that $\Delta(0) = I$. Then $\Delta(T)$ is the *monodromy matrix* of (2). It is a symplectic matrix, i.e. it holds that $\Delta^T\Omega\Delta = \Omega$, so its eigenvalues λ, which are the *Floquet multipliers*, come in reciprocal pairs. The *characteristic exponents* (CE) of the periodic solution η_0 are defined as

$$\sigma = \frac{1}{T}\ln\lambda \mod i\frac{2\pi}{T} \tag{3}$$

and appear in opposite pairs. Every simple σ provides a solution of (2), of the form

$$\xi(t) = e^{\sigma t} f(t) \tag{4}$$

where f is T-periodic, while exponents with multiplicity m correspond in general to solutions of the form

$$\xi(t) = e^{\sigma t} \sum_{k=0}^{m-1} f_k(t) t^k \tag{5}$$

where again f_k are T-periodic. A T-periodic solution of (2) corresponds to a zero characteristic exponent. If at least one σ is different from zero, the periodic solution η_0 is *isolated*.

Let ξ_1, ξ_2 be solutions of (2). Then it holds that (Poincaré, 1892)

$$\xi_1^T \Omega \xi_2 = \text{const.} \tag{6}$$

Let Φ be an integral of motion of (1). Then $\Omega\nabla\Phi$, evaluated on η_0, is a T-periodic solution of (2), so it is obvious that if Hamiltonian H possesses k integrals, independent on a periodic solution, then this solution possesses k zero CE.

Theorem 1 (Poincaré): If H possesses n integrals in involution Φ_i, independent on a periodic solution, then it is *non-isolated*, i.e. all its CE are zero.

Proof: Consider the $2n$ independent solutions of (2),

$$\xi_i = \Omega\nabla\Phi_i \quad \text{and} \quad \xi_{i+n} = e^{\sigma_i t} \sum_{k=0}^{m-1} f_{ik}(t) t^k \tag{7}$$

where σ_i are n presumably non-zero CE, and form the fundamental solution $\Xi(t) = (\xi_1, .., \xi_{2n})$. Then, according to (6), $\Xi^T\Omega\Xi = C$, where C is a constant matrix, which is non-singular, since Ξ, Ω are non-singular. On the other hand, for $i, j \leq n$,

$$C_{ij} = \xi_i^T \Omega \xi_j = [\Phi_i, \Phi_j] = 0,$$

so C assumes the form

$$C = \begin{pmatrix} 0 & B \\ -B^T & D \end{pmatrix} \tag{8}$$

where the elements of the $n \times n$ matrix B are, due to (7), of the general form

$$B_{ij} = \xi_i^T \Omega \xi_{n+j} = e^{\sigma_j t} \sum_{k=0}^{r} g_k(t) t^k \tag{9}$$

with T-periodic g_k. Since C is non-singular, every column of B must have at least one non-zero element, i.e. for at least one i, $B_{ij} = \text{const.} \neq 0$, which implies $\sigma_j = 0 \, \forall j$. □

Isolated periodic orbits are of course present in integrable systems, the integrals Φ_i are, however, dependent on them. So the above theorem actually proves non-integrability *if the Hamiltonian H possesses a dense or at least a key set of isolated periodic orbits*, in which case n integrals in involution, independent in an open domain cannot exist.

A generalization of the above theorem for the case of hyperbolic tori has been proved by Treshchëv (1991).

3. Properties of the Monodromy Group in Complex Time

The demand for the existence of a dense set of isolated periodic orbits has been overcome by Ziglin (1983) by proving that if Φ is an integral of (1) then the VE always possess an integral, homogeneous in ξ, of the form

$$I = D^m \Phi \tag{10}$$

for some integer m, where D is the operator $D = \xi^T \nabla$.

Consider a straight-line solution (SLS) of (1). The VE (2) around the SLS decouple to the tangential (TVE) and normal variational equations (NVE). It can be shown that DH is an integral of the TVE, while by (10) one shows that if a second integral Φ exists, then the NVE possess an integral which is homogeneous with respect to the normal variations and T-periodic in time. A monodromy matrix M of the NVE is *non-resonant* if no eigenvalue is a root of unity. Consider two independent monodromy matrices M_1, M_2 of the NVE in complex time. By acting M_1, M_2 on the above integral, the following theorem can be proved:

Theorem 2 (Ziglin): If two non-resonant, non-commuting monodromy matrices for the system of NVE exist, the Hamiltonian H does not possess a second independent integral of motion, holomorphic in a neighbourhood of the SLS.

For a proof see Ziglin (1983), Yoshida (1987a).

The application of Ziglin's theorem in concrete Hamiltonians implies the knowledge of at least two monodromy matrices, see for example Ziglin (1983), Ito

(1985, 1987, 1990), Françoise and Irigoyen (1993). For the case of motion in a two-dimensional planar potential of degree $k \neq 0$, i.e

$$H = \frac{1}{2}(p_x^2 + p_y^2) + V(x, y), \quad V(cx, cx) = c^k V(x, y), \tag{11}$$

Yoshida (1987a) has shown that the NVE transform to Gauss hypergeometric equation, whose monodromy group is known. Then, applying Ziglin's theorem, the simple algorithm of the following theorem is obtained:

Theorem 3 (Yoshida): Let $k \neq 0, \pm 2$ and c_1, c_2 be a solution of the system $c_1 = V_x(c_1, c_2), \;\; c_2 = V_y(c_1, c_2)$. Define the *integrability coefficient* λ as

$$\lambda = (V_{xx} + V_{yy})_{c_1, c_2} - (k - 1). \tag{12}$$

The Hamiltonian H does not possess a second independent integral if

$$\cos\left(\frac{2\pi}{k}\right) + 2\cos^2\left[\frac{\pi}{2k}\sqrt{(k-2)^2 + 8k\lambda}\right] > 1. \tag{13}$$

The above theorem has been generalized for potentials which are sums of homogeneous terms and has been applied to the Toda lattice (Yoshida, 1988), to perturbed Kepler problems (Yoshida, 1987b), to Yang-Mills potentials (Ichtiaroglou, 1989), the Störmer problem (Almeida *et al.*, 1992) and the J_2 problem (Irigoyen and Simó, 1993).

Example: Applying theorem 3 to the Yang-Mills potential

$$V(x, y) = \frac{1}{2}x^2 y^2, \tag{14}$$

we find $c_1^2 = c_2^2 = 1, \; \lambda = -1$ and inequality (13) is satisfied, so the potential (14) is non-integrable.

Based on Ziglin's theorem, Yoshida (1989) proved also the following theorem for the non-integrability of Hamiltonians of n degrees of freedom, of the form

$$H = \frac{1}{2}\sum p_i^2 + V(q_i)$$

with a homogeneous potential of degree $k \neq 0, \pm 2$. Let again $(c_1, c_2, \ldots, c_n)$ be a solution of

$$c_i = \frac{\partial V}{\partial q_i}(c_1, \ldots, c_n)$$

and μ_i the eigenvalues of the matrix

$$\left(\frac{\partial V}{\partial q_i \partial q_j}\right)_{(c_1,\ldots,c_n)}$$

For each μ_i, the roots ρ_i, ρ_{i+n} of the equation

$$\rho^2 - \frac{k+2}{k-2}\rho + (1-\mu_i)\frac{2k}{(k-2)^2} = 0$$

are the *Kovalevski exponents*, related to the branching of solutions of the system in complex time (e.g. Yoshida, 1983).

Theorem 5 (Yoshida): If the differences $\Delta\rho_i = \rho_i - \rho_{i+n}$ are rationally independent, then H does not possess a second analytic integral of motion.

The above theorem has been recently applied by Ferrándiz and Sansaturio (1995) in the J_{22}-problem. For recent results on singularity analysis and non-integrability see Goriely (1996) and for a review see Bountis (1992).

4. The Theorem of Poincaré

In the end of the previous century, Poincaré (1892) proved his famous non-integrability theorem, showing that, under general assumptions, a perturbed n-degree-of-freedom Hamiltonian of the form

$$H = H_0(J_i) + \varepsilon H_1(J_i, w_i) \tag{15}$$

where (J_i, w_i) are action-angle variables of the integrable part H_0, is generically non-integrable for $\varepsilon \neq 0$.

Assume that H_0 is *non-degenerate*, i.e. it holds that

$$\det\left|\frac{\partial^2 H_0}{\partial J_i \partial J_j}\right| = \det\left|\frac{\partial \omega_i}{\partial J_j}\right| \neq 0 \tag{16}$$

in an open domain of phase space, where $\omega_i = \partial H_0/\partial J_i$ are the *frequencies* of H_0. It can be shown that this assumption can be relaxed to *isoenergetic non-degeneracy*, i.e.

$$\det\begin{vmatrix} \frac{\partial \omega_i}{\partial J_j} & \omega_i \\ \omega_j & 0 \end{vmatrix} \neq 0.$$

Assume also the existence of an analytic integral Φ of H for $\varepsilon \neq 0$ and expand

$$\Phi = \Phi_0 + \varepsilon\Phi_1 + O(\varepsilon^2). \tag{17}$$

Since Φ is an integral of H,

$$[\Phi, H] \equiv 0. \tag{18}$$

Identity (18) at zeroth order in ε becomes $[\Phi_0, H_0] \equiv 0$, i.e. Φ_0 is an integral of H_0. It can be shown that Φ_0 depends only on the actions, since its dependence on the angles leads to linear dependence of the frequencies, contrary to the non-degeneracy condition (16). Suppose that $k+1$ independent integrals $H, \Phi^1, \Phi^2, ...\Phi^k$, analytic in ε exist for the perturbed Hamiltonian H and expand them with respect to ε as

$$\Phi^i = \Phi_0^i + \varepsilon\Phi_1^i + O(\varepsilon^2). \tag{19}$$

Lemma 1 (Poincaré): Φ^i can always be selected such that the zeroth order terms $H_0, \Phi_0^1, ..., \Phi_0^k$ are $k+1$ *independent* integrals of H_0.

This lemma, which is proved inductively, is fundamental for the proof of Poincaré's theorem and the theorem we will present in the next section.

Since H_1, Φ_1 must be 2π-periodic in the angles, they can be expanded in Fourier series as

$$\begin{aligned} H_1 &= \textstyle\sum A_{\mathbf{m}}(J) \exp i(\mathbf{m}\cdot\mathbf{w}), \\ \Phi_1 &= \textstyle\sum B_{\mathbf{m}}(J) \exp i(\mathbf{m}\cdot\mathbf{w}). \end{aligned} \tag{20}$$

Identity (18) in first order yields

$$[\Phi_1, H_0] \equiv [H_1, \Phi_0]. \tag{21}$$

Taking into account (20), identity (21) reads

$$B_{\mathbf{m}} \sum_{j=1}^{n} m_j \omega_j = A_{\mathbf{m}} \sum_{j=1}^{n} m_j \frac{\partial \Phi_0}{\partial J_j} \tag{22}$$

for *every* integer vector $\mathbf{m}$. Since the system is non-degenerate, a dense set $\mathcal{R}$ of points exists, where the relations

$$\frac{\omega_1}{p_1} = \frac{\omega_2}{p_2} = \ldots = \frac{\omega_n}{p_n} \tag{23}$$

hold for $p_i \in Z \setminus \{0\}$, in general with different integers p_i at each point. Then, one can define through (23), in every point of $\mathcal{R}$, $n-1$ independent integer vectors $\mathbf{k}$, such that $\mathbf{k}\cdot\omega = 0$. Let $\mathbf{m}$ be any integer linear combination of the vectors $\mathbf{k}$. We define the set $\mathcal{P}$ as the set of points of $\mathcal{R}$, where both equations

$$\sum_{i=1}^{n} m_i \omega_i = 0. \tag{24}$$

and

$$A_{\mathbf{m}} \neq 0, \tag{25}$$

hold, for $n-1$ independent integer vectors $\mathbf{m}$.

Theorem 6 (Poincaré): If the set $\mathcal{P}$ is dense in a domain of phase space, H cannot possess a second integral of motion, analytic in an open interval of ε around zero.

Proof: According to (22) and (25), on the points of $\mathcal{P}$ both equations (24) and

$$\sum_{i=1}^{n} m_i \frac{\partial \Phi_0}{\partial J_i} = 0$$

hold at the same time, for $n-1$ independent integer vectors $\mathbf{m}$. Since $\mathcal{P}$ is dense, functions H_0 and Φ_0 are dependent in an open domain, which contradicts lemma 1. □

For the application of this theorem in the three-body problem, see Poincaré (1892). See also Kozlov (1983), Giorgilli (1990) and Jackson (1990) for some interesting remarks.

5. A Theorem on the Allowed Number of Integrals of Motion for Perturbed Hamiltonians of n Degrees of Freedom

Most of the theorems on non-integrability of Hamiltonian systems supply criteria for showing either the non-existence of a complete set of independent integrals, or the absence of any other integral besides H. In this section we present a criterion for partial non-integrability, that supplies an upper bound for the maximum allowable number of integrals of motion (Meletlidou and Ichtiaroglou, 1994, Meletlidou, 1995).

We consider a perturbed Hamiltonian of n degrees of freedom, of the form (15), with non-degenerate or isoenergetically non-degenerate integrable part and assume the existence of $n-1$, in addition to H, analytic integrals of the form (19) for $\varepsilon \neq 0$.

For every integral Φ_i, an identity of the form (21) holds

$$[\Phi_{i1}, H_0] \equiv [H_1, \Phi_{i0}] \tag{26}$$

and parametrizing the above identity on the orbits of the unperturbed part, we obtain

$$\frac{d\Phi_{i1}}{dt} = [H_1, \Phi_{i0}]. \tag{27}$$

On a torus of H_0, defined by a set of constant values of J_i, the motion is described by the equations

$$w_i = \omega_i t + \vartheta_i \mod 2\pi \tag{28}$$

where ϑ_i are arbitrary initial conditions. Consider a resonant torus of H_0, defined by equations (23) for a set of non-zero and relative prime integers p_i and define the *vertical angles* by

$$\varphi_j = p_{j+1}\vartheta_1 - p_1\vartheta_{j+1} \mod 2\pi s_j, \quad j = 1, \ldots, n-1 \tag{29}$$

where s_j is the common divisor of the pair p_{j+1}, p_1. This torus is foliated by periodic orbits with period

$$T = \frac{2\pi p_i}{\omega_i}. \tag{30}$$

Note that a set of values for φ_j defines a particular periodic orbit on this torus. We integrate equations (27) on these periodic orbits for one period and, since $\mathbf{\Phi}_{i1}$ are single-valued and $\mathbf{\Phi}_{i0}$ depend only on J_i, we obtain

$$\sum_{j=1}^{n} \frac{\partial \Phi_{i0}}{\partial J_j} \int_0^T \frac{\partial H_1}{\partial w_j} dt = 0. \tag{31}$$

Let

$$< H_1 >= \frac{1}{T} \int_0^T H_1 dt$$

be the average value of the perturbing function along this periodic orbit. Taking into account equations (28) and the fact that $< H_1 >$ depends only on the particular orbit and not on its initial point, equations (31) become

$$\sum_{j=1}^{n-1} D_j^{(i)} \frac{\partial < H_1 >}{\partial \varphi_j} = 0 \tag{32}$$

where

$$D_j^{(i)} = p_{j+1}\frac{\partial \Phi_{i0}}{\partial J_1} - p_1 \frac{\partial \Phi_{i0}}{\partial J_{j+1}}.$$

Note that every integral $\mathbf{\Phi}_i$ of H defines through $\mathbf{\Phi}_{i0}$ a vector $\mathbf{D}^{(i)}$, which depends only on the actions. The following lemma can be easily proved:

Lemma 2: The s vectors $\mathbf{D}^{(1)}, \ldots, \mathbf{D}^{(s)}$ are linearly dependent on a resonant torus of H_0 iff the $s+1$ gradients $\nabla H_0, \nabla \mathbf{\Phi}_{i0}, \ldots, \nabla \mathbf{\Phi}_{s0}$ are linearly dependent on this torus.

Assume now that, for at least one periodic orbit on this resonant torus and for at least one j, it holds that

$$\frac{\partial < H_1 >}{\partial \varphi_j} \not\equiv 0. \tag{33}$$

Then, from (32)

$$\det\left|D_j^{(i)}\right| = 0$$

on the whole torus, and, according to lemma 2, the n integrals H_0, Φ_{i0} are dependent on it.

Differentiating (32) with respect to φ_k we obtain

$$\sum_{j=1}^{n-1} \frac{\partial^2 < H_1 >}{\partial\varphi_k \partial\varphi_j} D_j^{(i)} = 0. \tag{34}$$

If for at least one orbit

$$\mathrm{rank}\left(\frac{\partial^2 < H_1 >}{\partial\varphi_k \partial\varphi_j}\right) \geq r \tag{35}$$

then, according to lemma 2, only $n - r$ independent integrals of the unperturbed Hamiltonian, including H_0, can exist on this torus. So, combining these results with lemma 1, we have proved the following:

Theorem 7: If inequality (33) is valid on a dense or a key set of resonant tori of H_0, the perturbed Hamiltonian H does not possess a complete set of analytic integrals of motion and if (35) is valid for some r on such a set, H cannot possess more than $n - r$ analytic integrals, including the Hamiltonian, for an open interval of $\varepsilon \neq 0$ around zero.

For the relation of this result to the continuation of periodic orbits to the perturbed system, see Ichtiaroglou and Meletlidou (1996).

Examples: (1) The two-degree-of-freedom Hamiltonian

$$H = \sum_{i=1}^{2}\left(\frac{1}{2}P_i^2 + (e^{-x_i} - 1)^2\right) + \varepsilon P_1 P_2$$

has been used as a model for chaotic scattering in triatomic molecules (Sibert *et al.*, 1982; Bruhn and Koch, 1993; Meletlidou, 1995). The solution of the integrable part, for $E_i \in (0, 1)$, is

$$x_i = \ln\left\{\frac{1 - \sqrt{E_i}\cos\left[\sqrt{2(1 - E_i)}t - \vartheta_i\right]}{1 - E_i}\right\}, \quad P_i = \dot{x}_i \tag{36}$$

where E_i is the energy of the ith degree of freedom. By using (36), for the resonant torus defined by (23) for positive integers p_i, we find

$$< H_1 >= 4\omega^2 p_1 p_2 \sum_{m=1}^{\infty} e^{-m(p_2 a_1 + p_1 a_2)} \cos(m\varphi)$$

where

$$\omega = \frac{\sqrt{2(1-E_i)}}{p_i}, \quad \cosh a_i = \sqrt{E_i}.$$

Since $< H_1 >$ depends on the vertical angle φ, the criterion (33) is satisfied on a dense set of tori and the system is non-integrable for $\varepsilon \neq 0$.

(2) Consider the n-degree-of-freedom anharmonic oscillator

$$H = \sum_{i=1}^{n} \left(\frac{1}{2}P_i^2 + \frac{1}{4}d_i x_i^4\right) + \varepsilon \sum_{i=2}^{n} \alpha_i x_1 x_i.$$

where d_i, α_i are constants. The solution of H_0 is

$$x_i = \frac{\lambda_i}{d_i}\mathrm{cn}(\lambda_i t - \vartheta_i, 1/\sqrt{2})$$

and the average value of H_1 on the dense set of resonant tori with positive and odd p_i is (Meletlidou and Ichtiaroglou, 1994)

$$< H_1 >= \sum_{i=1}^{n-1} \sum_{j=1}^{\infty} b_{ij} \cos\left(\frac{2j+1}{s_i}\varphi_i\right)$$

where b_{ij} are non-zero constants whose explicit expression is of no importance here. It is easily seen that criterion (33) is satisfied and no complete set of integrals exists for $\varepsilon \neq 0$. On the other hand, the rank of the matrix in (35) equals $n-1$, so that no other analytic integral exists beside the Hamiltonian. These results are valid $\forall \varepsilon \neq 0$, due to the scale invariance of H with respect to the transformation

$$x_i \to c x_i, \ \ t \to c^{-1} t, \ \ \varepsilon \to c^2 \varepsilon$$

with arbitrary c. If for some $k \geq 2$ it holds that $\alpha_k = 0$, then the rank of the matrix in (35) becomes $n-2$ and H cannot possess more than two integrals. In this case, there exist two independent integrals indeed, i.e. the Hamiltonian and the energy along the kth degree of freedom. In general, every zero coefficient α_i in the perturbation supplies H with an additional integral of motion and the rank in (35) changes accordingly.

References

Almeida M.A., Moreira I.C. & Yoshida H., 1992, J. Phys. A: Math. Gen. **25**, L227-L230.
Arnold V.I., 1978, *"Mathematical Methods of Classical Mechanics"*, Springer, NY.
Bountis T., 1992, Int. J. of Bifurcation and Chaos **2**, 217-232.
Bruhn B. & Koch B.P., 1993, Int. J. of Bifurcation and Chaos **3**, 999-1012.
Ferrándiz J.M. & Sansaturio M.E., 1995, Phys. Lett. A **207**, 180-184.
Françoise J.-P. & Irigoyen M., 1993, Geom. Phys. **10**, 231-243.

Giorgilli A., 1990, in D.Benest & C.Froeschle (eds.) *"Modern Methods in Celestial Mechanics"*, Editions Frontières, Gif-sur-Yvette, 249-284.
Goriely A., 1996, J. Math. Phys. **37**, 1871-1893.
Guckenheimer J. & Holmes P., *"Nonlinear Oscillations, Dynamical Systems and Bifurcations of Vector Fields"*, Springer, NY.
Ichtiaroglou S., 1989, J. Phys. A: Math. Gen. **22**, 3461-3465.
Ichtiaroglou S. & Meletlidou E., 1996, in A.Delshams & C.Simó (eds.), *"Hamiltonian Systems with three or more degrees of freedom"*, NATO ASI Series, Plenum Press, NY (to appear).
Irigoyen M. & Simó C., 1993, Celest. Mech. Dyn. Astr. **55**, 281-287.
Ito H., 1985, Kodai Math. J. **8**, 120-138.
Ito H., 1987, ZAMP **38**, 459-476.
Ito H., 1990, Bol. Soc. Bras. Mat. **21**, 95-120.
Jackson E.A., 1990, *"Perspectives of nonlinear dynamics"*, Vol. 2, Cambridge Univ. Press, Cambridge.
Kozlov V., 1983, Russian Math. Surveys **38**, 1-76.
Meletlidou E. & Ichtiaroglou S., 1994, J. Phys. A: Math. Gen. **27**, 3919-3926.
Meletlidou E., 1995, *"Non-integrability criteria in Hamiltonian dynamical systems"*, Doctor's Thesis, Univ. of Thessaloniki (in Greek).
Poincaré H., 1892-1899, *"Les Méthodes Nouvelles de la Mécanique Céleste"*, Vol. I-III, Gauthier-Villars, Paris; English translation: D.L. Goroff (ed.), 1993, *"New Methods in Celestial Mechanics"*, American Institute of Physics.
Sibert E.L., Reinhard W.P. & Hynes J.T., 1982, J. Chem. Phys. **77** 3583- 3594.
Treshchëv D.V., 1991, Math. USSR Sbornik **68**, 181-203.
Wiggins S., 1990, *"Introduction to applied Nonlinear Dynamical Systems and Chaos"*, Springer, NY.
Yoshida H., 1983, Celest. Mech. **31**, 363-379.
Yoshida H., 1987a, Physica D **29**, 128-142.
Yoshida H., 1987b, Phys. Lett. A **120**, 388-390.
Yoshida H., 1988, Commun. Math. Phys. **116**, 529-538.
Yoshida H., 1989, Phys. Lett. A **141**, 108-112.
Ziglin S.L., 1983, Func. Anal. Appl. **16**, 181-189; **17**, 6-17.

REGULAR AND IRREGULAR PERIODIC ORBITS

G.CONTOPOULOS and E.GROUSOUZAKOU
Department of Astronomy,University of Athens
GR-15784 , Athens ,Greece

Abstract. We distinguish between regular orbits, that bifurcate from the main families of periodic orbits (those that exist also in the unperturbed case) and irregular periodic orbits, that are independent of the above. The genuine irregular families cannot be made to join the regular families by changing some parameters. We present evidence that all irregular families appear inside lobes formed by the asymptotic curves of the unstable periodic orbits. We study in particular a dynamical system of two degrees of freedom, that is symmetric with respect to the x-axis, and has also a triple resonance in its unperturbed form. The distribution of the periodic orbits (points on a Poincaré surface of section) shows some conspicuous lines composed of points of different multiplicities. The regular periodic orbits along these lines belong to Farey trees. But there are also lines composed mainly of irregular orbits. These are images of the x-axis in the map defined on the Poincaré surface of section. Higher order iterations of this map , close to the unstable triple periodic orbit, produce lines that are close to the asymptotic curves of this unstable orbit. The homoclinic tangle, formed by these asymptotic curves, contains many regular orbits, that were generated by bifurcation from the central orbit, but were trapped inside the tangle as the perturbation increased. We found some stable periodic orbits inside the homoclinic tangle, both regular and irregular. This proves that the homoclinic tangle is not completely chaotic, but contains gaps (islands of stability) filled with KAM curves.

Key words: periodic orbits – regular periodic orbits – irregular periodic orbits – Farey trees – homoclinic tangle

1. Regular Orbits

The distinction between regular and irregular periodic orbits (Contopoulos 1970) is very useful in studies of the geneology and distribution of periodic orbits.

Regular orbits are produced by bifurcation from the orbits of the unperturbed system, that continue to exist as the perturbation increases. The unperturbed system contains only a few families of periodic orbits. E.g. the Hamiltonian system

$$H = \frac{1}{2}(\dot{x}^2 + \dot{y}^2 + \omega_1^2 x^2 + \omega_2^2 y^2) - \epsilon x y^2 \tag{1}$$

with $\omega_1 = \sqrt{1.6}$, $\omega_2 = \sqrt{0.9}$ has all its orbits triply periodic for $\epsilon = 0$,i.e. all the orbits except the axes $x = 0$, $y = 0$ make 3 oscillations up and down and at the same time 4 oscillations right and left. But when $\epsilon \neq 0$ only four of these orbits remain; the axis $y = 0$, an orbit close to $x = 0$ perpendicular to the x-axis ($\dot{x} = 0$), and two triply periodic orbits (Fig.1), one stable and one unstable. The central periodic orbit is a continuation of the orbit $x = 0$, while the orbit $y = 0$ is represented by the boundaries $\bar{x} = \omega_1 x = \mp\sqrt{2H}$. If the ratio $\frac{\omega_1}{\omega_2}$ is slightly larger than $\frac{4}{3}$ the triple families are generated from the central family (for $\epsilon > 0$), while if the ratio $\frac{\omega_1}{\omega_2}$ is slightly smaller than $\frac{n}{m} = \frac{4}{3}$ the triple families are generated from the boundary $y = 0$. As ϵ increases many other families are generated from the central periodic orbit, from the boundary $y = 0$, and from the stable triple periodic orbit (Figs.1,2). E.g. if the rotation number of the central periodic orbit (i.e. the

Celestial Mechanics and Dynamical Astronomy **65**: 33–56, 1997.

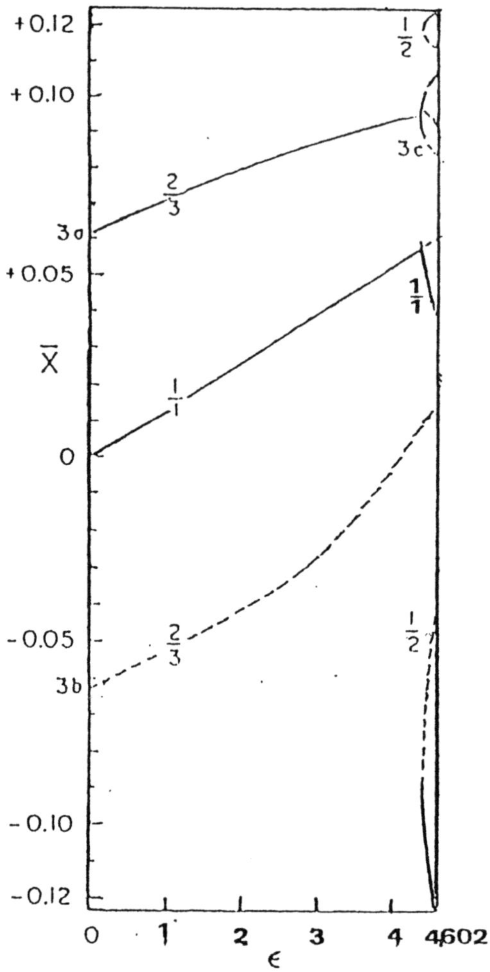

Fig. 1. Characteristics of the main families of periodic orbits generated at $\epsilon = 0$ in the case of the Hamiltonian (1) with $H = 0.00765$. $\bar{x} = \omega_1 x$ as a function of ϵ.(——) stable, (– – –) unstable families.$\epsilon = 4.602$ is the escape perturbation.

limit of the rotation numbers of nearby nonperiodic orbits) becomes rational $\frac{n'}{m}$, it generates by bifurcation a family of periodic orbits of multiplicity m.

These orbits make m oscillations up and down and at the same time n oscillations right and left. The ratio $\frac{n}{m}$ is equal to $(2 - \frac{n'}{m})$, where $\frac{n'}{m}$ is the rotation number (Contopoulos 1970).

The stable bifurcating families produce higher order bifurcations (Fig.2), that exist even for ϵ larger than the escape perturbation $\epsilon = 4.602$. E.g. the family $\frac{n'}{m} = \frac{4}{5}$ generates a double period family $\frac{8}{10}$, with the same rotation number, and an equal period family $\frac{4}{5}$ (and after this bifurcation the original $\frac{4}{5}$ family becomes

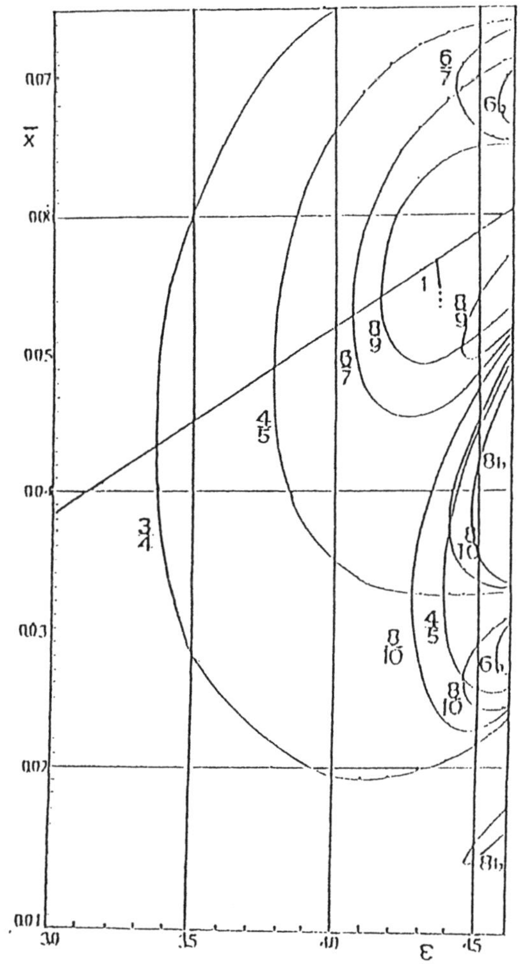

Fig. 2. Regular families of periodic orbits, generated by bifurcation from the central periodic family, and the irregular families 6b and 8b generated in a pair of stable and unstable orbits at a "tangent bifurcation"

unstable). The new family $\frac{4}{5}$ generates a family $\frac{8}{10}$, by a period doubling bifurcation, followed by a cascade of period doubling bifurcations $\frac{16}{20}, \frac{32}{40}$ etc.

All these families that bifurcate from the unperturbed families, or from their bifurcations, are called regular.

The central family becomes unstable for $\epsilon = 4.306$. At this point we have the bifurcation of two stable period 1 families (Fig.1) with the same x and opposite $\dot{x} \neq 0$. These become also unstable at a period doubling bifurcation for $\epsilon = 4.769$, followed by a cascade of period doubling bifurcations. The boundary $\bar{x} = \pm\sqrt{2H}$ ($y = 0$) becomes unstable at ϵ=7.0. The stable triple family becomes

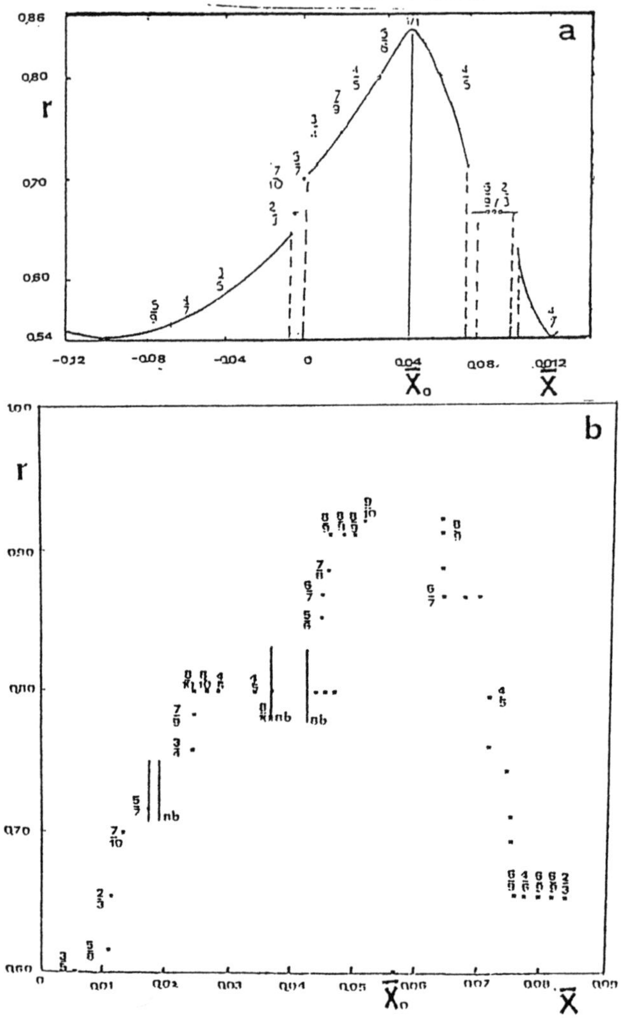

Fig. 3. Rotation numbers of periodic and nonperiodic orbits with respect to the central periodic orbit $\bar{x}_0$ as a function of $\bar{x} = \omega_1 x$. When the central orbit is stable its rotation number is maximum ($\epsilon = 4.0$, Fig. 3a), but when it is unstable its rotation number is not defined ($\epsilon = 4.5$, Fig. 3b).

unstable at ϵ=4.389, at an equal period bifurcation, followed by a cascase of period doubling bifurcations. All the period doubling cascades have the universal asymptotic bifurcation ratio of successive intervals between bifurcations $\delta = 8.72$ (Benettin et al. 1980).

2. Irregular Orbits

There are also other families, that are not produced by bifurcation from the unperturbed families, or from families bifurcating from them. Such families are called

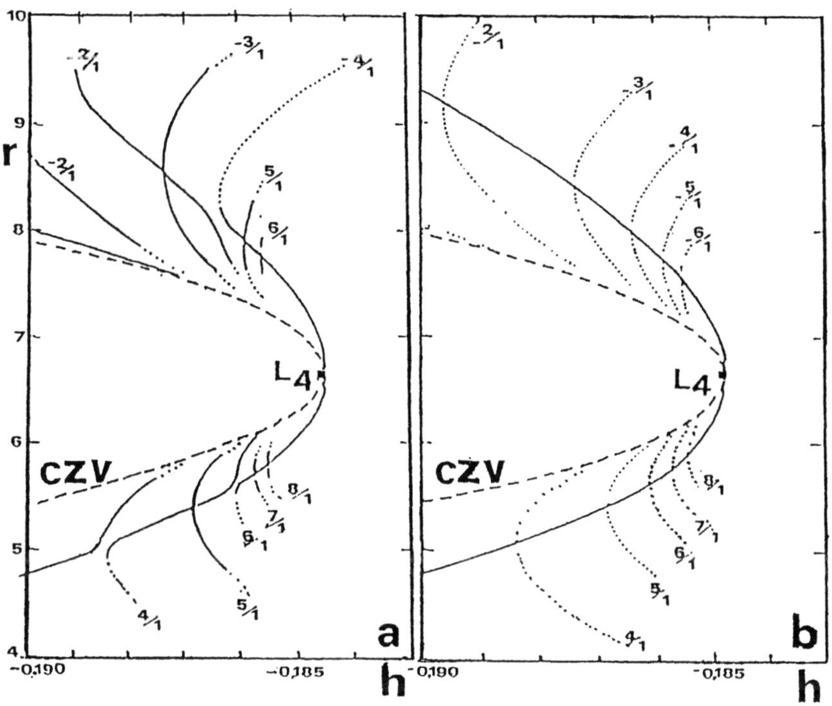

Fig. 4. Bifurcations and gaps from the central periodic orbit of a barred galaxy with a Hamiltonian of the form $H_0 = \bar{H}_0(r, \dot{r}) + \epsilon \bar{V}_1(r, \theta)$ (Contopoulos 1983) . The distance r along the axis $\theta = 0$ is given as a function of the energy H (a) for $\epsilon \neq 0$, (b) for$\epsilon = 0$.

irregular. E.g. in Fig.2 the families 6b and 8b do not bifurcate from any regular family.

The irregular families appear in pairs (one stable-one unstable) and each pair exists only for ϵ above a minimum value ϵ_{min}.

One way to see the pecular behaviour of the irregular families is by studying their rotation numbers (if they exist). Namely we consider the rotation number as a function of the coordinate $\bar{x}$ (Fig.3). When the central orbit is stable its rotation number is maximum (Fig.3a). The other stable periodic orbits are followed by an interval of values of $\bar{x}$ that have the same rational rotation number. E.g. the periodic orbit $\frac{n}{m}=\frac{4}{3}$ has rotation number $\frac{2}{3}$ (=2 − $\frac{4}{3}$)and the whole plateau around it has the same rotation number $\frac{2}{3}$. Near the unstable periodic orbits there is a chaotic region, where the rotation number is not defined (the region between the vertical dashed lines surrounding the unstable orbit with rot=$\frac{2}{3}$ in Fig.3a). Chaotic domains appear also near the boundaries of the islands (Fig. 3a).

But the irregular families have rotation numbers that may be quite different from the rotation numbers of nearby regular orbits. E.g. in the plateau $\frac{4}{5}$ (Figs. 2

and 3b) there are orbits with multiplicities 6 and 8. sometimes the rotation number of the irregular periodic orbits cannot be defined at all (Contopoulos 1970). E.g. the family 8b (Fig.2) has one branch in the plateau $\frac{4}{5}$ but its other branch is close to two families $\frac{5}{7}$, outside any plateau. Therefore one cannot assign a definite rotation number to it.

Even the family 6b, that has a rotation number $\frac{5}{6}$ (Fig.2), joins two plateaus of very different regular orbits, one with rot=$\frac{4}{5}$ and one with rot=$\frac{6}{7}$. This shows that the 5 islands around the stable periodic orbit $\frac{4}{5}$, that exist for $\epsilon = 4$, have been destroyed for $\epsilon = 4.6$, so that the orbit $\frac{5}{6}$ is not inside such an island.

One ambiguity about the irregular orbits is whether they can be joined to the regular orbits by changing another parameter besides ϵ. This happens, e.g., in the case of models of barred galaxies (Contopoulos 1983). The central family is split into an infinity of families by gaps, that appear at all even resonances $\frac{2n}{1}$, between the radial and the rotational frequencies, while at the odd resonances $\frac{(2n+1)}{1}$ we have bifurcations of resonant families.

Near the $\frac{4}{1}$ gap (Fig.4a) the central family x_1 deviates abruptly upwards, while a new pair of families appears beyond the gap. One of them is a continuation of the central family x_1 beyond the gap, and it is stable. The other family is unstable and joins the stable family at a minimum energy. similar gaps appear at the resonances $\frac{6}{1}, \frac{8}{1}$, etc. The appearance of gaps can be explained theoretically (Appendix of Contopoulos 1983). such gaps appear whenever the ratio of the frequencies $\frac{\omega_1}{\omega_2}$ is equal to an even number $\frac{2n}{1}$.

According to the original definition, the new families beyond the gaps are all irregular, because they do not join the original central family x_1 . But if we decrease ϵ the gaps become smaller and in the limit $\epsilon = 0$ they disappear (Fig.4b). In this last case we have bifurcations both at the even and the odd resonances. We notice that the galactic model depends on two parameters, the energy H and the strength of the bar ϵ. The fact that the gaps disappear and there are no irregular families for $\epsilon = 0$ leads us to consider the families beyond the gaps as not "genuine" irregular families. Thus from now on we will consider as (genuine) irregular families only those families that do not join the regular families with any change of parameters.

Now, however, the difficulty appears that one may not be sure that all possible changes of a parameter will not result in a joining of a presumably irregular family with the regular families.

Nevertheless there are generic cases of irregular families that cannot be joined to the regular ones by a continuous change of any parameter. Such are the families that are generated inside the lobes of the perturbed problem.

In the case of the Hamiltonian (1) for $\epsilon \neq 0$ the unstable triple periodic orbit forms lobes that generate the corrresponding homoclinic tangle (Fig.5).We define the homoclinic tangle as consisting of the lobes of a given unstable periodic orbit together with the areas between the intersecting lobes.The size of the tangle is found by calculating the asymptotic curves for long enough time in order to estimate the

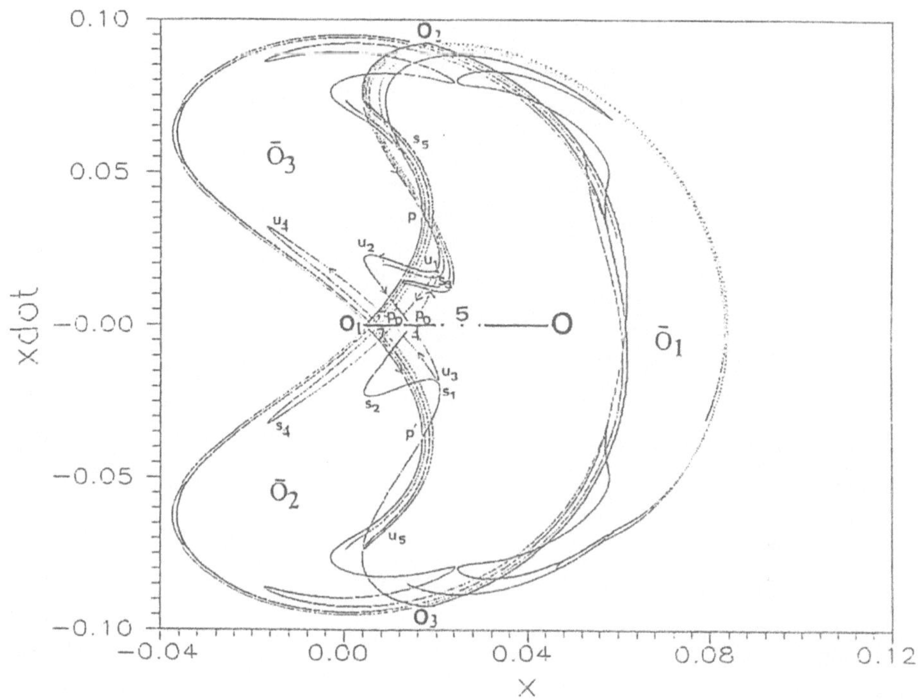

Fig. 5. The homoclinic tangle of the triple unstable periodic orbit $O_1O_2O_3$ of the Hamiltonian (1) for $h = 0.00765$ and $\epsilon = 4.3$.

maximum extent of the lobes.E.g the homoclinic tangle of the triple unstable orbit $O_1O_2O_3$,as given in Fig.7,covers a much larger area than what we could estimate by a short time calculation,as in Fig.5.

If ϵ decreases and tends to zero the lobes become smaller and for $\epsilon = 0$ they disappear completely. In fact the system (1) for $\epsilon = 0$ is an integrable Hamiltonian. The unstable triple periodic orbit does exist in this case also, but its asymptotic curves form separatrices and not a homoclinic tangle (Fig.6).

The regular periodic orbits are generated from the central periodic orbit (for $\epsilon \neq 0$), or from the boundary, or from the stable triple periodic orbit. But the periodic orbits inside the lobes cannot cross on asymptotic curve (if they did, they would be asymptotic orbits and not periodic orbits). Thus these periodic orbits cannot join the regular orbits that are all outside the lobes, and must disappear as ϵ decreases. Thus all the orbits inside the lobes are irregular. They are generated at a tangent bifurcation at a minimum $\epsilon = \epsilon_{min}$ and in pairs (stable-unstable). We conjecture that all (genuine) irregular periodic orbits are inside lobes. This is true for all the irregular orbits that we have calculated. A most remarkable case is a pair of irregular periodic orbits of multiplicity 19 in the case $\epsilon = 4.3$ (Fig.7). These orbits are far from the point O_1 and close to the central periodic orbit O. They are

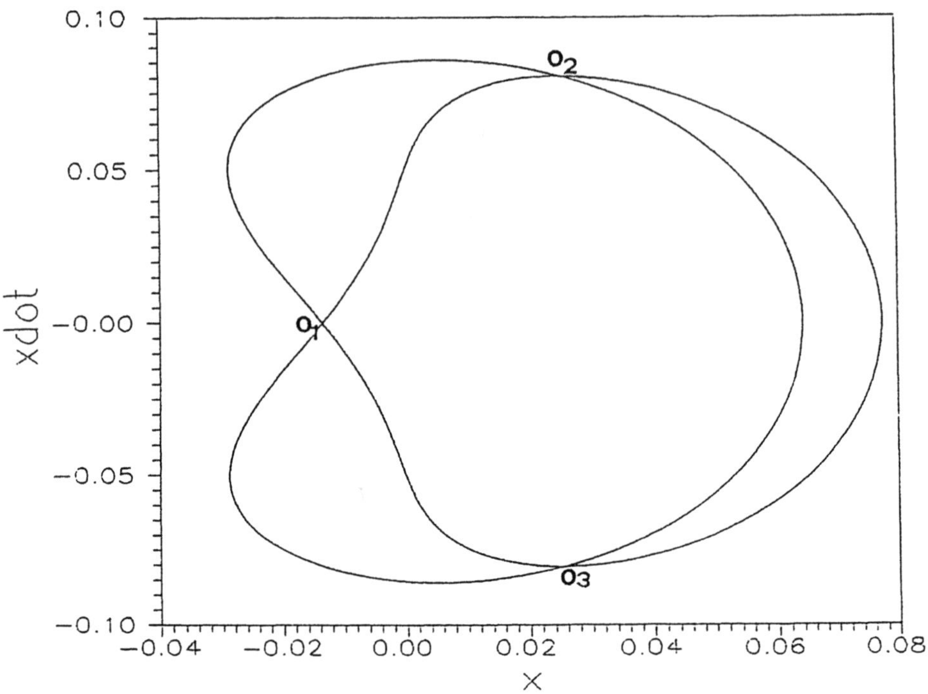

Fig. 6. The separatrices from the unstable orbit $O_1O_2O_3$ in the case $\epsilon = 0$.

on the right of the stable periodic orbit 5 and by looking at Fig.5 we cannot guess that the lobes of the triple periodic orbit $O_1O_2O_3$ can extend so far away from O_1 , i.e. beyond the orbit 5. Nevertheless if we calculate the asymptotic curves of the orbit $O_1O_2O_3$ for longer periods (Fig.7) we find lobes that reach the periodic orbits 19, well beyond the stable orbit 5. This is an unexpected verification (but not a proof) of our conjecture that all irregular orbits are inside lobes.

3. Distribution of Periodic Orbits

The easiest periodic orbits to calculate are those intersecting the x axis perpendicularly. If two orbits, starting perpendicularly at x and $(x + \Delta x)$, intersect the x-axis again at a positive and a negative angle, then between these two orbits there is an orbit intersecting the x-axis again perpendicularly, i.e. a (symmetric) periodic orbit. By using a small step Δx we can be sure that we have found all such orbits up to a given multiplicity.

An example of the distribution of perpendicular periodic orbits for $\epsilon = 4.0$ and $x > 0$ is given in Fig.8, up to multiplicity $m = 50$. In this figure we mark all the intersections of the orbits by the Poincaré surface of section $y = 0$ (i.e. the plane $(x, \dot{x})$) upwards (with $\dot{y} > 0$). We see the regular orbits that have bifurcated from

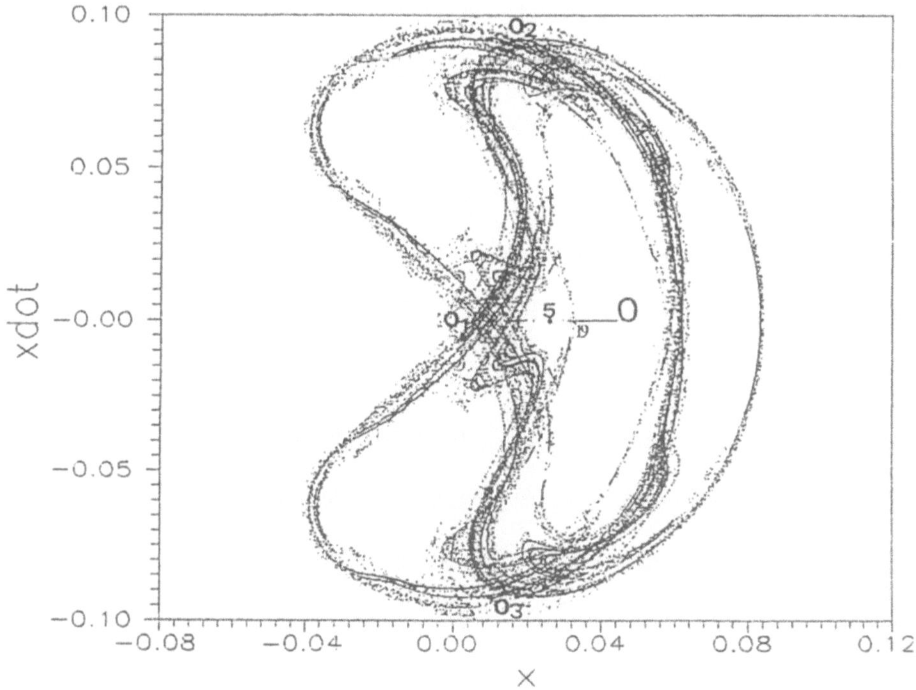

Fig. 7. Continuation of the asymptotic curves of Fig.5 of the case $\epsilon = 4.3$ for a long time. We see two lobes reaching the x-axis beyond the stable orbit 5, at the points where we have a pair of irregular orbits 19.Thus the orbit 5 is inside the homoclinic tangle.

the central periodic orbit O and from the stable triple periodic orbit $\bar{O}_1\bar{O}_2\bar{O}_3$. We see also some second order bifurcations, e.g. the period 45 orbit around the stable period 5 orbit.

The closest orbit to the center has a multiplicity $m = 19$, while the closest orbit to the stable triple orbit $\bar{O}_1\bar{O}_2\bar{O}_3$ has a multiplicity $m = 18$. Further away from the center we see a stable period 6 orbit, a stable and an unstable orbit of period 5, and further out a set of 8 points representing two unstable orbits of period 4. The orbits of period 6 and 5 intersect the surface of section downwards (i.e. with $\dot{y} < 0$) at the same points where they intersect it upwards (i.e. with $\dot{y} > 0$). But the orbits 4 have different points of intersection upwards (every second point of the ring of 8 points) and downwards (the other points of this ring). The second orbit of period 4 has $\dot{y} > 0$ ($\dot{y} > 0$) at the points where the first orbit has $\dot{y} > 0$ ($\dot{y} > 0$). The orbit marked 4 in Fig.8 has two points of perpendicular intersection upwards ($\dot{x} = 0, \dot{y} > 0$). Between the 8 dots of Fig.8 (unstable orbits 4 and 4') there are 8 gaps in the distribution, that contain 2 stable periodic orbits of period 4, one the reverse of the other. These orbits do not intersect perpendicularly the x-axis anywhere, but they are easy to find (by a Newton-Raphson method) because

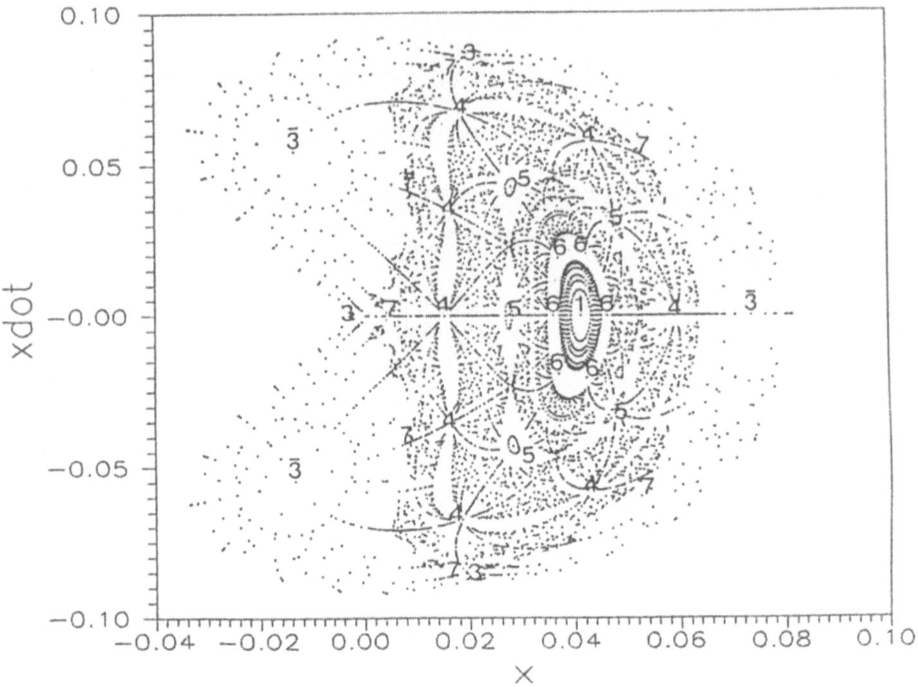

Fig. 8. The distribution of periodic orbits intersecting perpendicularly the x-axis (with $\dot{y} > 0$), on a Poincaré surface of section $(x,\dot{x})$, for $\epsilon = 4.0$.

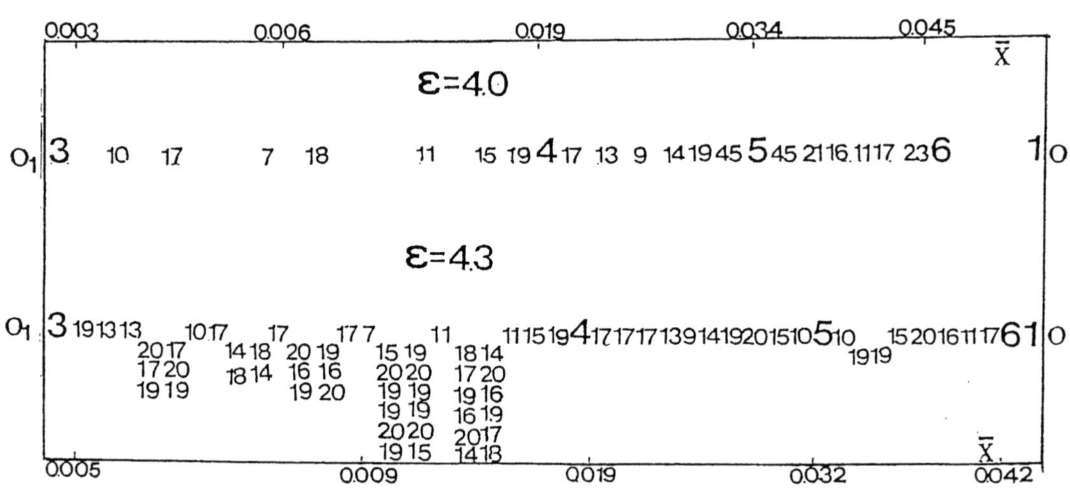

Fig. 9. The sequence of periodic orbits up to multiplicity $m = 25$ along the x-axis from O_1 to O for $\epsilon = 4$ and up to $m = 20$ for $\epsilon = 4.3$.The positions of the periodic orbits are only approximate. In the case $\epsilon = 4.3$ we have the regular orbits along the main line and the irregular orbits below it.

we can locate them approximately. In the same way, between the 6 dots closer to the center, there are 6 more points that represent an unstable periodic orbit.

In all cases the regular orbits appear in pairs (one stable - one unstable). All odd multiplicity orbits have perpendicular intersections with the x-axis, but only half of the even period orbits have perpendicular intersections and appear in Fig.8. Nevertheless the missing regular, even period, orbits can be easily located, because they have their points roughly halfway between the existing points of the same period, that form a ring around the center. But the missing irregular orbits are difficult to locate.

In Fig.8 we see some conspicuous lines composed of periodic orbits of different multiplicities. Such is the line that starts at the central orbit O (of multiplicity 1), passes through the upper left point of multiplicity 6 (stable), then goes through the upper left point of multiplicity 5 (stable), then the upper left point of multiplicity 4 (unstable) and terminates at the point O_2 of the unstable orbit of multiplicity 3.

This line is the image of the x-axis, therefore the dots that form it are easy to locate. The successive orbits along the x-axis from O_1 (orbit 3) and O (central orbit) are shown in Fig.9 (upper line). The orbits that we have mentioned above (1,6,5,4 and 3) are called "basic", and between them we have a large number of higher order peiodic orbits. All these orbits form Farey trees (Niven and Zuckerman 1960). Namely the rotation numbers of two orbits $\frac{n_1}{m_1}$ and $\frac{n_2}{m_2}$ are called "neighbors" if $m_1n_2 - n_1m_2 = \pm 1$. (such are the orbits above that have rotation numbers $\frac{5}{6}$, $\frac{4}{5}$,$\frac{3}{4}$ and $\frac{2}{3}$. Then the orbit with rotation number $\frac{(n_1+n_2)}{(m_1+m_2)}$ is a "neighbour" with both its "parents". By repeating this process we can find all the rationals between the original rotation numbers $\frac{n1}{m1}$ and $\frac{n_2}{m_2}$, arranged in the form of a tree (a Farey tree).

The periodic orbits of the Farey trees are generated by bifurcation from the central periodic orbit. In fact, as the perturbation increases, the rotation number of the central periodic orbit increases, passing through all the rational numbers of the Farey tree, and when the rotation number is rational the corresponding resonant family bifurcates from the central family.

The multiplicities of the intermediate orbits form arithmetic progressions. E.g. between m_1 and m_2 we have the progression of orbits of multiplicities $m_1 + m_2$, $m_1 + 2m_2$, $m_1 + 3m_2$ ... with increment m_2 , that approach the orbit m_2 , and the orbits $m_1 + m_2$, $2m_1 + m_2$, $3m_1 + m_2$ .... with increment m_1 that approach the orbit m_1 . These progressions of orbits are called "basic sequences". But between any two successive orbits of these sequences we have secondary sequences that form also Farey trees, so that finally all the orbits of multiplicities $s_1m_1 + s_2m_2$ appear in the Farey trees. As an example consider the orbits between 5 and 6 in Fig. 9 (case $\epsilon = 4.0$). We see the basic sequences 11,16,21... and 11,17,23,... . But we have also between 11 and 16 orbits not marked in Fig.9 like 27(11+16), 38(27+11), 49(38+11), and also 43(27+16) etc. Even the basic orbits in the present case form an arithmetic progression with increment 1, starting from 3 and approaching the central orbit (of multiplicity 1). Thus in the present case we can start the Farey

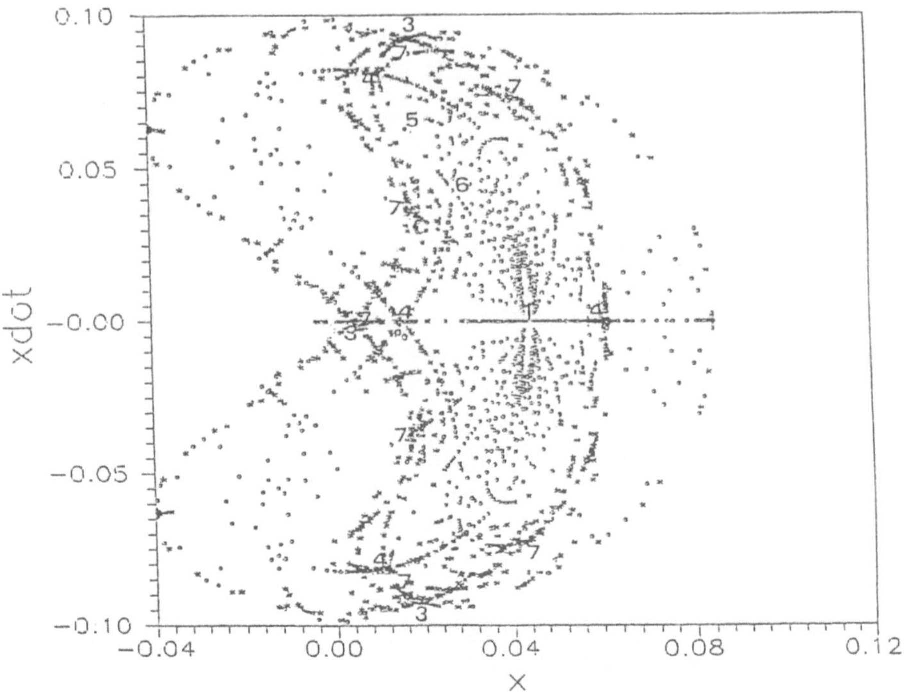

Fig. 10. The distribution of periodic orbits for $\epsilon = 4.3$.

trees with only 2 basic orbits, 3 and 1.

Besides these orbits we have the orbits bifurcating from every stable orbit of each sequence. such are the orbits 45 around the orbit 5 (Fig.9 for $\epsilon = 4$ and Fig.8). If we increase the perturbation to $\epsilon = 4.3$ (Fig.9) we find the same regular orbits as in the case $\epsilon = 4.0$, plus some lower order multiples of 5, namely the orbits 10,15 and 20, and some equal period bifurcations of the orbits 11 and 17. But there are also many more orbits that do not belong to the Farey trees. These are the irregular orbits shown in the lower line of Fig.9 for $\epsilon = 4.3$.

These irregular orbits have perpendicular intersections with the x-axis. Therefore they appear between the regular orbits on the x-axis. The successive images of the x-axis, in the mapping defined on the Poincaré surface of section, contain both regular and irregular orbits.

In the case $\epsilon = 4.3$ (Fig.10) we have the same basic orbits 1,6,5,4,3 and the same basic and secondary sequences,as in the case $\epsilon = 4$. In particular the first mapping gives a line that contains the basic orbits 1,6,5,4,3. But now we have many new irregular orbits that do not belong to Farey trees. Most of the irregular points are between the points 4 and 3. But we see also a pair of irregular orbits 19 between 6 and 5.

In Figs.8 and 10 we see many lines containing Farey tree (regular) orbits. The

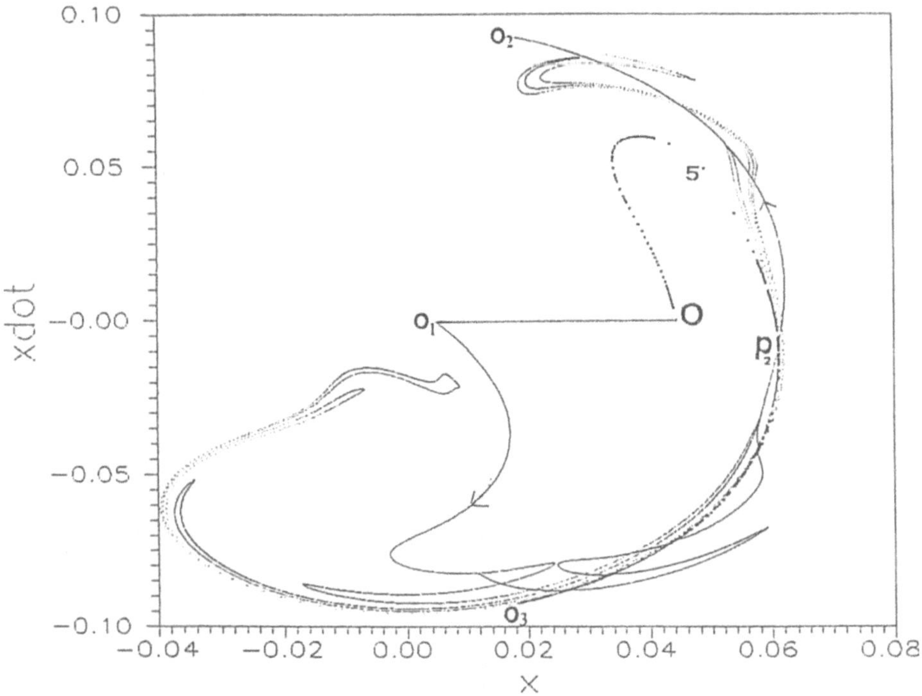

Fig. 11. The second image O_3p_2O of the set of periodic orbits of Fig.5 (x-axis) is a long thick line from O_3 to O with some gaps, mainly around the orbit 5. The point p_2 is the intersection of the unstable asymptotic curve from O_1 and the stable asymptotic curve from O_2.Some lobes formed by the unstable asymptotic curve from O_1,are given for comparison.

lines of Farey tree orbits do not contain irregular orbits in general except when they are images of the x-axis. But in Fig.10 we see also many more lines composed mainly of irregular orbits. These lines are close to the asymptotic curves of the unstable periodic orbit $O_1O_2O_3$. The appearance of these lines is explained as follows. Close to the unstable points $O_1O_2O_3$ their asymptotic curves form an intricate homoclinic tangle (Fig.5). E.g. from O_1 we have an unstable asymptotic curve, u, downwards, and a stable one, s, upwards (see the arrows). These curves intersect the asymptotic curves from the points O_2 and O_3 and form an infinity of lobes close to O_1 and almost parallel to the asymptotic curves from O_1 . Namely an unstable asymptotic curve from O_2 (arrows downwards) intersects the stable asymptotic curve s from O_1 at the point p, almost halfway between O_2 and O_1 , and from then on it forms the lobes u_1 ,u_2 ,u_3 , etc. similarly a stable asymptotic curve from O_3 intersects the unstable asymptotic curve u from O_1 at the point p' and forms the lobes s_1 ,s_2 ,s_3 etc, symmetric to the lobes u_1,u_2 ,u_3 ..... In particular the lobes u_3 and s_3 intersect at two homoclinic points p_0 and p_0'.

If we mark the periodic orbits along the x-axis from O_1 to p_0 , and beyond it, up

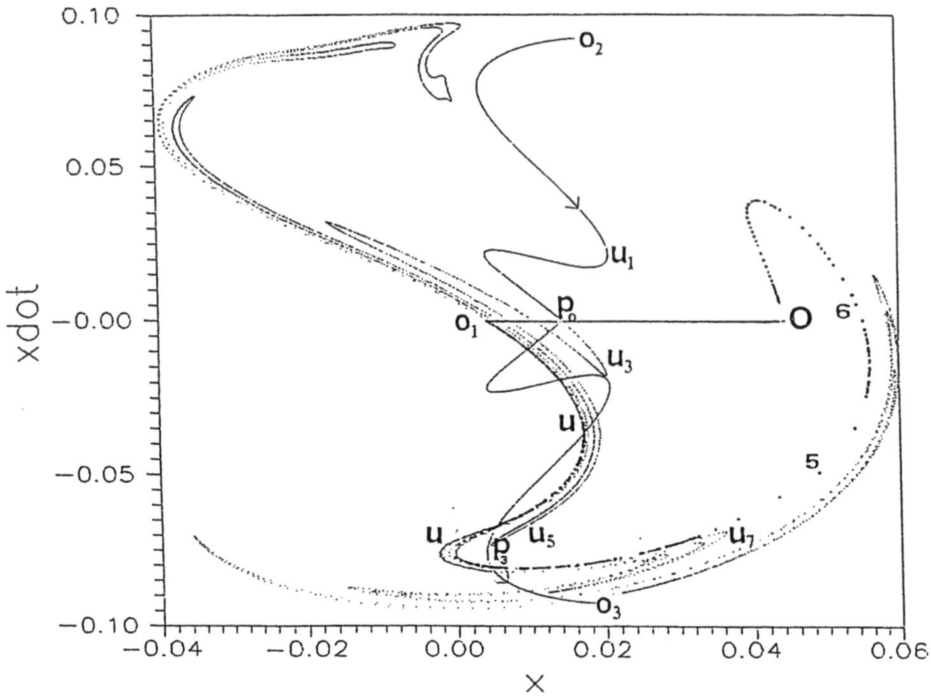

Fig. 12. The third image of the set of periodic orbits of Fig.5 is a thick line with many gaps, from O_1 to p_3 and O. The point p_3 is the intersection of the unstable asymptotic curve from O_2 (lobe u_7) and the stable asymptotic curve from O_3 . The line of periodic orbits O_1p_3 is sandwiched between the asymptotic curve u from O_1 and the lobe u_7 from O_2 . We see several unstable lobes from O_2 but only the initial arcs of the unstable asymptotic curve from O_1 and of the stable asymptotic curve from O_3.

to the central periodic orbit O (Fig.5) we can find their successive images on the Poincaré surface of section. E.g. the first image is the line 34561 that we can see in Figs. 8 and 10. But the higher order images of this line become more and more elongated along the unstable asymptotic curves from O_1 ,O_2 and O_3 , and close to them.

The second image of the line O_1p_0O is the thick line and dots O_3p_2O of Fig.11.In this figure we have marked also a few lobes,formed by the unstable asymptotic curve from O_1 (the exact form of these lobes is not important) and a part of the stable asymptotic curve from O_2 for comparison.These curves intersect at the point p_2 from which passes the thick line O_3p_2O.We notice that the thick line is close to the asymptotic lobes from O_1,except for the interval between the orbit 5 and the central periodic orbit O.

The third image of O_1p_0O is the line O_1p_3O (Fig.12) which is even longer and closer to the asymptotic curves than the line O_3p_2O of Fig.10 (except for the interval from the orbit 6 to the central point O). The point p_3 is the intersection of

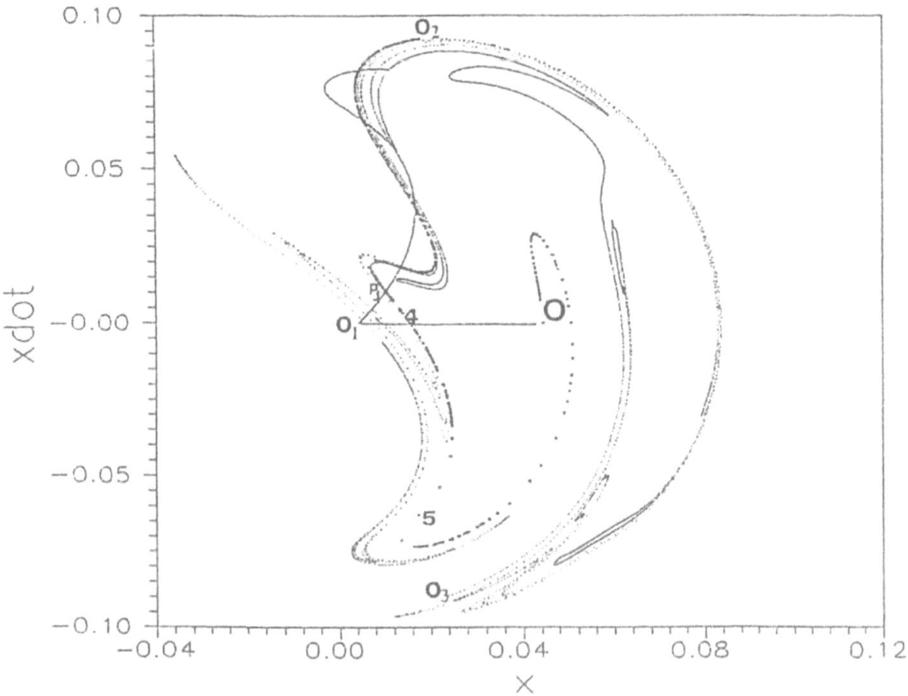

Fig. 13. The position of the orbit 7 on the x-axis in comparison with the lobes $u_3 s_3$ of the homoclinic tangle of the unstable periodic orbit $O_1 O_2 0_3$ for various values of e : (a)$\epsilon = 4.0$, (b)$\epsilon = 4.1$, (c)$\epsilon = 4.2$, (d)$\epsilon = 4.3$.

the lobe u_7 from O_2 with the stable asymptotic curve from O_3.But u_7 is very close to the unstable asymptotic curve u from O_1 (on the right of u).The curve of periodic orbits $O_2 p_3$ is on the left of u_7 and on the right of u,therefore it is sandwiched between u and u_7 ,i.e. it is very close to both these curves. The extension of the line of periodic orbits $O p_3$ is above the line u_7 , but again very close to it (and it is also sandwiched between u_7 and other higher order lobes above and close to u_7) up to the orbit 6. The region from the orbit 5 to 6 is no more close to the lobe u_7 but to higher order lobes from O_2 close to the right limit of Fig.12. From the orbit 6 up to the point O, the line of periodic orbits makes a turning to the left and then downwards.

This behaviour of the third image of the x-axis explains the appearance of lines of irregular orbits along (i.e. very close to) the asymptotic curves of the unstable periodic orbit $O_1 O_2 O_3$ in the general neighbourhood around O_1,O_2 and O_3 (but not around the central point O). The lines of periodic orbits close to O_1 ,O_2 and O_3 contain also some regular orbits. E.g. the point 7 in Fig.5 is on the left of the intersections $p_0' p_0$ of the lobes u_3,s_3 and close but on the right of the lobes u_5,s_5 . On the other hand the orbit 4 is on the right of p_0 . The segment 3-4 of the x-axis

(Fig.9 forϵ = 4.3) contains the regular orbit 7 and several other regular orbits. Some of them are close to 4, like the points 19,15 and 11 of Fig.9, then the points 11, (11,18,7,17), 17, (17,10) and many points like 13,16,19 close to 3. The points 11,18,7,17 and 10 are all in the interval between p'_0 and the lobes u_5 ,s_5 in Fig.5.

We conclude that the lines of periodic orbits in Fig.10 are of two basic forms. (a) Lines containing regular (Farey tree) orbits, and (b) Lines close to the asymptotic curves of the unstable periodic orbit $O_1O_2O_3$. Many lines are of composite type, i.e. they start close to the asymptotic curves of the orbit $O_1O_2O_3$ but continue as Farey tree lines that approach the central periodic orbit O.

4. Stable Orbits in the Homoclinic tangle

The orbit of multiplicity 7 on the left of the point p'_0 in Fig.9 was generated at the central periodic orbit O as a regular bifurcation, but in the present case it is separated from O by the lobes u_3 and s_3 . As we have pointed out above,a periodic orbit cannot cross an asymptotic curve as it evolves with changing ϵ. What happens is that the orbit 7 starting at O moves to the region between u_3 and s_3 (Fig.13a) and passes to the left of the lobes u_3 ,s_3 approaching the lobes u_5 , s_5 (Fig.13b). In these cases the lobes u_3 ,s_3 do not intersect and the homoclinic points p_0 ,p'_0 do not exist. Thus there is no barrier for the orbit 7 to pass to the left of the lobes u_3 , s_3 and approach (but not reach) the intersecting lobes u_5 ,s_5.

However for $\epsilon \simeq 4.2$ the lobes u_3 ,s_3 become tangent (Fig. 13c) and for$\epsilon = 4.3$ (Fig. 13d) they cross each other at p_0 ,p'_0, thus trapping the orbit 7 between the lobes u_3 ,s_3 and u_5 ,s_5 and away from the central orbit O, which is on the right of u_3, s_3. We note here that the orbit 7 is regular and does not belong to a lobe. According to our definition of the homoclinic tangle (Section 2),the homoclinic tangle contains the whole area inside the intersecting lobes around the unstable periodic orbit $O_1O_2O_3$ and not only the asymptotic curves or their lobes. Thus the orbit 7 is trapped inside the homoclinic tangle,although it does not belong to a lobe.

Several other orbits are trapped in the same way between the lobes of the triple resonant orbit $O_1O_2O_3$, as we have seen in the previous section, namely between $u_3(s_3)$ and $u_5(s_5)$, or between $u_5(s_5)$ and $u_7(s_7)$, etc.

The situation becomes more complicated when the central orbit O becomes unstable for $\epsilon = 4.305$. Then a new homoclinic tangle is formed around the central periodic orbit O and no more bifurcations occur at O. The regular periodic orbits, that bifurcated for smaller values of ϵ, are pushed away from O. For still larger ϵ the orbits O and $O_1O_2O_3$ have heteroclinic intersections and *all* the regular periodic orbits are then trapped inside the heteroclinic tangle. In Fig.14 ($\epsilon = 4.5$) we notice that the lobes of the orbit $O_1O_2O_3$ are much longer than the lobes of the case $\epsilon = 4.3$ (Fig.5). In particular the lobes u_1 ,s_1 are much longer and intersect on the x-axis. But we have also many intersections of the lobes of the orbit $O_1O_20_3$ with the lobes of the central periodic orbit.

The orbits trapped in the homoclinic tangle are sometimes stable. An example

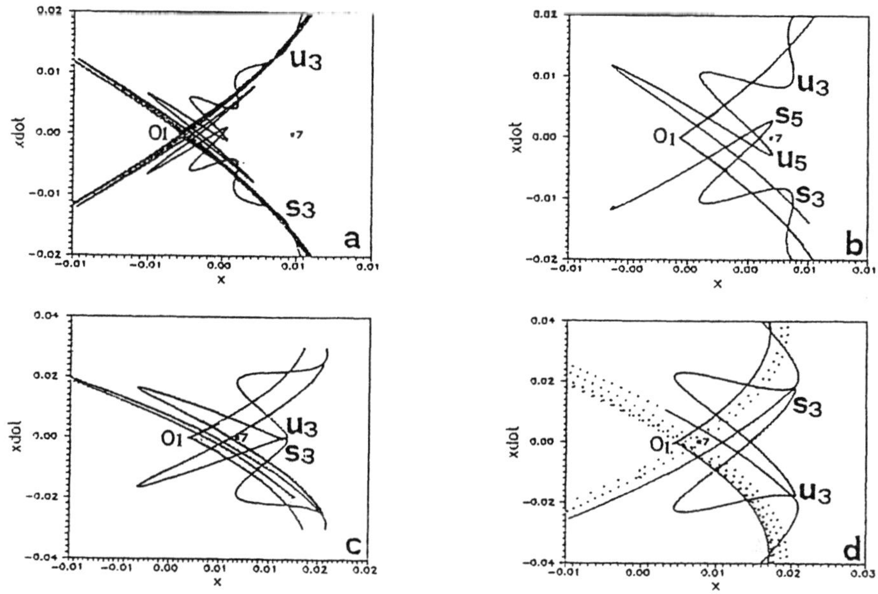

Fig. 14. Heteroclinic tangle between the asymptotic curves of the central periodic orbit O and the triple unstable periodic orbit $O_1O_2O_3$ for $\epsilon = 4.5$.

is shown in Fig.7, where the stable regular orbit 5 is trapped inside the lobes of the triple unstable periodic orbit $O_1O_2O_3$. Another example is shown in Fig. 16 ($\epsilon = 4.4$). In this case the central periodic orbit is marked by O and it is unstable, but its asymptotic curves do not have heteroclinic intersections with the triple periodic orbit. The lobes of $O_1O_2O_3$ are much longer than in the case $\epsilon = 4.3$ (Fig.5). The lobes u_1 and s_1 do not intersect, but higher order lobes intersect further to the right and trap several regular orbits away from the central periodic orbit O.

In Fig.15 we mark some stable regular periodic orbits along the x-axis that have bifurcated from the central periodic orbit O for smaller values of ϵ. Such are the orbits 7,9,13, and 10. Among them the orbit 10 is trapped in the homoclinic tangle of the triple periodic orbit.

Further on to the right we have the central triple periodic orbit $\bar{O}_1$ which is now unstable, but its own homoclinic tangle is very small and does not intersect the homoclinic tangle of the main triple unstable orbit $O_1O_2O_3$. On both sides of the central triple orbit $\bar{O}_1$ there are two stable triple orbits (marked 3) and further to the left there is a stable periodic orbit 6 that has bifurcated from the central triple orbit $\bar{O}_1$ for smaller ϵ. This orbit is now trapped inside the homoclinic tangle of the unstable orbit $O_1O_2O_3$.

We conclude that the homoclinic tangle of the unstable periodic orbit $O_1O_2O_3$

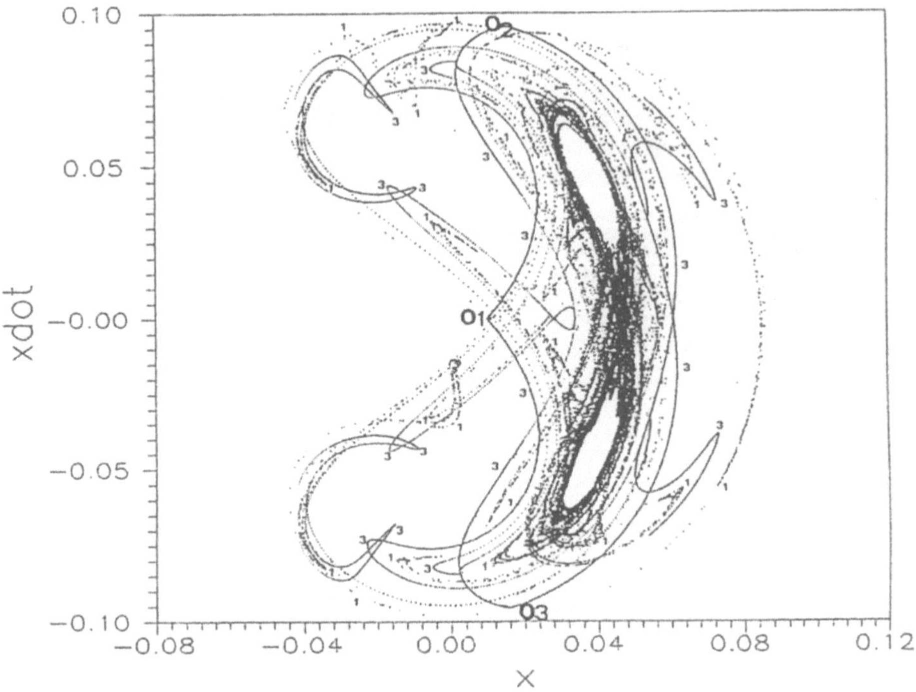

Fig. 15. Stable orbits in the homoclinic tangle, between the lobes of the triple orbit $O_1O_2O_3$ marked by stars (case $\epsilon = 4.4$).

has trapped inside it many regular periodic orbits (probably an infinity of them). Many such orbits are unstable, but some of them are stable, like the orbits 10 and 6 of Fig.15. These orbits are surrounded by small but finite islands of stability. Therefore the homoclinic tangle does not form a simply connected set. Although it contains a $\Lambda - set$ which is isomorphic to a Bernoulli shift (Guckenheimer and Holmes 1983), it contains also islands of stability.

This case seems to be a general one. In fact, as ϵ increases, the length of the lobes of the orbit $O_1O_2O_3$ (and of O) becomes larger in a finite total area. Thus we have new intersections between the various lobes. Such intersections appear first as tangencies between the lobes. Then the theorem of Newhouse (1983) proves the existence of nearby stable periodic orbits. Thus the stable periodic orbits are not exceptional, but probably generic.

This fact has been proven in the particular case of the logistic map (Jacobson 1981). Namely if the control parameter λ goes beyond the limit λ_∞ of the Feigenbaum sequence one can find infinite intervals of λ containing stable periodic orbits. The set of values of λ that do not have any stable periodic orbit has a positive measure, but this set is nowhere continuous. Thus between any two given values of λ there are islands of stability. This phenomenon is probably quite general.

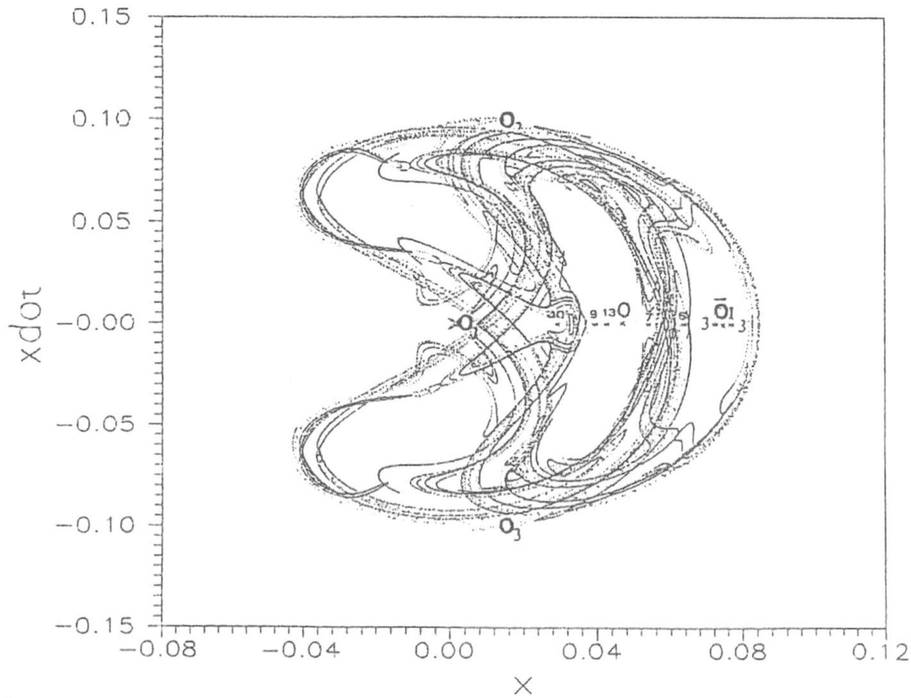

Fig. 16. An irregular stable periodic orbit of multiplicity $m = 16$ inside the lobes u_3 ,s_3 of the triple unstable periodic orbit $O_1O_2O_3$ for $\epsilon = 4.3879$.

The stable periodic orbits are both regular and irregular. An example of an irregular stable periodic orbit is given in Fig. 16. We know that the irregular families appear, as ϵ increases, in pairs (one stable and one unstable), at a tangent bifurcation. But the stable orbit soon becomes unstable, as ϵ increases further, followed by an infinity of period doubling bifurcations. Thus the interval of values of ϵ in which an orbit is stable is very small. The stable periodic orbit of Fig.15 was found by starting the opposite way, i.e. by following an orbit with decreasing ϵ until the orbit became stable. In particular the orbit 16 is an irregular family that exists for $\epsilon = 4.4$, but it is then unstable. By decreasing ϵ by steps $\Delta\epsilon = 0.001$ we found that this family becomes less and less unstable and disappears for $\epsilon = 4.387$. Then exploring with a smaller step $\Delta\epsilon = 0.0001$ we found that the orbit 16 which is unstable for $\epsilon = 4.388$, does not exist for $\epsilon = 4.3878$, but exists as stable for $\epsilon = 4.3879$.

In Fig. 16 we see that this orbit is inside the lobes u_3 and s_3.Further stable periodic orbits inside lobes were found in the same way. We conclude that even the regions inside lobes, which do not contain any regular periodic orbits, are not quite chaotic, but may have islands of stability. We expect that very small islands of stability exist in general inside the lobes.

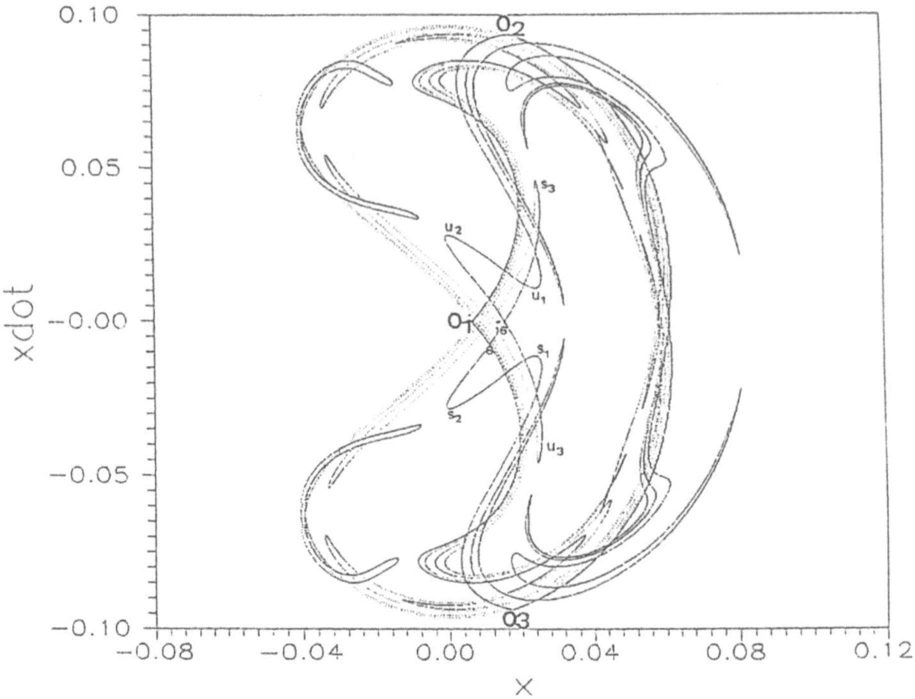

Fig. 17. The lobes of the unstable central periodic orbit O (for $\epsilon = 4.5$) are called u_1', u_2'... and s_0, s_{-1}... (outer lobes), or u_1 ,u_2 ... and $s_{0'}$,$s_{-1'}$ ... (inner lobes). All these lobes have equal area.

5. Minimum Recurrence Times

The (irregular) periodic orbits inside lobes cannot be of very low order, because a necessary (but not sufficient) condition in order to have such orbits is that a higher order image of a lobe intersects the original lobe. Only then there is a chance that some points of these lobes may be mapped into themselves, i.e. they define periodic orbits.

The intersection of images of a given set (of finite area) with the original set can be estimated by the Poincaré recurrence theorem (Poincaré 1899, Caratheodory 1919). This theorem specifies that an orbit in a compact phase space will come arbitarily close to its initial position an infinite number of times. The proof is well known. The images of a small circle of area M_0 inside a finite total area M on a Poincaré surface of section, have equal areas M_0 . If N such images do not itersect, their total area is NM_0 . But if N is the first integer for which $NM_0 > M$ then some images necessarily intersect, say M_a and M_b $(0 \leq a < b < N)$. Then, the images of M_0 and M_{b-a} also intersect, and $0 < b-a < N$. This theorem applies also to any area M_0 , which is not a circle.

The recurrence time may be smaller than the maximum N, but it cannot be

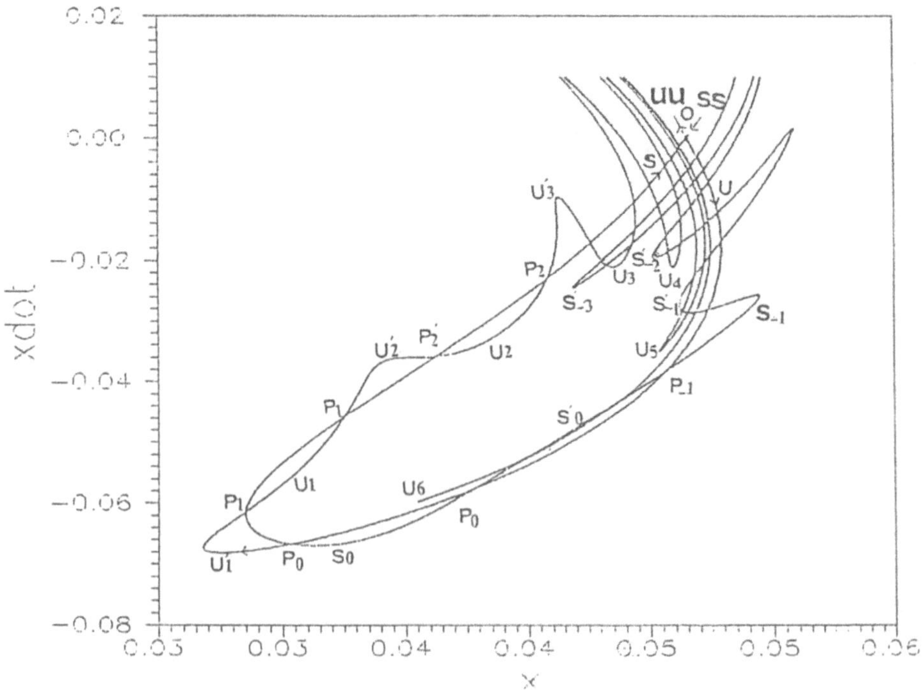

Fig. 18. The inner lobe u_6 does not intersect the outer lobe u'_1, but the lobes u_7 ,u_8 ... do.The arroes mark the end points of the lobes u'_{18},u'_{19} inside the lobes u_6,u_7,u_8. The lobe u'_{18} has parts inside the lobes u_6 and u_7 , but it does reach the lobe u'_1. The lobe $u_{19'}$ has some parts inside the lobes u_7 and u_8 and crosses the lobe u'_1.

arbitrarily small. In fact the minimum Poincaré recurrence time depends on the structure of the homoclinic tangle (Contopoulos and Polymilis 1996). In particular the irregular periodic orbits inside lobes cannot be of arbitrarily low order. They must be of sufficiently large order so that the images of the lobes intersect themselves.

Let us consider the lobes of the central periodic orbit when it is unstable (Fig.18). The asymptotic curves u (unstable) and s (stable) intersect at infinite homoclinic points P_0 ,P'_1, P_1 ,..., and P'_0 ,P_{-1} ... .At the points P_0 ,P_1 ,P_2 the curve u proceeds outwards from the resonance region O_1 and at the points P'_0 ,P'_1 ,P'_2 ... it proceeds inwards (for details of the nomenclature see Contopoulos and Polymilis 1993). The curves u and s between two successive homoclinic points form lobes. E.g. between P_0 and P'_1 we have the outer lobe u'_1 (also called s'_1), between P'_1 and P_1 we have the inner lobe u_1 (also called s_1), etc.similarly we have the asymptotic curves uu and ss around the resonance region O_2 , that are symmetric to s and u respectively with respect to the axis $\dot{x} = 0$, and form symmetric lobes.

All the lobes have equal areas. This is always the case for the outer lobes and the inner lobes separately; but in the present case the outer and inner lobes have also equal areas because of the symmetry of the problem.

The intersections of the lobes follow a simple rule: if the lobes u_n and s'_m intersect, the lobes u_{n+s} and s'_{m+s} also intersect (and the areas of intersection are the same). E.g. in Fig.17 the lobes u_3 and s_{-3} intersect. Then the lobes u_4 and s_{-2} also intersect, etc.

The outer (and inner) lobes become longer as their order increases. E.g. the lobes u'_4 and u'_5 are equal in area, but the base of u'_5 along the curve s is λ times smaller than the base of u'_4, where λ is the largest eigenvalue of the periodic orbit O. Thus the length of u'_5 is approximately λ times larger than the length of u'_4.

The unstable asymptotic curves u and uu cannot intersect themselves, or each other (and the stable asymptotic curves s and ss also cannot intersect). Thus in order that an image u'_n of the outer lobe u'_1 should intersect itself, u'_n should become very elongated and enter inside an inner lobe, which intersects the lobe u'_1 from its inner side s'_1. This cannot be done unless a higher order lobe is long enough to surround the whole resonance O_2 and most of the resonance O_1, and then enter the inner lobes. This is seen in Fig.18, where the lobe u'_{18} enters the inner lobes u_6 and u_7 , and the lobe u_{19} enters the lobes u_7 and u_8 . The inner lobe u_6 does not reach the outer lobe u'_1, but the lobes u_7 , u_8 and all higher order inner lobes do enter the lobe u'_1. On the other hand the outer lobe u'_{18} enters the lobes u_6 and u_7 , but does not reach the outer lobe u'_1, although the lobe u_7 does reach u'_1. But the lobe u'_{19} enters twice into the lobe u'_1, inside the inner lobes u_7 and u_8 .

We conclude that the first higher order outer lobe that enters the lobe u'_1 is u'_{19} . Therefore inside the lobe u'_1 there cannot be a periodic orbit of order smaller than 19-1=18. This is the minimum Poincaré time for recurrence of the lobe u'_1 with itself. The same minimum recurrence time applies to all other (outer and inner) lobes. E.g. the lobe u'_2 is intersected for the first time by the lobe u'_{20} , etc.

On the other hand it is not certain that inside the lobes there are periodic orbits of order 18, but it is expected that for a somewhat larger multiplicity we do have periodic orbits inside the lobes. In fact we found that in the present case the lobes contain periodic orbits of order $m \leq 24$.

If we would apply the Poincaré recurrence theorem without due consideration of the structure of the lobes we would find a much larger recurrence time. Namely the total chaotic area M, which contains all the images of u'_1 is about 4000 times larger than the area M_0 of u'_1. Thus the Poincaré recurrence time is about $N = \frac{M}{M_0} = 4000$, i.e. 200 time larger that the minimum recurrence time found above.

If we increase the perturbation ϵ the size of the lobes becomes larger, while the available space M remains almost constant. Thus the Poincaré recurrence time $N = \frac{M}{M_0}$ is smaller. But at the same time the eigenvalue λ of the unstable orbit O is larger. Therefore the lobes become longer fast and they can enter the original lobe u'_1 after a smaller number of iterations.

It is remarkable that the minimum recurrence time is defined also in cases above the escape perturbartion, when a Poincaré recurrence time cannot be defined

(Contopoulos and Polymilis 1996).

6. Conclusions

We make a distinction between regular periodic orbits, that bifurcate from the families of periodic orbits that exist in the unperturbed case, and irregular periodic orbits that are generated at tangent bifurcations from each other. We prove that all the orbits inside the lobes of the unstable periodic orbits are irregular.

On the other hand some orbits generated at a tangent bifurcation are not necessarily irregular, because they may join the regular families by changing some parameter. We present evidence (but not a proof) that all genuine irregular families, that cannot be generated from a regular family by any change of parameters, are generated inside lobes.

The distribution of the periodic orbits of a 2-D dynamical system of a Poincaré surface of section shows some conspicuous lines composed of orbits of various multiplicities. There are two types of lines. (a) Lines containing regular orbits that form Farey trees, and (b) lines that are close to the asymptotic curves of the main unstable periodic orbits. We explain the formation of both types of lines. some lines contain parts of both types, i.e. they start near the central periodic orbit as Farey tree lines, when the central orbit is stable, but as they come close to a resonant unstable periodic orbit they approach the asymptotes of this orbit.

Many regular orbits are trapped inside the homoclinic tangle of the main unstable periodic orbit (a triple orbit in our particular example).These orbits do not belong to the lobes of the triple unstable periodic orbit.The trapping is due to the fact that the lobes of this orbit increase in length, as the perturbation increases, and intersect each other, thus trapping several regular orbits away from the region around the central periodic orbit, where these regular orbits were generated.

Some orbits trapped in the homoclinic tangle are stable, therefore the homoclinic tangle is not completely chaotic, despite the fact that it contains a $\Lambda - set$, which is isomorphic to a Bernoulli shift. We found also stable irregular orbits inside some lobes, thus chaos is not complete inside the lobes either.

The irregular orbits inside the lobes cannot be of arbitrarily small multiplicity. Their multiplicity must be larger than the number of iterations needed for a lobe to intersect itself. This we call a minimum Poincaré recurrence time, and we show how such intersections can be realised.

Acknowledgements. We thank Dr. C.Polymilis for useful discussions.The present research was supported in part by the EEC Human Capital and Mobility project ERB4050PL930337.

References

Benettin, G., Cercignani, G., Galgani, L. and Giorgilli, A.: 1980, *Lett.Nuovo Cim.* **28**, 1.
Caratheodory, C.: 1919, *Sitzungsberichte Preuss. Akad. Wiss.* **580**.
Contopoulos, G.: 1970, *Astron.J.* **75**, 96.

Contopoulos, G.: 1971, *Astron.J.* **76**, 47.
Contopoulos, G.: 1983, *Physica* **D8**, 142.
Contopoulos, G. and Polymilis, C.: 1993, *Phys.Rev.* **E47**, 1546.
Contopoulos, G. and Polymilis, C.: 1996, *Celest.Mech.Dyn.Astron.* **63**,189 (in press).
Contopoulos, G., Grousousakou, E, and Polymilis, C.: 1996 *Celest.Mech.Dyn.Astron.* (in press).
Coullet, P. and Tresser, C.: 1978, *J. de Physique* **39**, C5.
Feigenbaum, M.: 1978, *J. Stat. Phys.* **19**, 25.
Guckenheimer, J. and Holmes, P.: 1983, Nonlinear Oscillations, Dynamical Systems and Bifurcations of Vector Fields, Springer Verlag, N. York.
Jacobson, M.I.: 1981, *Comm. Math. Phys.* **81**, 39.
Newhouse, S.E.: 1983, in G. Iooss, R.R.G. Helleman and R. stora (eds), Chaotic Behavior of Deterministic Systems, North Holland, Amsterdam, p. 443.
Niven, I. and Zuckerman, H.: 1960, An Introduction to the Theory of Numbers, Wiley, N. York.
Poincaré, H.: 1899, Les Méthodes Nouvelles de la Mécanique Céleste III, Gauthier Villars, Paris.

UNIVERSAL PROPERTIES OF ESCAPE IN DYNAMICAL SYSTEMS

CHRISTOS SIOPIS* , HENRY E. KANDRUP** and G. CONTOPOULOS*
Department of Astronomy, University of Florida, Gainesville, Florida, USA
and
RUDOLF DVORAK
Institut für Astronomie, Universität Wien, Türkenschanzstrasse 17, A-1180 Wien, Austria

Abstract. This paper summarizes a numerical study of the escape properties of three two-dimensional, time-independent potentials possessing different symmetries. It was found, for all three cases, that (i) there is a rather abrupt transition in the behaviour of the late-time probability of escape, when the value of a coupling parameter, ϵ, exceeds a critical value, ϵ_2. For $\epsilon > \epsilon_2$, it was found that (ii) the escape probability manifests an initial convergence towards a nearly time-independent value, $p_0(\epsilon)$, which exhibits a simple scaling that may be universal. However, (iii) at later times the escape probability slowly decays to zero as a power-law function of time. Finally, it was found that (iv) in a statistical sense, orbits that escape from the system at late times tend to have short time Lyapounov exponents which are lower than for orbits that escape at early times.

Key words: Chaotic Phenomena

1. Introduction

The aim of this paper is to review work that has been done in recent years on possibly universal escape properties of two-dimensional time-independent Hamiltonian systems ((Contopoulos, 1990), (Contopoulos and Kaufmann, 1992), (Contopoulos *et al.*, 1993), (Siopis *et al.*, 1995a), (Siopis *et al.*, 1995b)) as well as to report on some of the latest results. Three different systems have been studied, characterized respectively by the following Hamiltonians:

$$\begin{aligned} H_1 &\equiv \frac{1}{2}\left(\dot{x}^2+\dot{y}^2+x^2+y^2\right)-\epsilon x^2y^2=h_1, \\ H_2 &\equiv \frac{1}{2}\left(\dot{x}^2+\dot{y}^2+x^2+y^2\right)-\epsilon xy^2=h_2, \qquad (1) \\ H_3 &\equiv \frac{1}{2}\left(\dot{x}^2+\dot{y}^2+x^2+y^2-\frac{2}{3}y^3\right)+\epsilon x^2y=h_3. \end{aligned}$$

The first two systems each correspond to two harmonic oscillators, coupled via a quartic or cubic term which is multiplied by a coupling parameter ϵ. The last system is characterized by a modified Hénon-Heiles potential that allows for control of the coupling strength between the two space dimensions via the coupling parameter ϵ. When $\epsilon = 1$, H_3 becomes the standard Hénon-Heiles potential.

The isopotential surfaces that correspond to these Hamiltonians are shown in Figure 1. For a fixed energy h_i, $i = 1, 2, 3$, no orbit can escape if the value of ϵ is less than a critical $\epsilon_{esc} = \epsilon_{esc}(h_i)$. For $\epsilon \geq \epsilon_{esc}$, the isopotential surfaces

* Also: Astronomy Department, University of Athens, Panepistimiopolis, 157 84 Athens, Greece.
** Also: Department of Physics and Institute for Fundamental Theory, University of Florida.

Celestial Mechanics and Dynamical Astronomy **65**: 57–68, 1997.

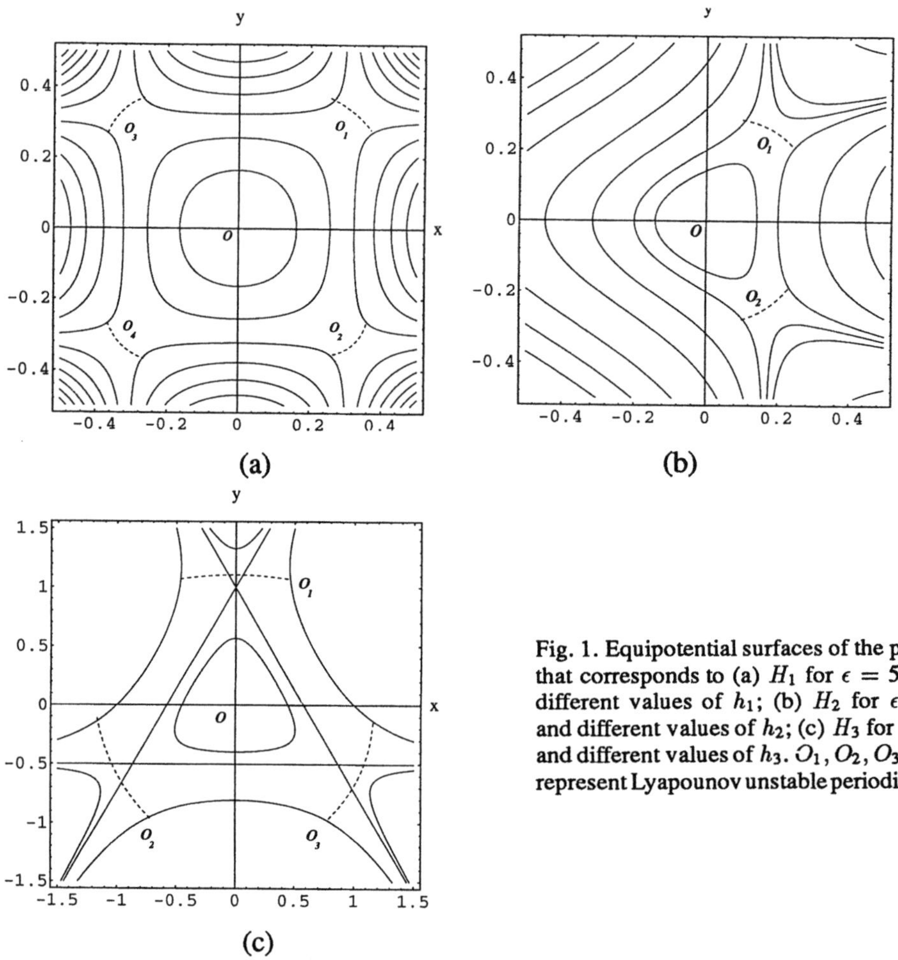

Fig. 1. Equipotential surfaces of the potential that corresponds to (a) H_1 for $\epsilon = 5.26$ and different values of h_1; (b) H_2 for $\epsilon = 3.0$ and different values of h_2; (c) H_3 for $\epsilon = 1.0$ and different values of h_3. O_1, O_2, O_3 and O_4 represent Lyapounov unstable periodic orbits.

form channels of escape, through which orbits can leak out. Any orbit that crosses, moving outward, one of the so-called Lyapounov unstable periodic orbits (see Figure 1), will escape from the system (Churchill *et al.*, 1979).

It will be shown in the following section that the escape properties that were found proved to be the same for all three Hamiltonians, despite their different symmetries. This result suggests strongly that the escape properties depend on generic phase-space characteristics rather than the details of individual potentials, and has motivated the claim for universality.

2. Short-Term Behaviour

The objective here is to study the escape properties of ensembles of orbits moving under the influence of the potentials associated with H_1, H_2 or H_3. To that end, a large number of orbits were computed numerically and followed up to a maximum

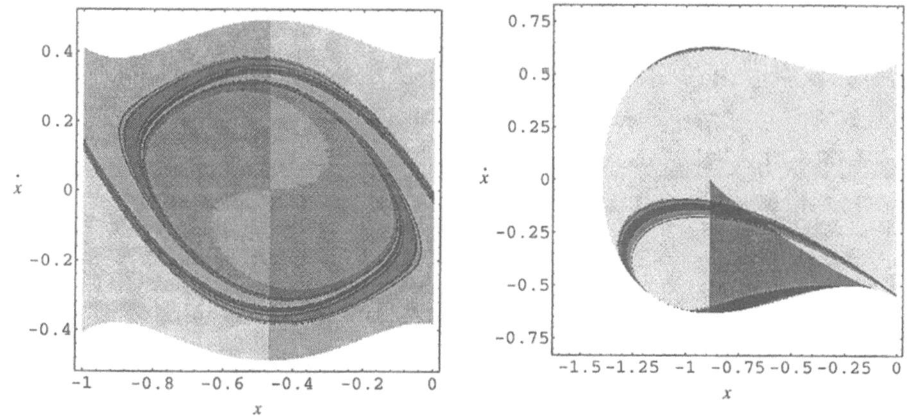

(a) (b)

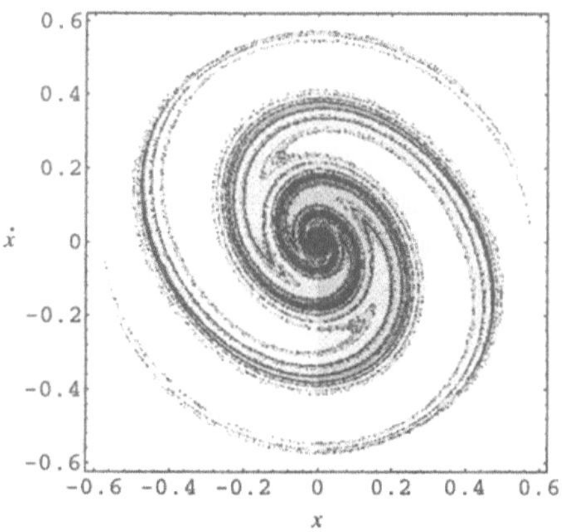

(c)

Fig. 2. The plane $(x-\dot{x})$ from which the initial conditions were sampled, for orbits characterized by the Hamiltonian (a) H_1 for $\epsilon = 5.26$; (b) H_2 for $\epsilon = 3.0$; (c) H_3 for $\epsilon = 1.3$. Darker shading implies longer escape times. Resolution is 400×400.

time or until they escaped, whichever happened first. In all experiments, time was discretised in terms of "consequents", defined as the number of intersections of the orbit with the $x = 0$ plane with $\dot{x} < 0$. The initial conditions were chosen to sample uniformly a square in the $(x - \dot{x})$ plane with $y = 0$, and $\dot{y} \geq 0$ uniquely determined from Eqs. (1). Figure 2 uses variable shading to illustrate the dependence of escape time on the initial conditions for the three Hamiltonians considered here. One can see that, whereas escape time is a rather smooth function of initial conditions in certain regions of phase space, elsewhere there is an extremely sensitive dependence on initial conditions. In fact, the latter regions exhibit the nearly self-similar structure characteristic of fractals (see (Contopoulos *et al.*, 1993) for more details).

For each Hamiltonian, ten 0.05×0.05 square cells were chosen to sample some of the "interesting" regions of phase space, *i.e.*, those exhibiting strong fractal structure. Scaling properties of the flow were also examined by studying smaller

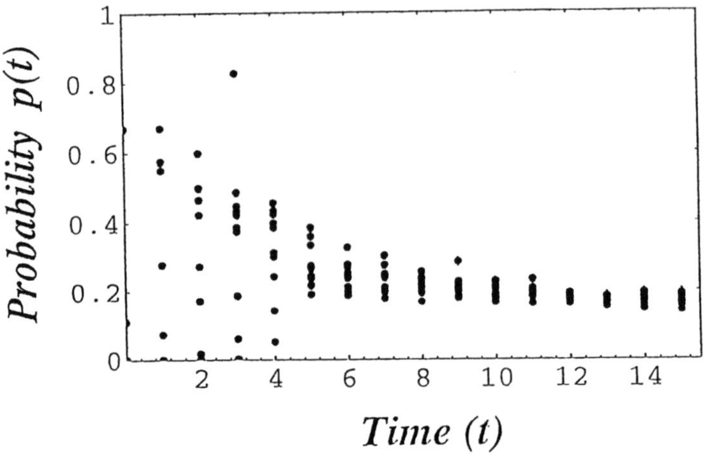

Fig. 3. The escape probability as a function of time for H_3, for $\epsilon = 1.3 > \epsilon_2$ (the error bars are too small compared to the dot size). Results of experiments from ten different cells of initial conditions with size 0.05×0.05 have been superimposed to produce the graph.

phase space regions, in some cases allowing for square cells with sides smaller by a factor of 500. For H_1, the initial $(x, \dot{x})$ pairs were sampled using a grid with an approximate resolution of 200×200 and a time-series orbit integrator. Using a Lie integration scheme (Hanslmeier and Dvorak, 1984) and an improved code to integrate the orbits, resolutions of 400×400, 800×800 and 1600×1600 as well as longer integration times were feasible for H_2 and H_3. The above procedure was repeated for several values of the coupling parameter ϵ.

The basic quantity of interest in this work is the probability $p(\epsilon, t)$ that an orbit will escape at time t from a Hamiltonian system with a coupling parameter ϵ, defined naturally via the relation

$$p(\epsilon, t) \pm \Delta p(\epsilon, t) = \frac{N_{esc} \pm \sqrt{N_{esc}}}{N_{tot}}. \tag{2}$$

Here N_{esc} is the number of trajectories from a given cell that escape between $t - 1$ and t, N_{tot} is the total number of trajectories still present at $t - 1$, and $\Delta p(\epsilon, t)$ represents an estimate of the uncertainty in the computed escape probability.

For $\epsilon < \epsilon_{esc}$ no orbit can escape and thus $p(\epsilon, t) = 0$ for all times. When $\epsilon \geq \epsilon_{esc}$, it is energetically *possible* for the orbits to escape. However, until a given orbit is actually calculated, it is not known if or when that orbit will escape. The numerical experiments show that, for a fixed energy, some orbits escape relatively fast whereas some others remain in the system for considerably longer times. In fact, slow escapers can remain bound for intervals more than two orders of magnitude longer than fast escapers. The escape probability behaves rather irregularly during the first several crossing times, in a fashion that reflects the specific choice of initial conditions (Figure 3). However, after these initial transients, $p(\epsilon, t)$ tends towards a constant value $p_0(\epsilon)$, the value of which depends on ϵ but not on the choice of the sampling square. More specifically, it appears that there is a critical value $\epsilon_2 > \epsilon_{esc}$ such that for $\epsilon_{esc} \leq \epsilon < \epsilon_2$, $p_0(\epsilon) \equiv 0$, whereas for $\epsilon > \epsilon_2$, $p(\epsilon, t)$ evolves towards a constant, nonzero value, $p_0(\epsilon)$. This transition occurs quite abruptly, within a narrow interval bracketing ϵ_2, and is strongly reminiscent of a phase transition.

It should be noted at this point that, as will be shown in the next section, the probability of escape $p_0(\epsilon)$ in the $\epsilon > \epsilon_2$ regime does *not* actually remain constant. Rather, it decays slowly with time. The rate of decay depends on the particular choice of Hamiltonian, but for the three systems considered here it is slow enough to justify the approximation that $p_0(\epsilon) \approx$ const. at intermediate (not too long) time scales.

The remainder of this section, as well as the following section will investigate properties of escape in the regime where $\epsilon > \epsilon_2$.

For all three Hamiltonians, a simple scaling behaviour was found, involving three distinct quantities ((Contopoulos *et al.*, 1993), (Siopis *et al.*, 1995a),(Siopis *et al.*, 1995b)):

i. the escape probability, $p_0(\epsilon)$;
ii. the time, $T(\epsilon)$, that is required for $p(\epsilon, t)$ to converge towards $p_0(\epsilon)$, for a fixed subset S of phase space volume, from which the initial con 'itions were sampled; and
iii. the time, $T(r)$, that is required for $p(\epsilon, t)$ to converge towards $p_0(\epsilon)$, as a function of the characteristic size r of S, for a fixed value of ϵ.

The following relations characterize the scaling:

$$\begin{aligned} p_0(\epsilon) &\propto (\epsilon - \epsilon_2)^{\alpha}, \\ T(\epsilon) &\propto (\epsilon - \epsilon_2)^{-\beta}, \\ T(r) &\propto r^{-\delta}. \end{aligned} \tag{3}$$

The numerical values of ϵ_2, the exponents α, β and δ as well as other relevant information for the three Hamiltonians can be found in Table I. It is significant that, within statistical uncertainties, the respective exponents of all three Hamiltonians could be identical. Also, again within statistical errors, $\alpha - \beta - \delta = 0$. This is more convincing for H_1 and H_3 than it is for H_2, which may be due to the superposition of short- and long-term behaviour.

3. Long-Term Behaviour

The results presented in the previous section, and in particular the evolution of $p(\epsilon, t)$ towards a constant $p_0(\epsilon)$, were obtained by integrating the orbits for roughly 20 crossing times. Thus, a natural question to ask is whether $p(\epsilon, t)$ remains constant at much later times. In order to address this question, a set of longer simulations were run for the system H_3. The results (Siopis *et al.*, 1995b) show that $p(\epsilon, t)$ is

TABLE I
System parameters of relevance to escape properties.

	H_1	H_2	H_3
h	0.12	0.125	$1/6 \approx 0.167$
ϵ_{esc}	$1/(4h) \approx 2.08$	$1/\sqrt{8h} = 1.00$	1.00
ϵ_2	4.90 ± 0.01	$1.15^{+0.02}_{-0.05}$	1.10 ± 0.05
α	0.49 ± 0.05	0.46 ± 0.05	0.45 ± 0.05
β	$0.39^{+0.14}_{-0.06}$	$0.50^{+0.15}_{-0.10}$	$0.37^{+0.10}_{-0.07}$
δ	0.08 ± 0.03	0.11 ± 0.03	0.12 ± 0.03

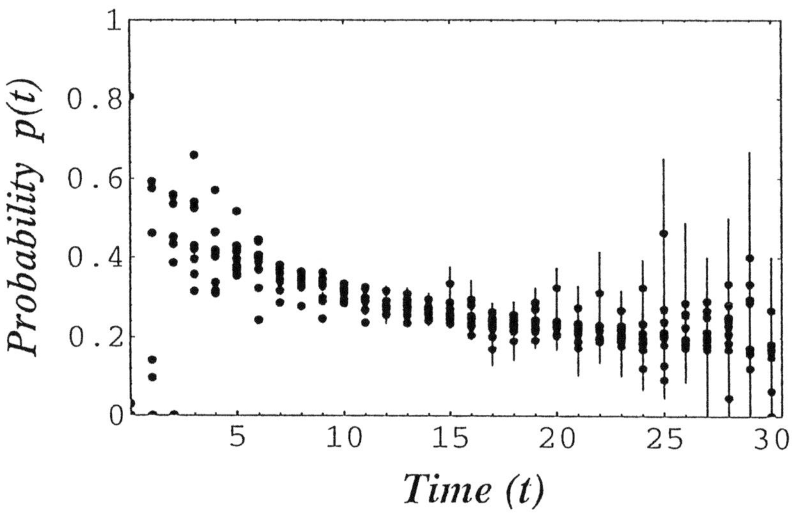

Fig. 4. The long-term decay for the Hamiltonian H_3 and $\epsilon = 1.6$.

not in fact constant. It decays, albeit slowly, with time (Figure 4). The total escape probability at late times was found to be well fit by the empirical relation

$$p(t) \sim t^{-\mu}. \tag{4}$$

Experiments with several values of ϵ indicate that $\mu = 0.31 \pm 0.05$, independent of ϵ.

4. Physical Interpretation

The numerical results obtained in the previous sections can be used to test some simple models of phase-space transport mechanisms. Perhaps the simplest model would be to assume that there is an absolute distinction between two populations of orbits, namely those that can escape and those that cannot. It can then be shown (Contopoulos *et al.*, 1993) that, at late times, $p(t) \propto \exp(-\lambda t)$, where $\lambda = -\ln(1 - p_\infty)$. However, as stated above, the late-time decay scales as a power-law, not exponential, function of time. Therefore, the assumption that there can only be two orbit populations, with no transitions possible between them, does not seem to be a valid one.

An alternative, more plausible, interpretation, can be achieved by invoking the presence of *cantori* around the islands of regularity. It can be shown that modeling of the structure of cantori as self-similar fractals can reproduce the observed late-time power-law decay ((Chirikov and Shepelyansky, 1984), (Hanson *et al.*, 1985), (Bleher *et al.*, 1990)). More specifically, according to the KAM theorem, regular orbits are surrounded in phase space by a hierarchy of invariant topological tori. These KAM tori act as impenetrable barriers that cannot be crossed by orbits, thus segregating phase space into disconnected regions that cannot communicate with each other. However, if somehow the system starts to deviate from integrability, the KAM tori become increasingly convoluted until they eventually lose their integrity and fragment into a Cantor set of holes. Orbits can now leak through these remnants of KAM tori, called cantori ((Contopoulos, 1971), (Aubry and Andre, 1978), (Mather, 1982), (Shirts and Reinhart, 1982)), and diffuse into the surrounding stochastic sea. The diffusion time scale depends on the size of the holes: if the system is sufficiently close to being integrable, the holes are small and diffusion proceeds slowly. However, as deviation from integrability becomes stronger, the holes very rapidly increase in size to the point where they practically no longer impede phase-space transport.

The fragmentation of phase space by KAM tori and cantori leads naturally to the conjecture (*cf.* (Mahon *et al.*, 1995)) that, at any given time, the orbits populating a dynamical system divide approximately into three different classes, namely *regular, sticky* or *temporarily confined chaotic,* and *unconfined chaotic* orbits. The *regular* orbits are trapped forever inside KAM tori. The *sticky,* or *temporarily confined,* orbits can spend a substantial time trapped inside cantori near regular islands, but eventually diffuse through the cantori and find their way out to the stochastic sea. If the phase space were compact, and escape to infinity impossible, the remaining *unconfined chaotic* orbits would simply travel unimpeded throughout the surrounding stochastic sea, although occasionally becoming temporarily confined near the regular islands. Given, however, that escape *is* possible, these orbits will eventually pass through a Lyapounov curve and escape to infinity. (Strictly speaking, in general there will also exist a finite measure of chaotic orbits inside the KAM tori, but these will never be able to enter the surrounding stochastic sea).

The preceding numerical results could then be interpreted if one makes the Ansatz that the transition occurring around ϵ_2 corresponds to the fracture of some dominant KAM tori. In this context, when ϵ exceeds ϵ_2, the last major KAM torus is destroyed, leaving behind one or more cantori which now dominate phase-space transport. Unlike KAM tori, cantori are leaky barriers, and a chaotic orbit moving around inside a cantorus will eventually find its way out in the surrounding stochastic sea through one of the holes of the cantorus. Nevertheless, if the size of the holes is small enough (which is presumably the case for values of ϵ only slightly greater than ϵ_2), an orbit can remain trapped inside the cantorus for extremely long times compared to the natural crossing time ("stickiness" effect). In any case, once in the stochastic sea at large, it will sooner or later escape from the system.

Ensembles of unconfined orbits that populate the stochastic sea inside the Lyapounov curves exhibit a tendency to evolve *towards* a near-invariant distribution (although if the orbits escape or diffuse through the cantori too fast compared to the time scale of approach towards the near-invariant distribution, no real sense of near-invariance may ever be achieved). The unconfined population initially consists only of that subset of the ensemble of initial conditions that lies outside the major cantorus. However, this population is continuously augmented by the orbits that leak through the cantorus. At the same time, orbits are drained from the system as they cross the Lyapounov curves and escape to infinity.

The situation is quite different when $\epsilon < \epsilon_2$. Now a substantial fraction of KAM tori still exist, surrounded by a hierarchy of cantori in the shallows of the stochastic sea. In this case, ensembles of orbits with initial conditions inside the KAM tori are trapped and cannot escape. However, the remaining orbits, even if located initially inside cantori, will eventually escape from the system in a manner similar to the $\epsilon > \epsilon_2$ case described above. In other words, at late times $N_{esc} \to 0$ [see Eq. (2)] but $N_{tot} \to N_{tot}^0 \neq 0$, since N_{tot} is bounded from below by the population of orbits inside the KAM tori that never escape. Therefore, the probability of escape as a function of time decays to zero faster than it does in the $\epsilon > \epsilon_2$ case, and does not exhibit the scaling behaviour observed for $\epsilon > \epsilon_2$ [Eqs. (3)].

5. Escape Time and Short Time Lyapounov Exponents

In the previous section, the distinction was made between three different types of orbits that can populate a dynamical system, namely regular, sticky or temporarily confined chaotic, and unconfined chaotic. This distinction was motivated by the presence of KAM tori and cantori, with the former acting as absolute and the latter as leaky barriers, together determining the dynamics of phase-space transport. Moreover, an interpretation was offered as to how such a model could explain the numerical results, *i.e.*, the initial convergence of the escape probability towards $p_0(\epsilon)$ and the subsequent power-law decay to zero.

An independent tool for gaining insight into phase-space transport and differentiated orbit populations is the (maximal) short time Lyapounov exponent $\chi(\Delta t)$, defined as

$$\chi(\Delta t) = \lim_{\delta z(0) \to 0} \frac{1}{\Delta t} \ln \left(\frac{\| \delta z(\Delta t) \|}{\| \delta z(0) \|} \right),$$

where $\delta z(0)$ is the initial phase-space perturbation and $\|\|$ represents a suitable (*e.g.*, Euclidean) norm (*cf. e.g.* (Grassberger *et al.*, 1988), (Kandrup and Mahon, 1994), (Voglis and Contopoulos, 1994)). This is the finite-time analogue of the well-known asymptotic maximal Lyapounov exponent $\chi \equiv \lim_{\Delta t \to \infty} \chi(\Delta t)$, which is

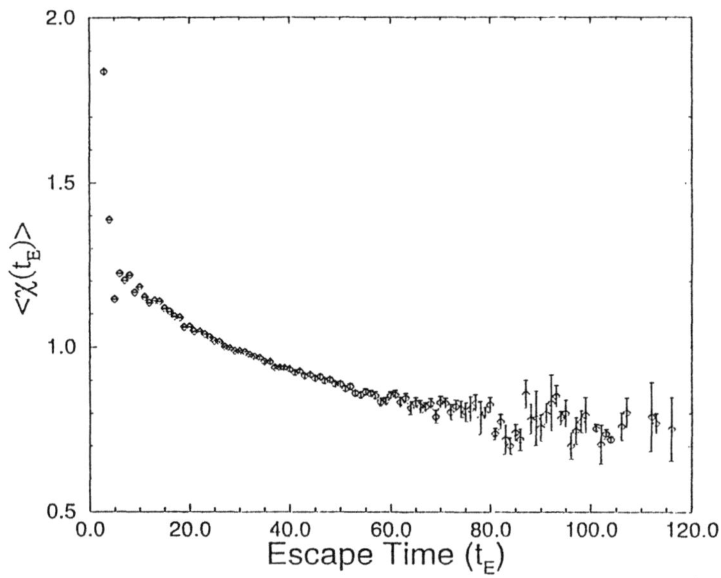

Fig. 5. The average short time Lyapounov exponent $\langle\chi(t_E)\rangle$ for ensembles of orbits that escape at time t_E versus t_E.

a measure, in an asymptotic sense, of the average rate of exponential instability of an orbit.

The usefulness of short time Lyapounov exponents for the problem at hand stems from the following two observations:

- First, one would expect that, for a given time interval Δt_0, the average of the short time Lyapounov exponents $\langle\chi(\Delta t_0)\rangle$ of a randomly chosen ensemble of sticky orbits restricted inside a major cantorus will be *smaller* than the corresponding $\langle\chi(\Delta t_0)\rangle$ of a randomly chosen ensemble of orbits moving around in (but without escaping from) the stochastic sea (*cf.* (Mahon *et al.*, 1995)).
- Second, in the spirit of the discussion of the previous section, the majority of the orbits that escape from the system at early times (say, before $t = 15$ or so) belong to that subset of the initial ensemble of orbits whose initial conditions placed them in the stochastic sea. By contrast, most of the orbits that escape at later times started as sticky orbits inside one of the major cantori wherein they spent most of their time until they crossed the cantorus through one of its holes and entered the stochastic sea, thus becoming unconfined. Escape followed relatively soon thereafter.

With these two observations in mind, one would expect that the average short time Lyapounov exponent for an ensemble of orbits that escape at early times will be *greater* than for an ensemble of orbits that escape at later times. This is so because the later an orbit escapes from the system the more time it will have spent

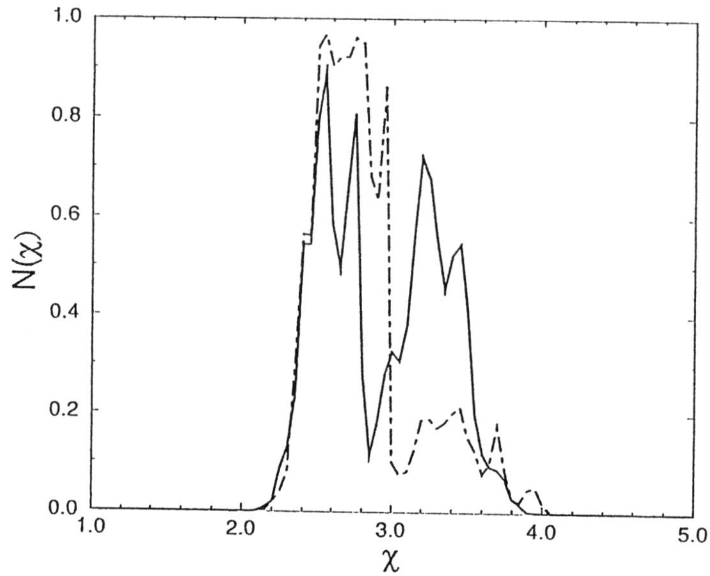

Fig. 6. The distribution $N[\chi(t)])$ at $t = 10$ of the short time Lyapounov exponents of the orbits that escape (i) at $10 \leq t_E < 15$ (solid line) and (ii) at $t_E \geq 40$ (dot-dashed line).

inside the lower-χ environment of a major cantorus relative to the time it spent in the higher-χ environment of the stochastic sea before it escaped. In other words, the lower Lyapounov exponents associated with the region inside the cantorus will weigh more towards the final average short time Lyapounov exponent.

In order to check the validity of this qualitative description, the following numerical experiment was performed. For the Hamiltonian H_3 with $\epsilon = 1.30$, a 0.05×0.05 square of initial conditions was sampled with a resolution of 1000×1000 and the orbits were evolved as in the previous sections until they escaped. In order to compute the short time Lyapounov exponent, a perturbation of magnitude $\| \delta z \| = 10^{-12}$ was introduced in the x coordinate of each orbit at $t = 0$. The perturbation was renormalised each time $\| \delta z \|$ exceeded 10^{-8}. At very early times, the computed exponent will depend sensitively on the initial δz, but for $t > 5$ or so it will provide a reasonable approximation to the exponent in the most unstable direction.

Some of the results are shown in Figures 5 and 6 (a more complete discussion will be presented in a future paper). Figure 5 plots the average short time Lyapounov exponent, $\langle\chi(\Delta t)\rangle = \langle\chi(t_E - 0)\rangle = \langle\chi(t_E)\rangle$, of an ensemble of orbits that escape at a given time t_E as a function of t_E. It is important to emphasize that, for any given escape time along the abscissa, only the Lyapounov exponents of those orbits that escape at that particular time are considered in the average. All the other orbits that are present, but escape at later times, are excluded. It is evident that orbits escaping at later times tend to have short time Lyapounov exponents that are smaller than orbits escaping at early times, in agreement with the interpretation given above.

In order to further test the validity of this simple model, a more detailed dissection of the system was made at a fixed early time. Figure 6 shows the distribution $N[\chi(t)])$ at $t = 10$ of the short time Lyapounov exponents of the orbits that escape (i) at $10 \leq t_E < 15$ (solid line) and (ii) at $t_E \geq 40$ (dot-dashed line). It is obvious that, despite the presence of some structure, there is an overall correlation between the computed χ and the time of escape. Orbits that escape from the system at early times tend systematically to have higher short time Lyapounov exponents than orbits that escape at later times.

Acknowledgements

The authors would like to thank Heinrich Eichhorn for his comments on the statistical interpretation of the numerical data.

H.E.K. was supported in part by the National Science Foundation Grant PHY92-03333. The remaining authors were supported in part by the European Community Human Capital and Mobility Program ERB4050 PL930312 (FWF project 10212-AST).

C.S. would like to thank the Department of Astronomy, the Graduate School and the Graduate Student Council of the University of Florida as well as the Organizing Committee for the financial support.

Some of the computations in this paper were effected using computer time made available through the Research Computing Initiative at the Northeast Regional Data Center (Florida) by arrangement with IBM.

References

S. Aubry and G. Andre. In A. R. Bishop and T. Schneider, editors, *Solitons and Condensed Matter Physics*, Springer Lecture Notes in Physics, Berlin, 1978. Springer.

S. Bleher, C. Grebogi, and E. Ott. *Physica D*, 46:87, 1990.

B. V. Chirikov and D. L. Shepelyansky. *Physica D*, 13:395, 1984.

R. C. Churchill, G. Pecelli, and D. L. Rod. In G. Casati and J. Ford, editors, *Como Conference Proceedings on Stochastic Behavior in Classical and Quantum Hamiltonian Systems*, volume 93 of *Springer Lecture Notes in Physics*, page 76, Berlin, 1979. Springer.

G. Contopoulos. *AJ*, 76:147, 1971.

G. Contopoulos. *Astron. Astrophys.*, 231:41, 1990.

G. Contopoulos and D. E. Kaufmann. *Astron. Astrophys.*, 253:359, 1992.

G. Contopoulos, H. E. Kandrup, and D. E. Kaufmann. *Physica D*, 64:310, 1993.

P. Grassberger, R. Badii, and A. Poloti. *J. Stat. Phys.*, 51:135, 1988.

A. Hanslmeier and R. Dvorak. *Astron. Astrophys.*, 132:203, 1984.

J. D. Hanson, J. R. Cary, and J. D. Meiss. *J. Stat. Phys.*, 39:327, 1985.

H. E. Kandrup and M. E. Mahon. *Astron. Astrophys.*, 290:762, 1994.

M. E. Mahon, R. A. Abernathy, Bradley B. O., and H. E. Kandrup. *Mon. Not. R. Astron. Soc.*, 275:443, 1995.

J. N. Mather. *Topology*, 21:457, 1982.

R. S. Shirts and W. P. Reinhart. *J. Chem. Pjus.*, 77:5204, 1982.

C. V. Siopis, G. Contopoulos, and H. E. Kandrup. In H. E. Kandrup, S. T. Gottesman, and J. R. Ipser, editors, *Three-Dimensional Systems*, volume 751 of *Annals of the New York Academy of Sciences*, pages 205–212, 1995.

C. V. Siopis, G. Contopoulos, H. E. Kandrup, and R. Dvorak. In J. H. Hunter, R. E. Wilson, and J. R. Ipser, editors, *Waves in Astrophysics*, volume 773 of *Annals of the New York Academy of Sciences*, pages 221–230, 1995.

N. Voglis and G. Contopoulos. *J. Phys. A*, 27:4899, 1994.

MAPPING AND DYNAMICAL SYSTEMS

MILOŠ ŠIDLICHOVSKÝ
Astronomical Institute of the Academy of Sciences of the Czech Republic, Boční II 1401, 141 31 Praha 4, Czech Republic

Abstract. This paper reviews various mapping techniques used in dynamical astronomy. It is mostly dealing with symplectic mappings. It is shown that used mappings can be usually interpreted as symplectic integrators. It is not necessary to introduce any δ functions it is just sufficient to split Hamiltonian into integrable parts. Actually it may be shown that exact mapping with δ function in the Hamiltonian may be non-symplectic. The application to the study of asteroid belt is emphasised but the possible use of mapping in planetary evolution studies, cometary and other problems is shortly discussed.

Key words: Dynamical Systems – Mapping – Symplectic Mapping

1. Introduction

Mapping as a rule which associates to each element of one set A one element in another set B is a very general concept widely used in mathematics. Sets A and B are usually endowed with some structure. The mappings respecting these structures are given special names, for instance homomorphism for mappings between groups respecting the group operation, homeomorphism for mapping between topological spaces respecting open sets and this concept is still sharpened for diffeomorphism of manifolds which must be infinitely often differentiable (together with its inverse).

The Hamiltonian dynamical systems are usually described in a phase space (cotangent bundle of configuration space). This cotangent bundle is a manifold where **q**, **p** represent $2n$ local coordinates $q_1, \ldots, q_n, p_1, \ldots, p_n$.

Dynamical system is a manifold M together with one parameter family of diffeomorphisms $g : M \times R \mapsto M$ and $g(x,t)$ is often denoted $g_t(x)$. In addition g_0 must be identity and $g_{t+s} = g_t \circ g_s$. The diffeomorphisms g_t form a group which is called a flow. Usually we imagine that parameter t is time and g_t for given t is a mapping $g_t : M \mapsto M$, which maps each x to its position after evolution for time t. The flow g_t is not usually given explicitely. For the Hamiltonian dynamical systems the flow is given by the differential Hamilton's equations

$$\frac{dq_i}{dt} = \frac{\partial H}{\partial p_i}, \quad \frac{dp_i}{dt} = -\frac{\partial H}{\partial q_i}, \tag{1}$$

where $H(q_i, p_i)$ is the Hamiltonian function. The cotangent bundle M has the natural symplectic structure given by a closed nondegenerate 2-form

$$\omega^2 = \sum dp_i \wedge dq_i. \tag{2}$$

Such manifolds M are called symplectic manifolds. For each differentiable map $f : M \mapsto N$ (M, N are manifolds) we may define (see for instance Arnold, 1989) f_* and f^* where f_* is the differential of f; $f_* : T(M) \mapsto T(N)$ where $T(M)$ and

Celestial Mechanics and Dynamical Astronomy **65**: 69–84, 1997.

$T(N)$ are corresponding tangent bundles (f_* simply maps the tangent vector of a curve to tangent vector of its image. On the other hand f^* takes the differential form ω defined on $T(N)$ to form $f^*\omega$ defined on T(M) so that e.g. for 2-forms

$$(f^*\omega)(x_1, x_2) = \omega(f_*x_1, f_*x_2), \tag{3}$$

where $x_1, x_2 \in T(M)$. For a Hamiltonian dynamical system defined on a symplectic manifold M it is easy to show (see for instance Arnold 1989)

$$g_t^*\omega^2 = \omega^2, \tag{4}$$

so that a Hamiltonian phase flow g_t preserves the symplectic structure. The consequence of Eq. (4) is that 2-forms $\omega^2, (\omega^2)^2 = \omega^2 \wedge \omega^2, \ldots, (\omega^2)^n$ are integral invariants of g_t. The existence of integral invariant $(\omega^2)^k$ ensures that $2k$-dimensional continuous swarm of representative points in the phase space evolves conserving the sum of $2k$-dimensional volumes of projections into coordinate planes. For $k = n$ there is only one term in this sum and we have conservation of the $2n$-dimensional volume (Liouville theorem).

If Φ is a chart mapping the neighborhood V of some $x \in M$ to some neighborhood U of $(q_1, \ldots, q_n, p_1, \ldots, p_n) \in R^{2n}$ then $\Phi^{-1} g_t \Phi$ gives the evolution of the dynamical system in local coordinates. Formally this evolution is given by the Lie operator

$$\exp tL_H = \sum_{j=0}^{\infty} \frac{1}{n!} t^n L_H^n, \tag{5}$$

where the Lie derivative L_H is operating on functions $F(\mathbf{q}, \mathbf{p})$ and is defined by the Poisson bracket $L_H\ F = [F; H]$. We can write the evolution from inital coordinates (**q**,**p**) to (**Q**(**q**,**p**,t), **P**(**q**,**p**,t))

$$(\mathbf{Q}, \mathbf{P}) = (\exp tL_H)(\mathbf{q}, \mathbf{p}). \tag{6}$$

It is easily shown (see for instance Deprit 1969) that

$$\exp tL_H\, [F; G] = [(\exp tL_H)F; (\exp tL_H)G] \tag{7}$$

and

$$F((\exp tL_H)\mathbf{q}, (\exp tL_H)\mathbf{p}) = (\exp tL_H)F(\mathbf{q}, \mathbf{p}). \tag{8}$$

If we introduce the $2n \times 2n$ matrix J

$$J = \begin{pmatrix} 0, & I \\ -I, & 0 \end{pmatrix}, \tag{9}$$

then the $2n \times 2n$ matrix L is symplectic if $L^T JL = J$ and the transformation $(\mathbf{q}, \mathbf{p}) \mapsto (\mathbf{Q}, \mathbf{P})$ is symplectic if its Jacobian is symplectic. From Eq. (7) it follows that (6) is symplectic. We know (see for instance Siegel and Moser 1971) that symplectic transformation $(\mathbf{q}, \mathbf{p}) \mapsto (\mathbf{Q}, \mathbf{P})$ leads to the canonical equations (1) even for new variables.

2. Mapping as Approximation of the Flow

When studiyng a Hamiltonian dynamical system, one possibility is to approximate the coordinate representation of the flow Eq. (6) by some approximation for fixed small $t = \tau$. For example we can use only the first k-terms of the series

$$
\begin{aligned}
Q_i &= q_i + \tau\,[q_i; H] + \ldots + \frac{1}{k!}\tau^k\,[\ldots[q_i; H]\ldots; H]\,,\\
P_i &= p_i + \tau\,[p_i; H] + \ldots + \frac{1}{k!}\tau^k\,[\ldots[p_i; H]\ldots; H]\,.
\end{aligned}
\tag{10}
$$

This mapping can be calculated and programmed. Then it can be applied many times and the evolution of the system followed with discrete time step τ. Mapping (10) is in fact a numerical integrator of order k as the error during one step τ is $O(\tau^{k+1})$. As we are dealing here with Hamiltonian systems it would be much nicer if the mapping (10) were symplectic. Then all the integral invariants would be conserved. The disadvantage of a non-symplectic mapping will be illustrated on the example of the linear harmonic oscillator (Yoshida, 1993).
Example:

$$
H = \frac{1}{2}(p^2 + q^2)
$$

Phase trajectories are circles. Taking k=1 (Euler method) Eq. (10) yields

$$
Q = q + \tau p, \quad P = p - \tau q. \tag{11}
$$

As $P^2 + Q^2 = (1 + \tau^2)(p^2 + q^2)$, the particles will be spiralling out from the exact phase trajectory. This error is present even in more sophisticated methods. Runge-Kutta integration step for the fourth-order method yields $P^2 + Q^2 = (1 - \frac{1}{72}\tau^6 + \ldots)(p^2 + q^2)$. By adding terms of the second order we can keep the first order integrator but change it into symplectic one.

$$
Q = q + \tau p, \quad P = (1 - \tau^2)p - \tau q. \tag{12}
$$

Then $[Q; P] = 1$. It can be checked that with (12) the value $\frac{1}{2}(p^2 + q^2 + \tau pq)$ is conserved. The calculated trajectory is now an ellipse with semimajor axes rotated about 135° from the q-axis. Its eccentricity e is

$$
e = \left(\frac{2\tau}{2 + \tau}\right)^{\frac{1}{2}}. \tag{13}
$$

As the volume is preserved there is no spiralling now and the calculated trajectory keeps close to the real one.

In what follows we will restrict ourselves to symplectic mappings. Then the Hamiltonian properties of the trajectories (existence of integral invariants) will be respected even if the trajectory will in fact be governed by a slightly disturbed

Hamiltonian from the original one. This ensures that the energy on the calculated trajectory can have no secular change as it is kept all time on the surface of constant disturbed Hamiltonian which is close to the surface of the original one.

If we construct a symplectic integrator of higher order the difference between those Hamiltonians will decrease.

3. Higher Order Symplectic Integrators

Such integrators were studied by Neri (1988) who proposed a general idea how to construct an explicit symplectic integrator for a Hamiltonian

$$H = T(\mathbf{q}, \mathbf{p}) + V(\mathbf{q}, \mathbf{p}), \tag{14}$$

where both T and V are integrable. We write both $\mathbf{q}$ and $\mathbf{p}$ in V, T to stress that V and T are not necessarily kinetic and potential energies. He calculated a fourth order symplectic integrator. For higher order integrators his method is very difficult to handle. Yoshida (1990) proposed a more managable indirect method and he was able to give the symplectic integrators of the sixth and eighth order. The general method is to find coefficients $c_1, \ldots, c_m, d_1, \ldots, d_m$ such that the flow operator for the Hamiltonian (14) is approximated by a multiple product of the flow operators of the individual integrable Hamiltonians T and V with accuracy $O(\tau^{k+1})$.

$$\exp\left[\tau\left(L_T + L_V\right)\right] = \prod_{i=1}^{m} \exp\left(c_i \tau L_T\right) \exp\left(d_i \tau L_V\right) + O(\tau^{k+1}). \tag{15}$$

The most simple symplectic integrators are $\exp(\tau L_T)\exp(\tau L_V)$ (first order) and $\exp(\frac{1}{2}\tau L_T)\exp(\tau L_V)\exp(\frac{1}{2}\tau L_T)$ (second order). Coefficients c_i, d_i for 4th ($k = 4, m = 4$), 6th ($k = 6, m = 8$) and 8th ($k = 8, m = 16$) order symplectic integrators ca be found in Yoshida (1990). An important role is played by the Baker-Campbell-Hausdorf (BCH) formula (see for instance Varadarajan , 1974). Schematically we can say that for any non-commutative operators X and Y, the product of two exponential functions is

$$\exp X \exp Y = \exp\left[X + Y + \sum_{j=2}^{\infty} C_j(X, Y)\right], \tag{16}$$

where $C_j(X, Y)$ is a linear combination of commutators of jth order. Here the second order commutator is $[X; Y]$, the third order commutators are commutators of X or Y with second order commutators and the $(s+1)$th order commutators are commutators of X or Y with the sth order commutators. Applying this formula to the product on the right hand side of Eq. (15) we get

$$\prod_{i=1}^{m} \exp\ \left(c_i \tau L_T\right) \exp\left(d_i \tau L_V\right) =$$

$$\exp\left[\tau\left(L_T + L_V + \sum_{j=k}^{\infty} \tau^j S_{j+1}(L_T, L_V)\right)\right], \tag{17}$$

where $S_j(L_T, L_V)$ is again a linear combination of jth order commutators of L_T, L_V. Using the fact that $[L_T; L_V] = -L_{[T;V]}$ we can see that

$$\prod_{i=1}^{m} \exp(c_i\tau L_T)\exp(d_i\tau L_V) = \exp(\tau L_{\bar{H}}), \tag{18}$$

where

$$\bar{H} = T + V + \sum_{j=k}^{\infty}(-1)^j\tau^j S_{j+1}(T, V). \tag{19}$$

The symplectic integrator of kth order for the Hamiltonian (14) actually calculates the trajectory for the Hamiltonian (19) if the series in Eq.(19) is convergent. The difference of these two Hamiltonians is of order $O(\tau^k)$. The value of $\bar{H}$ is actually conserved. So the particle in phase space keeps close to the energy surface and no secular deviations from this surface are possible. If one of the integrable parts, let us assume $V = O(\epsilon)$, where ϵ is small parameter, then $H - \bar{H} = O(\tau^k\epsilon)$. In problems with such small parameter the stepsize τ may be chosen $\epsilon^{-1/k}$ times longer to achieve the same accuracy in the Hamiltonian.

The problem is that the series (10) is usually divergent. First terms may still be used as asymptotic approximation which gives mathematical foundation of using divergent series in approximating a function.

Symplectic integrators were tested by Kinoshita et al. (1991) on two body problem and it was demonstrated that they did not show a secular trend in energy as the 4th order Runge-Kutta integrator, 6th and even 13th order Störmer's integrator. A 4-th order symplectic integrator was used for the three body problem by Šidlichovský (1992). The disadvantage is that they are slower. Kinoshita et al. 1991 found the Yoshida's symplectic integrator of the 6th order about five time slower than corresponding Störmer's integrator. Another weak point is that it is difficult to build symplectic integrators with adaptable stepsize, which is quite necessary for some kind of dynamical problems with close approaches of bodies.

4. Chirikov's Method Applied to the Pendulum

Chirikov (1969, 1979) was the first who introduced the kicks from periodic $\delta_{2\pi}$ function as the mean of transition between continuous evolution and mapping. It is best illustrated on the pendulum as it leads to the so called standard mapping widely discussed in the literature. Let us start with the Hamiltonian

$$H(\theta, I) = \frac{1}{2}I^2 + \sum_{j=-\infty}^{\infty} K_j \cos(\theta - j\Omega t), \tag{20}$$

where we assume $\Omega \gg K_0^{1/2}$. In this case we have a pendulum disturbed by high frequency terms (terms with $j \neq 0$). Using the averaging principle we can omit these disturbing terms or more exactly kill them by canonical transformations introducing new averaged variables $\bar{\theta}, \bar{I}$ and the new Hamiltonian $\bar{H}$

$$\bar{H}(\bar{\theta}, \bar{I}) = \frac{1}{2}\bar{I}^2 + K_0 \cos(\bar{\theta}). \tag{21}$$

Now we will change

$$\cos(\bar{\theta}) \mapsto 2\pi K_0 \cos(\bar{\theta}) \delta_{2\pi}(\Omega t) \tag{22}$$

This change is nothing but again reintroduction of some new high frequency terms as from the Fourier transform of $\delta_{2\pi}$ it follows

$$2\pi \cos(\bar{\theta}) \delta_{2\pi}(\Omega t) = \sum_{j=-\infty}^{\infty} \cos(\bar{\theta} - j\Omega t). \tag{23}$$

The mapping giving the change of $\bar{\theta}, \bar{I}$ during a high frequency period $T = 2\pi/\Omega$ consists of a kick on $\delta_{2\pi}$ at $t = 0$ (only $\bar{I}$ is changed) and evolution due to $\bar{I}^2$ for time T. The resulting mapping gives the $\bar{\theta}_1, \bar{I}_1$ after time T in terms of the initial $\bar{\theta}, \bar{I}$

$$\begin{aligned} \bar{I}_1 &= \bar{I} + T K_0 \sin(\bar{\theta}), \\ \bar{\theta}_1 &= \bar{\theta} + T \bar{I}_1. \end{aligned} \tag{24}$$

If we exchange the order of the $\delta_{2\pi}$–kick and the evolution we get another mapping

$$\begin{aligned} \bar{\theta}_1 &= \bar{\theta} + T \bar{I}, \\ \bar{I}_1 &= \bar{I} + T K_0 \sin(\bar{\theta}_1). \end{aligned} \tag{25}$$

For $T = 1$ ($\Omega = 2\pi$) we will have two versions of the well known standard mapping also known as the Chirikov-Taylor mapping. Its properties were studied by many authors, especially the transition of the system to global stochasticity when K_0 increases. A good reference on this topic can be found in (Lichtenberg and Lieberman, 1983). Chirikov (1979) showed that the standard mapping describes approximately some real physical systems such as a charged particle in a magnetic bottle. Also some other mappings as the Fermi mapping (for the point bouncing between fixed and oscillating walls) and the separatrix mapping (called also sometimes the whisker mapping) can be converted to the standard mapping by linearization (Lichtenberg and Lieberman, 1983). The standard mapping was recently studied by Laskar et. al (1992). The power of Laskar Modified Fourier Transform (MFT) was demonstrated. They watched the destruction of the last invariant curve with frequency corresponding to the golden mean (the number least easily approximated by rationals).

To avoid two levels of subscripts, let us introduce the operator E

$$E(\tau, H) = \exp(\tau L_H) . \tag{26}$$

Then the two symplectic integrators of the first order with step T which we found in Section 2 applied to the Hamiltonian (21) yield the transformations

$$\begin{pmatrix} \bar{I}_1 \\ \bar{\theta}_1 \end{pmatrix} = E\left(T, K_0 \cos \bar{\theta}\right) E\left(T, \bar{I}^2/2\right) \begin{pmatrix} \bar{I} \\ \bar{\theta} \end{pmatrix} \tag{27}$$

and

$$\begin{pmatrix} \bar{I}_1 \\ \bar{\theta}_1 \end{pmatrix} = E\left(T, \bar{I}^2/2\right) E\left(T, K_0 \cos \bar{\theta}\right) \begin{pmatrix} \bar{I} \\ \bar{\theta} \end{pmatrix} . \tag{28}$$

It is easy to show that Eqs (27) and (28) yield exactly the mappings (24) and (25). The standard mapping is obtained as the symplectic integrator of the first order applied to the pendulum Hamiltonian (21) with step $T = 1$. Points obtained by multiple application of the mapping should be on the level curves of (19), if it was convergent. The actual chaotic behavior, well known from standard mapping investigations, demonstrates the divergence of the series (19). Some orbits may be, however, well described by several first terms of (19).

5. Wisdom Mapping for Asteroid in Resonances

Wisdom (1982,1983) applied the Chirikov method to the problem of asteroids in the 3/1 resonance with Jupiter. He started with the elliptical planar three-body problem averaged over fast variables. Generally for the resonance $(p + q)/p$ we can introduce the canonical variables

$$\begin{aligned} \Phi &= \sqrt{\mu_1 a}/p, \\ \phi &= p\lambda - (p + q)\lambda', \\ x &= \sqrt{2\rho} \cos \tilde{\omega}, \\ y &= -\sqrt{2\rho} \sin \tilde{\omega}, \end{aligned} \tag{29}$$

where

$$\rho = \sqrt{\mu_1 a}\left(1 - \sqrt{1 - e^2}\right) \tag{30}$$

and $a, e, \lambda, \tilde{\omega}$ denote the asteroid semimajor axis, eccentricity, mean longitude and longitude of perihelion (primed quantities are for Jupiter) and μ, μ_1 are the masses of Jupiter and the Sun ($\mu + \mu_1 = 1$). Wisdom wrote the averaged Hamiltonian for the 3/1 resonance in the form

$$H = H_0(\Phi) + H_s(x, y) + H_r(\phi, x, y), \tag{31}$$

and he neglected terms of higher than second degree in eccentricities in H_s and H_r. Then H_s is explicitly integrable (we can write down the solution of the problem with H_s in a simple closed form describing secular evolution) and

$$H_r = F(x,y)\cos\phi + G(x,y)\sin\phi. \tag{32}$$

Wisdom wrote F and G as a sum of simple explicitely integrable Hamiltonians

$$F(x,y) = \sum_{i=1}^{n} F_i(x,y), \quad G(x,y) = \sum_{i=1}^{m} G_i(x,y) \tag{33}$$

and used the trick with $\delta_{2\pi}$-function by adding high frequency terms (Ω is the frequency of Jupiter mean longitude)

$$\begin{aligned} F_i(x,y)\cos\phi &\mapsto 2\pi\delta_{2\pi}(\Omega(t-\xi_i))F_i(x,y)\cos\phi, \\ G_i(x,y)\sin\phi &\mapsto 2\pi\delta_{2\pi}(\Omega(t-\gamma_i))F_i(x,y)\sin\phi. \end{aligned}$$

It was assumed $0 < \xi_1 < \ldots < \xi_n$, $\pi/2\Omega < \gamma_1 < \ldots < \gamma_m$ and $\xi_n \to 0$,$\gamma_m \to \pi/2\Omega$. The change of the elements Φ, ϕ, x, y during the time period $T = 2\pi/\Omega$ is then calculated as a series of n jumps at δ-functions followed by secular evolution together with change of ϕ due to H_0 from $t = 0$ to $t = T/4$, then m jumps at δ-functions followed by evolution due to $H_0 + H_s(x,y)$ from $T/4$ to T. In Wisdom treatment of the 3/1 resonance $n = 3, m = 2$ so that the resulting mapping consists of 7 mappings. The jump through δ-function has been calculated by representing δ-functions by rectangular peak and using limiting procedure (narrowing down the peak while keeping its area conserved). However, this procedure is not correct. The jumps calculated by exact direct integration of equations (yielding implicit relations for elements before and after jumps) are represented by different mappings which generally are not even symplectic (Henrard, 1997). Wisdom obtained with his limiting procedure the jump through δ-function at ξ_i expressed by mapping $\phi_0, x_0, y_0, \Phi_0 \mapsto \phi_1, x_1, y_1, \Phi_1$

$$\begin{aligned} \phi_1 &= \phi_0, \\ x_1 &= X(x_0, y_0, T), \\ y_1 &= Y(x_0, y_0, T), \\ \Phi_1 &= -F_i(x_0, y_0)T, \end{aligned} \tag{34}$$

where $X(x_0, y_0, t), Y(x_0, y_0, t)$ are the known solutions of equations

$$\frac{dx}{dt} = \frac{\partial F_i(x,y)\cos\phi_0}{\partial y}, \quad \frac{dy}{dt} = -\frac{\partial F_i(x,y)\cos\phi_0}{\partial x}, \tag{35}$$

with constant ϕ_0 and initial $x = x_0, y = y_0$. The jumps through δ-functions at γ_i is calculated similarly. It is very easy to see that this sequence of $m+n+2$ mappings is equivalent to the action of the following operator

$$\prod_{i=1}^{n} E \ \left(T, cos\phi \, F_i(x,y)\right) E(T/4, H_0 + H_s(x,y)) \times$$
$$\prod_{i=1}^{m} E \ \left(T, \sin\phi \, G_i(x,y)\right) E(3T/4, H_0 + H_s(x,y)) \tag{36}$$

As to the first order it agrees with the operator $E(T, H)$ it is a symplectic integrator of the first order. Without introducing δ functions the Wisdom mapping may be obtained just by splitting the Hamiltonian into integrable parts and applying the mappings corresponding to evolutions of individual Hamiltonians for time interval T corresponding to the period of the high frequency terms which were discarded by averaging. Of course we obtain different first order integrators if we change the order of the individual mappings similarly as we had two versions of the standard mapping.

Wisdom mapping is nothing but the symplectic integrator of the first order applied after splitting of the Hamiltonian into integrable parts. It is a legitimate approach. The derivation of this mapping by means of introducing δ-function into Hamiltonian may be misleading as the correct approach for such Hamiltonian would yield a different non-symplectic mapping.

The Wisdom mapping is very fast compared to the numerical integration and Wisdom (1982) could integrate for millions of years. He showed that particles starting in the 3/1 gap could keep low eccentricity ($e \leq 0.05$) for about one million years and then jump to a large eccentricity ($e \geq 0.3$). The particles can switch between the high and low eccentricity modes quite irregularly in unpredictable moments and this switching is quite different for very close initial positions of two test particles. Such irregular behaviour was found already by Scholl and Froeschlé (1974, 1975) but it was then understood as exceptional. This behavior was explained by Wisdom (1985) who was able to calculate guiding trajectories employing the adiabatic approximation (frozen x, y) while averaging over the faster oscillations in the resonant variable ϕ. The transition of trajectory from one guiding trajectory to another is possible only in a small region where adiabatic approximation fails. In this critical region the oscillations in ϕ are slow, so it must be close to the separatrix of the ϕ, Φ motion. The critical curves in the x, y plane can be calculated and it was shown that the chaotic region is formed practically just by trajectories intersecting this curve. Wisdom (1983) was also the first to show that the observed 3/1 Kirkwood gap coincides with this chaotic region.

The Wisdom mapping was soon applied to other resonances. Murray (1986) used a mapping for the 2/1 and 3/2 resonances, Šidlichovský and Melendo (1986) and Šidlichovský (1988) for the 5/2 resonance. Šidlichovský (1991) calculated all necessary coefficients for the most important resonances and suggested a general method of constructing Wisdom mapping which made it possible to calculate mappings for various resonances with one program with p and q as input (for

$(p+q)/p$ resonance). We follow the same approach as Wisdom but in general the resonant part H_r for the resonance of the q-th order starts with terms of the q-th degree in eccentricities. For the first order resonances terms of the second degree in eccentricities contribute to the resonant Hamiltonian with terms containing $\cos 2\phi$ and $\sin 2\phi$ but the spliting of the Hamiltonian and the construction of the symplectic integrator may be used again. For the fourth order resonances the resonant part is of the 4-th degree in eccentricities and to be consistent also terms of the 4-th degree in the secular part H_s should be taken into accout. It makes the calculations of the mapping more difficult. Šidlichovský (1992) presented the results for the 7/3 resonance calculated with his universal program for mappings. All these calculations show a similar picture of a chaotic zone with two modes of motion with high and small eccentricities. Wisdom (1985) perturbative treatment may be applied to other resonances giving good explanation of the chaotic zone using guiding trajectories and critical curves (Šidlichovský, 1987) for the 5/2 resonance and (Yokoyama and Balthazar, 1992) for the 5/2 and 7/3 resonances.

The accuracy of the 3/1 mapping was studied by Murray and Fox (1984). They compared three numerical methods: (a) integrating the full equations of motion, (b) integration of averaged equations of motion, (c) Wisdom mapping. They showed that in regular regions methods b and c give an excellent agreement with each other and that provided the eccentricity $e_{\max} < 0.4$ differences with the exact solution in $e_{\max}$ are $< 6\%$. In chaotic region it is not possible to compare individual solutions but all methods reflect the qualitative behaviour of chaotic trajectories and give good agreement on the bound of the motion. The situation is not so good with the 2/1 resonance. Henrard and Lemaître (1987) showed that Murray's "central region" (which exhibits large excursions and could lead to collisions with planets from moderate eccentricities $e \sim 0.15$) is an effect of the truncation of the perturbation function. This disagreement is not a failure of the mapping but the failure of early truncation. In fact it was shown (Henrard and Lemaître, 1987) that we should take into account the 8th degree terms in eccentricity to have the right qualitative behaviour for $e \sim 0.35$. Lemaître and Henrard (1988) showed that for the 3/2 resonance the situation is still worse and the mapping results are not reliable for e higher that 0.15 but probably even for $e > 0.07$. Šidlichovský (1993) compared the results of the numerical integration of the full system with the mapping for the 5/2 resonance and showed that exact excursion of eccentricity are still higher in the full system. These calculation indicated the existence of a very high eccentricity ($e \sim 0.52$) unstable equilibrium solutions similar to those studied for the 3/1 resonance (Ferraz-Mello, 1993). Trajectories starting at $e \sim 0.1$ can approach this unstable equilibrium point and then go to $e \sim 0.78$. It is not possible to get such trajectories with mapping. It may be shown (Ferraz-Mello, 1994) that the convergence radius for Laplacian expansion of the disturbing potential of Jupiter at the resonances 5/2, 7/3, 2/1 and 3/2 is $e \sim 0.28$, 0.26, 0.20 and 0.09, respectively. These values are less than eccentricities required for collisions with Mars which are 0.41, 0.44, 0.49 and 0.58, respectively. Wisdom mapping played a very important and stimulating role in our understanding of asteroidal resonances. Its great disadvantage is that it is based on a truncated perturbation function and it cannot be reliable for eccentricities as high as needed for the asteroid to become Mars crosser.

6. Improvements of Wisdom Mapping

We will mention two attempts to improve the mapping method for resonant asteroid in order to get better results for high eccentricity. One was suggested by Hadjidemetriou (1992), the second by Wisdom and Holman(1991).

6.1. Hadjidemetriou's Method

Hadjidemetriou (1991) constructed a mapping using the fact that the function

$$W(\mathbf{p}_1, \mathbf{q}) = \mathbf{p}_1\mathbf{q} + \tau H(\mathbf{q}, \mathbf{p}_1) \tag{37}$$

is the generating function for the infinitesimal transformation from initial values $\mathbf{q}, \mathbf{p}$ to $\mathbf{q}_1, \mathbf{p}_1$ after an infinitesimal time τ. If τ is not infinitesimal Eq. (37) still generates a symplectic integrator of the first order. In (Hadjidemetriou, 1993) this method was applied to the 3/1 resonance with variables (σ, ν, S, N) and the Hamiltonian $H(\sigma, \nu, S, N)$ as used by Henrard and Caranicolas (1990). Hadjidemetriou (1992) compared fixed points of the averaged Hamiltonian (terms of the third and higher order neglected) with the 3/1 resonant families of periodic orbits for the elliptical problem. He found four families I_c, II_c, I_e, II_e. Families I_c, II_c bifurcate from the unstable region of the family of periodic orbits of the first kind of the circular problem and I_e, II_e from the stable resonant 3/1 family of the second kind. For the eccentricity of Jupiter ($e' = 0.048$) the points on I_c, II_c correspond to periodic trajectories with low eccentricities (about 0.1) but the eccentricities of trajectories on I_e, II_e are high (about 0.8). The truncated averaged Hamiltonian yields fixed points which are in good agreement with low eccentricity periodic trajectories but there is poor agreement for the high eccentricity periodic trajectories. Hadjidemetriou (1992) suggested heuristic correction term in the Hamiltonian which would put the high eccentricity fixed points to correct place. This term

$$H_c = \mu F' \left(1 - F'' e' cos(\sigma + \nu)\right) S^3 \tag{38}$$

is of the sixth degree in e and, therefore, will change the low eccentricity fixed points insignificantly. The value of the constants F' and F'' were tuned (Hadjidemetriou, 1992) to give the correct position of the fixed points. Hadjidemetriou (1993) constructed the mapping for the Hamiltonian with correction term using the mapping generated by Eq. (38) for this problem (putting $\tau = 2\pi$). In variables given by Eq. (29) Hadjidemetriou correction term is

$$H_c = \mu F' \rho^3 (1 - F'' e' \cos\tilde{\omega}). \tag{39}$$

so that another term $E(T, H_c)$ might be added to Eq. (36) to generalize explicit Wisdom mapping. The actual transformation corresponding to this Hamiltonian is not so easy to write, even if we split H_c into two simpler parts. It is possible to give a solution in quadratures so we have to include some numerical way of calculating integral. This is not necessary in Hadjidemetriou's formulation. His mapping is, however, implicit so that eg. equation for S_1 must be solved and again some numerical methods have to be employed.

6.2. Wisdom and Holman Generalization

Wisdom and Holman (1991) tried to improve Wisdom method so that it worked without truncation of the perturbing function. The original Hamiltonian of the n-body problem $H = H_0 + H_1$ was replaced (again using addition of high frequency terms) by $H_{\mathrm{Map}} = H_0 + \Phi(\Omega t)H_1$, where

$$\Phi(t) = 2\pi \sum_{i=0}^{N-1} a_i \delta_{2\pi}(t - 2\pi d_i), \tag{40}$$

where constants a_i, d_i are introduced to match the Taylor series of a mapping step. The unperturbed Hamiltonian H_0 is the sum of the two body Kepler Hamiltonians (in Jacobi coordinates). The time step of the mapping should be much smaller than the period of the fastest moving object. This method avoids truncation as no averaging is made but without averaging the short time step is (as usually) required.

Wisdom and Holman obtained mapping where the evolution is governed by H_0 between δ functions and by H_1 which is obtained from kicks at δ-functions (as obtained by limiting procedure). Coefficients a and d determine for how long this individual evolutions are applied and are chosen so that the best accuracy is achieved. It is clear that this method is similar to Neri and Yoshida's symplectic integrator. Wisdom and Holman performed numerous tests of their maps obtained for the n-body problem. Finally they have chosen the second order version (leap frog) $E(T/2, H_0)E(T, H_1)E(T/2, H_0)$ and used Jacobi cordinates and integrated outer planets with Pluto for 1.1E9 years. The evolution was solved using Gauss f and g functions to avoid repeating conversion between Kepler and Jacobi cordinates. The step was one year so that the map was used in qualitative mode. The map reproduced all the principal results obtained previously using Digital Orrery (Sussman and Wisdom, 1988), where the step size was 32.7 days.

7. Some other Mappings used in Celestial Mechanics

Mapping techniques were recently applied to various problems. It is impossible to give here a complete description of these methods. We will mention just four fields of interest. Mappings for systems with dissipation, synthetic mappings, Keplerian maps and mapping approach to the stability of orbits in the solar system.

7.1. Mappings for Systems with Dissipation

(Malhotra, 1994) studied the non-Hamiltonian system

$$\frac{dq_i}{dt} = \frac{\partial H}{\partial p_i}, \quad \frac{dp_i}{dt} = -\frac{\partial H}{\partial q_i} + \varepsilon F(t, q, \dot{q}), \tag{41}$$

where

$$H = H_{\mathrm{Kep}}(q, p) + \mu H'(q). \tag{42}$$

Here H_{Kep} is the two body problem Hamiltonian (describing the unperturbed Keplerian orbit of the j-th planet), $\mu H'$ describes the gravitational perturbation of the planets and εF is a small dissipative force. Symplectic integrator of the second order for $\varepsilon = 0$

$$E(\tau/2, H_{\mathrm{Kep}})E(\tau, \mu H')E(\tau/2, H_{\mathrm{Kep}}) \tag{43}$$

can be modified (for $\varepsilon \neq 0$) to

$$\bar{E}(\tau/2, H_{\mathrm{Kep}}, F)E(\tau, \mu H')\bar{E}(\tau/2, H_{\mathrm{Kep}}, F), \tag{44}$$

where $\bar{E}(\tau, H_{\mathrm{Kep}}, F)$ describes the evolution of the Keplerian orbit with drag F. For this case a modification of the Gauss f and g function was presented. Malhotra has shown that the error for one step τ is $O(\mu\tau^3) + O(\varepsilon\tau^3)$ for positions and $O(\mu\tau^3) + O(\varepsilon\tau^2)$ for velocities. This method is not of the second order exactly. A modificaton yielding exact second order method where orbit is advanced twice for $\tau/2$ using just H_{Kep} and one simple jump of velocities is calculated in between these steps (where both dissipative and perturbing term come into play) was constructed in (Cordeiro et al., 1996). Hadjidemetriou implicit mapping was used in obtaining another method for near conservative system (Zhou et al., 1994). This method was applied to Duffing's equation and to the study of grain motion with PR drag near the 3/1 resonance.

7.2. Synthetic Mapping

The Poincaré mapping for two degrees of freedom conservative system leads to a mapping of a two dimensional surface (with coordinates x, y) into itself

$$x' = f(x, y), \quad y' = g(x, y). \tag{45}$$

Usually we are willing to calculate the value of functions f and g only in some grid of points as this calculating amounts to time consuming integration of the trajectory starting at x, y on the surface up to the next intersection x', y' with the surface. The basic idea of the synthetic mapping is to interpolate the image of a point, when given the images of a set of points located at the vertices of the grid. Such mapping was presented in (Froeschlé and Petit, 1990). For method of the 5th order they use grid of 36 neighbouring points, for the 3rd order they need only 16 points. When partial derivativatives f_x and f_y are introduced together with methods of their calculation (Petit and Froeschlé, 1994) only 12 values of f, f_x and f_y at four corners of the grid are necessary for calculation of the image of the point located inside these four corners (in the 3rd order calculations) (Petit and Froeschlé, 1994).

7.3. Keplerian Map

This name is used for the mapping giving the evolution of near parabolic orbits in the restricted three body problem (e.g. the Sun, Jupiter, comet), where we are interested in possible transition to unbounded orbits. Petrosky and Broucke (1988) obtained the mapping

$$\begin{aligned} P_{n+1} &= P_n + 2\sigma\mu s \sin g_n \\ g_{n+1} &= g_n - 2\pi\sigma/(-P_{n+1})^{3/2} \end{aligned} \tag{46}$$

where g_n is the phase angle of Jupiter at the moment of the perihelion passage of the comet, P_n is the energy, μ the mass of Jupiter, $\sigma = \pm 1$ is the parity of direction of revolution of the comet. The change of energy from negative to positive value corresponds to the transition from bounded to unbounded state. Using Melnikov integral Liu and Sun (1994) obtained similar mapping (sines of multiples of g_n appeared in their mapping) and investigated fixed points and their stability.

7.4. The Long-Term Evolution of Orbits in the Solar System

A mapping approach to this problem was suggested by Duncan et al. (1989) under the assumption that the perturbation to the test particle are localized near conjunctions with the neighbouring planets. The planetary orbits are circular. The behaviour of test particles is investigated in various regions of the solar system using this mapping. While many orbits between Uranus and Neptune and most orbits between Jupiter and Saturn become planet crossing and hence unstable within the age of the solar system, most orbits between Saturn and Uranus, between Venus and Earth and in the asteroid belt seem to be stable over this time scale. As the planetary orbits are constant circles no secular resonances are present in this model.

8. Conclusion

It may be concluded that mappings has proved to be efficient and stimulating tools in the study of model dynamical system we meet in celestial mechanics. Wisdom papers about Kirkwood gaps inspired much work in this field. Some maps (as the standard map) for which the behaviour is well known, are used as 'laboratory' for testing new methods for chaos. Symplectic integrators as exact integrators are still rather slow and some possibilities of changing their stepsize would be desired for systems with close approaches. They can, however, be very good in qualitative mode with large stepsizes, as the interesting features of a system may be found quite fast. Of course it is a good idca always to check such interesting results with numerical integration. Failure of mapping for large e in asteroidal resonances showed by numerical check is a failure of models with truncated Hamiltonian and not a failure of the mapping itself.

Acknowledgements

The author acknowledges support provided by grant 205/95/0184 of GAČR (Grant Agency of the Czech Republic) and he thanks to Jacques Henrard for his critical comments which helped very much to improve the paper.

References

Arnold, V. I.: 1989 Mathematical methods of classical mechanics, Springer-Verlag, New York, Berlin, Heidelberg, London, Paris, Tokyo

Chirikov, B.V.: 1969, ' Research concerning the theory of nonlinear resonance and stochasticity', Preprint 267 of the Institute of Nuclear Physics, Novosibirsk (English Translation in it CERN Trans. 1971)

Chirikov, B.V.: 1979, 'A universal instability of many-dimensional oscillator systems', *Physics Reports* **52**, pp. 264–379

Cordeiro, R. R., Gomes, R. S., Martins, R. V.: 1996 'A mapping for nonconservative systems', *Celest. Mech*, in print

Deprit, A. :1969, 'Canonical transformations depending on a small parameter', *Celest. Mech.* **1**, pp. 12–30

Duncan, M., Quinn, T., Tremaine, S.: 1989, 'The long-term evolution of orbits in the solar system: A mapping approach', *Icarus* **82**, pp. 402–418

Ferraz-Mello S.: 1993, 'Kirkwood gaps and resonant groups', in Asteroids, Comets, Meteors, IAU Symp. No. 160, pp. 175–188 (eds. Milani, Di Martino, Cellino), Kluwer Academic Publishers, Dordrecht, Boston, London

Ferraz-Mello, S.: 1994, 'The convergence domain of the Laplacian expansion of the disturbing function', *Celest. Mech. Dyn. Astron.* **58**, pp.37–52

Froeschlé, Cl., Petit, J. M.: 1990, 'Polynomial approximation of Poincaré maps for Hamiltonian systems', *Astron. Astrophys.* **238**, pp. 413–423

Hadjidemetriou, J. D.: 1991, 'Mapping models for Hamiltonian systems with application to resonant asteroid motion', in Predictability, Stability and Chaos in N-body Dynamical Systems, ed. A. Roy, pp. 157–175, Plenum Press, New York, London

Hadjidemetriou, J. D.: 1992, 'The elliptic restricted problem at the 3:1 resonance', *Celest. Mech Dyn. Astron.***53**, pp. 151–183

Hadjidemetriou, J. D.: 1993, 'Asteroid motion near the 3:1 resonance', *Celest. Mech.* **56**, pp. 563–599

Henrard, J.: 1997, 'Symplectic integrators', to be published in Analysis and modelling the discrete dynamical systems–with application to dynamical astronomy, eds. D. Benest and C. Froeschlé, Gordon and Breach Henrard, J., Caranicolas, D.: 1990, 'Motion near the 3:1 resonance of the planar elliptic restricted three body problem', *Celest. Mech.***47**, pp. 99–121

Henrard, J., Lemaître, A.: 1987, 'A perturbative treatment of the 2/1 Jovian resonance', *Icarus* **69**, pp. 266–279

Kinoshita, H., Yoshida, H., Nakai, H.: 1991, 'Symplectic integrators and their application to dynamical astronomy', *Celest. Mech.* **50**, pp. 59–71

Laskar J., Froeschlé, Cl., Celleti, A., 'Numerical analysis of the fundamental frequencies. Application to standard mapping', *Physica* **56D**, pp. 253–269

Lemaître, A., Henrard, J.: 1988, ' The 3/2 resonance', *Celest. Mech.* **43**, pp.91–98

Lichtenberg, A. J., Lieberman, M. A.: 1983, Regular and stochastic motion, Springer-Verlag, New York, Heidelberg, London

Liu, J., Sun, Y. S.: 1994, 'Chaotic motion of comets in near-parabolic orbit: mapping approaches', *Celest. Mech.* **60**, pp. 3–28

Malhotra, R.: 1994, 'A mapping method for the gravitational few-body problem with dissipation', *Celest. Mech.***60**, pp. 373–385

Murray , C. D.: 1986, 'Structure of the 2:1 and 3:2 Jovian resonances', *Icarus* **65**, pp. 70–82

Murray, C. D., Fox, K.: 1984, 'Structure of the 3:1 Jovian resonance: A comparison of numerical methods', *Icarus* **59**, pp. 231–233

Neri, F.: 1988, 'Lie algebras and canonical integration', *Preprint of Department of Physics*, University of Meryland

Petit, J. M., Froeschlé, Cl.: 1994, 'Polynomial approximations of Poincaré maps for Hamiltonian systems. II', *Astron. Astrophys.* **282**, pp. 291–303

Petrosky T. Y., Broucke, R.: 1988, 'Area-preserving mappings and deterministic chaos for nearly parabolic motions', *Celest. Mech.* **42**, pp. 53–79

Scholl, H., Froeschlé, Cl.: 1974, 'Asteroidal motion at the 3/1 commensurability', *Astron. Astrophys.* **33**, pp. 455–458

Scholl, H., Froeschlé, Cl.: 1975, 'Asteroidal motion at the 5/2, 7/3 and 2/1 resonances', *Astron. Astrophys.* **42**, pp. 457–463 Šidlichovský, M.: 1987, 'Comparison of perturbation and Wisdom methods for 5/2 resonance', In *Proceedings 10th ERAM 3.- Dynamics and the solar system*, ed. M. Šidlichovský, Praha, pp. 571–580

Šidlichovský, M.: 1992, 'Mapping for the asteroidal resonances', *Astron. Astrophys.* **259** pp. 341–348

Šidlichovský, M.: 1993, 'Chaotic behaviour of trajectories for asteroidal resonances', *Celest. Mech.* **56**, pp. 143–152

Šidlichovský, M., Melendo, B.: 1986, 'Mapping for 5/2 asteroidal commensurability', *Bull. Astron. Inst. Csechosl.* **37**, pp. 65–80

Siegel L. S., Moser G.: 1971, Lectures on celestial mechanics, Springer-Verlag, Berlin, Heidelberg, New York, Berlin

Sussman, G. J., Wisdom J.: 1988, *Science*241, pp. 433–437

Varadarayan V. S., 1974, Lie groups, Lie algebras and their representations, Prentice-Hall, Engelwood Cliffs

Wisdom, J.: 1982, 'The origin of the Kirkwood gaps', *Astron. J.* **87**, pp.577–593

Wisdom, J. 1983, 'Chaotic behaviour and origin of the 3/1 Kirkwood gap, *Icarus* **56**, pp.51–74

Wisdom, J.: 1985, ' A perturbative treatment of motion near the 3/1 commensurability', *Icarus* **63**, pp.272–289

Wisdom, J.: 1992, 'Long term evolution of the solar system', in Chaos, Resonance and Collective Dynamical Phenomena in the solar system, IAU Symp. No. 152, pp. 17–24 (ed. S. Ferraz-Mello), Kluwer Academic Publishers, Dordrecht, Boston, London

Wisdom, J., Holman, M.: 1991, ' Symplectic maps for the N-body problem', *Astron. J.* 1991, pp. 1528–1538

Yokoyama , T., Balthazar, J. M.: 1992, 'Application of Wisdom's perturbative method for 5/2 and 7/3 resonances', *Icarus* **99**, pp. 175–190

Yoshida, H.: 1990, 'Construction of higher order symplectic integrators', *Physics Letters A* **150**, pp. 262–268

Yoshida, H.: 1993, 'Recent progress in the theory and application of symplectic integrator' , *Celest. Mech.* **27**, pp. 27–43

Zhou, J. L., Fu, Y. N., Sun, Y. S.: 1994, 'Mapping model for near-conservative system with applications', *Celest. Mech.* **60**, pp. 471–487

A NEW FAST FOURIER METHOD FOR EVALUATING FOURIER SPECTRA AT ARBITRARY FREQUENCIES

KARL WODNAR
Institute of Astronomy, University of Vienna
Türkenschanzstraße 17, A-1180 Vienna, AUSTRIA
email: wodnar@astro.ast.univie.ac.at

Abstract. This paper deals with some recent problems of frequency determination in dynamical systems. As a main result a new *Fast Fourier Transform (FFT)* method, the *High Resolution Fast Fourier Transform (HRFFT)* is presented, which allows for economic evaluation of discrete Fourier spectra at arbitrary frequencies, not just at the evenly spaced points of the rigid frequency grid the ordinary FFT is restricted to (continuous evaluation). Comparison is made to existing methods in what regards continuous evaluation, computational speed, memory requirements and precision, evidencing the high efficiency of the HRFFT. Possible applications of the new technique are indicated, showing that existing methods, as e.g. that of Laskar (1990, Icarus **88**) and its extension by Šidlichovský and Nesvorný (1996, this volume), may strongly benefit from the HRFFT.

Key words: Fourier methods — Fast Fourier Transform — frequency analysis — Hamiltonian dynamical systems

1. Introduction

1.1. Background

The behaviour of fundamental frequencies in Hamiltonian dynamical systems, especially in celestial mechanics, has been the subject of recent interest. Just to mention a few, the work of Laskar (1990, 1993, 1995), Laskar & al. (1992), Voyatzis & Ichtiaroglou (1992), Šidlichovský & Nesvorný (1996) has been devoted to this topic. Other current applications of Fourier methods in astronomy are e.g. the resolution of the spiral arm structure of galaxies and image processing in general, as well as the frequency analysis of pulsating stars. See e.g. Ferraz-Mello (1981) and Foster (1996ab) for the solution of orthogonality problems that arose in this context and have a close connection to frequency methods developed for dynamical systems.

This study grew out from the analysis of fundamental frequencies in the orbital behaviour of Pluto and Pluto-like planets as obtained by extensive numerical integrations in various models (Dvorak & al. 1995). The amount of data required efficient methods regarding the use of memory, computational speed and precision. Therefore we developed a method — the *High Resolution Fast Fourier Transform (HRFFT)* — extending the existing *Fast Fourier Transform (FFT)*, such that values of the *Discrete Fourier Transform (DFT)* can be obtained at *any* frequency point (continuous evaluation), as far as accessible through computer precision, with high speed and economic use of memory.

The mutual relationship of the parameters determining the HRFFT method wrt. precision, speed and use of memory is displayed through simple asymptotic formulas; two free integer parameters allow to adjust the method to specific requirements.

Celestial Mechanics and Dynamical Astronomy **65**: 85–94, 1997.

The method is compared to existing ones, again using asymptotic relations, proving its efficiency.

Finally, the applicability of the HRFFT in the context of Laskar's method is briefly discussed.

1.2. The Fourier Transform — Notation and Definitions

In view of the vast amount of literature on the subject — I especially recommend the excellent book of Bracewell (1978) — basic definitions and conventions will be referred to only briefly.

We consider a complex function f_t — the 'signal' — to be given and may think of some physical quantity in dependence of time[1]. We are interested in the question of how f_t may be linearly composed — over the field of complex numbers — of single harmonic oscillations $e^{2\pi i\nu t}$ with frequency ν. We thus define the *Continuous Fourier Transform (CFT)* $\mathcal{F}_\nu f_t$ of f_t, as a function of ν, as well as its inverse $\mathcal{F}_t^{-1}$ applied to F_ν and being a function of t as

$$\mathcal{F}_\nu f_t := \int_{-\infty}^{\infty} f_t e^{-2\pi i\nu t} dt, \qquad \mathcal{F}_t^{-1} F_\nu := \int_{-\infty}^{\infty} F_\nu e^{2\pi i\nu t} d\nu,$$

wherein $i = \sqrt{-1}$. Likewise, we define its discrete counterpart, the *Discrete Fourier Transform (DFT)* and its inverse, assuming the knowledge of f_t at N evenly spaced points $(t_k = t_0 + k\delta t)_{k=0,\ldots,N-1}$ on time axis, as

$$\mathcal{F}_l^D f_{t_k} := \sum_{k=0}^{N-1} f_{t_k} e^{-2\pi ikl/N}, \qquad \mathcal{F}_k^{D^{-1}} F_l := \frac{1}{N} \sum_{l=0}^{N-1} F_l e^{2\pi ikl/N}. \tag{1}$$

In the classical form the latter transformation is being evaluated only at integer values of l, transforming an N-vector $(f_{t_k})_{k=0,\ldots,N-1}$ into an N-vector $\left(\mathcal{F}_l^D f_{t_k}\right)_{l=0,\ldots,N-1}$, although in (1) the evaluation is possible at *any* frequency point l, not necessarily integer, which has important applications in determining precisely the maxima of the DFT. This is of significance in order to find estimates of frequency components and their complex amplitude not confined to the rough grid of integer multiples $l\delta\nu$ of $\delta\nu = 1/(N\delta t)$, apart from the well known problems of *leakage* and *aliasing* (see e.g. Bracewell 1978). Formula (1) is, however, fairly slow to be applied in computer programs; on the other hand the known, classical FFT is restricted to integer values of l, only. In the following an extension to the FFT is constructed, which essentially keeps the speed properties of the FFT, but is able to cope also with non-integer l.

[1] In order to emphasize the vector character of a function f as an element of an infinite-dimensional linear space, its argument t shall be written as a subscript.

2. Realisation of a High Resolution FFT Algorithm

2.1. CLASSICAL FFT

The classical *Fast Fourier Transform (FFT)* (s. Brigham 1974) has been widely used as an efficient means for the calculation of a DFT. It assumes N to be an integer power of 2 and defines a finite triple sequence

$$\left(f_{k,l}^{(j)}\right)_{k=0,\dots,2^{n-j}-1;\, j=0,\dots,n;\, l=0,\dots,N-1}, \qquad N = 2^n,$$

through the recursion

$$\begin{aligned}
&\text{for } l = 0,\dots,N-1:\\
&\quad \text{for } k = 0,\dots,N-1:\\
&\qquad f_{k,l}^{(0)} := f_{t_k}\\
&\quad \text{for } j = 0,\dots,n-1:\\
&\qquad \text{for } k = 0,\dots,2^{n-j-1}-1:\\
&\qquad\quad f_{k,l}^{(j+1)} := f_{k,l}^{(j)} + e^{-2\pi i 2^{-j-1} l} f_{k+2^{n-j-1},l}^{(j)}.
\end{aligned}$$

Then the DFT comes up as

$$\mathcal{F}_l^D f_{t_k} = f_{0,l}^{(n)}.$$

The algorithm results from a hierarchically continued grouping of the summation in (1) into two halves and appropriate extraction of exponential factors. The speed of this method comes from two facts: First from a drastic reduction of the number of multiplications, and second from calculating the exponential factors $E_{jl} := e^{-2\pi i 2^{-j-1} l}$ in advance:

$$\begin{aligned}
\text{ExpiFast}(m) &:= \exp(-2\pi i \operatorname{mod}(m,N)/N); \quad m = 0,\dots,N-1\\
\Longrightarrow \quad E_{jl} &= \text{ExpiFast}(2^{n-j-1} l),
\end{aligned}$$

wherein part of the notation has been lent from common computer-languages. For the typically used conventions in the context of computer-oriented description of algorithms, as employed here, see e.g. Golub & van Loan (1986).

2.2. STANDARD APPROACHES TO HIGH FREQUENCY RESOLUTION

By straightforward inspection of the classical FFT algorithm, one counts $N\,\mathrm{ld}N$ complex multiplications, where 'ld' denotes the binary logarithm. However, because of well known symmetries, one may achieve $(N/2)\,\mathrm{ld}N$ for the arithmetic complexity instead (s. Rabiner & Gold 1975, pp. 361). In the present exposition emphasis is given to those cases in which transforms at *selected* frequencies are of interest, because construction of an algorithm is intended that allows for a locally high density of frequency values in order to study close spectral features or to find the precise location of maxima[2]. Thus, it is not desirable to use an algorithm which

[2] Although use of the algorithm at a *complete* grid of equidistant frequencies in a Nyquist interval is not excluded by this assumption.

prescribes large numbers of frequencies to be evaluated, even if not needed by the user, as does the classical FFT, which assumes the number of frequencies M to be equal to N, so as to yield $(N/2)\,\mathrm{ld}N$ as computational effort. An increase of the local density requiring the calculation of large amounts of spectral points which might belong to redundant features is highly undesirable. At the same time, this form of the FFT procedure would consume enormous amounts of memory and computing time. Therefore, we decided to refer to that form of the FFT, where just a selected number $M \leq N$ of frequencies is being calculated. In this case one cannot take advantage of the factor two reduction in complexity, unless one implements certain case distinctions into the algorithm, which would result in a random variable for the computational effort taking values between $(M/2)\,\mathrm{ld}N$ and $M\,\mathrm{ld}N$. Thus, for simplicity and consistency of comparison, once and for all in this paper, we assume the complexity of the FFT to be $M\,\mathrm{ld}N$. The only exception of this rule occurs when the FFT appears as part of the chirp z-transform (see below), where the $(M/2)\,\mathrm{ld}N$ complexity is an inherent feature of the FFT in this method.

Compared to the definition of the DFT (1), which in principle allows for an evaluation at arbitrary non-integer values of l, the classical FFT algorithm lacks such a property. Although, by concatenating a sequence of $(2^r-1)N$ $(r = 1,2,3,\ldots)$ zero ordinates to the original data vector, one may evaluate — by re-scaling the resulting FFT — the DFT at points $l = L + m2^{-r}$ $(L = 0,\ldots N-1;\ m = 0,\ldots,2^r-1)$, resulting in the fairly good complexity value $M\,\mathrm{ld}(N/\varepsilon)$, where $\varepsilon = 2^{-r}$ is the frequency grid step in units of $\delta\nu$. This procedure — referred to as *FFT with zeroes* — however, will consume exceedingly much memory ($2^r N = N/\varepsilon$ complex storage locations), if high resolution is required. As mentioned above, in many applications the evaluation is not necessary at all points of a fine grid, but only at some of them. In this case the latter variant of the FFT, for high frequency resolution, still needs to store an extraordinarily large vector of exponentials, as is well known.

Another standard method to increase grid density in frequency space is the so-called *chirp z-transform*. This method, which is of similar complexity order $\geq (N+M-1)\,\mathrm{ld}(N+M-1)$ as the FFT with zeroes, not only faces similar storage problems as the latter, if high resolution of frequencies is required, but also enforces evaluation at *all* grid points, which is a major drawback of this method. This technique, however, serves as one of the main tools for the computation of general z-transforms. Note that the expression for the complexity of the chirp z-transform strongly depends on the specific implementation and constraints to be put on M and N. The above complexity formula therefore represents a lower bound for the 'best' form in which it may appear in, assuming efficient computer equipment and considering only the leading terms, for the sake of a fair comparison with other methods where the complexity is to be determined easier.

For an illustration of the addressed memory problem occurring with a standard method, one may think of the determination of maxima of a DFT spectrum up to computer precision. This task — using an FFT with zeroes or a chirp z-transform

— may require more random access memory, than presently available on any computer on earth for fairly realistic practical examples. Therefore, in this paper, the classical FFT has been modified and extended to serve this purpose.

2.3. Generalisation of the FFT to Continuous Evaluation

In the following we assume l to be a real of the form

$$l = L + \sum_{q=1}^{q_{\max}} \alpha_q a^{-q}, \quad \alpha_q \in \{0, \ldots, a-1\}, \quad a = 2^K,$$

where L is the largest integer less or equal to l, and $q_{\max}$, K are respectively two free positive integer parameters, which serve to adjust the method to the specific requirements of the problem, as explained in detail below. The α_q are just the digits of the non-integer part of l in an a-adic expansion. With this setting the exponential factor E_{jl} becomes

$$E_{jl} = e^{-2\pi i 2^{-j-1} l} = e^{-2\pi i 2^{-j-1} L} \prod_{q=1}^{q_{\max}} e^{-2\pi i 2^{-j-1-Kq} \alpha_q}.$$

In order to compute as much as possible of this expression in advance, define

$$\text{Expi2}(p, \alpha) := e^{-2\pi i 2^p \alpha}; \quad p = -n - Kq_{\max}, \ldots, -1-K; \quad \alpha = 0, \ldots, a-1$$

as the second array, besides 'ExpiFast', to be calculated in advance. Then we may compute E_{jl} for a single pair (j, l) during the data analysis as

$$E_{jl} = \text{ExpiFast}(2^{n-j-1} L) \prod_{q=1}^{q_{\max}} \text{Expi2}(-j-1-Kq, \alpha_q), \tag{2}$$

which involves just slightly more effort than that of the classical FFT, because of the additional $q_{\max}$ factors per evaluation of E_{jl}.

The resulting algorithm now assumes a quite similar form to the usual FFT with only minor modifications. One has just to replace the first line of the FFT algorithm by, say,

for $l = l_{\min}, \ldots, l_{\max}$ step ε :,

where ε is the fine grid step to be chosen arbitrarily small, but positive and $l_{\min}$, $l_{\max}$ are bordering values of a frequency interval of interest, and finally to calculate E_{jl} according to (2) during the data analysis. This defines the new algorithm completely, apart from the problem of the involved conversion to an a-adic number-system, which will be treated below. The two free integer parameters of the method, $q_{\max}$ and K, relate to parameters

s	...	memory requirement (no. of complex storage locations)
ε	...	frequency grid resolution (in units of $\delta\nu$)
T	...	$\approx$ time $\approx$ computational effort (no. of multiplications)
M	...	number of frequencies to be evaluated
N	...	number of data points

in the following manner

$$s = (n + K(q_{\max} - 1))2^K \tag{3}$$

$$\varepsilon = a^{-q_{\max}} = 2^{-Kq_{\max}} \tag{4}$$

$$T = M(1 + q_{\max})\,\mathrm{ld}N. \tag{5}$$

Thus, using these asymptotic relations, one may optimize the method by selecting $q_{\max}$ and K according to amount of data, computer capability, speed and precision requirements. The standard strategy will be the following: Given ε, one selects K maximal and $q_{\max}$ minimal along the hyperbola defined through (4), such that — according to (3) — s does not exceed available memory. >From (5) it is evident that with minimal $q_{\max}$, speed is optimized. In most practical cases, contemporary computer resources will allow to work with the minimum value of $q_{\max} = 1$, achieving maximum computational speed, which, in this case, is just half the speed of that form of the classical FFT with complexity $M\,\mathrm{ld}N$. Note that with this new algorithm the frequency grid may be taken as fine as required without the need to actually evaluate at all the grid points.

The basic heuristic principle underlying this method is to achieve close compatibility with the classical FFT, which is strongly related to binary expansions, through introducing an a-adic number-system in the non-integer part of l, where a is a power of two. There remains however the problem of conversion to an a-adic system. In practice two situations of application will be typical to deal with: A certain interval of frequencies should be evaluated on a fine grid or maxima are to be found through successive bisection in a given interval, necessitating evaluation at these points. In the first case we can circumvent the problem of transforming the non-integer parts of frequencies measured in units of $\delta\nu$ to the a-adic number-system, by properly defining the increment in steps of ε in terms of changes of the digits α_q, which just involves a chain of $q_{\max}$ 'if'-clauses and possible increment of some digits by one or setting them equal to zero, depending on the amount carried forward by less significant digits. In the second case, since the basis a of the number-system is a power of two, division by two is an easy task to be carried out on the digits α_q, as long as the borders of the interval are suitably chosen (e.g. the non-integer parts must be negative powers of two). In both of these cases, conversion may be performed by directly accessing the bits of the non-integer part of the frequency variable and using partitions of length K of the corresponding bit string, in order to dispose of the digits in the a-adic system. Most of the modern programming languages allow for that. Following these suggestions, the problem of conversion thus does not increase essentially the computational effort, because no additional multiplications occur. This has been verified also through computer tests.

For further reference this method will be named the *High Resolution Fast Fourier Transform (HRFFT).*

2.4. Comparison of Computational Effort of Various Methods

In the following we give an overview on the respective computational efforts of various existing methods compared to the HRFFT in terms of asymptotic formulas, also indicating whether continuous evaluation is possible in the last column, see Rabiner & Gold (1975, pp. 392) for the related facts about the Bluestein algorithm and the chirp z-transform.

method	effort	cont. eval.
classical DFT	NM	yes
Bluestein algorithm	$M^{3/2}$	no
classical FFT	$M \,\mathrm{ld}\, N$	no
chirp z-transform	$\geq (N+M-1)\,\mathrm{ld}(N+M-1)$	yes
FFT with zeroes	$M\,\mathrm{ld}(N/\varepsilon)$	yes
HRFFT	$M(1+q_{\max})\,\mathrm{ld}N$	yes

We note that the Bluestein algorithm needs $N = M = M'^2$, with some integer M' in order to work. We remind on the huge amount of memory consumed by the FFT with zeroes, if high precision is required, and that, using this method, $-\mathrm{ld}\varepsilon$ needs to be a positive integer. Furthermore, the chirp z-transform faces similar memory problems, requires evaluation at all grid points and has thus a strict coupling $\varepsilon = N/M$. Remember also that $q_{\max} \geq 1$ and $-\mathrm{ld}\varepsilon$ must be a positive integer for the HRFFT to work. Note, that if we would allow for the mentioned factor two speeding up of the FFT (also of that with zeroes), by requiring the evaluation of the respective full grid of frequencies, we could do also the same for the HRFFT, because of the close relationship between those methods. Thus the above comparison is consistent within itself.

We conclude that out of the three methods allowing for continuous evaluation with acceptable computational effort, the HRFFT is strongly preferable, if either many frequencies or selected sets of those are to be evaluated, especially if high precision is required and even if much memory is available. In addition, out of the three, it is the only one to be adjusted according to the necessities of the specific problem through two parameters.

3. Application to Laskar's Method

The HRFFT method, as defined in the preceding, of course allows efficient application to *any* problem involving continuously evaluated DFT spectra, whenever

N equidistant data points are given and high resolution in frequency domain is required. In particular, application to several variants of Laskar's (1990, 1993, 1995) *Numerical Analysis of Fundamental Frequencies (NAFF)*, which was especially designed for the study of orbital behaviour in dynamical systems, seems promising, as outlined in the following.

Laskar introduces a continuous scalar product, wherein a signal vector is multiplied with a single harmonic-oscillation model vector, the absolute value of which has to be maximized wrt. the model frequency, in order to obtain an estimate for the most dominant frequency, yielding thus one more model vector. The signal is then projected onto the orthogonal complement of the model vectors found so far and the projection is taken as the new signal onto which the same procedure of finding the maximum of the mentioned scalar product is then performed. This is done until reasonable conditions for interruption are encountered. The integral involved in this step is calculated by means of a standard interpolation method, applied to the set of discrete data points. The reason for taking a continuous inner product rests on the fact that the implicit periodicity assumption on the spectrum due to sampling in time domain can be thus overcome, which amounts to circumvent the annoying superposition of aliasing and leakage, appearing even within a Nyquist interval; leakage itself is of course not removed using this technique. There is, however, a drawback connected with the use of an arbitrary interpolation method: Since in most applications little is known about the behaviour of the signal in-between the data points, error estimates of the deviation in determination of frequency components introduced by the interpolation are difficult, if not impossible, unless there is an error bound of the interpolation itself at hand, which will be the case only in artificially constructed examples. The oscillations of the interpolation polynomial between data points — unless the signal is of trivial form — cause the appearance of unreal frequency components strongly pronounced at half the value of the sampling rate, involving all harmonics and even sub-harmonics and continua with magnitudes decaying with the sequence of enumeration. Though the error thereby inferred may be small in absolute value, it cannot be controlled, because of the only finite density of data points available.

If one now replaces the scalar product of Laskar's method by the natural discrete one, based just on the data points, all the other steps become obviously more simple and the finding of the maximum just turns out to be a problem of straightforward continuous evaluation of a DFT. This suggests to take advantage of the HRFFT for this task. Moreover, the search for the maximum can be done particularly efficient using a bisection algorithm, due to the a-adic expansion involved in the HRFFT, with a being a power of two.

Constructing a sequence of signals with increasing length, obtained from the original signal by adding that number of zeroes in its trailing part which is given by the sequence index, one obtains a sequence of corresponding DFT spectra, which, properly re-scaled, just represents a sequence of equidistant interpolations of the DFT with increasing density of points in frequency domain. One might introduce now a discrete scalar product in frequency space according to the increased number of frequency points and again search for the maxima of these scalar products when pairing the re-scaled DFT spectra with the corresponding ones of elementary harmonic oscillations, analogous to Laskar's method. In the limit of infinite zeroes one thus obtains a continuous scalar product in frequency domain. On the other hand, for all finite indices, the values of the discrete inner products coincide with the value of the plain DFT of the original signal at the respective model frequency. Thus, by continuity, the equivalence between the finding of maxima of the DFT through continuous evaluation and a continuous scalar product applied in frequency domain is demonstrated. The difference to Laskar's approach now essentially rests on the fact that no artificial effect is introduced by this method, apart from the fact that one cannot get rid of the periodicity in frequency space. It is worth noting that such an approach amounts to formally use the interpolation supplied by Shannon's theorem, though one should be aware that the essential hypothesis of band limitation of this theorem in order to be applicable, is not fulfilled, because of the only finite length of the signal.

Thus, if one employs an iterative correction scheme, as the one devised by Šidlichovský and Nesvorný (1996), or one based on the *Determinant Spectral Analysis (DSA)* method developed by the author (to be presented in a forthcoming paper), one might start using Laskar's interpolation in order to get fairly good initial values for the iteration and then proceed with the HRFFT-based scalar product.

Extensive practical tests will be necessary to clarify the situation and improve existing procedures, however, the essential point to be stated here is, that there occurs a *conceptual problem* with the classical form of the NAFF. This is a fact that can be established on a rigorous mathematical basis and there are ways to avoid the errors introduced thereby, using e.g. the new method HRFFT.

4. Summary

In the present study a new method, named HRFFT, for obtaining the accurate values of the DFT of a signal at any point of frequency domain has been developed with an extraordinarily efficient balance between computation speed and memory effort. Applications to the precise description of the frequency structure of dynamical systems have been indicated, leaving much future work for their practical implementation, though the method itself is applicable to a wide class of problems far beyond the field of dynamical system theory and has already been employed successfully to subtle practical problems.

As a future perspective we note that application of the same principles as used here in frequency domain in order to yield the HRFFT, when applied also in time domain, results in a new, very fast algorithm to evaluate DFT spectra also of *unevenly* spaced data, as latest investigations revealed.

Acknowledgements

I am indebted to R. Dvorak, S. Ferraz-Mello, S. Ichtiaroglou, J. Laskar, D. Nesvorný, M. Šidlichovský and J. Waldvogel for valuable suggestions and discussions related to this work. Unknown referees helped to remove serious mathematical errors of an earlier draft. Part of this work was carried out at the Department of Physics of Aristotle University in Thessaloniki, Greece. Support by the Austrian 'Fonds zur Förderung der wissenschaftlichen Forschung' (FWF), contract P 9921-AST, as well as P 10212-AST associated to EU contract ERB 4050 PL930312, is acknowledged.

References

Bracewell, R.N.: 1978, The Fourier Transform and its Applications, McGraw-Hill, New York

Brigham, E.O.: 1974, The Fast Fourier Transform, Prentice Hall, Englewood Cliffs, New Jersey

Dvorak, R., Lohinger, E. & Maindl, T.: 1995, 'On the Chaoticity of Pluto's Orbit', in: Perturbation Theory and Chaos in Nonlinear Dynamics with Emphasis to Celestial Mechanics, J. Hagel, M. Cunha & R. Dvorak (eds.), Universidade da Madeira, Funchal, Madeira

Ferraz-Mello, S.: 1981, 'Estimation of Periods from Unequally Spaced Observations', *Astron. J.* **86/4**, p. 619

Foster, G.: 1996a, 'Time Series Analysis by Projection. I. Statistical Properties of Fourier Analysis', *Astron. J.* **111/1**, p. 541

Foster, G.: 1996b, 'Time Series Analysis by Projection. II. Tensor Methods for Time Series Analysis', *Astron. J.* **111/1**, p. 555

Golub, G. & van Loan, Ch.F.: 1986, Matrix Computations, North Oxford Academic

Laskar, J.: 1990, 'The Chaotic Motion of the Solar System: A Numerical Estimate of the Size of Chaotic Zones', *Icarus* **88**, p. 266

Laskar, J., Froeschlé, C. & Celletti, A.: 1992, 'The Measure of Chaos by the Numerical Analysis of the Fundamental Frequencies. Application to the Standard Map.', *Physica D* **56**, p. 253

Laskar, J.: 1993, 'Frequency Analysis of a Dynamical System', in: Qualitative and Quantitative Behaviour of Planetary Systems, R. Dvorak & J. Henrard (eds.), Kluwer, Dordrecht

Laskar, J.: 1995, 'Introduction to Frequency Map Analysis', in: Hamiltonian Systems with Three or More Degrees of Freedom, A. Delshams & C. Simo (eds.), pre-print

Rabiner, L.R. & Gold, B.: 1975, Theory and Application of Digital Signal Processing, Prentice Hall Inc., Englewood Cliffs, New Jersey

Šidlichovský, M. & Nesvorný, D.: 1996, 'Frequency Modified Fourier Transform and its Application to Asteroids', this volume

Voyatzis, G. & Ichtiaroglou, S.: 1992, 'On the Spectral Analysis of Trajectories in Near-Integrable Hamiltonian Systems', *J. Phys. A* **25**, p. 5931

ON THE USE OF ANALYTICAL THEORIES IN THE CALCULATION OF PRECESSION-NUTATION

P. BRETAGNON
Bureau des longitudes
77, avenue Denfert-Rochereau, 75014, Paris FRANCE

Abstract. We present the use of the analytical solutions of the planets and of the Moon's motion in the determination of the quantities which relate the barycentric and the geocentric coordinate systems and of the expressions of precession-nutation. The computation of the precession and nutation quantities are built with the analytical theories of the motion of the Moon, the Sun and the planets of the Bureau des longitudes. We take into account the influence of the Moon, the Sun and all the planets on the potential of the Earth limited to $C_{j,0}$ for j from 2 to 5, $C_{2,2}$, $S_{2,2}$, $C_{3,k}$, $S_{3,k}$ for k from 1 to 3 and $C_{4,1}$, $S_{4,1}$. We determine the 3 Euler angles ψ, ω, and φ calculating the components of the torque of the external forces with respect to the geocenter in the case of the rigid Earth. The equations are solved by iterations and so are taken into account the nutations-on-the-nutations effects. We have determined the analytical variations of the angles ψ and ω fixing the equator with respect to the ecliptic J2000. We find, in ω, a secular term of -26.5026 mas per century. The analytical solution of the precession-nutation has been compared to a numerical integration over the time span 1900–2050. The differences do not exceed 16 μas for ψ and 12 μas for ω.

Key words: precession – nutation – analytical theory

1. Introduction

Many problems need a solution of the planets and of the Moon's motion. The solutions are used either in analytical form or in numerical integration form. Among these problems, we have for instance the determination of the expressions of transformation between the barycentric, geocentric and topocentric reference systems for the space coordinates and the time coordinate.

Another problem which needs to know the motion of the Moon and of the planets is the study of the orientation of the Earth about the geocentre. For all these studies, the motion of the planets and of the Moon are given by the numerical integration DE200/LE200 of the JPL or by the analytical solutions VSOP and ELP of the Bureau des longitudes and we have to know an estimation of the accuracy of the numerical and analytical solutions.

2. Accuracy of the Numerical and Analytical Solutions

The precision of the VSOP87 solutions (Bretagnon, Francou, 1988) is obtained by comparison with internal numerical integrations and by comparison with DE200 (Standish, 1982). Table I gives differences for the longitudes between VSOP and internal numerical integration over one century (1900-2000). The differences are about 0.001″ for the inner planets and about 0.01″ for the outer planets. In the comparison with DE200, the residuals are a little more important because the model is slightly different. There are important residuals in the longitudes of Saturn and Neptune because of the utilization of a wrong value of the mass of

Celestial Mechanics and Dynamical Astronomy **65**: 95–106, 1997.

Uranus in DE200. As for the longitude of Jupiter, the residuals come from the mass of Uranus as well, but also from the Saturn one. For the inner planets, the differences of masses take a less important part and the difference between the analytical solution and DE200 are about 0.001″.

TABLE I
Comparison of VSOP with internal numerical integration (INI) and with DE200 over one century.

	Mercury	Venus	Earth	Mars	Jupiter	Saturn	Uranus	Neptune
INI	0″.001	0″.002	0″.002	0″.007	0″.01	0″.02	0″.01	0″.01
DE200	0″.001	0″.005	0″.006	0″.020	0″.07	0″.14	0″.02	0″.07

The numerical integrations have a very great internal accuracy. Their uncertainties come from the model which is used and from the fitting to the observations. The actual accuracy may be obtained by comparison of the numerical integrations DE200 and DE245. Table II shows the differences, over the time span 1900-2000, for the heliocentric right ascensions and declinations for the eight planets. The third body is the Earth-Moon barycentre. A part of the differences of the table II comes from differences in the orientations of the two ephemerides and the absolute accuracy errors for the inner planets through Mars should be less than 0″.02 for DE200.

TABLE II
Differences between DE245 and DE200 for the heliocentric right ascensions and declinations.

	Mercury	Venus	Earth	Mars	Jupiter	Saturn	Uranus	Neptune
α	0″.08	0″.09	0″.10	0″.07	0″.12	0″.27	0″.57	0″.82
δ	0″.03	0″.03	0″.04	0″.03	0″.07	0″.22	0″.13	0″.15

We see that the errors of DE200 are greater than the differences between DE200 and VSOP and, finally, the main limitation of the accuracy of the planetary solutions comes from the quality of the observations. It is useful to precise that the integration constants of the analytical planetary theories have been determined by fitting to DE200. So, the VSOP solutions cannot give a more precise ephemeris than DE200.

The lunar ephemeris ELP2000 (Chapront-Touzé, Chapront, 1983) gives the position of the Moon with an accuracy of 0.01″ for the chosen model. For LE200 and ELP, the uncertainty essentially comes from the tidal term which involves in an error of 1″ per century squared. So, over 1980-2000, the uncertainty of the lunar ephemerides is 0.01″ and, over 1900-2100, 1″.

3. Relations between Barycentric and Geocentric Coordinates

The solutions of the planets and of the Moon's motion are used to determine the transformation between the barycentric and the geocentric coordinate systems. Neglecting the c^{-4} terms, the Barycentric Coordinate Time (TCB) and the Geocentric Coordinate Time (TCG) are related by

$$\begin{aligned}\mathrm{TCB} &= \mathrm{TCG} + c^{-2}\int_{TCB_0}^{TCB} (U_E + \tfrac{1}{2}v_E^2)\,\mathrm{d}(TCB') \\ &= \mathrm{TCG} + L_C \times TCB + P(TCB) \qquad (1)\end{aligned}$$

with :

$$L_C = 1.480\,826\,8475 \times 10^{-8}$$

$$\begin{aligned}\mathrm{P(TCB)} = {} & 1\,656.674\,564\,\mu\mathrm{s}\,\sin(6\,283.075\,850\,TCB + 6.240\,054) \\ & + \ldots\end{aligned}$$

In the expression (1), the determination of L_C from the analytical solutions VSOP and ELP is easier than with a numerical integration. In this latter case, we have to perform a Fourier analysis, the result of which depends on the time span taken into account.

The use of analytical solutions ensures an accuracy of the relation (1) of 2 nanoseconds over (1980-2000). In this connection, a problem is the choice of the time scale TCB or TDB in calculation of periodic and Poisson terms P.

4. Rotation of the Rigid Earth

Using the VSOP and ELP solutions, we studied the rotation of the rigid Earth by determining the variations of the Euler angles ψ, ω, and φ by the calculation of the torque on the oblate Earth.

For years, there have been many contributions improving the rigid Earth precession-nutation series.

Kinoshita (1977) and Kinoshita and Souchay (1990) used Hamiltonian methods. From VSOP and ELP, they took into account step by step the perturbations by the disturbing bodies : for each body, at first, they consider a Keplerian orbit then the different perturbations.

Roosbeek and Dehant (1996) compute the torque on the equatorial bulge by using analytical solutions and proceeding step by step.

Williams (1994, 1995) proceeds in the same way by using VSOP and ELP or DE200/LE200.

Hartmann and Soffel (1995) make a similar work by solving numerically the equations and by using DE200/LE200.

In this paper, the rotation of the rigid Earth is obtained by using analytical theories for the motion of the solar system bodies including the perturbations as a whole.

4.1. Coordinate systems

Let us define the used coordinate systems. The coordinate system $Oxyz$ is defined by the ecliptic and the equinox J2000. A non-rotating equatorial coordinate system $OXYZ$ is obtained by a rotation of ψ about the z axis and of ω about the X axis. The terrestrial coordinate system $O\xi\eta\zeta$ is obtained by a rotation of φ about the Z axis. This coordinate system has the axes $O\xi$ and $O\eta$ in the plane of the equator with the $O\xi$ axis pointing toward the intersection of the equator with the plane containing the Greenwich Meridian. Let α be the longitude of the major axis of the equatorial ellipse corresponding to the principal moment of inertia A and $\tilde{\varphi} = \varphi + \alpha$. At last, we note : $\varphi = \varphi_0 + \varphi_1\, t + \Delta\varphi$.

5. Equations of Motion

The differential equations for the variables ψ, ω and φ are written in the form

$$\begin{aligned} \ddot{\omega} + \tfrac{C}{A}\, \sin\omega_0\, \varphi_1\, \dot{\psi} &= \tfrac{L}{A} + F_2 + \tfrac{B-A}{A}\, F_1 \\ \sin\omega_0\, \ddot{\psi} - \tfrac{C}{A}\, \varphi_1\, \dot{\omega} &= \tfrac{M}{A} + G_2 + \tfrac{B-A}{A}\, G_1 \\ \ddot{\varphi} &= \tfrac{N}{C} + H_2 + \tfrac{B-A}{C}\, H_1 \end{aligned} \tag{2}$$

with :

$$\begin{aligned} F_1 &= \tfrac{1}{2}\ddot{\psi}\sin 2\tilde{\varphi}\sin\omega + \dot{\psi}\dot{\varphi}\cos 2\tilde{\varphi}\sin\omega - \dot{\omega}\dot{\varphi}\sin 2\tilde{\varphi} \\ &\quad -\tfrac{1}{2}\ddot{\omega}(1-\cos 2\tilde{\varphi}) + \tfrac{1}{4}\dot{\psi}^2(1+\cos 2\tilde{\varphi})\sin 2\omega \\ G_1 &= -\tfrac{1}{2}\ddot{\psi}(1+\cos 2\tilde{\varphi})\sin\omega + \dot{\psi}\dot{\varphi}\sin 2\tilde{\varphi}\sin\omega + \dot{\omega}\dot{\varphi}\cos 2\tilde{\varphi} \\ &\quad +\tfrac{1}{2}\ddot{\omega}\sin 2\tilde{\varphi} + \tfrac{1}{4}\dot{\psi}^2\sin 2\tilde{\varphi}\sin 2\omega \\ H_1 &= \tfrac{1}{2}\dot{\omega}^2\sin 2\tilde{\varphi} - \dot{\psi}\dot{\omega}\cos 2\tilde{\varphi}\sin\omega - \tfrac{1}{4}\dot{\psi}^2\sin 2\tilde{\varphi}(1-\cos 2\omega) \end{aligned} \tag{3}$$

$$\begin{aligned} F_2 &= -\tfrac{C}{A}\,\varphi_1\,\dot{\psi}\,(\sin\omega - \sin\omega_0) - \tfrac{C}{A}\,\Delta\dot{\varphi}\,\dot{\psi}\,\sin\omega - \tfrac{1}{2}\,\tfrac{C-A}{A}\,\dot{\psi}^2\,\sin 2\omega \\ G_2 &= \tfrac{C}{A}\,\Delta\dot{\varphi}\,\dot{\omega} - \ddot{\psi}\,(\sin\omega - \sin\omega_0) - \tfrac{A+B-C}{A}\,\dot{\psi}\,\dot{\omega}\,\cos\omega \\ H_2 &= -\ddot{\psi}\,\cos\omega + \dot{\psi}\,\dot{\omega}\,\sin\omega \end{aligned} \tag{4}$$

In the expressions (2) and (4), A, B, C are the moments of inertia. L, M, N represent the components, in the equatorial coordinate system $OXYZ$, of the torque of the external forces with respect to the geocentre. F_2, G_2, H_2 are small quantities. The factors $(B-A)/A$ and $(B-A)/C$ are about 2×10^{-5}.

6. Solution of the Equations

The integration of the third equation (2) presents no difficulties. For the two first equations, the solution of the reduced system is

$$\begin{aligned} \dot{\omega} &= K_S\,\sin\sigma t + K_C\,\cos\sigma t \\ \sin\omega_0\,\dot{\psi} &= K_C\,\sin\sigma t - K_S\,\cos\sigma t \end{aligned} \tag{5}$$

where :

$$\sigma = \frac{C}{A}\varphi_1 = 2\,308\,800.579\,285\,173 \text{ rd/thousand Julian years.}$$

σ is the Eulerian frequency in the reference frame J2000. This term has a period of 0.997 day in the reference frame J2000 and of 305 days in the terrestrial coordinate system. K_S and K_C are two integration constants.

A particular solution of the equations (2) is built by the method of the variation of the constants :

$$\begin{aligned}\dot{K}_S &= \tfrac{L}{A}\sin\sigma t - \tfrac{M}{A}\cos\sigma t + (F_2 + \tfrac{B-A}{A}F_1)\sin\sigma t - (G_2 + \tfrac{B-A}{A}G_1)\cos\sigma t\\ \dot{K}_C &= \tfrac{L}{A}\cos\sigma t + \tfrac{M}{A}\sin\sigma t + (F_2 + \tfrac{B-A}{A}F_1)\cos\sigma t + (G_2 + \tfrac{B-A}{A}G_1)\sin\sigma t\end{aligned} \quad (6)$$

and the solution is obtained by integration.

In the equations (2-4), the quantities F, G, H depend on ψ, ω, φ. To compute L, M, N, the disturbing body must be referred to the terrestrial coordinate system by a rotation also depending on ψ, ω and φ. So, it is necessary to compute again the right-hand sides of the equations (2) and to proceed by iteration.

6.1. Integration constants

The two integration constants K_S and K_C are issued from observation. The period of these terms is the Euler period of 305 days but no periodic term is observed for a non rigid Earth with this period. So, one gets $K_S = K_C = 0$.

For the two first equations, the other integration constants are the values ψ_0 and ω_0 of the secular parts of the angles ψ and ω at t = J2000.

The two last integration constants are φ_0 and φ_1. Let R be the intersection of the equator of date with the ecliptic J2000 and γ_D the equinox of date. The rotation of the Earth is measured by the angle φ from R or by the sidereal time from γ_D. Therefore φ is related to the sidereal time by

$$\varphi = \text{temps sidereal} + \chi,$$

where χ is the arc $R\gamma_D$. Taking the secular part of χ from Simon et al (1994), we obtain the two integration constants φ_0 and φ_1.

6.2. Torque L, M, N

The components L, M, N of the torque depend on the derivatives of the potential with respect to the longitude Λ and the latitude β of the disturbing body reckoned in the terrestrial coordinate system $O\xi\eta\zeta$.

$$\begin{aligned}L &= -m\sin(\Lambda+\varphi)\tfrac{\partial U}{\partial\beta} + m\tan\beta\cos(\Lambda+\varphi)\tfrac{\partial U}{\partial\Lambda}\\ M &= m\cos(\Lambda+\varphi)\tfrac{\partial U}{\partial\beta} + m\tan\beta\sin(\Lambda+\varphi)\tfrac{\partial U}{\partial\Lambda}\\ N &= -m\tfrac{\partial U}{\partial\Lambda}\end{aligned} \quad (7)$$

with :

$$\frac{\partial U}{\partial \beta}=\frac{Gm_E}{r}\sum_{n=2}^{\infty}\frac{a_E^n}{r^n}C_{n,0}\frac{\partial}{\partial \beta}[P_{n,0}(\sin\beta)]$$
$$+\frac{Gm_E}{r}\sum_{n=2}^{\infty}\sum_{k=1}^{\infty}\frac{a_E^n}{r^n}(C_{n,k}\cos k\Lambda+S_{n,k}\sin k\Lambda)\frac{\partial}{\partial \beta}[P_{n,k}(\sin\beta)]$$
$$\frac{\partial U}{\partial \Lambda}=\frac{Gm_E}{r}\sum_{n=2}^{\infty}\sum_{k=1}^{\infty}\frac{a_E^n}{r^n}k(S_{n,k}\cos k\Lambda-C_{n,k}\sin k\Lambda)P_{n,k}(\sin\beta) \tag{8}$$

where $P_{n,k}$ are the Legendre polynomials.

In these expressions, the position of the Sun and the planets are given by the VSOP87A solution which represents the heliocentric motions in rectangular coordinates with respect to the ecliptic and equinox J2000. The Moon is also represented by rectangular coordinates with respect to the ecliptic and equinox J2000.

6.3. Use of the ELP solution

The ELP solution contains the perturbations by the Sun, the direct and indirect planetary perturbations, the perturbations due to the potential of the Earth and the tidal terms. It is expressed in Fourier and Poisson series with the mean longitudes of the planets, the Delaunay angles D, F, l, l' and the longitude $\mathcal{L}$ of the Moon reckoned from the equinox of date.

The angles $\mathcal{L}$, λ_3 and D are related by :

$$\mathcal{L}=\lambda_3+D+180^\circ+\mathcal{P}t$$

where $\mathcal{P}$ is the constant of the general precession in longitude. Combinations of $\mathcal{L}$, λ_3 and D can raise the angle with the frequency $\mathcal{P}$ and the period 26 000 years. So, we replace $\mathcal{L}$ by λ_3+D and develop the quantity $\mathcal{P}t$ in Poisson series.

The angles λ_3 and l' are related by :

$$\lambda_3-l'=\varpi_3$$

and in the same way, we can obtain a term the argument of which is the longitude ϖ_3 of the Earth perihelion with a 111 600 year period. l' is replaced by λ_3 and ϖ_3 is developed in Poisson series.

So, in our solution, the argument of the most important term of the nutation is not the longitude Ω_D of the node of the Moon referred to the equinox of date but the longitude Ω_{J2000} of the node referred to the equinox J2000 given in this form :

$$\Omega_D\rightarrow\lambda_3+D-F=\Omega_{J2000}+180^\circ$$

TABLE III
Most important perturbations of the nutation. Amplitudes are in μas, periods in days.

Origin	Argument	Amplitude in ψ	Amplitude in ω	Period
Moon $C_{2,0}$	$\lambda_3 + D - F$	17 292 173.10	9 227 878.02	6 793.48
$C_{3,0}$	$\lambda_3 + D - l$	104.05	88.95	3 232.61
$C_{4,0}$	$\lambda_3 + D - F$	0.73	6.84	6 793.48
$C_{5,0}$	$\lambda_3 + D - l$	0.01	0.00	3 232.61
$C_{2,2} - S_{2,2}$	$2\lambda_3 + 2D - 2\varphi$	29.44	11.71	0.52
$C_{3,1} - S_{3,1}$	$\lambda_3 + D + \varphi$	38.44	15.25	0.96
$C_{3,2} - S_{3,2}$	$\lambda_3 + D - 2\varphi$	0.39	0.14	0.51
$C_{3,3} - S_{3,3}$	$\lambda_3 + D - 3\varphi$	0.41	0.20	0.34
$C_{4,1} - S_{4,1}$	φ	1.68	0.67	1.00
Sun $C_{2,0}$	$2\lambda_3$	1 276 710.99	552 389.67	182.63
$C_{3,0}$	λ_3	0.26	0.22	365.26
$C_{2,2} - S_{2,2}$	$2\lambda_3 - 2\varphi$	12.32	4.90	0.50
$C_{3,1} - S_{3,1}$	$\lambda_3 + \varphi$	2.79	1.11	1.00
Mercury $C_{2,0}$	$\lambda_1 - 4\lambda_3$	1.03	0.43	2 432.11
Venus $C_{2,0}$	$3\lambda_2 - 5\lambda_3$	216.71	90.76	2 959.21
Mars $C_{2,0}$	$\lambda_3 - 2\lambda_4$	11.55	0.95	5 764.01
Jupiter $C_{2,0}$	$2\lambda_5$	104.41	45.69	2 166.29
Saturn $C_{2,0}$	$2\lambda_6$	12.15	5.16	5 379.61
Uranus $C_{2,0}$	$2\lambda_7$	0.65	0.29	15 344.24
Neptune $C_{2,0}$	$2\lambda_8$	0.40	0.16	30 091.15
Complements	$\lambda_3 + D - F$	15 361.12	0.04	6 793.48

With a such representation, we find no arguments of similar periods in the series. In the classical tables of the nutation, we find the arguments :

$$\begin{array}{ll} \Omega_D & \text{(period 6798.38 days)} \\ -2l' + 2F - 2D + \Omega_D & \text{(period 6786.32 days)} \end{array}$$

which are in phase every 10 468 years. They are replaced by $\lambda_3 + D - F$, longitude of the node referred to the equinox J2000 with a 6793.48 day period. The arguments

$$\begin{array}{ll} l - l' - D & \text{(period 3232.86 days)} \\ l - F - \Omega_D & \text{(period 3231.50 days)} \end{array}$$

are in phase every 20 937 years and are replaced by $\lambda_3 + D - l$, longitude of the Moon perigee with a 3232.61 day period.

Many couples of such arguments can be found in the classical solutions and it is not necessary to keep them in expressions of nutation which are valid over only some thousand years.

In the first step of this work, the solution used for the motion of the Moon is truncated to 0.″001 and thus ensures an accuracy of only 0.″09.

7. Results

In the construction of the torque, we have taken into account the harmonics gathered in table III. This table gives the most important periodic perturbation of each effect in the nutation.

For the Moon, the effect on $C_{2,0}$ of course gives the 18.6 year period term with an amplitude of $17''$. The contributions on the zonal harmonics $C_{3,0}$, $C_{4,0}$ are not negligible.

As for the tesseral harmonics, we find the half-day period term with an amplitude of 30 μas due to the Moon and 12 μas due to the Sun but also the influence of the harmonics $C_{3,1} - S_{3,1}$ and $C_{4,1} - S_{4,1}$ which give one day period terms with amplitudes of 40 μas and 2 μas. The importance of these terms comes from the resonance with the Euler period. In the case of the planets, only the harmonic $C_{2,0}$ is to take into account. The complementary part of the equations (2) gives 15 mas in the nutation in longitude.

Table IV contains the different contributions to the secular terms for ψ and ω. The moment of inertia C is fixed in order to obtain $\dot{\psi} = 50\,384''\!.5649$ which corresponds to $50\,287.7''$ per thousand years for the constant of the general precession in longitude.

TABLE IV
Secular term of ψ and of ω in arcseconds per thousand years.

Origin	$-\psi$	$-\omega$
Moon $C_{2,0}$	34 454.955 434	−0.254 414
$C_{3,0}$	−0.000 057	−0.000 011
$C_{4,0}$	0.025 192	
Sun $C_{2,0}$	15 948.701 336	0.002 923
$C_{3,0}$	−0.000 026	−0.000 005
Mercury $C_{2,0}$	0.003 698	−0.000 088
Venus $C_{2,0}$	0.181 581	−0.016 814
Mars $C_{2,0}$	0.005 999	0.000 357
Jupiter $C_{2,0}$	0.117 059	0.002 804
Saturn $C_{2,0}$	0.005 208	0.000 220
Uranus $C_{2,0}$	0.000 100	0.000 001
Neptune $C_{2,0}$	0.000 029	0.000 001
Complements	−0.231 853	
p_g	−19.198 800	
	50 384.564 900	−0.265 026

Then, the value of the dynamical ellipticity is

$$H_d = \frac{2C - A - B}{2C} = 0.003\,273\,7670.$$

TABLE V
Secular terms of ψ in mas/yr.

Planet	Williams 94	Williams 95	this work
Mercury	0.003 651	0.003 697	0.003 698
Venus	0.187 273	0.181 565	0.181 581
Mars	0.005 393	0.005 998	0.005 999
Jupiter	0.116 665	0.117 068	0.117 059
Saturn	0.005 177	0.005 188	0.005 208
Uranus	0.000 100	0.000 100	0.000 100
Neptune	0.000 029	0.000 029	0.000 029
Total	0.318 287	0.313 645	0.313 674

TABLE VI
Secular terms of ω in mas/yr.

Planet	Williams 94	Williams 95	this work
Mercury	−0.000 090	−0.000 088	−0.000 088
Venus	−0.017 372	−0.016 813	−0.016 814
Mars	0.000 255	0.000 356	0.000 357
Jupiter	0.002 782	0.002 804	0.002 804
Saturn	0.000 217	0.000 219	0.000 220
Uranus	0.000 001	0.000 001	0.000 001
Neptune	0.000 001	0.000 001	0.000 001
Total	−0.014 207	−0.013 520	−0.013 519

This result is very close of the values obtained by Williams (1994) and by Souchay and Kinoshita (1996) :

$$\text{Williams (1994)} = 0.003\,273\,7634$$
$$\text{Souchay-Kinoshita (1996)} = 0.003\,273\,7580$$

Williams (1994, 1995) has published the secular perturbations due to the planets. His results of 1994 were not very precise. His values of 1995 are very close of ours, specially for ω. We give in table V and VI the planetary secular terms of Williams (1994), Williams (1995), and of our solution for ψ and ω. The last results of Souchay and Kinoshita (1996) are very close as well.

The comparison to the Kinoshita-Souchay solution (1990) shows, in the periodic part, important discrepancies (Bretagnon, 1996). In their recent solution, they have introduced the long period terms of 900, 1800 and 239 year period and corrected some mistakes such as the 122 day period term with a residual of 1 mas. But, we still have a difference of 300 μas between our solution and the Souchay-Kinoshita solution for the 18.6 year term.

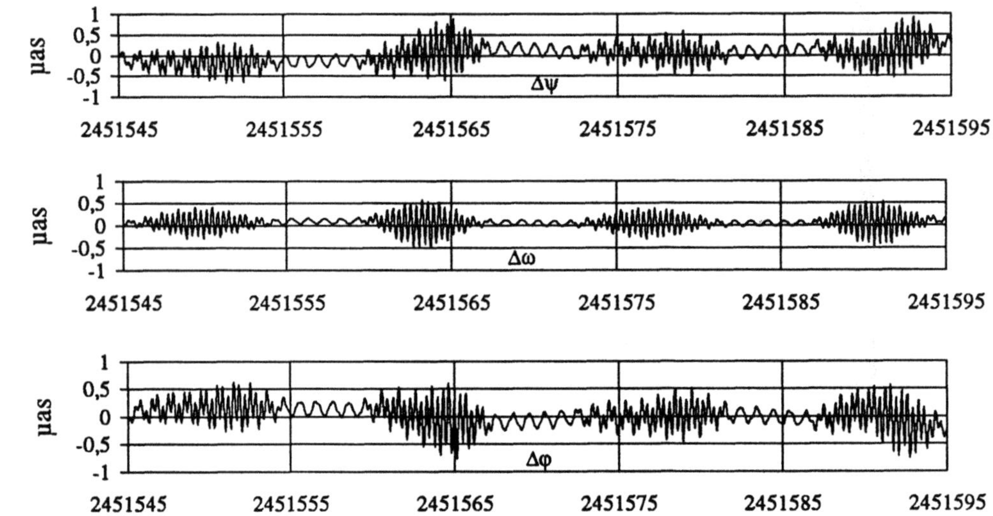

Fig. 1. Theory – numerical integration over 50 days

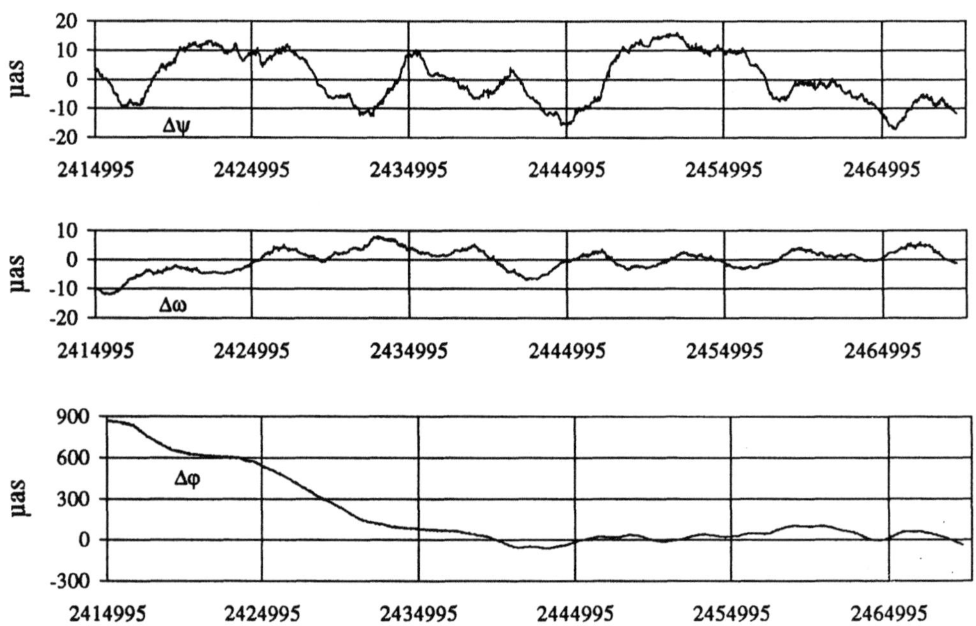

Fig. 2. Theory – numerical integration over 1900-205

8. Comparison to Internal Numerical Integrations

We have numerically integrated the 3 angles ψ, ω and φ with the Stoer-Bulirsch (1980) method taking into account all the harmonics of the analytical solution, but the numerical integration use the full ELP solution.

To check for the diurnal terms of the analytical solution, we have performed an integration over 50 days from J2000. Figure 1 shows the differences in the sense theory - numerical integration. These differences are smaller than 1 μas for the 3 angles ψ, ω and φ, but there are discrepancies of 0.5 μas with an 8 hour period not yet explained.

We have also performed an integration over 150 years between 1900 and 2050. Figure 2 shows that the residuals reach 16 μas in the precession-nutation in longitude and 12 μas in obliquity. For the angle φ, the discrepancies are inferior to 80 μas between 1950 and 2050 but reach 1 mas about 1900. These discrepancies seem to come from numerical difficulties, the variation of the angle φ being $5 \times 10^{10''}$ per century.

9. Conclusion

In this method, we use the whole solution of the Sun, Moon and planets motion. In this manner, we consider :
- the direct planetary attraction on the Earth,
- the indirect planetary effects,
- the non-Keplerian terms induced by the Sun on the Earth-Moon gravitational interaction and by the Moon on the Earth-Sun gravitational interaction,
- the effect of the Earth's equatorial bulge on the Moon which perturbs the Earth-Moon gravitational attraction.

Proceeding by iteration, we consider the effect of the precession-nutation changing the Earth's orientation and so, changing Moon-Earth-Sun gravitational interaction : the nutations on the nutations effects.

The present solution is 50 times as precise as the solution of Souchay-Kinoshita, however we have to improve it. For that, we are now going to use the full ELP solution and to keep a greater number of terms in the series of the nutation. Indeed, the VLBI observations reach an accuracy of some $10^{-5''}$ and it is necessary to obtain a solution of the precession-nutation for the rigid Earth with a precision of some μas to reveal the effects of the non-rigidity of the Earth by comparison to observations.

References

Bretagnon, P., Francou, G. (1988) *Astron. Astrophys.*, **Vol. no. 202**, p. 309–315.
Bretagnon, P. (1996) *IAU Symposium no. 172*, to be published.
Chapront-Touzé, M., Chapront, J. (1983) *Astron. Astrophys.*, **Vol. no. 124**, p. 50.
Hartmann, T., Soffel, M. (1994) *Astron. J.*, **Vol. no. 108**, p. 1115.
Kinoshita, H. (1977) *Celest. Mech.*, **Vol. no. 15**, p. 277.

Kinoshita, H., Souchay, J. (1990) *Celest. Mech.*, **Vol. no. 48**, p. 187.
Roosbeek, F., Dehant, V. (1996) *Celest. Mech.*, in preparation.
Simon, J.L., Bretagnon, P., Chapront, J., Chapront-Touzé, M., Francou, G., Laskar, J. (1994) *Astron. Astrophys.*, **Vol. no. 282**, p. 663-683.
Souchay, J., Kinoshita, H. (1996) *Astron. Astrophys.*, to be published.
Standish, E.M. (1982) *Astron. Astrophys.*, **Vol. no. 114**, p. 297–302.
Stoer, J., Bulirsch, R. (1980) Introduction to Numerical Analysis. *Springer-Verlag, New-York.*
Williams, J. G. (1994) *Astron. J.*, **Vol. no. 108** (2), p. 711.
Williams, J. G. (1995) *Astron. J.*, **Vol. no. 110** (3), p. 1420.

THE NEKHOROSHEV THEOREM AND THE ASTEROID BELT DYNAMICAL SYSTEM

ALESSANDRO MORBIDELLI
CNRS, Observatoire de la Côte d'Azur
B.P. 229, 06304 Nice Cedex 4, France
E–Mail: morby@obs-nice.fr

and

MASSIMILIANO GUZZO
Dipartimento di Matematica Pura ed Applicata
Università degli Studi di Padova
Via Belzoni 7, 35131 Padova, Italy
E–Mail: guzzo@pdmat1.math.unipd.it

Abstract. The present paper reviews the Nekhoroshev theorem from the point of view of physicists and astronomers. We point out that Nekhoroshev result is strictly connected with the existence of a specific structure of the phase space, the existence of which can be checked with several numerical tools. This is true also for a degenerate system such as the one describing the motion of an asteroid in the so called main belt. The main difference is that in some parts of the belt, the Nekhoroshev result cannot apply a priori. Mean motion resonances of order smaller than the logarithm of the mass of Jupiter and first order secular resonances must be excluded. In the remaining parts, conversely, the Nekhoroshev theorem can be proved, provided some parameters, such as the masses, the eccentricities and the inclinations of the planets are small enough. At the light of this result, a massive campaign of numerical integrations of real and fictitious asteroids should allow to understand which is the real dynamical structure of the asteroid belt.

Key words: Nekhoroshev theorem – frequency analysis – Lyapunov exponents – asteroids – degenerate systems

1. Introduction

The Nekhoroshev theorem (Nekhoroshev, 1977, 1979) is one of the most outstanding results in the field of Hamiltonian dynamics.

In order to achieve a *global* result, valid for *every* initial condition over a given domain, the Nekhoroshev theorem substitutes to the usual mathematical concept of stability (which implies confinement of motion for infinite times), the more practical concept of stability over *long times*. Since the stability times T_S grow exponentially with respect to the inverse of the non–integrability parameter ε (i.e. $T_S \sim \exp(1/\varepsilon)$), the concept of long time stability is often equivalent for physicists to the concept of *effective* stability. Indeed, every dynamical system, associated to a realistic physical model, is characterized by a limited lifetime; the latter could be easily exceeded by the long stability times given by Nekhoroshev theorem, provided ε is small enough.

This is the reason why the Nekhoroshev theorem has become very popular among dynamical physicists and astronomers. Unfortunately, most attempts of application of Nekhoroshev results have turned to frustration. Indeed it is very hard to check if the conditions for the application of Nekhoroshev theorem are

Celestial Mechanics and Dynamical Astronomy **65**: 107–136, 1997.

fulfilled (in particular the one imposing the non–integrability parameter to be small enough), and to compute analytically the value of the stability time. The results are often unrealistic. To our knowledge the best result along this line is the one obtained by Giorgilli and Skokos (1996) concerning the stability of the Lagrange triangular equilibrium points of Jupiter: they prove the existence of a region of stability over the age of the Solar System, which is large enough to include a good fraction of the Trojan asteroids; however this is done at the price of using an oversimplified model (the circular restricted three body problem).

The purpose of this review paper is to revisit the Nekhoroshev theorem and to show that it has still to be considered as a mile stone for understanding the dynamical behaviour of a given system. To this end, in section 2 we will recall the original result by Nekhoroshev and we will stress that such a result is strictly related to a particular structure of the phase space, the core of which is made of invariant KAM tori. In section 3 we will show that the existence of such a structure can be checked numerically by many means: using frequency analysis or the sup–action map method, as well as looking at the distribution of Lyapunov exponents. This gives to physicists a precious numerical tool for checking whether a Nekhoroshev–like stability result can apply to a given system.

Unfortunately, the Hamiltonian dynamical systems which are of astronomical interest are very specific, since they are *degenerate*, so that the description given in sections 2 and 3 cannot apply directly. Then, in the second part of this paper we will discuss how a Nekhoroshev–like result can be extended to degenerate systems and which are the associated structures of the phase space, which can be observed numerically.

This could give to astronomers the theoretical background for understanding, through a massive campaign of numerical investigations, which is the real structure of the asteroid belt dynamical system.

More precisely, section 4 will be devoted to investigate the secular dynamics in the asteroid belt, outside of all mean motion resonances. This is a non–degenerate dynamical system, but global perturbation parameters cannot be easily identified. Therefore, we will explore the *local* geography of perturbation parameters, and show that, in most parts of the belt, these are considerably small. Moreover we will discuss the existence of coordinate singularities corresponding to circular motions and some convexity problems which appear in the specific asteroid case. Section 5 will discuss how to control "à la Nekhoroshev" the dynamics in mean motion resonances of order $K \sim |\ln \varepsilon|$, where ε is the mass of Jupiter. Mean motion resonances of lower order cannot be controlled a priori using a Nekhoroshev–like construction, and most of them correspond indeed to the well known Kirkwood gaps. Finally, section 6 will discuss the connection between Nekhoroshev stability and invariant KAM tori, as well as the effects of the chaotic motion of the inner planets on the dynamical structure of the asteroid belt.

The conclusions will follow in section 7.

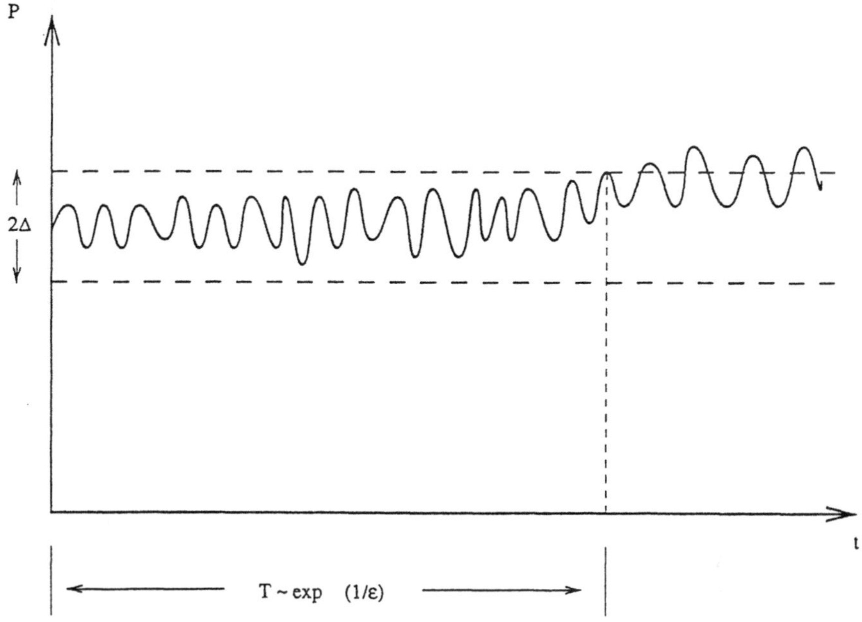

Fig. 1. Sketch of the evolution of the action p with respect to time, according to Nekhoroshev theorem. The action changes (possibly chaotically), but is confined within a strip of size Δ around the initial value. The action can escape from the strip only after a time T, which is exponentially long.

2. Nekhoroshev Theorem, Non–Overlapping Resonances and Invariant Tori

The Nekhoroshev theorem concerns autonomous Hamiltonian systems with Hamilton function $H(p,q)$, (p,q) being action angle variables defined in a domain $\mathcal{D} \equiv \mathcal{G} \times \mathbf{T}^n$, where $\mathcal{G}$ (space of actions) is a domain of $\mathbf{R}^n$ and $\mathbf{T}^n$ (space of angles) is the n–dimensional torus. The Hamiltonian is assumed to be analytic, i.e. can be expanded in local convergent series around any point $(\overline{p},\overline{q}) \in \mathcal{D}$. This implies that H can be extended to some complex neighbourhood of $\mathcal{D}$. In the following we denote by $\mathcal{G}-\Delta$ the set of points p which are contained in $\mathcal{G}$ together with a Δ–neighbourhood. With these premises, we can state in the following form the

NEKHOROSHEV THEOREM: 1. *Let $H(p,q) = H_0(p) + \varepsilon H_1(p,q)$ be real analytic in $\mathcal{D} \equiv \mathcal{G} \times \mathbf{T}^n$, with $\mathcal{G} \subset \mathbf{R}^n$ open and bounded, and $\|H_1\| \leq 1$. Consider the matrix $C(p)$ defined by $C_{ij}(p) = \frac{\partial^2 H_0}{\partial p_i \partial p_j}(p)$, and assume that there exist positive constants M and m such that:*

$$\|C(p)v\| \leq M\|v\| \quad \forall p \in \mathcal{G} \text{ and } \forall v \in \mathbf{R}^n \tag{1}$$

$$|C(p)v \cdot v| \geq m v \cdot v \quad \forall p \in \mathcal{G} \text{ and } \forall v \in \mathbf{R}^n \tag{2}$$

Then, there exist positive constants ε_*, α, β, a and b such that for any $\varepsilon < \varepsilon_*$ one has

$$\|p(t) - p(0)\| \leq \Delta \equiv \alpha\varepsilon^a ,$$

for all $p(0) \in \mathcal{G} - \Delta$ and for all $|t| \leq T(\varepsilon)$, where

$$T = \beta\left(\frac{1}{\varepsilon}\right)^{\frac{1}{2}} \exp\left(\frac{1}{\varepsilon}\right)^b .$$

A few comments are in order. The Nekhoroshev theorem does not exclude the possibility of chaotic motions. Indeed, the actions p can possibly change in a chaotic way: the theorem just states that these changes are bounded by a quantity Δ up to the time T. Slow drifts can force the actions to change more than Δ with respect to the initial conditions only after a time larger than T, as sketched in figure 1. The important fact is that the stability time T grows exponentially with respect to $1/\varepsilon$. Therefore, as soon as ε is somewhat small, the stability time is extremely long and can possibly exceed the physical lifetime of the system in study, thus providing a result of effective stability, as mentioned in the introduction. We stress that such an important stability result is achieved *uniformly* for every initial condition (p, q) with p in $\mathcal{G} - \Delta$. Obviously, orbits closer than Δ to the borders of the action domain are excluded, since they can escape from $\mathcal{G}$ in a short time. Concerning the hypotheses of the theorem, we remark that it is crucial to deal with analytic Hamiltonians. For the reasons which will be explained below, one should expect that differentiable Hamiltonians have a stability time T which is a power of $1/\varepsilon$ instead of having an exponential dependence. Conversely, the convexity condition (2) can be weakened, using a condition of finite tangency between resonant flows and corresponding resonant manifolds. We will explain below the role of convexity, and in section 4 we will discuss how to use weaker conditions, suitable for the asteroid belt case.

In our opinion, the Nekhoroshev theorem is much more than the celebrated long time stability result reported in the statement above: it proves the existence of a specific structure of the phase space from which the stability result is derived. However this fact is usually hidden in the proof of the theorem, so that it has attracted little attention. In order to bring this structure into light, we sketch in the following the basic construction of Nekhoroshev theorem. For sake of simplicity, we consider a three degrees of freedom case, and, following Poschel (1993), we refer to the space of frequencies $\omega = \partial H_0/\partial p$ to design the structure of the phase space, thus allowing to draw nice explanatory pictures.

The starting point is that one can remove from the perturbation by means of canonical transformations the harmonic term $\varepsilon H_{1,k}(p)e^{ik\cdot q}$ only outside a suitable neighborhood of its resonant plane defined by $k \cdot \omega = 0$, at the price of introducing a small remainder. Conversely, near the resonant plane the term can't be removed. It is also well known that the set of resonances related to all integer vectors $k \in \mathbf{Z}^3\backslash 0$ is dense in $\mathbf{R}^3$ and so, given any open subset of $\mathcal{G}$, it is not possible to remove from

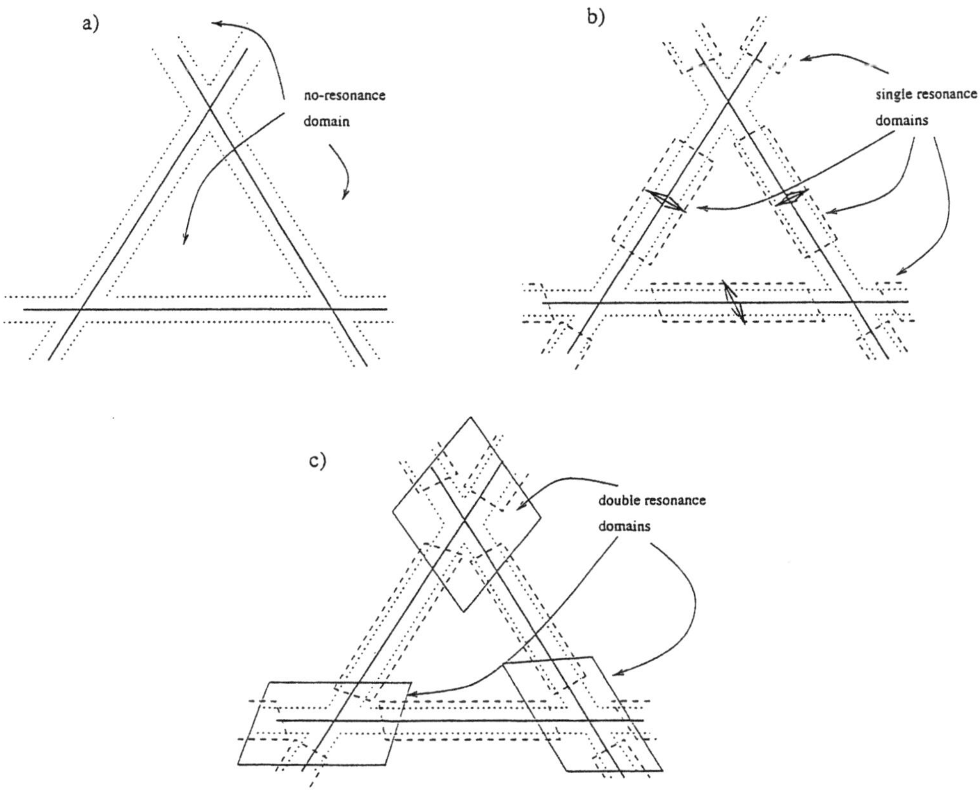

Fig. 2. Sketch of the geometric construction of Nekhoroshev theorem. See text for explanation.

the perturbation an infinite number of harmonics. This fact prevents in general the integrability of the system (Poincaré, 1892). The idea of Nekhoroshev is to consider resonances only up to a threshold order K. We recall that the order of a resonance of type $k \cdot \omega = 0$ (where $k \cdot \omega \equiv k_1\omega_1 + \ldots + k_n\omega_n$) is defined as $|k| = \sum_{i=1}^{n} |k_i|$, n being the number of degrees of freedom (3 in the present case) and $\omega_i, i = 1, \ldots, n$ being the unperturbed frequencies of the system. This approach is very important. Indeed, on one hand, the number of resonances up to a given order is finite, and so each open subset of the space of actions contains only a finite number of resonant planes, as in figure 2; on the other hand the neglected terms, corresponding to resonances of order larger than K, have a size not exceeding $\exp(-K\sigma)$, where σ is a positive parameter such that the Hamiltonian is analytic for any q with $|\mathrm{Im} q| \leq \sigma$. It will be shown at the end that K cannot be chosen greater than $1/\varepsilon^b$ with some positive $b < 1$, so that the neglected terms turn out to be exponentially small in $1/\varepsilon^b$. Note that in case the Hamiltonian is only r times differentiable (instead of analytic), one can expect that the neglected terms are as large as K^{-r}. As a consequence, the exponential dependence on $1/\varepsilon^b$ which characterizes Nekhoroshev theorem would be destroyed.

Following Nekhoroshev, we first define a *no–resonance domain*, as the set of points which are far enough from all resonances up to order K. In figure 2a it is the non connected domain, bounded by the dotted lines. More precisely, it is defined as the set of frequencies ω such that $|k \cdot \omega| > \sqrt{\varepsilon}$ for all k with $|k| \leq K$. In the no–resonance domain, one can eliminate all harmonic terms in the perturbation εH_1 of order smaller than K, i.e. the terms of type $e^{ik \cdot q}$ with $|k| \leq K$. Indeed, by construction, all these terms are non–resonant in such a domain, since the denominators $k \cdot \omega$ have a lower bound (see also Morbidelli and Giorgilli, 1996). As a consequence, the Hamiltonian $H_0 + \varepsilon H_1$ can be integrated, apart from a remainder $\mathcal{R}_K$, made (essentially) of harmonic terms of order larger than K, which are exponentially small. Therefore, neglecting at most exponentially slow diffusion forced by $\mathcal{R}_K$, one can conclude that the frequencies of the system don't change with time in the no–resonance domain.

As a second step, we consider the *single–resonance domains*, bounded by dashed lines in figure 2b, which are characterized by the presence of only one resonance of order smaller than K. As in the previous case, all non–resonant terms of order smaller than K can be eliminated. Then one reduces the Hamiltonian to have only one resonant term (of order $|k| \leq K$) and a remainder $\mathcal{R}_K$ made of terms of order larger than K, which is, again, exponentially small. Neglecting the remainder, such Hamiltonian is still integrable, but frequencies are no longer fixed. Frequencies change along what is usually called the *fast drift direction*, sketched in figure 2b by an arrow. The convexity hypothesis (2) guarantees that the fast drift direction is transversal to the resonance line. As a consequence, the motion along this direction must be confined. Indeed, following indefinitely the fast drift direction, the motion would enter the no–resonance domain. But this is impossible, since in the no–resonance domain frequencies are fixed, as explained above. On the other hand, transversal motion with respect to the fast drift direction can be forced only by the non–integrable remainder $\mathcal{R}_K$, so that it is exponentially slow.

As a third step, we consider the *double resonance domains*, centered around the resonance crossings, bounded by light continuous lines in figure 2c. In such domains, the reduced Hamiltonian still has two independent resonant terms of order smaller than K. Then, one can expect that these domains are characterized by strongly chaotic motions and that frequencies can move in any direction of the plane around the resonance crossing. However, this chaotic motion is bounded. Indeed, if frequencies moved far enough from the double resonance point, they would enter either the no–resonance domain or one of the single resonance domains. But this is impossible, since in the no–resonance domain frequencies are fixed, and in the single resonance domains frequencies can change only along the fast drift direction.

In conclusion, neglecting the exponentially small remainder $\mathcal{R}_K$, for each initial condition, the motion is confined within one of the resonance domains. As a consequence, frequencies (and actions) can change at most by a quantity equal to the radius of the double resonance domains (in n degrees of freedom the worst case corresponds to the domains characterized by $n-1$ independent resonances). Such radius is proportional to ε^a for some positive $a < 1$, (a decreasing with increasing number of degrees of freedom).

Moreover, in order to have a consistent picture as in fig. 2, the number of resonances of order smaller than K must be not too large, otherwise there would be no place for the no–resonance domain and the construction of figure 2 would be impossible. The fact that the largest resonance domains are of order ε^a gives an upper bound of type $1/\varepsilon^b$ to the choice of K, as anticipated above.

Finally, we take into account the exponentially small remainder $\mathcal{R}_K$. It is evident that this remainder can force diffusion in every direction of the frequency space, but only with exponentially small speed. Then the result concerning bounded motion, deduced neglecting $\mathcal{R}_K$, will be true in principle only up to exponentially long times.

The Nekhoroshev theorem is therefore proved.

The scheme sketched in this section is described with all mathematical details in Benettin et al. (1985). The reader can consult the papers by Nekhoroshev (1977), Poschel (1993), Lochak (1992) for alternative approaches.

2.1. NEKHOROSHEV STRUCTURE OF THE PHASE SPACE

The construction of Nekhoroshev theorem can be iterated in order to explore the dynamical structure of the phase space also with respect to the resonances of order larger than Nekhoroshev's threshold $K \sim 1/\varepsilon^b$ (see Morbidelli and Giorgilli, 1995 and Giorgilli and Morbidelli, 1996).

Indeed, in the no–resonance domain defined above (denoted hereafter by $\mathcal{G}_K$), after the elimination of the resonances up to order K, in the new action angle variables (p', q') the Hamiltonian has the form:

$$H'(p',q') = H_0'(p') + \varepsilon' H_1'(p',q') \,, \quad \text{with } \varepsilon' H_1'(p',q') \equiv \mathcal{R}_K(p',q') \,. \tag{3}$$

This is again an autonomous convex system, but the new perturbation is exponentially small with respect to the original one, being $\varepsilon' \sim \exp(-1/\varepsilon^b)$. Then, applying the Nekhoroshev theorem to this Hamiltonian, (i.e. considering resonances up to a new cut–off $K' \sim 1/\varepsilon'^b \sim \exp[b/\varepsilon^b]$) one proves the global stability of motion in $\mathcal{G}_K$ for superexponentially long times, namely up to $T \sim \exp[\exp(1/\varepsilon^b)]^b$. Moreover, one finds a new no–resonance domain $\mathcal{G}_{K'}$, characterized by the absence of resonances up to order K'. On $\mathcal{G}_{K'}$ one can introduce new action angle variables such as to transform the Hamiltonian into an integrable part and a remainder $\mathcal{R}_{K'}$, which is superexponentially small. This procedure can be iterated, and is proved to be convergent to a set of invariant KAM tori of relative large volume (Giorgilli and Morbidelli, 1996). The fact that, at each step of the iteration, the no–resonance domain is fragmented into smaller and smaller pieces never prevents the application of Nekhoroshev theorem: indeed the most relevant parameter, i.e. the ratio

TABLE I
Scheme of the iteration of Nekhoroshev theorem. The first line gives the number of iterations N, the second one the size ϱ of the domain of definition of the Hamiltonian, the third one the size of the perturbation $||H_1||$, the fourth one the cutoff order K of the considered resonances, the fifth one the Nekhoroshev stability time T_S, the sixth one the size ϱ_K of each connected component of the no–resonance domain, the last one gives the size of the remainder $\mathcal{R}_K$ of the normal form constructed on the no–resonance domain. The iteration consists in applying the Nekhoroshev theorem subsequently on the no–resonance domain defined at the previous step, i.e. substituting lines 6 and 7 to lines 2 and 3. All powers of ε are not indicated

N	1	2	3	...	n
ϱ	1	ε	$1/\exp(1/\varepsilon)$	...	$1/\exp^{n-2}(1/\varepsilon)$
$\|\|H_1\|\|$	ε	$1/\exp(1/\varepsilon)$	$1/\exp^2(1/\varepsilon)$	...	$1/\exp^{n-1}(1/\varepsilon)$
K	$1/\varepsilon$	$\exp(1/\varepsilon)$	$\exp^2(1/\varepsilon)$	...	$\exp^{n-1}(1/\varepsilon)$
T_S	$\exp(1/\varepsilon)$	$\exp^2(1/\varepsilon)$	$\exp^3(1/\varepsilon)$	...	$\exp^n(1/\varepsilon)$
ϱ_K	ε	$1/\exp(1/\varepsilon)$	$1/\exp^2(1/\varepsilon)$	...	$1/\exp^{n-1}(1/\varepsilon)$
$\|\|\mathcal{R}_K\|\|$	$1/\exp(1/\varepsilon)$	$1/\exp^2(1/\varepsilon)$	$1/\exp^3(1/\varepsilon)$	...	$1/\exp^n(1/\varepsilon)$

between the size of the perturbation and the size of each connected component of the domain, decreases at each step. The procedure of iteration of Nekhoroshev theorem is summarized in table I.

In conclusion, the global picture that we obtain at the light of Nekhoroshev theorem and of its successive iterations is the one sketched in figure 3. When the perturbation parameter ε is *sufficiently small* the Nekhoroshev theorem can be applied. The phase space is *structured* (fig. 3a). Resonances of a given order cross over at resonant nodes, in the neighbourhood of which one can find well defined chaos, but do not overlap, in the sense that there is always a no–resonance domain in each mesh of the resonant web. Moreover, this scenario is true at every scale, i.e. looking successively at resonances of increasing order, in a sort of renormalization picture. The core of this structure, which is the limit of convergence of the "renormalization" algorithm provided by the iteration of Nekhoroshev theorem, is made of invariant KAM tori. The latter form a Cantor–like set, since one has to remove strips associated to all resonances, which are dense (Arnold, 1963). We will say hereafter that a dynamical system characterized by such a structure is a *Nekhoroshev system.*

The converse situation is the one sketched in figure 3b. At some order, resonances overlap. There is no place for a no–resonance domain, and invariant tori do no longer exist. The overlapping of resonances allows orbits to pass from one resonance to another one, in a chaotic "fast" diffusion. The phase space is affected by *large scale chaos*. Since the width of resonant strips scales as $\sqrt{\varepsilon}$ this situation happens if ε is not *sufficiently small.*

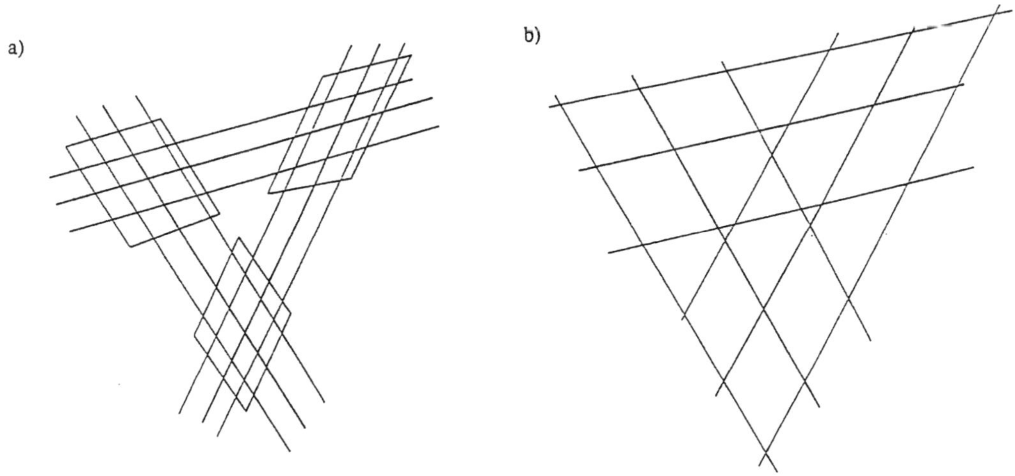

Fig. 3. Figure a: Sketch of Nekhoroshev structure. The resonances of a given order cross over at resonant nodes but do not overlap: there is always a no–resonance domain within each mesh of the resonance web. This picture repeats at every scale, i.e. looking subsequently to resonances with increasing order. Figure b: resonances of some order overlap. The no–resonance domain cannot be defined. Invariant tori are destroyed. Motion can pass from one resonance to another.

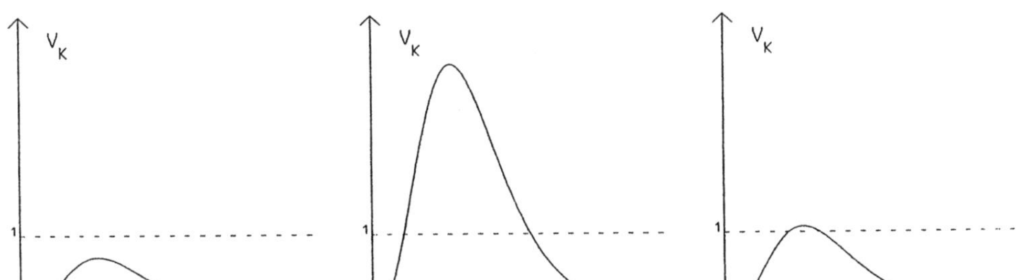

Fig. 4. The relative volume V_K filled by all resonances of order K, as given by formula (4). In the first picture, ε is small, so that for any K the volume V_K is always smaller than the volume of the phase space (here normalized to 1). Therefore resonances cannot overlap; invariant tori exist and the system has the Nekhoroshev structure. In the second and the third pictures, the volume V_K is larger than 1 for some K. Therefore resonances must overlap, and invariant tori cannot exist. The order at which resonances overlap depends on ε. If ε is very large (second picture), the order is small; otherwise (third picture) the order can be very large, close to the one corresponding to the maximum of V_K.

3. Exploring the Dynamical Structure of a Given System

At the light of the previous section, the correct attitude of a physicist, for approaching the investigation of a given dynamical system, is to search for an answer to the following questions: is the system a Nekhoroshev one? In the negative case, at which order do resonances overlap?

Unfortunately, it is not an easy task to answer to these questions in an analytic way. This can be understood with the following qualitative argument, inspired by the work of Arnold (1963). The width of a resonant strip associated to a resonance of order K scales as $\sqrt{\varepsilon}\exp(-K\sigma)$; the number of resonances of order K grows as $2^n K^{n-1}$, n being the number of degrees of freedom. Therefore, the relative volume of the phase space covered by all resonances of order K is

$$V_K \sim 2^n \sqrt{\varepsilon} K^{n-1} \exp(-K\sigma) \,. \qquad (4)$$

The volume V_K has a maximum value for $K = (n-1)/\sigma$, and decays exponentially for larger K. Figure 4 plots V_K as a function of K, for different values of ε. If ε is small, then, for every K, the volume concerned by resonances of order K is always smaller than the volume of the full phase space, i.e. the relative volume is smaller than 1 (fig. 4a). This implies that, at all orders, resonances cannot overlap completely, and there is some volume free from resonances for the existence of invariant tori; the system has the Nekhoroshev structure. [1] If ε is larger, then for some K the relative volume V_K is equal to one. As a consequence, the resonances of order K must overlap completely, invariant tori cannot exist and the system looses its Nekhoroshev structure. The minimal order K at which resonances overlap depends on ε. If ε is very large, this can happen at very low order (fig. 4b), but if ε is smaller, this can happen for K closer to $K_{\max} = (n-1)/\sigma$ (fig. 4c).

Now, the analytic algorithms, in the limit of implementation imposed by modern technology (algebraic manipulators, semi–numerical computations etc.), allow to investigate the location and the amplitude of resonances only up to a very limited order. Conversely, if ε is not big, resonance overlapping can happen at orders close to $K_{\max}$. The latter can be very large if the number of degrees of freedom is large or the parameter of analyticity σ is small. For example, the dynamical system for a realistic description of the motion of an asteroid has at least 10 degrees of freedom, and σ can be estimated to be about 0.1. Therefore, resonance overlap can occur at order 100! As a consequence, the analytic approaches cannot allow to conclude whether a given system has the Nekhoroshev structure, unless $K_{\max}$ is small (small number of degrees of freedom, large σ). As a matter of fact, all applications of Chirikov criterion of resonance overlapping (see Contopoulos, 1966; Chirikov,

[1] Note that, from the rigorous mathematical viewpoint, in order to prove the existence of invariant tori one has to show that the sum of the V_K, for $K \in [1,\infty)$ is smaller than 1; however, since most of high order resonance strips are actually covered by low order resonance ones, the sum of V_K leads to an overestimate of the global resonance volume, so that we believe to be more intuitive to show in figure 4 the V_K distribution rather than its integral.

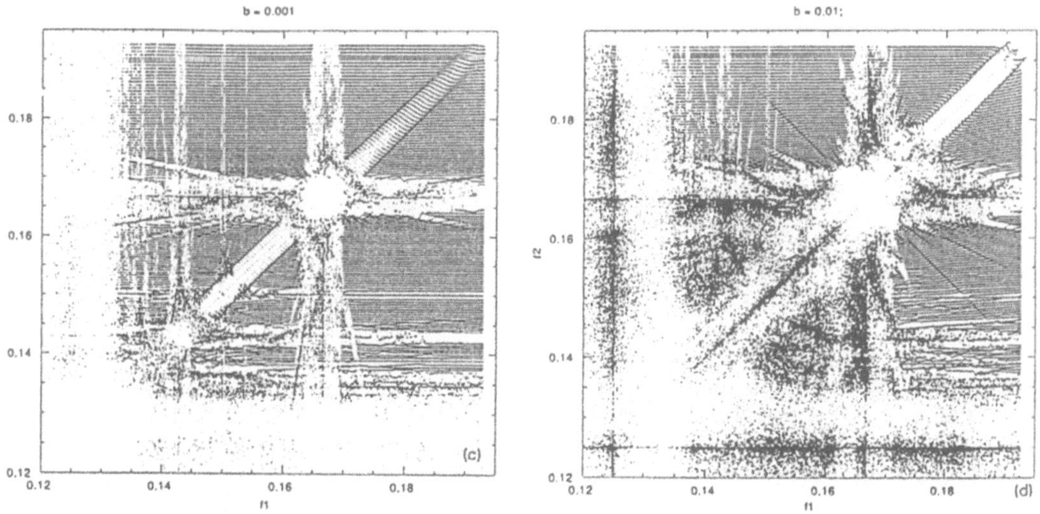

Fig. 5. The structure of the phase space of Froeschlé map (5) detected by frequency analysis (from Laskar, 1993). In the first picture the coupling parameter b is equal to 0.001. In the second picture b is equal to 0.01. The frequency map consists in computing, for each initial condition on a regular grid, the corresponding fundamental frequencies over a given timespan. The x–axis denotes the frequency of libration of q_1; the y–axis denotes the frequency of libration of q_2. See text for comments.

1979) have been carried on with success only on two degrees of freedom systems, where good results can be achieved at low to moderate order. To our knowledge Chirikov criterion has never been applied with realistic results on systems with larger number of degrees of freedom.

The situation is not so desperate if one turns to the numerical exploration of dynamical systems. Indeed, the existence of the Nekhoroshev structure can be checked numerically by many means. In the following we will discuss three useful tools: frequency analysis, sup–action map, Lyapunov exponents.

Frequency analysis has been introduced by Laskar (1990) (see also Laskar et al., 1992) and has been shown to be very efficient for obtaining the global picture of the dynamics of 3 degrees of freedom systems (Laskar, 1993). In the previous section we have described the Nekhoroshev structure directly in the space of frequencies, therefore there is no doubt that frequency analysis is one of the best adapted numerical methods for detecting such a structure. Figure 5 shows the picture provided by frequency analysis on the Froeschlé map of equations (see Froeschlé, 1972)

$$p_1' = p_1 + a_1 \sin(q_1 + p_1) + b \sin\left(\frac{q_1 + p_1 + q_2 + p_2}{2}\right)$$
$$q_1' = q_1 + p_1$$

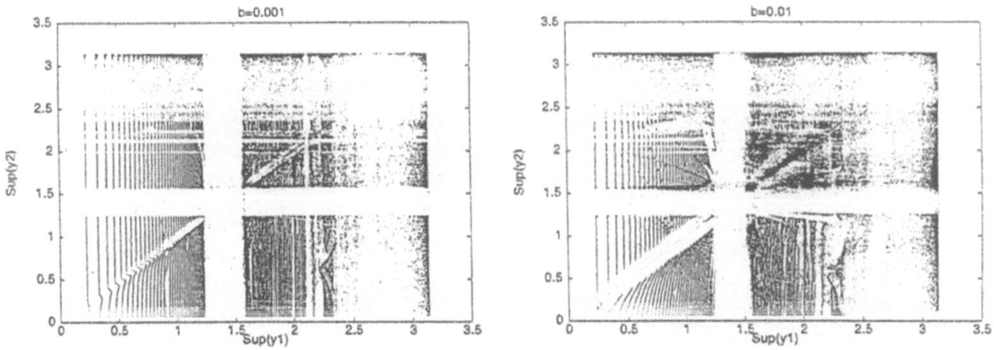

Fig. 6. The same as in figure 5, but the structure of the phase space is here detected through the sup–action map. The sup–action map consists in computing, for each initial condition on a regular grid, the maximal values assumed by the actions over a given timespan. The x–axis denotes the maximum assumed by p_1; the y–axis denotes the maximum assumed by p_2. Note that the axes are reversed with respect to figure 5. (Courtesy of E. Lega)

$$\begin{aligned} p_2' &= p_2 + a_2 \sin(q_2 + p_2) + b \sin\left(\frac{q_1 + p_1 + q_2 + p_2}{2}\right) \\ q_2' &= q_2 + p_2 \,, \end{aligned} \tag{5}$$

for two values of the coupling parameter b. In the first case (fig. 5a) where b is small, we can recognize the Nekhoroshev structure. All resonant strips with non–negligible width are well visible. These cross each other at resonant nodes, but do not overlap. The no–resonance domains, among resonance strips, are filled by dots disposed along regular lines. This is due to the fact that no–resonance domains are practically filled by invariant tori (the gaps associated to resonances are of negligible volume), and that the frequencies of invariant tori can be fitted by a very smooth C^∞ function (Lazutkin, 1973). Conversely, when b is larger (fig. 5b), resonances overlap. The region characterized by $f_1 < 0.17$ and $f_2 < 0.17$ is covered by large scale chaos; invariant tori do not exist and the structure visible in fig. 5a has disappeared. Orbits can diffuse over this region in a quite short time (see Laskar, 1993 for examples). It can be seen, however, that the region in the upper right corner of fig. 5b still conserves the Nekhoroshev structure. Indeed, a dynamical system can have the Nekhoroshev structure on some macroscopic parts of its domain of definition, and show large scale chaos on the complementary parts. As a matter of fact, this is the most typical realistic situation, as we will see concerning the asteroid belt dynamical system.

A second useful tool for exploring the dynamical structure of a given system is the so called "sup–action map". This is quite similar to the frequency map, and consists in computing, for every initial condition, the supremum value assumed by the actions over a sufficiently long timespan. This technique has been first used by Laskar (1995) in order to show the evolution of the planetary system over several billions of years, and has been extensively studied by Froeschlé and Lega (1995).

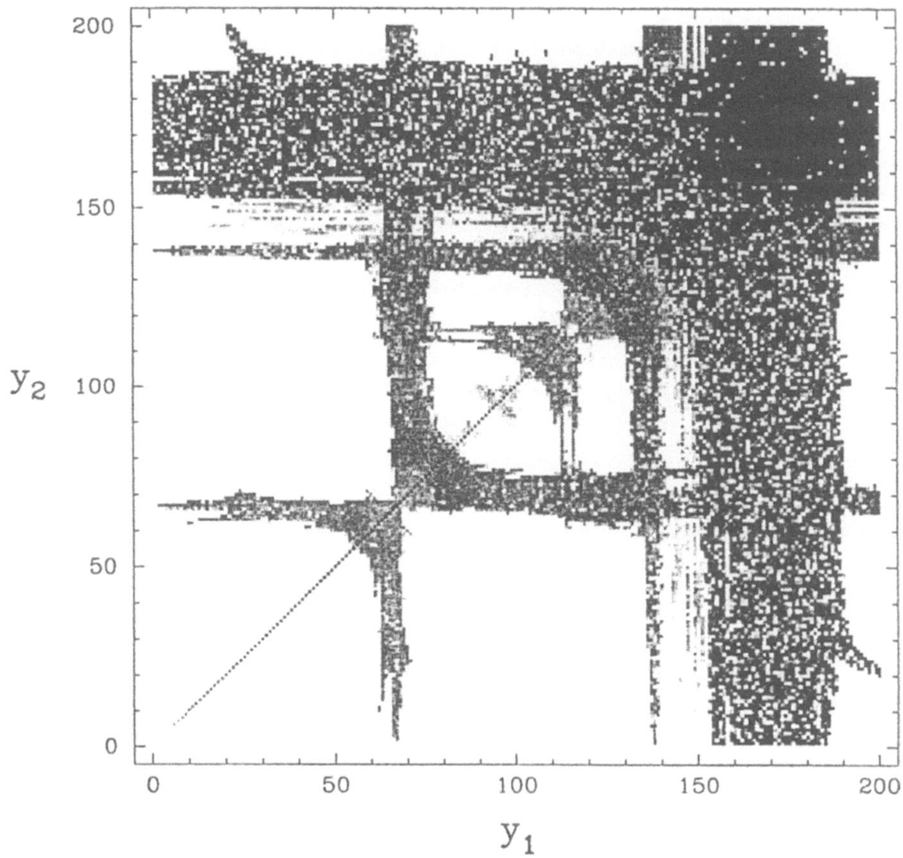

Fig. 7. The structure of the phase space of Froeschlé map (5) in the case $b = 0.001$, revealed by the distribution of Lyapunov exponents. Lyapunov exponents are computed, over a given timespan, for every orbit of a regular grid of initial conditions. The x and the y axes denote the values of the initial actions p_1 and p_2 (the initial q_1 and q_2 are set to 0). The value of the Lyapunov exponent is indicated by the darkness of the corresponding dot. The white domains are those for which the Lyapunov exponent is smaller than the level of significance due to the limited timespan of computation. (Courtesy of R. Gonczi)

Figure 6 shows the results provided by the sup–action method on the Froeschlé map (5), for the same values of the coupling parameter b used in fig. 5. We just stress that the axes are reversed (since in fig. 6 they refer to actions and in fig. 5 they refer to frequencies), so that the large chaotic region, which is visible in the left bottom corner in fig. 5b, is found in the right upper corner in fig. 6b. Again, the Nekhoroshev structure is clearly detectable (when it exists!) as well as large chaotic regions.

Last but not least, also the distribution of Lyapunov exponents is a useful numerical tool for detecting the dynamical structure of a given system. Indeed, such distribution has a clear correspondence with the Nekhoroshev structure outlined in the previous section (see Morbidelli and Froeschlé, 1996). Indeed, in the no–resonance domain, the Lyapunov exponents are equal to zero almost everywhere, since invariant KAM tori fill a very large volume of it. At most, the Lyapunov exponent can be exponentially small, under the effect of the exponentially small remainder $\mathcal{R}_K$ (see section 2). In the single resonance domains, the Lyapunov exponent can be significantly large (of order of $\sqrt{\varepsilon}$) only in the chaotic layer associated to the separatrices of the unperturbed resonance. Such chaotic layer, however, being generated by the effects of the exponentially small remainder $\mathcal{R}_K$, is exponentially thin, as proved first by Neishtadt (1984). On double resonance domains, conversely, the Lyapunov exponent turns out to be of order $\sqrt{\varepsilon}$ over a region of size $\sqrt{\varepsilon}$, as shown by Benettin and Gallavotti (1986) (see Morbidelli and Froeschlé (1996) for a description of Benettin and Gallavotti's argument). Figure 7 shows the distribution of Lyapunov exponents for the Froeschlé map with the small coupling parameter b corresponding to figures 5a and 6a. Darker domains denote the sets of initial conditions which give a larger Lyapunov exponents, while the white domain is the set of orbits for which the Lyapunov exponent is smaller than the level of significance attained over the integration timespan.

In conclusion, all these results show that it is possible to detect numerically the dynamical structure of a given system. The Nekhoroshev theorem allows the interpretation of these results and to conclude about the long time stability of motion on the simple basis of the observed geography of resonances. In our opinion, this symbiosis between mathematical results and numerical computations is very promising for the future developments of applied dynamical system science and, in particular, for Celestial Mechanics. Unfortunately, the dynamical systems which concern Celestial Mechanics are very specific, since they are *degenerate*, in the sense of the next section. It is therefore important to extend, as much as possible, to the case of degenerate systems the results on Nekhoroshev stability and phase space structure described above. This is the purpose of the second part of this paper.

4. Nekhoroshev Result for Asteroid Dynamics: the Secular System

An integrable system is said to be *degenerate* if it admits a number of independent prime integrals larger than the number of degrees of freedom, so that some angles don't move, whatever the initial condition over the phase space is. The prototype of such a system is one with Hamilton function $H(p,q) = H_0(p_1,\ldots,p_m)$, with $(p,q) \in \mathbf{R}^n \times \mathbf{T}^n$ and $m < n$: not only the actions $p_1,\ldots,p_n$ are independent prime integrals, but also the "angles" $q_{m+1},\ldots,q_n$. As a consequence, a degenerate perturbed system has fast angles (those which already move in the integrable approximation, the $q_1,\ldots,q_m$ in the example) and slow angles (those which stay fixed in the integrable approximation, the $q_{m+1},\ldots,q_n$). For example, the Keplerian problem admits 5 first integrals (semi–major axis a, eccentricity e, inclination

i, and the angles longitude of perihelion ϖ and longitude of node Ω), so that it is strongly degenerate, and its perturbation presents a fast angle (the mean longitude λ, denoting the revolution around the Sun) and slow angles (ϖ and Ω, denoting the secular precession of the orbit).

The existence of slow angles breaks down the perturbation approach sketched in the previous sections for the proof of Nekhoroshev theorem. Indeed, since in the integrable approximation the frequencies of the slow angles are equal to zero, the terms in the perturbation H_1 which depend only on the slow angles cannot be eliminated (the denominators $k \cdot \omega$ would be identically equal to zero). As a consequence, one could not define no–resonance domains and single resonance domains, where the Hamiltonian is integrable apart from an exponentially small remainder.

Therefore, the construction of the Nekhoroshev result must be modified. This cannot be done in general, but some specific properties of the system in study must be exploited. In the following we will concentrate on the asteroid belt case, which has been already the subject of a technical paper (Guzzo and Morbidelli, 1996). More precisely, the system that we deal with describes the motion of an asteroid in the *space* under the perturbations of the planets of the Solar System. The latter are assumed to be on their real orbits, which change quasi–periodically with time as described by the analytical theories of planetary motions (see Bretagnon, 1974; the effects of secular chaotic motion of the inner planets will be described later in section 6). This is a quasi–integrable degenerate system, with perturbation parameter given by the mass of Jupiter, denoted hereafter by ε.

The aim of this part of the present work is not to prove that the asteroid belt is stable in the sense of Nekhoroshev; the aim is to show which parts of the belt *could* be stable provided some parameters are sufficiently small. Moreover we will discuss the numerical indications that we already have concerning the nature of asteroids dynamics.

In order to construct a Nekhoroshev result on the asteroid belt dynamical system, we first consider mean motion resonances up to some large order $K \sim 1/\varepsilon^b$, and we investigate whether they overlap. Mean motion resonances are defined as usual as the resonances among the mean longitude λ of the asteroid and the mean longitudes λ' of the planets, i.e. among the fast angles of the system. Since the frequencies of λ' are fixed and the frequency of λ depends only on the semi–major axis of the asteroid, the geography of mean motion resonances is, in first approximation, a very simple one, made of parallel planes in the (a, e, i) space. If mean motion resonances overlap, then the system cannot have the Nekhoroshev structure and the semi major axis of an asteroid can diffuse, passing from one resonance to another; conversely, if mean motion resonances do not overlap, then one can define a no–mean–motion–resonance domain, where the system can be averaged with respect to mean longitudes up to some remainder term, exponentially small in K. The overlapping/non–overlapping of mean motion resonances depends simply on their widths, and therefore gives a condition on the value ε of the mass of

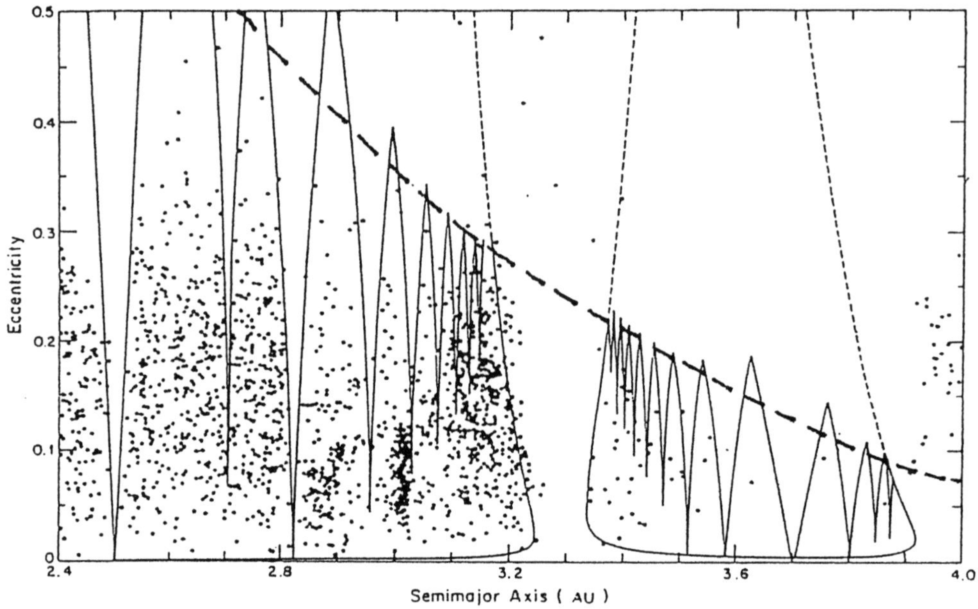

Fig. 8. The width of mean motion resonances of type $p/(2p+1)$, $p/(2p-1)$, $3p/(3p+1)$ and 3/8 in the asteroid belt (Dermott and Murray, 1983). Mean motion resonances seem not to overlap below the dashed curve. Most of asteroids are indeed in what seems to be the no–mean–motion–resonance domain.

Jupiter. Although we do not know precisely which is the situation for the real value of ε (10^{-3} the mass of the Sun), numerical computations seem to indicate that mean motion resonances do not overlap, at least for small to moderate values of eccentricity and inclination. Figure 8 shows the results of a computation made by Dermott and Murray (1983) on the width of mean motion resonances with Jupiter (the dominating ones) in the range 2.4–4.0 AU. The borders of each resonance are plotted by bold or partly–dashed lines (the resonant region is the one above such lines); the dashed curve marks what seems to be the limit for resonance overlapping in the (a, e) plane. The dots mark the values of a and e for the first numbered asteroids. Moreover, the limit for overlapping of mean motion resonances of type $(p+1)/p$ at $e = 0$ has been computed analytically by Wisdom (1980) to be equal to 4.5 AU.

Provided that we can introduce a no–mean–motion–resonance domain, we further proceed to study the stability of the averaged system on such a domain. (We will see in section 5 how to confine motion in mean motion resonances of order larger than $|\ln \varepsilon|$).

The averaged system is no longer degenerate (all remaining angles move now on the same timescale), so that we join the usual framework of Nekhoroshev result, explained in the previous sections. However, some difficulties must be

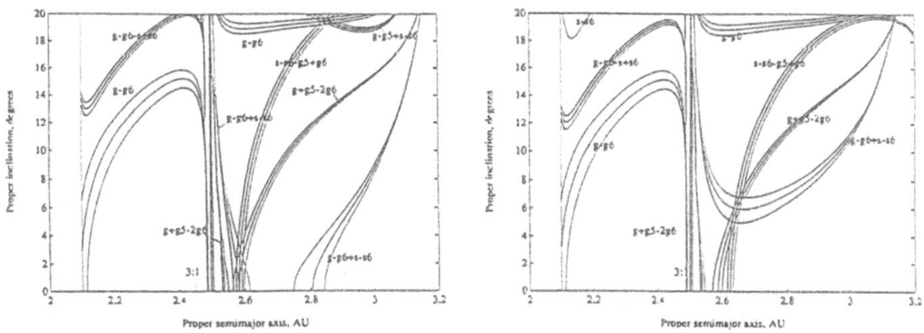

Fig. 9. The location of secular resonances up to order 3 in μ for $e = 0$ (picture a) and $e = 0.1$ (picture b). The three curves for each resonance denote the location of the resonance and the estimated borders of the resonant strip. From Milani and Knežević(1990). See text for comments.

overcome, such as the definition of new small perturbation parameters, the existence of singularities in the equation of motion, and the break–down of the convexity hypothesis.

4.1. Local Geography of Perturbation Parameters

As astronomers know, the problem with the averaged system (usually called also the secular system) is that there is no evident *small parameter* to divide the Hamiltonian into an integrable part and a perturbation. Different partitions of the secular Hamiltonian have been proposed. Here we follow the approach by Williams (1969), which consists in expanding the Hamiltonian in powers of the eccentricities and inclinations of the planets, which therefore play the role of small parameters. So, the secular Hamiltonian is written as

$$H_{\rm sec} = \varepsilon\left(\mathcal{H}_0 + \sum_{i\geq 1}\mu^i \mathcal{P}^{(i)}_{\rm sec}\right) , \tag{6}$$

where μ is of order of e', i', i.e. about $\sqrt{\varepsilon} \sim 10^{-3/2}$, and $\mathcal{H}_0$ is integrable and non–degenerate. $\mathcal{H}_0$ depends only on the actions, provided suitable action–angle variables are used (see Morbidelli and Henrard, 1991). Note that $H_{\rm sec}$ is globally of order ε. Indeed it describes the slow (secular) evolution associated to orbital precession, the fast angles being averaged out. Scaling time, i.e. using a time unit of order of thousands of years, the factor ε in $H_{\rm sec}$ can be eliminated.

The new "small parameter" μ is not very small indeed. However, we remark that each term $\mathcal{P}^{(i)}_{\rm sec}$ has a very limited Fourier spectrum, due to the well known D'Alembert characteristics (Tisserand, 1882). As a consequence, the usual situation is that, in a given domain, the first resonant terms appear only at some order

$i^* > 1$. Therefore all terms in $\mathcal{P}_{\rm sec}^{(i)}$, $i < i^*$ can be eliminated, which leads to a new Hamiltonian

$$H_{\rm sec} = H_0 + \eta H_1 \ ,$$

with a new effective small parameter $\eta \sim \mu^{i^*}$. So, we can conclude that the averaged problem has *local* small parameters which depend strongly on the local geography of secular resonances.

The local geography of low order secular resonances is now well known in the asteroid belt, so that we have a good estimate of the size of the effective perturbation parameters. As an example, figure 9 shows the location of secular resonances up to order 3, computed by Milani and Knežević(1990) on the (a, i) plane for $e = 0$ and $e = 0.1$. The resonances denoted by $g - g_6$ and $s - s_6$ are of first order, so that the corresponding terms are in $\mathcal{P}_{\rm sec}^{(1)}$; then in a neighbourhood of these resonances the local perturbation parameter is $\eta \sim \mu \sim \sqrt{\varepsilon}$, which is quite big. However, far from these two resonances, the perturbation parameter is significantly smaller. The resonances denoted by $g - g_6 + s - s_6$, $g - g_5 + s - s_6$ and $g - g_6 - s + s_6$ are of second order, so that in their neighbourhood, $\eta \sim \mu^2 \sim \varepsilon$. All other resonances marked in fig. 9 are of third order: in their vicinity $\eta \sim \mu^3$. Furthermore, in the remaining parts of the (a, i, e) space, the local perturbation parameter η is at least $\mu^4 \sim \varepsilon^2$, since secular resonances are at least of order 4.

Therefore, we can conclude that, apart from some domains characterized by the presence of secular resonances of very low order, the secular problem has perturbation parameters which are significantly small. Then, the first difficulty for the proof of a Nekhoroshev result on the stability of secular motion could be overcome.

4.2. Circular motion and singularities

It is well known that action angle variables (for example Delaunay variables) cannot be used to describe the dynamics at very small eccentricity. Indeed such variables are singular at $e = 0$ due to a very simple geometrical reason: the longitude of perihelion ϖ is not defined for circular motion, so that when e passes through 0 during its secular evolution, the motion of ϖ is discontinuous. The same happens to the longitude of node Ω when the inclination becomes equal to zero, but this can be overcome by a simple change of the reference frame. Conversely, in the neighbourhood of $e = 0$ one has to introduce rectangular coordinates, such as the Poincaré variables $(h = e\cos\varpi, k = e\sin\varpi)$.

The Nekhoroshev construction can still be done using rectangular coordinates (see Benettin et al., 1996, Guzzo and Morbidelli, 1996), however the radius ϱ of the neighbourhood $\mathcal{U}$ of $e = 0$ (i.e. $\mathcal{U} \equiv \{e : \ e < \varrho\}$) where rectangular coordinates are used, cannot be chosen in an arbitrary way. Indeed, on the one hand, ϱ^2 plays the role of a new perturbation parameter, so that one would like to choose ϱ as small as possible. On the other hand, ϱ must be large enough to contain the oscillations of the eccentricity, being this a standard requirement in the perturbative construction. This condition, makes the optimal choice of ϱ strongly dependent on the value of the local perturbation parameter η defined in section 4.1. Roughly speaking, in the vicinity of first order secular resonances the Hamiltonian is like $e^4 + e\mu\cos\sigma$ (where σ is the resonant angle), so that eccentricity has oscillations of size $\mu^{1/3} \sim 0.3$; therefore, ϱ must be at least $\mu^{1/3}$ and ϱ^2 cannot be considered as a "small parameter". Concerning resonances of second order, the worst case has Hamiltonian like $e^4 + ie\mu^2\cos\sigma$, where i is the inclination, which can be large; so, the oscillation of e is in this case of order $\mu^{2/3} \sim 0.1$. Away from second order resonances the coefficient of the resonant harmonic is not larger than $e\mu^3$, so that the oscillation of e is not larger than μ. On the other hand, μ is the magnitude of non–resonant oscillations forced by $\mathcal{P}^{(1)}_{\rm sec}$, since the leading Hamiltonian is in this case like $e^2 + e\mu\cos\ldots$. As a consequence, ϱ cannot be smaller than μ anywhere in the asteroid belt. However, the perturbation parameter ϱ^2 is in this cases of order $\mu^2 \sim \varepsilon \sim 10^{-3}$, which is quite small.

Finally, we need to require that the domains in which we proved the degenerate non–singular Nekhoroshev theorem overlap with $\mathcal{U}$, so that every domain in the phase space, singular or not, turns out to be covered by a Nekhoroshev like construction. This means essentially that ϱ must be chosen of the same order as the variation of the eccentricity predicted by the degenerate non–singular Nekhoroshev theorem; however this is not a significant further constraint.

In conclusion, as in sect. 4.1 we expect, away from secular resonances of low order, the perturbation parameters in game are significantly small. This is not enough, of course, to conclude that the Nekhoroshev theorem can be applied, but at least shows that a Nekhoroshev like result is not impossible, a priori.

4.3. Confinement of resonant motion and convexity problems

In section 2 we have seen that convexity plays an important role to confine resonant motion. This can be simply understood in the following way. Let us consider a 2 degrees of freedom convex Hamiltonian $H_0(p)$, such as, for example

$$H_0(p) = \frac{1}{2}(p_1^2 + p_2^2)\,.$$

The level curves of such Hamiltonian are shown in figure 10a; they are closed, so that they are transversal to the resonant lines, which come out from the origin $p_1 = p_2 = 0$. Since the perturbed motion of the actions must be ε–close to the level curves of H_0, it follows that the resonant motion must be transversal to the resonant lines. As a consequence, the motion cannot go indefinitely far, otherwise it would leave the resonance: resonant motion must be confined. For instance, let us consider the Hamiltonian

$$H(p,q) = \frac{1}{2}(p_1^2 + p_2^2) + \varepsilon\cos(q_1 - q_2)\,.$$

The 1/1 resonance is located at the line $p_1 = p_2$; the fast drift direction is $p_1 = -p_2$. Therefore, the fast drift direction and the resonant line are orthogonal. As a consequence the motion is bounded. Indeed, in the canonical coordinates $\varphi_1 = q_1 - q_2, \varphi_2 = q_2, I_1 = p_1, I_2 = p_1 + p_2$ the Hamiltonian has the pendulum like form

$$H(I,\varphi) = I_1^2 - I_1 I_2 + \frac{I_2^2}{2} + \varepsilon \cos(\varphi_1) ,$$

so that the motion of I_1 is always confined. Conversely, figure 10b concerns a non–convex Hamiltonian, such as

$$H_0(p) = \frac{1}{2}(p_1^2 - p_2^2) .$$

The level curves are open, and there are two escape directions along the lines $p_1 = -p_2$=constant and $p_1 = p_2$. The resonances which are transverse to the level curves of H_0 still have bounded motion as in the convex case. However, this is not the case for the resonances which coincide with the escape lines. For instance, let's consider the 1/1 resonance in the Hamiltonian

$$H_0(p) = \frac{1}{2}(p_1^2 - p_2^2) + \varepsilon \cos(q_1 - q_2) ;$$

the resonant line is $p_1 = -p_2$ and the fast drift direction is also $p_1 = -p_2$. With the same set of variables I, φ introduced above, the Hamiltonian reads

$$H(I,\varphi) = I_1 I_2 - \frac{I_2^2}{2} + \varepsilon \cos \varphi_1 ,$$

and, for almost all initial conditions (I_1, φ_1) on the resonance line $I_2 = 0$ the action I_1 escapes to infinity.

Now, the integrable approximation $\mathcal{H}_0$ of the secular Hamiltonian (6) of asteroid dynamics is non–convex. If one expands it locally around any point of the action space, the level curves of the quadratic form look like those sketched in figure 10b. This means that any secular resonance coinciding with one of the two escape directions would have unbounded motions! Luckily, if one considers also higher orders in p in the expansion of $\mathcal{H}_0$, the escape directions bend in the action space, as well as resonant lines. Then, the typical case is the one sketched in figure 11a: the resonant line and the escape line can be tangent *only locally*. This is enough to bound resonant motion, although in a larger strip. Indeed, assuming that for every small displacement in the actions δp the resulting change in frequencies $\delta\omega$ is such that the scalar product $|\delta\omega \cdot \delta p| > m||\delta p||^3$, then resonant motion can be confined within a strip $\eta^{1/3}$, for a perturbation of given size η. In the convex case, where $|\delta\omega \cdot \delta p| > m||\delta p||^2$ (see hypothesis [2] of Nekhoroshev's theorem), the motion would be confined in a strip of width $\eta^{1/2}$.

With the only exception of the ν_6 secular resonance we have checked that in the asteroid belt

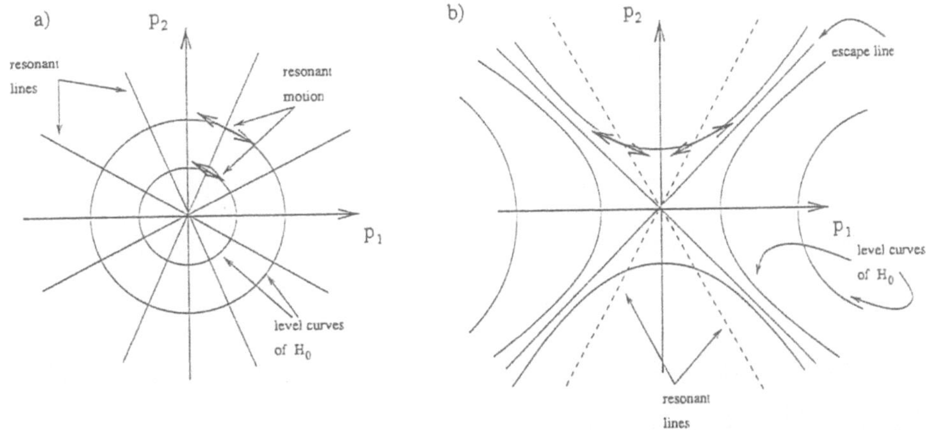

Fig. 10. Sketch of the dynamics in the action space of a quadratic 2D convex system (picture a) and of a quadratic 2D non–convex system (picture b). In the convex case, the energy level curves are transversal to the resonant lines. Therefore, resonant motion must be confined. In the non–convex case, if a resonance coincides with an escape line, the resonant motion is no longer bounded.

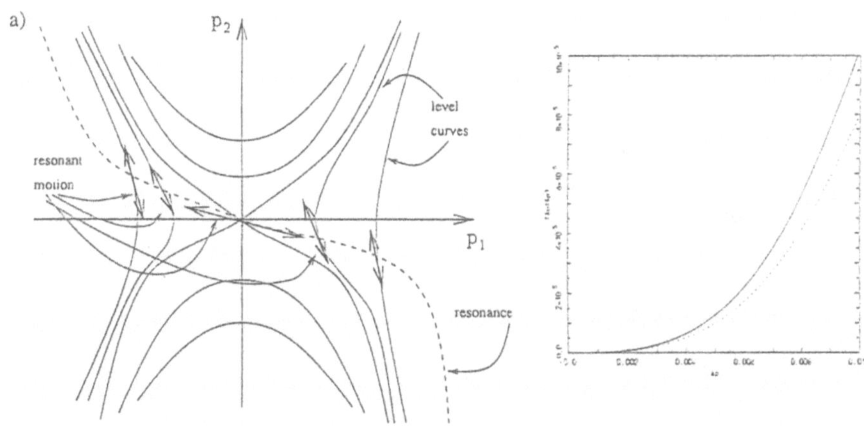

Fig. 11. Picture a: Sketch of the dynamics in the action space of a non–convex system when the terms of higher order than p^2 are considered. As a generic fact, resonances have only a finite tangency with the escape curves, so that resonant motion can be confined again, although in a larger strip. Picture b: Numerical evaluation of the scalar product $\delta\omega \cdot \delta p$ as a function of δp for $a = 3.0175$AU, $e = 0.15$ and $i = 10°$. The fact that $\delta\omega \cdot \delta p$ is bounded from below by a cubic function of δp (dashed curve) proves that resonances can have only a tangency of first order to the escape curves, as sketched in picture a.

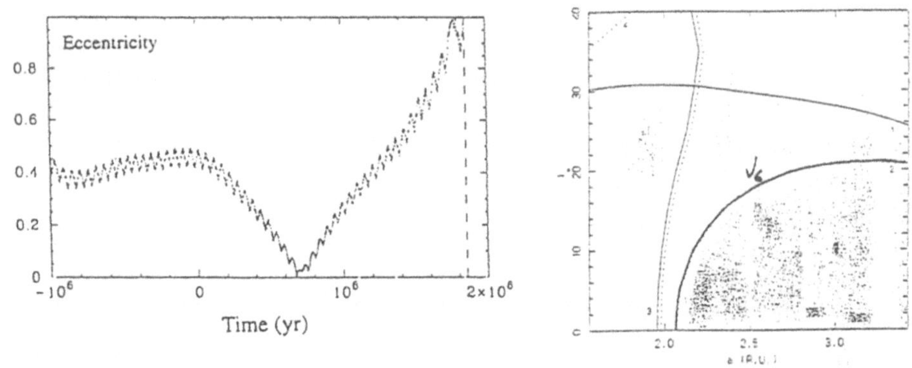

Fig. 12. Picture a: the dramatic effects of the ν_6 resonance in the asteroid belt. The eccentricity can be pumped from 0 to 1, forcing the resonant body to collide with the Sun (from Farinella et al., 1994). Picture b: the localization of the ν_6 resonance in the asteroid belt (bold line). The coordinates are the semi major axis and the inclination and the location is computed for $e = 0.1$. Note that the resonance marks the border of the main distribution of the asteroids in the belt. Other lines refer to other secular resonances.

i) the secular resonances of lowest order which can be locally tangent to an escape direction are of type $2\dot{\Omega} + \dot{\varpi} = \ldots$, $3\dot{\Omega} + \dot{\varpi} = \ldots$, $4\dot{\Omega} + \dot{\varpi} = \ldots$, which are at least of order 3, 4 and 5 respectively. Therefore these resonances must be associated to small local perturbation parameter η (see sect. 4.1);

ii) in these cases, the condition $|\delta\omega \cdot \delta p| > m\|\delta p\|^3$ is fulfilled for some positive m. For instance, at $a = 3.0175$ AU, $e = 0.15$ and $i = 10$ degrees, m is not smaller than 80 (see fig 11b).

The ν_6 secular resonance is the 1/1 resonance between the motion of the asteroid longitude of perihelion ϖ and the frequency g_6 of the planetary system. The latter corresponds to the average frequency of precession of Saturn's longitude of perihelion. This is the only resonance of low order which is tangent to an escape direction of $\mathcal{H}_0$. Moreover, the coefficient m is close to zero in this case, so that a condition of type $|\delta\omega \cdot \delta p| > m\|\delta p\|^\gamma$ is fulfilled only for γ much larger than 3. As a consequence, the resonant action (i.e. the eccentricity) can have extremely large changes, or even escapes, as in the non–convex case of figure 10b. Figure 12a shows the evolution of an orbit in the ν_6 resonance, integrated numerically taking into account the perturbations of all planets on their real orbits. As one sees, the eccentricity is equal to 0 at time 6.5×10^5 y and grows to 1 at time 1.83×10^6 y. From the mathematical viewpoint, this means that under the effect of the ν_6 resonance the action has crossed the whole domain of definition. From the astronomical viewpoint, this explains why the ν_6 resonance is the border of the asteroid belt, as shown in fig. 12b; indeed the eccentricity increases so much that resonant asteroids eventually cross the orbits of the planets or collide with the Sun (Morbidelli et al., 1993).

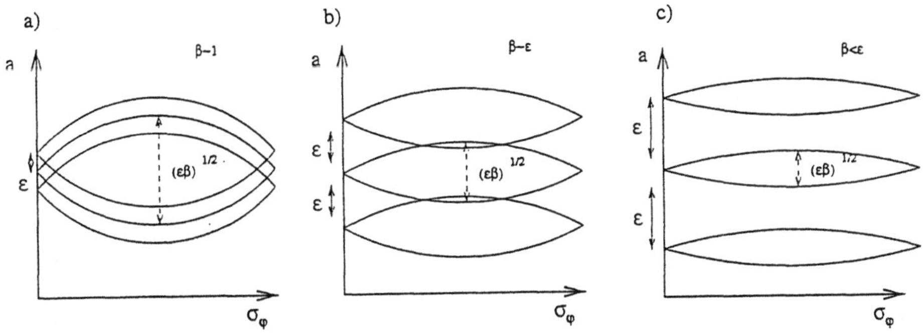

Fig. 13. Sketch of mean motion resonances in the asteroid belt. Due to degeneracy, a mean motion resonance splits into a multiplet of resonances. If the order of the resonance is small ($\beta \sim 1$, picture a) the resonances overlap almost completely, which gives a sort of modulated pendulum. If the order is close to $|\ln \varepsilon|$ ($\beta \sim \varepsilon$, picture b), resonances overlap marginally. The semimajor axis can diffuse passing from one resonance of the multiplet to another one. If the order of the resonance is larger than $|\ln \varepsilon|$ ($\beta < \varepsilon$, picture c) resonances do not overlap. Stability is possible.

We can conclude that, apart form the ν_6 resonance which is one of the most unstable places in the asteroid belt, the fact that the Hamiltonian of asteroids secular dynamics is not convex does not provide an a priori obstruction to the application of Nekhoroshev results.

5. Dynamics in Mean Motion Resonances

In the previous section we have considered the domain characterized by the absence of mean motion resonances of up to order $K \sim 1/\varepsilon^b$.

In this section we investigate the dynamics in mean motion resonances, looking for the necessary conditions which could allow the establishment of a Nekhoroshev like result of stability.

In a degenerate system, it is very difficult to control the dynamics near resonances among the fast angles. The reason is that, denoting by ω_0 and ω_1 respectively the fast and the slow frequencies, in the neighbourhood of a resonance $k_0 \cdot \omega_0 = 0$ several resonances of low order of type $k_0 \cdot \omega_0 + k_1 \cdot \omega_1 = 0$ are possible.

In Celestial Mechanics, it is well known that for a mean motion resonance of type $(p+q)/q$ several critical angles

$$\sigma_\varphi = \frac{(p+q)\lambda' - p\lambda}{q} + \varphi$$

can be introduced, where λ and λ' are the mean longitudes of the asteroid and of the resonant planet, and φ is any combination of the secular angles with algebraic sum of the coefficients equal to 1, such as ϖ, Ω, $2\varpi - \varpi'$ etc. Since the secular angles have frequencies of order ε, each mean motion resonance $p\dot{\lambda} = (p+q)\dot{\lambda}'$

splits into a multiplet of resonances (given by $\dot{\sigma}_\varphi = 0$, for different φ) which are ε–close in the semi major axis a.

If the size of the resonant harmonics $\cos \sigma_\varphi$ is ε times a coefficient β, then each resonance of the multiplet has a width proportional to $\sqrt{\varepsilon\beta}$. Therefore, for different values of β, the situation is the one sketched in figure 13. If $\beta \sim 1$ (fig. 13a) the resonances overlap completely. This gives a sort of modulated pendulum, with a large chaotic layer at its borders. If $\varepsilon < \beta << 1$ we have marginal overlapping (fig. 13b). The action, i.e. the semi major axis, can diffuse in the band concerned by the resonance multiplet, passing from one resonance to another. Finally, if $\beta << \varepsilon$, resonances split apart (fig. 13c). Each component of the multiplet is an independent resonance.

Because the perturbation is an analytic function of all the angles the value of the coefficient β decreases at exponential rate with respect to the order K of the mean motion resonance. Then, behaviours like those represented in fig. 13a, fig. 13b and fig. 13c can possibly take place in mean motion resonances with different order. Indeed, mean motion resonances of low order behave as in fig. 13a. In first approximation, they behave like modulated pendulums. Indeed, the model of the modulated pendulum has been applied with success by Morbidelli et al. (1995) in order to compute the extent of the chaotic zones in semi major axis, and compare with the distribution of the asteroids. Moreover, the resonant terms (which cannot be averaged out) dominate the secular Hamiltonian $\mathcal{H}_0$ in (6). As a consequence, the geography of secular resonances *inside* low order mean motion resonances is completely different from that *outside* mean motion resonances. Then, in general, it is not possible to establish a priori a Nekhoroshev result of stability. Each resonance is a specific dynamical system and must be studied in a separate way. For example, Moons and Morbidelli (1995) have shown that in the 3/1, 5/2, 4/1 and 7/3 resonances with Jupiter secular resonances overlap completely, leading to large scale chaotic motion with the consequence that the eccentricity of resonant orbits can increase to Earth–crossing or Sun–collision values. It's worth noting that most low order mean motion resonances with Jupiter correspond to the well–known Kirkwood gaps (see fig. 8); the notable exceptions are the 3/2 the 4/3 and the 1/1 resonances, inhabited respectively by the Hilda, the Thule, and the Trojan asteroids.

Up to now, we don't know much about the dynamics in mean motion resonances which are not of small order. Resonances of intermediate orders ($\sim 20 - 30$) could behave as in fig. 13b. As a matter of fact, the overlapping of resonances associated to different critical angles σ_φ has been invoked to explain the strange dynamical behaviours of some real asteroids (Milani et al, 1995). Conversely, resonances of larger order ($K >> |\ln \varepsilon|$) could be sable in the spirit of Nekhoroshev (Guzzo and Morbidelli, 1996). This can already be guessed from fig. 13c, since we have seen in section 3 that resonance non–overlapping is at the basis of the Nekhoroshev result. However, the picture complicates if one takes into account that the frequencies of the secular angles change with the actions (in particular with the eccentricity and

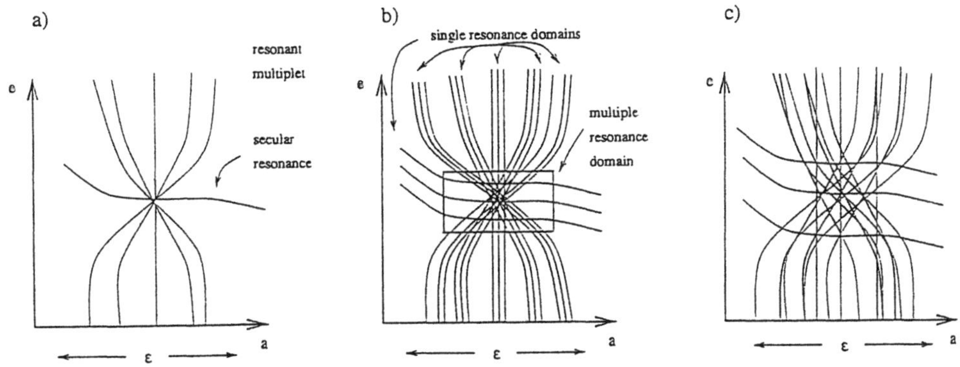

Fig. 14. Sketch of mean motion resonances in the asteroid belt for the planar elliptic 3 body problem, taking into account that the secular frequencies of ϖ change with e. The resonances of the multiplet must cross in a node due to the intersection with the secular resonances $\dot{\varpi} - \dot{\varpi}' = 0$. Then, if resonances are thin enough (large order), the structure of the phase space is Nekhoroshev–like (picture b). Conversely, if the width of the resonances in the multiplet is large (low order), resonances overlap and Nekhoroshev stability cannot be proved.

the inclination), and can be zero when some secular resonance occurs.

In order to understand this point and how the Nekhoroshev construction could still be implemented, let us consider for simplicity the planar elliptic problem Sun–asteroid–Jupiter and a n/n' mean motion resonance, splitting into a multiplet of resonances with critical angles

$$\sigma_k = n'\lambda' - n\lambda + (n - n')\varpi + k(\varpi - \varpi')\ .$$

On the (a, e) plane, the multiplet of resonances looks like the one in figure 14a. The lines correspond to the exact resonances $\dot{\sigma}_k = 0$, for different k. If one limits to resonances with $|k|$ smaller than a given cutoff $\mathcal{K}$, the multiplet concerns a band the width of which is of order ε in a. Indeed, since $\dot{\varpi}$ and $\dot{\varpi}'$ are of order ε, resonances can occur only for $|n'\lambda' - n\lambda + (n - n')\varpi| < \mathcal{K}\varepsilon$. Moreover, all resonances in the multiplet must intersect in a node, with the secular resonance $\dot{\varpi} - \dot{\varpi}' = 0$. Now, if the width of the resonances is small enough (i.e. $\beta < \varepsilon$) the picture that we get is like fig. 14b, which is similar to fig. 3a: resonances intersect at the node, but do not overlap completely, leaving space for the existence of a no–resonance domain. The Nekhoroshev construction of section 2 can therefore be repeated, regardless of the degeneracy of the problem. We can call this a *miniature Nekhoroshev*, in order to underline that it is a local construction concerning a part of phase space of size ε in semi major axis, centered around the unperturbed n/n' mean motion resonance. Conversely, if the width of resonant strips is too large (i.e. $\beta > \varepsilon$), as in fig. 14c, no long time confinement of motion is possible.

We can conclude that mean motion resonances of logarithmically large order with respect to the mass of Jupiter ε could have long time stability in the sense of Nekhoroshev.

6. Nekhoroshev Stability, Invariant Tori and Chaotic Perturbations in the Asteroid Belt

In sections 4 and 5 we have shown that, in principle, the dynamics in most parts of the asteroid belt could be governed by Nekhoroshev theorem. This has been done defining some cutoff on the order of the considered resonances. Indeed, the no–mean–motion–resonance domain is defined as the part of the belt where there are no mean motion resonances up to order $K \sim 1/\varepsilon^b$; moreover, in order to prove the Nekhoroshev theorem on such a domain, one has to consider, as usual, secular resonances up to a second cutoff $K_{\text{sec}} \sim 1/\eta^c$, where η is the local small parameter characterizing secular dynamics. As well, a miniature Nekhoroshev theorem can be established for mean motion resonances of order larger than $|\ln \varepsilon|$, provided the resonances in the multiplet are considered up to a cutoff $\mathcal{K} \sim 1/\eta^d$. ($b$, c, and d are to be considered here as some positive parameters). In order to prove the connection of Nekhoroshev result on long–time stability with the existence of invariant tori, we have to push all cutoff orders to infinity.

This is however quite easy. Indeed, in the no–resonance domains defined with respect to the cutoff thresholds K, K_{sec} and $\mathcal{K}$ indicated above, the perturbation can be reduced to be exponentially small with respect to these parameters, so that the system is reduced to

$$H(p_0, p_1, q_0, q_1) = H_0(p_0) + \varepsilon\mathcal{H}_0(p_0, p_1) + \mathcal{R}_{K,K_{\text{sec}},\mathcal{K}}(p_0, p_1, q_0, q_1) , \tag{7}$$

with

$$\|\mathcal{R}_{K,K_{\text{sec}},\mathcal{K}}\| < \max\left(\exp[-\frac{1}{\varepsilon^b}], \exp[-\frac{1}{\eta^c}], \exp[-\frac{1}{\eta^d}]\right) . \tag{8}$$

(For simplicity, we have denoted by q_0 the fast angles, q_1 the slow angles, and by p_0 and p_1 the respectively conjugate actions). The new system (7) can be considered as non degenerate, since the ratio ε between the frequencies of slow angles and fast angles is much bigger than the size of the perturbation $\mathcal{R}_{K,K_{\text{sec}},\mathcal{K}}$. From the physical point of view, this means that, at the scale of exponentially small effects, mean motions and secular motions are completely coupled. So, one can refer to the results on the iteration of Nekhoroshev theorem and the existence of KAM tori for non degenerate systems, explained in section 3.

Therefore, we can conclude that, in those privileged parts of the asteroid belt where a Nekhoroshev result of long–time stability can be established, the Nekhoroshev structure of resonance non–overlapping exists at every scale, up to the limit of invariant tori, as in the non–degenerate case. Therefore, similarly to non–degenerate systems, the existence of such a structure can be investigated numerically.

This is true if the planets perturbing the asteroid system have quasi–periodic motion, as assumed at the beginning of section 4. The fact that the inner planets have a weakly chaotic motion (Laskar, 1990) introduces non quasi–periodic perturbations to the asteroid motion. This must destroy at some level the Nekhoroshev structure, and prevents the existence of invariant tori, although some long–time stability result can still be derived.

This can be understood as follows. A non quasi–periodic perturbation has a continuous Fourier spectrum. In the case of weak chaos, such as the one of the inner planets, we can assume that the continuous Fourier spectrum is non–zero only on bounded intervals of width $\Delta\omega$ around the discrete values of the frequencies characterizing the quasi–periodic approximation of their motion (as given, for instance, by low order analytic theories). As a consequence of the continuous spectrum, each resonance splits into a continuous package of resonances. The width of this package in the space of frequencies is $\Delta\omega$, whatever the order of the resonance is. Then, the width of each resonance strip must be larger than $\Delta\omega$, instead of decaying exponentially with the order of the resonance as in the usual case. Referring to formula (4) the volume filled by resonances of order K becomes

$$V_K \sim 2^n \sqrt{\varepsilon} K^{n-1} \Delta\omega \,, \tag{9}$$

which increases indefinitely with K. Therefore, at some order K resonances must overlap, thus destroying the Nekhoroshev structure and the existence of invariant tori. However, if $\Delta\omega$ is small, this can happen only at a large order K, so that diffusion due to resonance overlapping has to be slow.

In the case of a main belt asteroid perturbed by the inner planets, the size ε of the chaotic component of the perturbation is very small (of order 10^{-8}) and the size of the intervals $\Delta\omega$, defined with respect to some cutoff level, is also quite small, since the frequencies of the inner planets do not change much over the age of the Solar System. Therefore, we can hope that the effects of chaotic perturbations do not affect the stability of asteroids' motion over a time significantly shorter than the age of the Solar System.

7. Conclusions

In the first part of this paper we have shown that the long–time stability result by Nekhoroshev for non–degenerate systems is strictly related with the existence of a specific structure of the phase space, where resonances of a given order do not overlap completely and leave space to some no–resonance domain in each mesh of the resonance web (fig. 3a). This structure repeats at every order in a sort of renormalization picture, the limit being the set of invariant KAM tori.

The existence of such a structure can be investigated with high accuracy using several numerical tools, such as frequency analysis, the sup–action map, the computation of Lyapunov exponents. This symbiosis between numerical investigations and theoretical results is a powerful approach for understanding the dynamical nature of a given system.

In the second part of the paper, we have devoted to the asteroid belt dynamical system, which is degenerate. We have shown that a Nekhoroshev–like long time stability result can still be applied in principle, except on some parts of the belt.

These parts correspond to mean motion resonances up to order $|\ln \varepsilon|$ (ε being the mass of Jupiter in solar units), and to secular resonances of very low order. Most of these resonances correspond indeed to gaps in the asteroid distribution. Moreover, if the perturbing planets move on quasi–periodic orbits, the Nekhoroshev stability result is strictly related to the specific structure of the phase space mentioned above, as in the case of non–degenerate systems.

This should give the theoretical background for understanding, through a big campaign of numerical integrations of real and fictitious asteroids, which is the dynamical structure of the asteroid belt. Indeed, we don't know at present whether the asteroid belt is a big chaotic see (although with diffusion times of order of billions of years) or is structured in the sense described by Nekhoroshev theorem for quasi–integrable Hamiltonian systems. This is not just an academic question, but can have some important astronomical implications. In the first case, the asteroid belt could be marginally stable, and the asteroids that we observe now could be those, among a much larger population, which have still not escaped. In the second case, the dynamics in the asteroid belt would be frozen for very long times, possibly largely exceeding the Solar System age. So, what we observe now could be, in the first case, a transient dynamical situation, and in the second case, a sort of "permanent" configuration of the asteroid belt.

Acknowledgements

This work has been developed with the support of the grant "EC contract ERBCHRXCT940460" for the project "Stability and universality in classical mechanics".

References

Arnold, V., I.: 1963, Proof of A. N. Kolmogorov's theorem on the conservation of conditionally periodic motions with a small variation in the Hamiltonian., *Russian Math. Surv.*, **18**, No. 5, 9–36.

Benettin, G., Galgani L., and Giorgilli A.: 1985, A proof of Nekhoroshev's theorem for the stability times in nearly integrable Hamiltonian systems., *Cel. Mech.*, **37**, 1.

Benettin, G. and Gallavotti G.: 1986, Stability of motions near resonances in quasi–integrable Hamiltonian systems.. *J. Stat. Phys.*, **44**, 293–338.

Benettin, G., Fassò F. and Guzzo M.: 1996, Fast rotations of the rigid body: a study by Hamiltonian perturbation theory. Part II: gyroscopic rotations., preprint.

Bretagnon, P.: 1974, Termes à longues périodes dans le système solaire., *Astron. Astrophys.*, **30**, 141–154.

Chirikov, B. V.; 1979, A universal instability of many dimensional oscillator systems *Physics reports*, **52**, 265–379.

Contopoulos, G.: 1967, in Nahon, F. and Hénon, M. (eds): *Les nouvelles méthodes de la dynamique stellaire*, CNRS, Paris (1966); see also Bull. Astron. Ser. 3, **2**, Fasc. 1, 233.

Dermott, F. S. and Murray, C. D.: 1983, Nature of the Kirkwood gaps in the asteroid belt., *Nature*, **301**, 201–205.

Farinella, P., Froeschlé, Ch. Froeschlé, C., Gonczi, R., Hahn, G., Morbidelli, A. and Valsecchi, G. B.: 1994, Asteroids falling onto the Sun, *Nature*, **371**, 315–317.

Froeschlé C.: 1972, Numerical study of a four dimensional mapping, *Astron. Astrophys.*, **16**, 172–189.

Froeschlé C. and Lega E.: 1995, On the measure of the structure around the last KAM torus before and after its break–up, *Celest. Mech.*, in press.

Giorgilli, A. and Morbidelli A.: 1996, Invariant KAM tori and global stability for Hamiltonian systems, *ZAMP*, in press.

Giorgilli, A. and Skokos C.: 1996, On the stability of the Trojan asteroids, *Astron. Astrophys.*, in press.

Guzzo M. and Morbidelli A.; 1996, Construction of a Nekhoroshev like result for the asteroid belt dynamical system, *Celest. Mech.*, submitted.

Laskar, J.,: 1990, The chaotic motion of the solar system: a numerical estimate of the size of the chaotic zones, *Icarus*, **88**, 266–291.

Laskar J., Froeschlé C. and Celletti A: 1992, The measure of chaos by the numerical analysis of the fundamental frequencies. Application to the standard mapping, *Physica D*, **56**, 253–269.

Laskar, J.: 1993, Frequency analysis for multi–dimensional systems. Global dynamics and diffusion., *Physica D*, **67**, 257–281.

Laskar, J.: 1995, Large scale chaos and marginal stability in the Solar System, *XIth ICMP Colloquium*, Paris.

Lazutkin, V. F.: 1973, The existence of caustics for a billiard problem in a convex domain, *Math. USSR Izis*, **7**, 185–214.

Lochak, P.: 1992, Canonical perturbation theory via simultaneous approximations, *Usp. Math. Nauk.*, **47**, 59–140. English transl in Russ. Math. Surv.

Milani, A., and Knežević, Z.: 1990, Secular perturbation theory and computation of asteroid proper elements., *Celest. Mech.*, **49**, 247–411.

Milani A, Nobili A. M. and KneževićZ.: 1995, Stable chaos in the asteroid belt, *Icarus*, submitted.

Morbidelli, A., and Henrard, J.: 1991, Secular resonances in the asteroid belt: theoretical perturbation approach and the problem of their location., *Celest. Mech.*, **51**, 131–167.

Morbidelli, A., Gonczi, R., Froeschlé, Ch. and Farinella, P.: 1993, Delivery of meteorites trough the ν_6 secular resonance, *Astron. Astrophys.*, **282**, 955–979.

Morbidelli, A., and Giorgilli, A.: 1995, On a connection between KAM and Nekhoroshev theorems, *Physica D*, **86**, 514–516.

Moons, M. and Morbidelli, A.: 1995, Secular resonances inside mean-motion commensurabilities: the 4/1, 3/1, 5/2 and 7/3 cases., *Icarus*, **114**, 33–50.

Morbidelli, A., Zappalá, V., Moons, M. Cellino, A. and Gonczi, R.: 1995, Asteroid families close to mean motion resonances: dynamical effects and physical implications, *Icarus*, **118**, 132–154.

Morbidelli A., and Froeschlé C.: 1996, On the relationship between Lyapunov times and macroscopic instability times *Celest. Mech.*, **63**, 227–239.

Morbidelli A. and Giorgilli A.: 1996. On the role of high order resonances in normal forms and in separatrix splitting, *Physica D*, **1454**, 1–13.

Neishtadt, A. I.: 1984, The separation of motions in systems with rapidly rotating phase, *Prikl. Matem. Mekan.* **48**:2, 197–204; english transl. in *J. Appl. Math. Mech.* **48**:2, 133–139 (1984).

Nekhoroshev, N., N.: 1977, Exponential estimates of the stability time of near–integrable Hamiltonian systems., *Russ. Math. Surveys*, **32**, 1-65.

Nekhoroshev, N., N.: 1979, Exponential estimates of the stability time of near–integrable Hamiltonian systems, 2., *Trudy Sem. Petrovs.*, **5**, 5-50.

Pöschel, J.: 1993, Nekhoroshev's estimates for quasi–convex Hamiltonian systems, *Math. Z.*, **213**, 187.

Poincaré, H.: 1892, *Les Méthodes Nouvelles de la Mécanique Céleste*, Gauthier–Villars, Paris.

Tisserand, M. F.: 1882, *Ann. Obs. Paris*, **16**, E1.

Williams, J. G.: 1969, Secular perturbations in the solar system., *Ph.D. dissertation*, University of California, Los Angeles.

Wisdom J.: 1980, The resonance overlap criterion and the onset of stochastic behavior in the restricted three body problem, *AJ*, **85**, 1122–1133.

FREQUENCY MODIFIED FOURIER TRANSFORM AND ITS APPLICATION TO ASTEROIDS

MILOŠ ŠIDLICHOVSKÝ
Astronomical Institute of the Academy of Sciences of the Czech Republic, Boční II 1401, 141 31 Praha 4, Czech Republic, sidli@ig.cas.cz

and

DAVID NESVORNÝ
Universidade de São Paulo, Instituto Astrônomico e Geofísico, Av. Miguel Stefano 4200, Agua Funda, CEP 04301, São Paulo, Brasil, david@vax.iagusp.usp.br

Abstract. Recently a method has been suggested to analyze the chaotic behaviour of a conservative dynamical system by numerical analysis of the fundamental frequencies. Frequencies and amplitudes are determined step by step. As the frequencies are not generally orthogonal, a Gramm-Schmidt orthogonalization is made and for each new frequency the old amplitudes of previously determined frequencies are corrected. For a chaotic trajectory variations of the frequencies and amplitudes determined over different time periods are expected. The change of frequencies in such a calculation is a measure of the chaoticity of the trajectory. While amplitudes are corrected, the frequencies (once determined) are constant. We suggest here simple linear corrections of frequencies for the effect of other close frequencies. The improvement of frequency determination is demonstrated on a model case. This method is applied to the first fifty numbered asteroids.

Key words: Asteroids – Chaos – Fourier Transform

1. Algorithm for Modified Fourier Transform

The Modified Fourier Transform (MFT) was introduced by Laskar (1988) for analysis of the numerical solution to his secular system for planetary motion. The method was described in more detail in Laskar (1990) and applied to the standard mapping (Laskar *et al.*, 1992) and to multi-dimensional systems (Laskar, 1993). For regular motion the method yields an analytical representation of the solutions. For chaotic system the frequency changes calculated for two time intervals give a measure of chaos.

In this section we give a recursive algorithm for MFT, while in Section 2 the modification of the algorithm to correct the frequencies and amplitudes will be presented. Let us assume a complex function $f(t) = k(t) + ih(t)$ ($k(t)$ and $h(t)$ are real) defined on the interval $[0, 2\tau]$. We shall denote $\nu_0 = \pi/\tau$. Usually $f(t)$ is sampled at evenly spaced intervals in time. We introduce the scalar products of two such functions

$$\langle f, g \rangle = \frac{1}{2\tau} \int_0^{2\tau} f(t)\bar{g}(t) \left(1 - \cos\frac{\pi t}{\tau}\right) dt \tag{1}$$

corresponding to introducing the Hanning window filter on the interval $[0, 2\tau]$. Such integrals can be calculated by the Simpson's rule or other more

Celestial Mechanics and Dynamical Astronomy **65**: 137–148, 1997.

sophisticated methods. Then

$$\langle e^{i\nu t}, e^{i\omega t}\rangle = e^{i(\nu-\omega)\tau} Q(\nu,\omega), \tag{2}$$

where $Q(\nu,\omega)$ is real valued function

$$\begin{aligned} Q(\nu,\omega) &= \frac{\sin(\nu-\omega)\tau}{(\nu-\omega)\tau}\frac{\pi^2}{\pi^2-(\nu-\omega)^2\tau^2} && \text{for } \nu\neq\omega, \\ Q(\nu,\omega) &= 1 && \text{for } \nu=\omega. \end{aligned} \tag{3}$$

If

$$f(t) = \sum_{j=1}^{\infty} A'_j \exp i\nu'_j t, \tag{4}$$

with

$$|A'_1| \geq |A'_2| \geq \cdots,$$

then the MFT tries to approximate $f(t)$ with the first N terms

$$f(t) = \sum_{j=1}^{N} A_j \exp i\nu_j t, \tag{5}$$

where A_j and ν_j are very close to the corresponding A'_j, ν'_j. In the first step of the MFT we determine ν_1 as the value σ, for which the absolute value of the function

$$\varphi_1(\sigma) = \langle f(t), \exp i\sigma t\rangle \tag{6}$$

takes on a maximum value. Then

$$A_1 = \varphi_1(\nu_1). \tag{7}$$

For the next step we use as input function

$$f_1(t) = f(t) - A_1 \exp i\nu_1 t. \tag{8}$$

To give the necessary formulas for the mth step in which the frequency ν_m is determined, let us make some preparatory definitions. We will define unit vectors

$$e_j = \exp i\nu_j(t-\tau). \tag{9}$$

As $\langle e_j, e_k\rangle = Q(\nu_j,\nu_k)$, the vectors e_j are not orthonormal. We will form an orthogonal basis

$$b_j = \sum_{k=1}^{j} \alpha_{jk} e_k. \tag{10}$$

In mth step e_m is determined first and then m coefficients α_{mk} for $k = 1, \ldots, m$ follow from the orthonormality conditions for vectors $b_1, \ldots, b_m$. We will choose $\alpha_{11} = 1$. Then we can choose all α_{ij} real. In mth step we calculate coefficient S_m in the development

$$f = \sum_{j=1}^{m} S_j b_j + f_m. \tag{11}$$

Here

$$S_j = \langle f, b_j \rangle, \tag{12}$$

so that

$$\langle f_m, b_j \rangle = 0 \quad \text{for } j \leq m. \tag{13}$$

For mth step we need to know the function $f_{m-1}(t)$, coefficients α_{nk} (for $n < m$ and $k \leq n$), and the coefficients S_k (for $k < m$). This input is certainly known for $m = 2$, since we know f_1, $\alpha_{11} = 1$, and $S_1 = A_1$. The algorithm for the mth step consists of following substeps:

1. Calculation of ν_m as the value σ for which the function
$$\varphi_m = \langle f_{m-1}, \exp i\sigma t \rangle \tag{14}$$
takes on a maximum value. This is achieved by applying the FFT to f_{m-1} for rough estimate (accuracy $\sim \nu_0$) followed by Brent's method for a much more accurate determination of ν_m. As our program is written in C++, we used a slightly modified routine 'brent' from Numerical Recipes in C (Press *et al.*, 1992). As a result we have not only ν_m but also value of φ at its maximum
$$\varphi_m(\nu_m) = F_m \exp i\delta_m \tag{15}$$
e.g. its amplitude F_m and phase δ_m.
2. Calculation of b_m, or equivalently the coefficients α_{mk} for $k \leq m$. The vectors b_j for $j < m$ are known from the previous steps. Writing
$$b_m = \alpha_{mm} \left(\sum_{j=1}^{m-1} B_j^{(m)} b_j + e_m \right), \tag{16}$$
we obtain the coefficients $B_j^{(m)}$ from the orthogonality of b_m to all b_j for $j < m$.
$$B_j^{(m)} = -\sum_{s=1}^{j} \alpha_{js} Q(\nu_m, \nu_s) \quad \text{for } j = 1, \ldots, m-1 \tag{17}$$
and coefficient α_{mm} from condition $\langle b_m, b_m \rangle = 1$ which yields
$$\alpha_{mm} = \left(1 - \sum_{j=1}^{m-1} |B_j^{(m)}|^2 \right)^{-\frac{1}{2}} \tag{18}$$

and finally

$$\alpha_{mj} = \alpha_{mm} \sum_{s=j}^{m-1} B_s^{(m)} \alpha_{sj} \quad \text{for } j < m. \tag{19}$$

3. Calculation of f_m. From Eq. (11)

$$f_m = f_{m-1} - \langle f_{m-1}, b_m \rangle b_m. \tag{20}$$

Employing Eqs. (16) and (13) we have

$$\begin{aligned} f_m &= f_{m-1} - \alpha_{mm} \langle f_{m-1}, e_m \rangle \sum_{j=1}^{m} \alpha_{mj} e_j \\ &= f_{m-1} - \alpha_{mm} F_m \sum_{j=1}^{m} \alpha_{mj} \exp i \left[\nu_j t + (\nu_m - \nu_j)\tau + \delta_m \right]. \end{aligned} \tag{21}$$

For the real and imaginary part of f_m we have the following recurrence relations

$$\begin{aligned} k_m &= k_{m-1} - \alpha_{mm} F_m \sum_{j=1}^{m} \alpha_{mj} \cos i \left[\nu_j t + (\nu_m - \nu_j)\tau + \delta_m \right], \\ h_m &= h_{m-1} - \alpha_{mm} F_m \sum_{j=1}^{m} \alpha_{mj} \sin i \left[\nu_j t + (\nu_m - \nu_j)\tau + \delta_m \right]. \end{aligned} \tag{22}$$

4. Calculation of S_m from Eq. (12).

$$\begin{aligned} S_m &= \langle f, b_m \rangle = \langle f_{m-1}, b_m \rangle = \\ &= \sum_{j=1}^{m} \alpha_{mj} \langle f_{m-1}, e_j \rangle. \end{aligned} \tag{23}$$

But f_{m-1} is orthogonal to $b_1, \ldots, b_{m-1}$ and, therefore, to $e_1, \ldots, e_{m-1}$.

$$S_m = \alpha_{mm} F_m \exp i(\nu_m \tau + \delta_m) \tag{24}$$

After N steps we have

$$f = \sum_{j=1}^{N} S_j b_j + f_N = \sum_{s=1}^{N} e_s \sum_{j=s}^{N} S_j \alpha_{js} + f_N. \tag{25}$$

Neglecting f_N and comparing Eq. (25) with (5) we find

$$A_s = \sum_{j=s}^{N} \alpha_{jj} \alpha_{js} F_j \exp i \left[(\nu_j - \nu_s)\tau + \delta_j \right]. \tag{26}$$

The program for calculation of ν_j and A_j can be written in the form of a loop with N steps. The function f_j is always calculated and written to a file with the same sampling period as f. Our program ort.cc is still more general in incorporating the frequency corrections described in following section.

2. The Frequency Modified Fourier Transform (FMFT)

If we increase the number N of calculated frequencies and amplitudes to $N+1$ it follows from Eq. (26) that amplitudes previously determined are corrected with each additional step. On the other hand frequencies, once determined, are not changed. It is, however, clear that the maximum amplitude of $\varphi_k(\sigma)$ is shifted from ν_k mostly because of the existence of nearby frequencies (several ν_0 apart from ν_k). We shall assume that no two frequencies with significant amplitudes are closer than ν_0. This could lead to calculation of a false frequency somewhere in between. Our assumption is that the error of the MFT calculation of most important frequencies is small, and that a linear correction of the calculated frequencies can be given.

Let us first assume we have only two frequencies

$$f = C_1 \exp i(\omega_1 t + \beta_1) + C_2 \exp i(\omega_2 t + \beta_2), \tag{27}$$

with real positive $C_1 > C_2$; then

$$\varphi(\sigma) = \langle f, \exp i\sigma t \rangle = \sum_{j=1}^{2} C_j Q(\omega_j, \sigma) \exp i\left[(\omega_j - \sigma)\tau + \beta_j\right]. \tag{28}$$

Calculating ν_1 as σ for which the amplitude of $\varphi(\sigma)$ takes on a maximum would give $\nu_1 = \omega_1$ only if $Q(\omega_2, \sigma)$ is negligible for $\sigma \sim \nu_1$. We determine ν_1 from the condition

$$\frac{d}{d\sigma}|\varphi(\sigma)|_{\sigma=\nu_1} = 0, \tag{29}$$

Inserting Eq. (28) in (29) we obtain an equation satisfied by ν_1:

$$\begin{aligned} &C_1^2\ Q(\omega_1, \nu_1)Q'(\omega_1, \nu_1) + C_2^2 Q(\omega_2, \nu_1)Q'(\omega_2, \nu_1) + \\ &C_1\ C_2 \left[Q(\omega_1, \nu_1)Q'(\omega_2, \nu_1) + Q(\omega_2, \nu_1)Q'(\omega_1, \nu_1)\right] \\ &\cos\ \left[(\omega_1 - \omega_2)t + \beta_1 - \beta_2\right] = 0, \end{aligned} \tag{30}$$

where $Q'(\omega, \nu)$ is the derivative of $Q(\omega, \nu)$ with respect to the second argument. Introducing the error ϵ in approximating ω_1 by ν_1

$$\nu_1 = \omega_1 + \epsilon, \tag{31}$$

we can make a linearization of Eq. (30) in ϵ and obtain

$$\epsilon = -\frac{C_2 Q'(\omega_2, \omega_1) \cos\left[(\omega_1 - \omega_2)\tau + \beta_1 - \beta_2\right]}{C_1 Q''(\omega_1, \omega_1)}, \tag{32}$$

where $Q''(\omega_1, \omega_1)$ is the second derivative of $Q(\omega_1, \omega_1)$ again with respect to the second argument. We introduce

$$Q(y) = \frac{\sin y}{y} \frac{\pi^2}{\pi^2 - y^2}. \tag{33}$$

Then

$$Q'(y) = \frac{1}{y}\frac{\pi^2}{\pi^2 - y^2}\left[\cos y + \frac{\sin y}{y}\frac{3y^2 - \pi^2}{\pi^2 - y^2}\right] \tag{34}$$

and

$$Q''(0) = \left(\frac{2}{\pi^2} - \frac{1}{3}\right). \tag{35}$$

The final algorithm for the calculation of the corrections ϵ_j to ν_j (for N frequencies) to produce new frequencies $\nu_j - \epsilon_j$ is to use

$$\epsilon_j = \sum_{s>j}^{N} \frac{C_s Q'(y_{sj})}{C_j Q''(0)\tau} \cos(y_{sj} + \beta_s - \beta_j), \tag{36}$$

where

$$y_{sj} = (\nu_s - \nu_j)\tau. \tag{37}$$

Real values C_s, β_s are determined from relation

$$A_s = C_s \exp i\beta_s \tag{38}$$

and $Q'(y)$, $Q''(0)$ are given by Eqs. (34), (35) and (37).

The FMFT method consists of the MFT as described in Section 1 followed by a correction of frequencies via Eq. (36) a and new determination of the amplitudes by again employing the algorithm of Section 1, with only the frequencies given a priori.

An alternative way to correct the frequencies determined by the MFT is provided by the following simple method. One first applies the MFT to the original function $f(t) = \sum C_j \exp i(\nu_j t + \beta_j)$ and gets its development $f'(t) = \sum C'_j \exp i(\nu'_j t + \beta'_j)$. The error in the frequency determination is the small quantity

$$\epsilon_k = \epsilon_k(\nu_j, C_j, \beta_j) = \nu_k - \nu'_k. \tag{39}$$

Similarly, the errors in the amplitudes and phases are also small quantities. A second MFT applied to $f'(t)$ leads again to a slightly different development $f''(t) = \sum C''_j \exp i(\nu''_j t + \beta''_j)$ with small errors in the frequencies

$$\epsilon'_k = \epsilon_k(\nu'_j, C'_j, \beta'_j) = \nu'_k - \nu''_k. \tag{40}$$

Substituting $\nu'_j = \nu_j - \epsilon_j$ into Eq. (40), and developing function $\epsilon_k()$ into a Taylor series (on the assumption that its first derivatives are small quantities), one can neglect the linear and higher order terms and write $\epsilon_k = \epsilon'_k$. Thus, in the above approximation, $\nu_k = \nu'_k + \epsilon'_k$, whose right-hand side can be simply evaluated since ν'_k and ϵ'_k are known. A similar approach provides the amplitudes $C_j = C'_j + (C'_j - C''_j)$ and phases $\beta_j = \beta'_j + (\beta'_j - \beta''_j)$.

In order to distinguish between the two FMFT methods described in this section, we call the first FMFT_1 and the second FMFT_2.

TABLE I
The reconstruction of frequencies ("/year) by MFT and FMFT

j	Original	MFT	$FMFT_1$	$FMFT_2$
1	4.2488163	4.2488183	4.2488163	4.2488163
2	28.2206942	28.2206916	28.2206942	28.2206942
3	3.0895148	3.0895150	3.0895149	3.0895148
4	52.1925732	52.1925732	52.1925732	52.1925732
5	27.0613982	27.0611601	27.0613983	27.0613957
6	29.3799573	29.3802248	29.3799577	29.3799611
7	28.8679427	28.8679114	28.8679409	28.8679426
8	27.5734578	27.5734593	27.5734593	27.5734579
9	5.4070444	5.4070414	5.4070413	5.4070444
10	0.6671228	0.6671228	0.6671228	0.6671228

3. Simple Test of the Method

Both $FMFT_1$ and $FMFT_2$ were programmed and first tested on the function

$$f(t) = k(t) + ih(t) = \sum_{j=1}^{10} C_j \exp i(\nu_j t + \beta_j), \tag{41}$$

where the coefficients ν_j, C_j, β_j were taken from a development of $f = e \exp i\tilde{\omega}$ for Jupiter ($\tilde{\omega}$ is its longitude of perihelion and e its eccentricity) as given by Laskar (1990).

The sampling period was 120 000 days and the number of sampling points was 32 768 so that the time interval 2τ for reconstruction of the frequencies was about 10.8 Myr. It is possible to distinguish frequencies $\nu_0 = \pi/\tau = 0.25$"/year apart, which is well sufficient as the closest frequencies, ν_8 and ν_5, are separated by 0.51"/year.

Table 1 shows the original and reconstructed frequencies using MFT, $FMFT_1$ and $FMFT_2$. It is clear that both $FMFT_1$ and $FMFT_2$ improve the result. The first frequency is determined exactly up to seven decimal digits and close frequencies 5 – 8 are computed by several orders more precisely than by MFT. This improvement is significant in amplitudes and phases as well, and leads to better decomposition of $f(t)$. $FMFT_1$ and $FMFT_2$ are roughly comparable in the quality of reconstruction. Frequencies 5 and 6 are better determined by $FMFT_1$, while frequencies 7, 8 and 9 are better from $FMFT_2$. There are slight differences in the computed amplitudes and phases as well, but both methods can be considered to work with similar precision.

TABLE II
The g and s for first 50 asteroids, for which a good decomposition was obtained

No.	g "/year	s "/year	No.	g "/year	s "/year
1	54.07464	-59.10736	24	131.99292	-103.31862
3	43.63334	-61.22194	26	55.26366	-55.66960
4	36.48091	-39.20237	27	38.09601	-43.25415
7	37.94909	-46.06354	28	57.32945	-66.64166
9	38.36402	-41.63721	29	49.43937	-46.98591
11	40.41837	-42.82523	30	38.18240	-40.13398
12	33.71960	-40.34126	34	55.14321	-58.69910
13	35.72530	-45.27553	37	54.53217	-57.80620
14	48.58297	-56.04458	39	53.70462	-56.40034
15	42.49709	-51.80465	40	34.27385	-34.75085
16	76.75683	-73.06070	42	39.67433	-46.66578
17	43.41116	-45.97697	44	40.89591	-45.98494
19	41.50964	-44.81832	45	56.46531	-58.19380
20	40.47798	-44.74191	46	58.62063	-48.61045
21	41.12942	-44.24252	47	70.20357	-69.61741
23	48.42881	-64.34235	48	105.30808	-81.97421

4. An Application to Asteroids

The time evolution of $e \exp \tilde{\omega}$ and $\sin I/2 \exp \Omega$ (I is the inclination and Ω is the longitude of node) for the first 50 asteroids under the force of four outer planets was obtained by numerical integration using the MSI integrator (Šidlichovský and Nesvorný, 1994). Initial conditions were taken from Milani's `oscelnum.92` at `gauss.dm.unipi.it` and corrected to the barycenter of the inner solar system. The integration step was 10 days, the output was filtered by sequential application of two low-pass filters (Nesvorný and Ferraz-Mello, 1996) and sampled at 120 000 days. The number of points (32 768) implies a time interval over which the decompositions were computed of about 10.8 Myr. During the integration, the distance to Jupiter was checked every 10 days. All the studied asteroids stayed farther than 1 AU from Jupiter, ensuring the accuracy of the integrator. The coefficients of decomposition were determined by FMFT_2 and in several cases checked by FMFT_1.

In Tab. 2, we give the main secular frequencies g in $e \exp \tilde{\omega}$ and s in $\sin I/2 \exp \Omega$ for those asteroids among the first 50 for which a good decomposition was obtained. The others, for which the frequencies were not shown, are likely to be in some secular resonance or at least in its close vicinity, and their decomposition can not be obtained in 10.8 Myr due to presence of very low frequencies. Similarly, chaoticity on a shorter time scale than 10.8 Myr would also not allow for a good orbital decomposition.

TABLE III
The coefficients for Ceres

	$e \exp \tilde{\omega}$			$\sin(I/2) \exp \Omega$		
j	ν_j "/year	C_j	β_j deg	ν_j "/year	C_j	β_j deg
1	54.07464	0.115573	152.273	-59.10736	0.080378	99.223
2	4.24465	0.030770	29.497	-60.94878	0.014701	338.786
3	28.23886	0.019704	300.863	-57.26561	0.014650	39.140
4	52.23318	0.008700	32.124	-0.00001	0.013683	107.648
5	-172.28929	0.003359	226.036	-58.42389	0.008938	129.248
6	53.39115	0.001435	302.504	-59.79006	0.008786	248.077
7	-174.13056	0.001307	105.332	-26.33878	0.005271	313.763
8	3.08658	0.001257	119.333	-57.95748	0.001899	201.575
9	55.91697	0.001077	272.615	-61.62944	0.001780	124.544
10	-170.44725	0.001017	165.590	-56.57933	0.001459	64.909

Nobili et al. (1989) discusses the effect of the inner planets on the fundamental frequencies of the outer solar system. If the quadrupole force term of the inner planets is not included in the model, the fundamental frequency g_5 of Jupiter differs by about 0.012 "/yr from the real value. Asteroid g will differ even more and the values in Tab. 2 must be regarded, although very precise in the frame of studied model, as preliminary and underestimating the real frequencies. Indeed, our recent integration of the first 20 asteroids together with seven planets (Venus to Neptune with the mass of Mercury added to Sun) led to $g = 54.25081$ "/yr and $s = -59.23250$ "/yr for Ceres, which in comparison with Tab. 2 indicates 0.18 "/yr and 0.13 "/yr differences. Similarly, the frequencies of other asteroids differ at order 0.1 "/yr. Including of the inner planets is thus necessary for good estimation of the real frequencies and thus computations which include the full effect of Venus, Earth, and Mars are in progress.

4.1. Asteroids with good decomposition

The coefficients for Ceres' $e \exp i\tilde{\omega}$ and $\sin(I/2) \exp i\Omega$ are shown in Tab. 3. We checked the quality of both decompositions by comparison with the filtered output of the integration over 10.8 Myr. In the case of $f(t) = k(t) + ih(t) = e \exp i\tilde{\omega}$, the maximum error in k was 0.009 and in h 0.008. We show the first 1 Myr in Fig. 1. We decided to consider the Ceres as a marginal case of an asteroid with good decomposition and listed in Tab. 2 only those asteroids, which have the maximum errors over 10.8 Myr in both k and h less than 0.01.

Additional prolonged integration showed that the maximum error of Ceres' decomposition over 43.2 Myr is about 0.03 in both h and k, which is

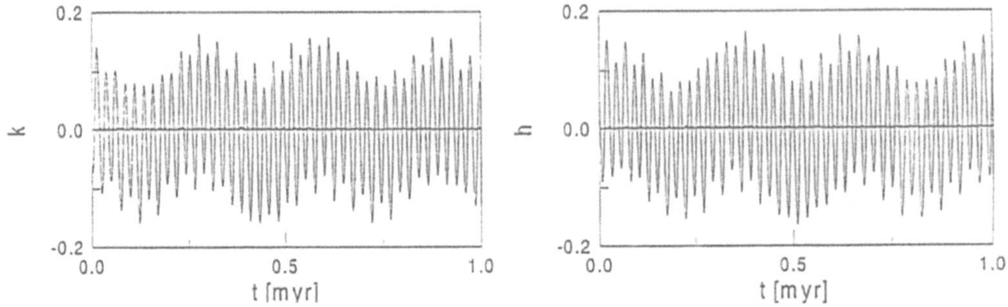

Fig. 1. Comparison between original and reconstructed k and h evolution for Ceres. Here the filtered output of the integrated k and h is plotted with its difference in the time evolution obtained from Tab. 3. The difference is the rough line at almost zero. The maximum errors on this interval in k and h are both about 0.006.

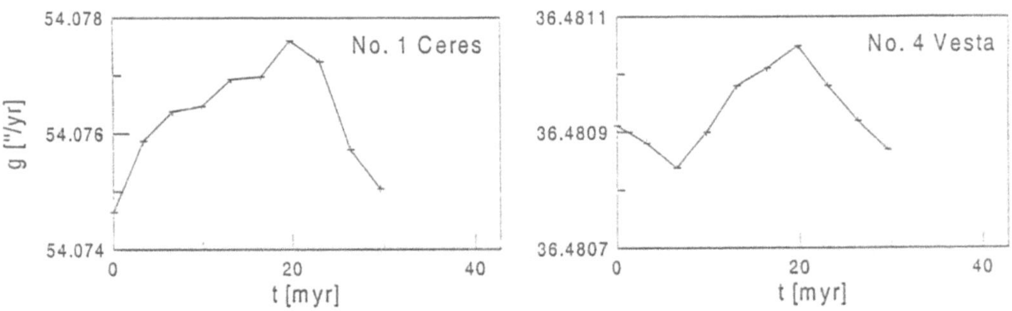

Fig. 2. Time dependence of g for Ceres (left) and No. 4 Vesta (right)

a rather large disagreement. Moreover, the decomposition obtained from the whole 43.2 Myr, which would better distinguish the low frequencies, does not represent the real filtered evolution any better. We infer that the motion of Ceres can't be properly represented by a decomposition with fixed frequencies and claim that the frequencies are changing due to orbital chaoticity. The trajectory of No. 4 Vesta (listed in Tab. 2) is much better represented by the decomposition obtained from the 10.8 Myr interval. Indeed, the error over 43.2 Myr is only 0.004.

In Fig. 2 we show the time dependence of g for Ceres and Vesta. We computed several FMFT_2 while shifting the interval for the frequency computation by 10 000 points (roughly by 3.28 Myr). Notice that g for Ceres changes by one order of magnitude more than Vesta's g. This is the reason why the decomposition for Vesta with fixed $g = 36.48091$ "/year represents its trajectory more precisely than fixed $g = 54.07464$ "/year of Ceres.

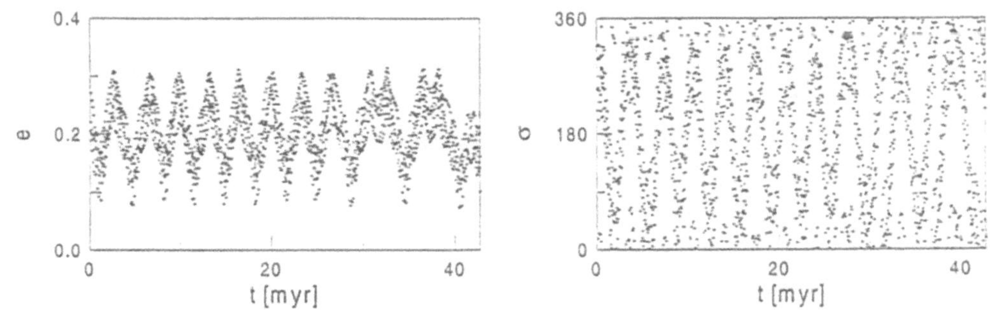

Fig. 3. Eccentricity and resonant angle $\sigma = \tilde{\omega} + \tilde{\omega}_J - 2\tilde{\omega}_S$ for No. 5 Astraea

4.2. Asteroids with bad decomposition

Asteroid No. 5 Astraea is, at least in our model neglecting the quadrupole moment of the orbits of the inner planets, is in the secular resonance $g + g_5 - 2g_6 \sim 0$, where g, g_5 and g_6 are the main secular frequencies of the asteroid's, Jupiter's and Saturn's $e \exp i\tilde{\omega}$. Fig. 3 (right) shows the resonant angle $\tilde{\omega} + \tilde{\omega}_J - 2\tilde{\omega}_S$ versus time. The period of libration is about 3.5 Myr. It is impossible to obtain, in this case, a good decomposition from only 10.6 Myr, since the low frequency forms important harmonics close to g, as discussed in Nesvorný and Ferraz-Mello (1996). Moreover, at about 30 Myr, the resonant angle begins to circulate indicating a strong orbital chaoticity.

The asteroids not shown in Tab. 2. likely have characteristic similar to Astraea; they are in or very near the secular resonances and low frequencies do not allow us to obtain a good decomposition of their trajectories.This was tested and verified for several of them. For instance, asteroids No. 6 Hebe and No. 8 Flora fulfil $g - g_6 - g_7 \sim 0$, where g_7 is the secular frequency of longitude of perihelion of Uranus.

5. Summary

Our preliminary calculations show what behaviour one may expect when investigating asteroids with the MFT. If the inner planets are taken into account the fundamental frequencies of the system change slightly and conclusions about secular resonances from a model with outer planets only might not be correct. The more exact calculations taking into account the inner planets are in progress. Such calculations are time consuming as the step of numerical integration must be much shorter.

Acknowledgements

The first author (M.Š.) acknowledges the financial support provided by grant 205/95/0184 of GAČR (Grant Agency of the Czech Republic) and Key Project AV ČR K1-003-601. The second author (D.N.) thanks the FAPESP (Research Foundation of the State of São Paulo) for sponsorship of his work.

References

Laskar, J.: 1988, 'Secular evolution of the solar system over 10 million years', *Astron. Astrophys.* **198**, pp. 341–362

Laskar, J.: 1990, 'The Chaotic Motion of the Solar System: A Numerical Estimate of the Size of the Chaotic Zones', *Icarus* **88**, pp. 266–291

Laskar, J., Froeschlé , Cl. and Celleti, A.: 1992 ,' The Measure of Chaos by the Numerical Analysis of the Fundamental Frequencies. Application to the Standard Mapping.', *Physica D* **56**, pp. 253–269

Laskar, J.: 1993, 'Frequency analysis for multi-dimensional systems. Global dynamics and diffusion', *Physica D* **67**, pp. 257–281

Nesvorný, D. and Ferraz-Mello, S.: 1996, 'Chaotic diffusion in the 2/1 asteroidal resonance', submitted to *Astron. Astrophys.*

Nobili, A.M., Milani, A. and Carpino, M.: 1989, 'Fundamental frequencies and small divisors in the orbits of the outer planets', *Astron. Astrophys.* **210**, pp. 313–336

Press, W. H., Teukolsky, S. A., Vetterling, W. T. and Flannery, B. P.: 1992, Numerical Recipes in C, Cambridge University Press, Cambridge, UK

Šidlichovský, M. and Nesvorný, D.: 1994, 'Temporary capture of grains in exterior resonances with the Earth: Planar circular restricted three-body problem with Poynting-Robertson drag', *Astron. Astrophys.* **289**, pp. 972–982

THE TROJAN PROBLEM

BÁLINT ÉRDI
Department of Astronomy, Eötvös University
Budapest, Hungary

Abstract. There has been a renewed interest in the Trojan problem in recent years. Significant progress has been made in discovering and understanding dynamical features of motion of Jupiter's Trojan asteroids. The dynamics of hypothetical Trojan-type asteroids of other major planets has also been the subject of several recent investigations. This paper offers an overview on the current status of researches on real and hypothetical Trojan asteroids of the major planets. Results of analytical and numerical works are surveyed. Questions of dynamical properties, long-term evolution of orbits, stability regions around the triangular Lagrangian points are discussed among other problems of the Trojans.

Key words: Lagrangian points – Trojan asteroids – resonances.

1. Introduction

The discovery of the triangular solutions of the three-body problem by Lagrange in 1772 provided celestial mechanics with a fascinating subject. The problem of motions around the Lagrangian points L_4 and L_5 is very interesting and it has been extensively studied in many papers. A major part of the problem concerns the existence and motions of celestial bodies around the Lagrangian points in the Solar System. For a long time Lagrange's solutions had been considered only as of theoretical interest until in 1906 M. Wolf discovered an asteroid, (588) Achilles, near Jupiter's L_4 point providing thus the first example in Nature for an approximate Lagrangian solution. By now many other examples have become known. The most prominent are the two groups of Trojan asteroids near Jupiter's L_4 and L_5 points. There are also a few examples in the satellite system of Saturn. Considering the ever increasing number of the known Trojans it has been a thoroughly studied question in recent years if Trojan-like asteroids of other planets than Jupiter exist. The answer is affirmative, now we know one Martian Trojan, (5261) Eureka, discovered by H. E. Holt and D. H. Levy in 1990 near Mars' L_5 point (Holt and Levy, 1990), and identified by E. Bowell as a possible Mars' Trojan.

The Trojan asteroids are in a 1:1 mean motion resonance with Jupiter. Their mean motions oscillate around that of Jupiter as a mean value and this is true for their semi-major axes as well. While moving around the Sun, the Trojans perform librating motion around Jupiter's L_4 or L_5 point, situated at 60° before and after Jupiter on Jupiter's orbit. This libration is asymmetric with respect to the Lagrangian points. The amplitude of libration can reach large values ($\approx 33^o$) thus the libration is non-linear. Despite the large amplitudes the Trojans fill only a narrow part of the libration region allowed by the restricted three-body problem. The Trojans have moderate orbital eccentricities. For most Trojans the proper eccentricity $e_p < 0.15$, the average is 0.0619 and only in two cases $e_p = 0.22$. The proper inclinations range from a few degrees up to 37° with a high average value.

Celestial Mechanics and Dynamical Astronomy **65**: 149–164, 1997.

A remarkable feature of the Trojans is that their basic periods are well separated. The period of revolution around the Sun is about 12 yr, corresponding to the 1:1 resonance with Jupiter. The period of libration around the Lagrangian points is 145.7 – 240 yr. The period of the free motion of the perihelion is between 3000 and 5600 yr. The proper period of the ascending node is in the interval 38000 – 2700000 yr. As a result of these different periods, no low order resonances are possible between the basic motions of the Trojans. Apart from the 1:1 resonance with Jupiter and the 5:2 near-resonance with Saturn, secular resonances can affect the orbital evolution of some of the Trojan asteroids.

There are classical studies on the Trojan problem, references can be found in Szebehely (1967). Garfinkel (1977, 1978, 1980, 1983, 1985) developed a theory of the Trojan asteroids in the planar, circular restricted three-body problem. Zagretdinov (1986) extended Garfinkel's solution to the spatial case. Kevorkian (1970) studied the planar motion of the Trojan asteroids in the restricted three-body problem using the two-timescale method. Érdi (1977) generalized this solution to the planar, elliptic restricted problem, and to the three-dimensional case (Érdi, 1978), and studied the Trojans' problem in several papers (Érdi, 1979, 1981, 1983a, 1983b, 1984, 1988, 1995; Érdi and Presler, 1980; Érdi and Váradi, 1983).

There have been numerical studies on the long-time evolution of the orbits of the Trojan asteroids. Chebotarev *et al* (1974) integrated the orbits of 29 Trojans for 400 yr. Bien and Schubart studied the motion of the Trojans in different models by numerical methods with characteristic time-spans of $10^4 - 10^5$ yr (Bien, 1978, 1980a, 1980b; Bien and Schubart, 1983, 1984, 1987; Schubart and Bien, 1984, 1986, 1987). Milani (1993, 1994) integrated the orbits of 174 Trojan asteroids for 1 Myr and in several cases for 5 Myr considering the perturbations of the four outer major planets. De la Barre *et al* (1996) integrated (hypothetical) Trojan orbits for 198 Myr.

In the last few years significant progress has been made in discovering and understanding dynamical features of motions around the L_4, L_5 points of the major planets (Zhang and Innanen, 1988 a, b, c; Innanen and Mikkola, 1989; Mikkola and Innanen, 1990, 1992, 1994; Holman and Wisdom, 1993; de la Barre and Kaula, 1995; de la Barre *et al*, 1996; Varadi *et al*, 1995). These points are stable in the restricted three-body problem. The question is, whether they remain stable when the perturbing effects of the other planets are taken into account. Except for the case of Jupiter, the external perturbations on the Lagrangian points of the planets are larger than those of the mother planet. The problem of stability can be studied only by carefully considering the specific conditions in the case of each planet.

2. Dynamics of the Trojan Asteroids

2.1. A THEORY OF MOTION

Garfinkel's (1977) Trojan theory is based on the restricted three-body problem. Even this simplest model needs very involved mathematical treatment, but for the understanding of the complex dynamics of the Trojans more realistic models are necessary. A better approximation is the three-dimensional elliptic restricted three-body problem (Érdi, 1978) in which Jupiter moves around the Sun in fixed elliptic orbit and the asteroid is not confined to Jupiter's orbital plane. Since the Trojans are influenced by the other major planets as well, directly and indirectly, it is necessary to account for these perturbations also. A part of the indirect perturbations was taken into account in (Érdi, 1995), where Jupiter's orbit was considered as a secularly changing elliptic orbit.

In the framework of the secularly changing, three-dimensional, elliptic restricted three-body problem the following equations of motion can be derived for an asteroid:

$$\begin{aligned}
&\frac{d^2r}{dv^2} - r\left(\frac{d\alpha}{dv}\right)^2 - 2r\frac{d\alpha}{dv} - 2\beta r\left(\frac{d\alpha}{dv}+1\right) = \\
&\quad = \frac{1}{1+e_J\cos v}\left[r - \frac{1-\mu}{R_1^3}r + \mu\left(\frac{\cos\alpha - r}{R_2^3} - \cos\alpha\right)\right], \\
&\frac{d}{dv}\left(r^2\frac{d\alpha}{dv} + r^2\right) + 2\beta r\frac{dr}{dv} = \frac{\mu r\sin\alpha}{1+e_J\cos v}\left(1 - \frac{1}{R_2^3}\right), \\
&\frac{d^2z}{dv^2} + z = \frac{z}{1+e_J\cos v}\left(1 - \frac{1-\mu}{R_1^3} - \frac{\mu}{R_2^3}\right),
\end{aligned} \tag{1}$$

where

$$R_1 = (r^2+z^2)^{1/2}, \quad R_2 = (1 + r^2 - 2r\cos\alpha + z^2)^{1/2}.$$

In Equations (1) r and α are coordinates of the asteroid in Jupiter's orbital plane (r is the distance from the Sun, α is the angle between the asteroid and Jupiter as seen from the Sun), z is the distance of the asteroid from Jupiter's orbital plane, v is the true anomaly of Jupiter which serves as independent variable, $\mu = m_J/(m_S+m_J)$, m_S, m_J being the mass of the Sun and Jupiter, e_J is the eccentricity of Jupiter's orbit. The coordinates r and z are dimensionless, the instantaneous Sun-Jupiter distance serving as unit distance. The parameter β is connected with the precession of the orbit of Jupiter: $\beta = \dot{\varpi}_J/n_J$, $\dot{\varpi}_J$ is the secular rate of the motion of the perihelion of Jupiter's orbit, and n_J is the mean motion of Jupiter. In the derivation of Equations (1), it has been assumed that the orbit of Jupiter is a precessing ellipse for which the relations of the two-body problem are valid with slowly changing e_J eccentricity. Equations (1) are only approximate, terms depending on $\dot{\varpi}_J^2$, $\ddot{\varpi}_J$,

$\dot{e}_J/n_J$ have been neglected. Equations (1) are accurate up to the third order of the small parameter $\varepsilon = \sqrt{\mu} = 0.030885$. The parameter β is of order ε^3, since the mean value of $\dot{\varpi}_J$ is 4.2"/yr and $n_J = 299.1$"/day.

Since the basic periods of the Trojans are well separated, Equations (1) are suitable for the application of the method of the multiple-variable expansions. The solution of Equations (1) can be assumed in the form of a four-variable asymptotic expansion, each variable v, $u = \varepsilon(v - v_0)$, $\tau = \varepsilon^2(v - v_0)$, $w = \varepsilon^3(v - v_0)$ (v_0 the epoch) corresponding to a time-scale of the motions of the Trojans (Érdi, 1995). The application of the method transforms Equations (1) into a system of partial differential equations. In the solution of this system arbitrary functions of the independent variables appear. These have been specified according to the principle that the solution for r, α, z be bounded. The perturbations of the orbital elements can be derived from the solution for r, α, z by means of the relations of the two-body problem.

2.2. Libration around L_4 and L_5

The leading term α_0 in the expansion of α describes the main part of the libration around L_4. The equation governing α_0 is

$$\frac{\partial^2 \alpha_0}{\partial u^2} + 3\sin\alpha_0[1 - 2^{-3/2}(1 - \cos\alpha_0)^{-3/2}] = 0. \tag{2}$$

This equation admits an integral

$$\frac{1}{2}\left(\frac{\partial \alpha_0}{\partial u}\right)^2 + U = h, \quad U = 3[2^{-1/2}(1 - \cos\alpha_0)^{-1/2} - \cos\alpha_0], \tag{3}$$

where h is constant. The function $U(\alpha_0)$ is shown in Figure 1. For a given h, libration is possible where $h \geq U$. The interval $3/2 \leq h < 9/2$ defines the tadpole region around L_4 and L_5, while $h = 9/2 = U$ yields the Brown's limit $\alpha_0 = 23.9°$, the limiting distance from Jupiter on tadpole orbit. For $h > 9/2$ the libration orbits are horseshoes.

The main variation in the semi-major axis a and in the mean motion n of the asteroid is

$$a = a_J\left(1 - \frac{2}{3}\varepsilon\frac{\partial \alpha_0}{\partial u}\right), \quad n = n_J\left(1 + \varepsilon\frac{\partial \alpha_0}{\partial u}\right). \tag{4}$$

The variation of a against α_0, obtained from (3) and (4), is shown in Figure 2 both for tadpole and horseshoe orbits (see also Schubart, 1964, Figs. 22-24).

To obtain the time-dependence of the libration one has to integrate Equation (3). This leads to a hyperelliptic integral $u(\alpha_0)$, the function $\alpha_0(u)$ can be obtained by inversion. This method of solution was followed by Garfinkel (1977, 1980) in his theory.

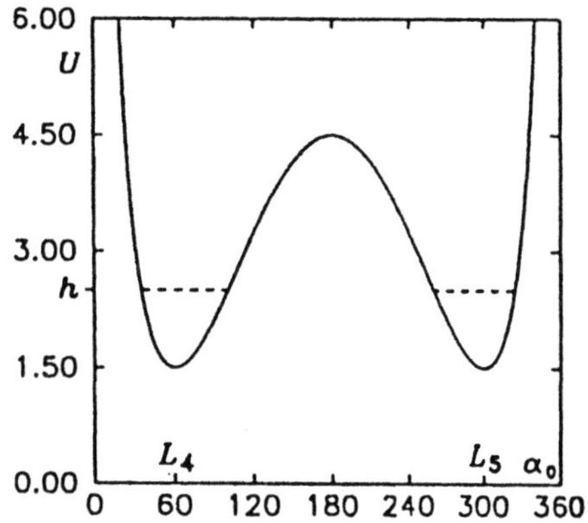

Fig. 1. The function $U(\alpha_0)$.

Instead of integrating Equation (3), Érdi (1981) derived a solution of Equation (2) in the form

$$\alpha_0 = \frac{\pi}{3} + \sum_{i\geq 0} a_i(l) \cos i\phi, \tag{5}$$

where $\phi = \omega(l)\, u + \delta(\tau)$, the coefficients $a_i(l)$ are polynomials of a constant of integration l (the approximate amplitude of libration), the frequency $\omega(l)$ depends also on l, and δ is a slowly changing function of τ.

The solution (5), valid for moderate amplitudes in the tadpole region, describes a periodic libration in ϕ around L_4. According to (5), the extremum values of the libration are: $\alpha_{0_{max}} = C + d$, $\alpha_{0_{min}} = C - d$,

$$C = \frac{\pi}{3} + \frac{\sqrt{3}}{2^2} l^2 + \frac{11\sqrt{3}}{2^7 3} l^4 + \mathrm{O}(l^6), \quad d = l + \frac{5}{2^6} l^3 + \frac{77}{2^{10} 3 \cdot 5} l^5 + \mathrm{O}(l^7). \tag{6}$$

C is the centre of the libration, d is the amplitude. The libration is asymmetric with respect to L_4: with increasing amplitude the centre shifts in opposite direction to Jupiter.

Using the relation between h and l, deducible from (3) and (5):

$$h = \frac{3}{2} + \frac{27}{2^3} l^2 - \frac{81}{2^8} l^4 + \mathrm{O}(l^6),$$

the extrema of a, reached at $\alpha_0 = \pi/3$, can be obtained from (3) and (4) as $a_{\max} = a_J + A$, $a_{\min} = a_J - A$,

$$A = a_J \varepsilon \sqrt{3}\, l \left(1 - \frac{3}{2^6} l^2 + \mathrm{O}(l^4)\right). \tag{7}$$

The semi-major axis oscillates around a_J as a mean value, with a phase shift with respect to the libration of α_0.

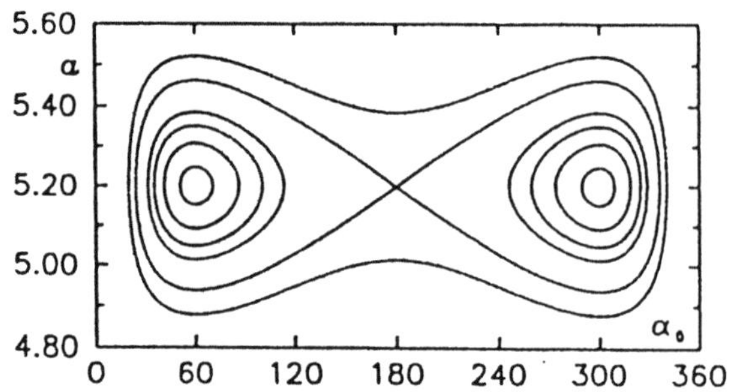

Fig. 2. The variation of a against α_0. The curves correspond to the values $h = 1.6, 2, 2.5, 3, 4.5, 6$, beginning with the innermost libration orbit.

The amplitudes A and d are in a quasi-linear connection. From (6) and (7) one obtains the approximate relation

$$A = da_J\varepsilon\sqrt{3}\left(1 - \frac{1}{8}d^2\right). \tag{8}$$

The contribution of the third order term is small (for $d < 0.5$). This quasi-linearity holds even when perturbations of the other planets are taken into consideration (see Schubart and Bien, 1987, Fig. 2). Milani (1993) also used this linear approximation between the amplitudes in his computation of the proper elements.

The approximate period of libration is

$$P_L = \frac{P_J}{\varepsilon\omega}, \quad \omega = \sqrt{\frac{27}{2^2}\left(1 - \frac{3}{2^3}l^2 - \frac{97}{2^9}l^4\right)}, \tag{9}$$

where $P_J = 12.86$ yr is Jupiter's orbital period. For $l = 0$, $P_L = 147.8$ yr and P_L increases with l.

The picture, described above, is somewhat idealistic, since it considers only the main part of the libration. In reality, several perturbations distort the smooth curves of Figure 2. Considering only the effect of Jupiter, several short and long periodic terms, appearing in the expansion of α are added to the main part α_0. The inclusion of the effects of the other planets makes the picture more complex. There appear short-period perturbations from the direct action of Saturn, and also long-period perturbations from the indirect action of the planets through the secular changes of the orbit of Jupiter. As a result, the curves of Figure 2 broaden and the libration periods change. For example, there are long periodic variations in the libration period with the secular frequencies $g_5 - g_6$ coming from the secular change of Jupiter's orbital eccentricity (Milani, 1993).

2.3. Perturbations of e and ϖ

The secular variation of Jupiter's orbit exerts its main effect in the variation of the eccentricity e and of the longitude of perihelion ϖ of the Trojans. The main perturbations of e and ϖ can be expressed as (Érdi, 1995)

$$\begin{aligned}\psi_1 &= (b - c\cos\chi)(1 + \frac{1}{4}\kappa + \frac{5}{8}\kappa\cos\gamma) + (a - c\sin\chi)\frac{5}{8}\kappa\sin\gamma,\\ \psi_2 &= (a - c\sin\chi)(1 + \frac{1}{4}\kappa - \frac{5}{8}\kappa\cos\gamma) + (b - c\cos\chi)\frac{5}{8}\kappa\sin\gamma,\end{aligned} \tag{10}$$

where $\psi_1 = e\cos(\varpi - \varpi_J)$, $\psi_2 = e\sin(\varpi - \varpi_J)$, a (not the semi-major axis) and b are proportional to e_J and depend also on l (the approximate amplitude of libration), c constant, κ is connected with the inclination, χ changes on the time-scale of the proper motion of the perihelion, γ changes on the time-scale of the motion of the ascending node.

The long-term variations of e_J and ϖ_J can be described by the theory of secular perturbations. Considering only the most important terms,

$$h_J = \sum_{j=5}^{6} m_{Jj}\sin(g_j t + \beta_j), \quad k_J = \sum_{j=5}^{6} m_{Jj}\cos(g_j t + \beta_j), \tag{11}$$

where $h_J = e_J\sin\varpi_J$, $k_J = e_J\cos\varpi_J$. The constants m_{Jj}, β_j and the secular fundamental frequencies g_j are given in Nobili *et al* (1989).

The variations of e and ϖ according to Equations (10) are shown in Figure 3 for several values of the parameters. In each case $l = 0.3$ (corresponding to $d = 17.2^o$ amplitude of libration), for other values of l the picture is qualitatively the same.

(a) *Planar case, Jupiter's orbit fixed*

In this case $i = 0$, $e_J = 0.048$. Figure 3(a) shows e versus $\varpi - \varpi_J$ for the values $c = 0.02, 0.04, 0.06, 0.08$. The inner closed curve corresponds to $c = 0.02$, with c increasing outwards. The centre $e = e_J$, $\varpi - \varpi_J = 60^o$ corresponds to the Lagrange's solution. The orbit of the asteroid can have two basic types of motion: libration or circulation. In both cases the approximate period of the perihelion is $P_\varpi = T_J/(\varepsilon^2 A_0)$. ($A_0$ given as a polynomial of l in Érdi, 1981). $P_\varpi = 3685$ yr for $l = 0$. The possibility of the libration of the perihelion is a consequence of the non-zero eccentricity of Jupiter's orbit. If e_J were 0, the perihelion could only circulate.

(b) *Planar case, Jupiter's orbit secularly changing.*

In this case $i = 0$ and e_J is changing according to Equations (11). Figures 3(b), 3(c), 3(d) are obtained for $c = 0.02, 0.04, 0.08$, respectively. The secular variation of e_J determines essentially the behaviour of e, and broadens the limits of its variation. For example, the innermost closed curve of Figure 3(a) evolves into the structure of Figure 3(b).

The value of e_J oscillates quasi-periodically between 0.025 and 0.061, the main period being 304000 yr. Since the proper period of the perihelion is about 3600 yr, during one revolution or libration of the perihelion e_J can be considered constant and the variation of e and ϖ can be described by curves similar to those of Figure 3(a). With the changing of e_J these curves change slowly and the structures

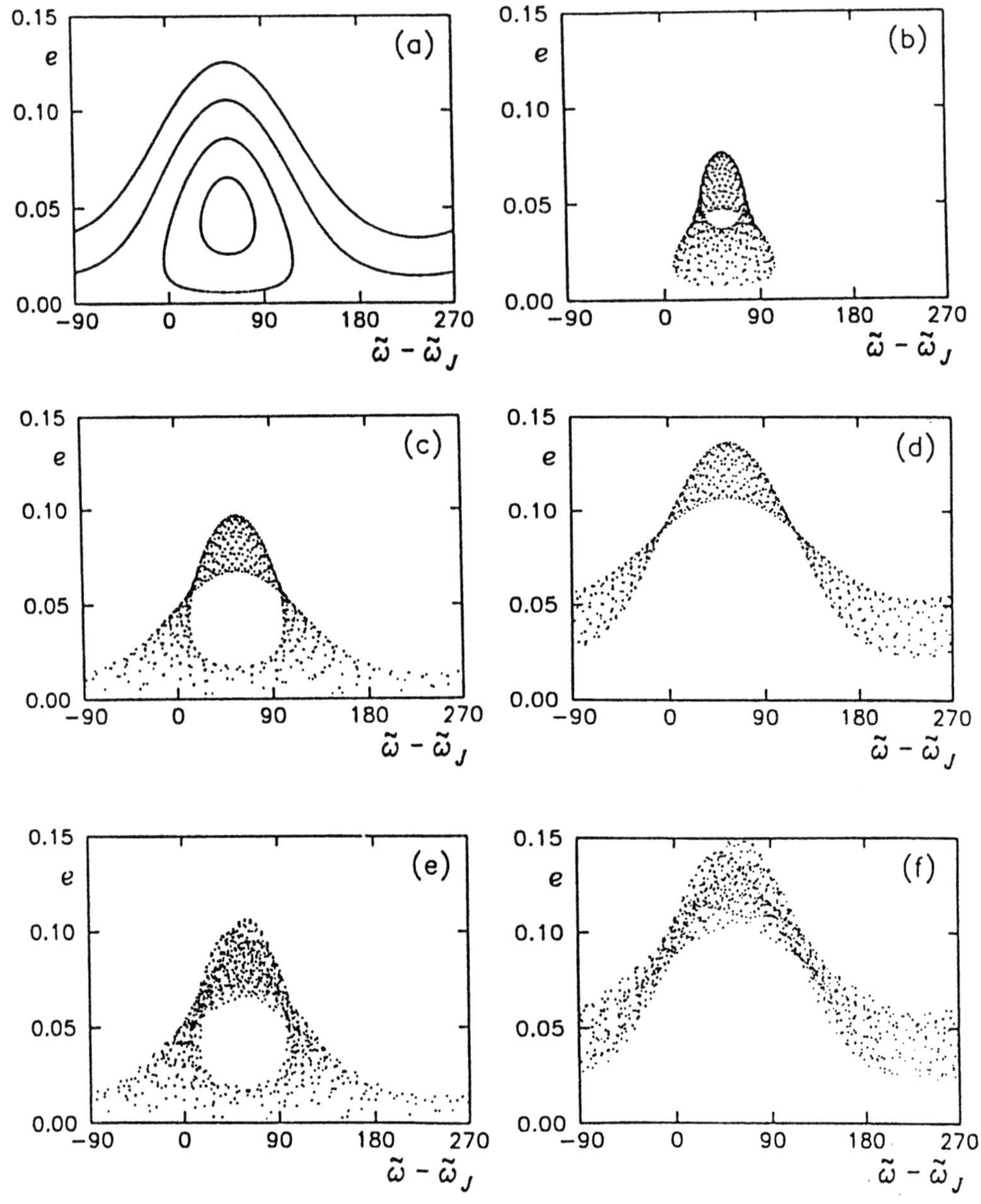

Fig. 3. Variation of e versus $\varpi - \varpi_J$. (a): zero inclination orbits, e_J constant; (b)-(d) zero inclination orbits, e_J secularly changing; (e)-(f): inclined orbits, e_J secularly changing.

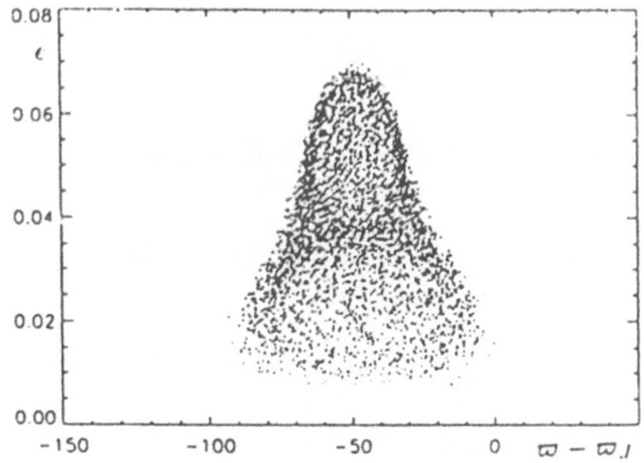

Fig. 4. The variation of e against $\varpi - \varpi_J$ in the case of (1871) Astianax.

of Figures 3(b)-3(d) are built up by these modified curves. In Figure 3(b) the perihelion librates, in Figure 3(d) it circulates. In Figure 3(c) the motion of the perihelion alternates between libration and circulation. This phenomenon is caused by the secular variation of Jupiter's orbit.

(c) *Spatial case, Jupiter's orbit secularly changing*

Figures 3(e), 3(f) are obtained with $i = 19^o$, and c=0.04, 0.08, respectively, e_J is changing according to Equations (11). The inclination dependent terms in Equations (10) bring in several long periods and the result is the broadening of the former structures of Figures 3(c)-3(d). The main long periodic perturbation corresponds to the frequency $2s$ (s is the frequency of the ascending node).

Figures 3(b)-3(f) offer a good description for the behaviour of the eccentricity and of the longitude of the perihelion of the real Trojans. Figures 4-6 show the variations of e against $\varpi - \varpi_J$ for (1871) Astianax, (2146) Stentor, and (1173) Anchises according to a numerical integration of the equations of motion for 1 Myr considering the perturbing effects of the four outer major planets. The good qualitative agreement with Figure 3 indicates that the indirect perturbations of the planets have the dominating effects on the Trojans' perihelion and eccentricity, since Figures 3(b)-3(f) have been obtained considering only the main indirect effects, while Figures 4-6 show the results of the full effect of the planets.

In the case when Jupiter's orbit is secularly precessing, the libration of the perihelion of the Trojans means that the asteroid's perihelion is locked to Jupiter's perihelion, the libration is about a direction differing from Jupiter's perihelion by 60°.

2.4. Proper Elements

The proper elements are important dynamical parameters of the asteroids which are almost constant for very long times. Bien and Schubart (1987) computed proper

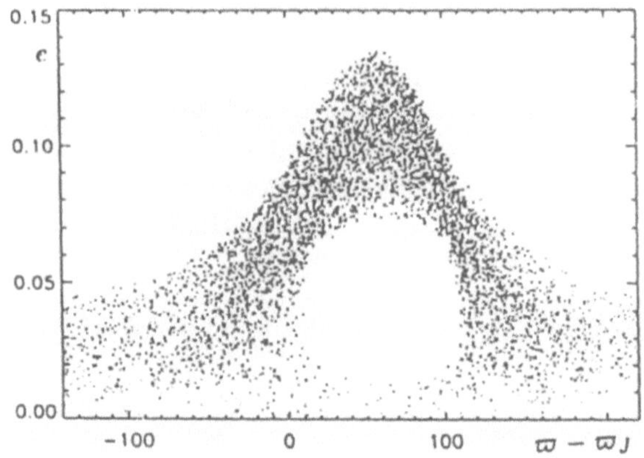

Fig. 5. The variation of e against $\varpi - \varpi_J$ in the case of (2146) Stentor.

elements for 40 Trojan asteroids based on numerical integrations for about 150000 yr. They introduced three proper elements, D, e_p, i_p, characterizing the amplitude of libration around L_4, and also the amplitude of free oscillations of the eccentricity and the inclination, respectively. This work was substantially extended by Milani (1993) who computed proper elements for 174 Trojans. He integrated the equations of motion for 1 million yr (in several critical cases for 5 million yr) considering the perturbations of the four outer major planets and referring the orbits to the barycenter of the inner solar system to account for most of the effect of the inner planets. He determined the stability range of the computed proper elements: in most of the cases the variation of da (the amplitude of the proper oscillation of the semi-major axis) was within the range 0.001 AU while the variation of e_p and $\sin i_p$ were smaller than 0.0015 (Milani, 1994).

The proper elements make it possible to search for families of asteroids. These are groupings of asteroids with very close proper elements. They can be considered as remnants of a parent body disrupted by collision. The identification of families among the Trojans is difficult for several reasons. One problem is that due to the dynamical properties of the Trojan belt, the fragments of an asteroid may leave the libration region after a collision and thus it is difficult to identify the members of the family. Another problem is that we do not know enough good orbits of small asteroids. A family typically consists of a larger fragment and several much smaller pieces. Among the known Trojans objects of larger size dominate, thus the lack of smaller objects makes it difficult to find the small members of the families. On the other hand, the number of the known Trojans is low for reliable statistical investigations.

On the basis of the proper elements computed by him, Milani (1993, 1994) confirmed the reliability of four families from among those proposed earlier (Shoemaker et al, 1989). These are the Menelaus family with 4-8 members and the Teucer family (4 to 6 members) at L_4, and the Sarpedon family (consisting of 3

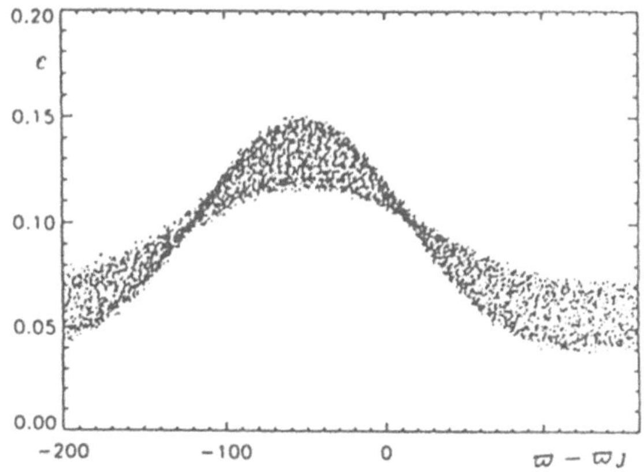

Fig. 6. The variation of e against $\varpi - \varpi_J$ in the case of (1173) Anchises.

very close members) at L_5. There is also a very close couple at L_4: the pair (1583) Antilochus-(3801) Thrasimedes. They are so close to each other that their relative velocity is only 10 m/s. The existence of Trojan families indicates that collisional evolution might have played a role among the Trojans in the same way as in the main belt of the asteroids.

2.5. Stability of the Trojans

The stability of the Trojan asteroids is a very difficult subject. Milani (1993, 1994) discussed the problems of understanding the connections between stability, chaos, and resonances of the Trojans. About 15% of the Trojans' orbits are chaotic in the sense that they have positive Liapunov exponents. However it is not easy to find the cause of the chaoticity. Since the frequencies of the known Trojans are far apart, no low order resonances are possible between them. There exist however secular resonances involving the nodes (Bien and Schubart, 1984, 1986; Nakai and Kinoshita, 1985). There are 5 Trojans near the $s = s_6$ resonance, and 3 near the $s = s_7$ resonance. The Liapunov exponents are positive in these cases, and the proper elements also show instabilities. The majority of the chaotic Trojans however do not show significant increase of instability in the proper elements. These asteroids belong to the class of stable chaos: asteroids with stable proper elements and positive Liapunov exponents.

The majority of the Trojans have stable proper elements and their behaviour can be considered quasi-periodic for millions of years. De la Barre *et al* (1996) found stable (hypothetical) Trojan orbits existing for 198 Myr. Though the motions in the vicinity of Jupiter's L_4, L_5 points seem to be stable for very long times, it is not clear where are and what are like the boundaries of the stability zones. In the plane $(da, \sin i_p)$ the distribution of the Trojans is bounded by the secular resonances $s = s_6$, $s = s_7$. However, the number of the known Trojans is too low to conclude from this that these boundaries are really the exact ones of dynamical stability.

Recent numerical investigations seem to indicate that the stability zones are not much wider than the regions occupied by the presently known Trojans. Levison *et al* (1991) studied the stability field in the plane of the proper eccentricity e_p and the libration amplitude D by integrating the orbits of 110 test particles for about 2 Myr. They found that with the time increasing the stability field shrinks in both e_p and D approaching the limits of the observed Trojan population. Holman and Wisdom (1993) integrated the orbits of about a thousand test particles for 20 Myr and found that the maximum libration amplitude in the stable region is 0.2 AU, only a little larger than for the known Trojans.

3. Motions around the L_4, L_5 Points of the Major Planets

3.1. Inner planets

Mikkola and Innanen (1992) studied the evolution of Trojan-type asteroids of the major planets. In the case of the inner planets they followed the evolution of a few test particles near the triangular Lagrangian points by integrating the equations of motion for 2.25 Myr considering the external perturbations of the other planets except Uranus and Neptune. The initial orbital elements corresponded to those of the orbit of the mother planet rotated by 60 degrees, except the initial semi-major axis which was varied in the range $\Delta a/a = 0 - 0.004$ (with respect to the planet's semi-major axis) with stepsize 0.001. In the case of *Mercury* all the orbits lost their Trojan-type character (tadpole or horseshoe) very rapidly, except the one starting from L_4. It followed a tadpole-like orbit with increasing amplitude, became a horseshoe orbit at 2.1 Myr, and since by this time its eccentricity reached the value 0.7, encounters with Venus became possible. It seems that Mercury's L_4 point is strongly unstable. In the case of *Venus* all the integrated orbits remained stable, preserved their Trojan-like character for the whole integration time-span with no essential evolution. Venus' triangular Lagrangian points may persist for very long times, possibly for a significant fraction of the age of the Solar System. In the case of *Earth* all the (planar) orbits remained stable with no essential variation in the orbital elements. Thus stable Trojan-type orbits may exist at Earth's triangular libration points for millions of years. Changing, however, the initial orbital inclination, exponential instabilities occur in the eccentricities for high inclination orbits ($i > 70°$) in a few thousand years due to close encounters with Venus (Mikkola and Innanen, 1990)

The study of the orbital behaviour of hypothetical *Mars*' Trojans is of utmost interest, since one Martian Trojan, (5261) Eureka, is already known. Eureka is a following Trojan at Mars' L_5 point. It has low eccentricity (0.065) and high inclination ($\approx 20°$). Mikkola and Innanen (1994) studied the dependence of the stability of motion on the orbital inclination. They integrated the equations of motion for more than 4 Myr considering the perturbations of the planets from Venus to Neptune and using the symplectic method of Wisdom and Holman (1991). The test particles were placed initially at L_5 and very close to it ($\Delta a/a = 0.00025$), and the inclination was varied between 0° and 80°. It turned out that long-term stability may exist only in two intervals of inclination (with respect to Jupiter's orbit), namely in $15^o \leq i \leq 30^o$, and in $32^o \leq i \leq 44^o$. For other values of i, the eccentricity grows above 0.3 and close encounters with Earth cause instabilities in the motion of the test particles. The reason for the increase of the eccentricity comes from several secular resonances. For low inclination orbits ($i < 12^o$) the perihelion of the asteroid precesses with nearly the same rate as Mars' perihelion. At $i \approx 12^o$, there is a 1:1 secular resonance between the rates $2\dot{\varpi} + \dot{\Omega}$ of the asteroid and $\dot{\varpi}_J$ of Jupiter. Finally, at $i \approx 32^o$ the orbits of the asteroid and Jupiter precess with the same rate.

Eureka is in the stable inclination window $15^o \leq i \leq 30^o$. Its orbital behaviour was studied by Mikkola *et al* (1994). They integrated the equations of motion for 2 Myr considering the perturbations of all the planets except Mercury. The numerical experiments showed no evidence of instability in the motion of Eureka for 2 Myr. The amplitude of libration in the difference between the mean longitudes of Eureka and Mars is about 5.7° (with small variations), the period changes between 1313-1383 yr in correlation with the variation of the inclination. The inclination varies between 17° and 28° (with respect to Mars' orbit) with an approximate period of 0.6 Myr. (With respect to Jupiter's orbit Eureka's inclination changes less, between about 21°-23°.) The eccentricity varies between 0.03-0.10 with a major period of 200000 yr. This is also the period of the perihelion which is presently aligned with that of Mars. The variations of the eccentricity are caused mainly by the other planets, and actually they seem to stabilize the variation of e at lower values than if only the perturbations of Mars existed.

3.2. Outer Planets

The stability of *Saturn*'s triangular Lagrangian points was investigated by Innanen and Mikkola (1989). They made an interesting discovery. Considering the perturbing effects of the outer major planets, they found that a test asteroid starting from Saturn's L_5 point became unstable in about 100000 yr. Changing, however, the initial semi-major axis by $1 - 2\%$, the resulting large amplitude orbits remained stable over 1 Myr. This behaviour was confirmed by Holman and Wisdom (1993) who found unstable "holes" at the centers of the stability regions around Saturn's L_4, L_5 points. Innanen and Mikkola (1989) pointed out that the instability of the orbits near the Lagrangian points is caused by the critical growth of the eccentricity above 0.15 due to the effect of the great inequality (GI), the major perturbation in the Jupiter-Saturn system due to the 5:2 near resonance between the mean motion of these planets.

The role of the GI in the instability of the orbits near the triangular Lagrangian points of Saturn was also discussed by de la Barre *et al* (1996). They carried out long-time numerical integrations for 350 hypothetical low-inclination ($i < 12^o$)

Saturnian Trojans, considering the perturbations of the outer major planets. They found that orbits with amplitudes of libration less than 80° (difference between the maximum and minimum value of the libration angle) become unstable in less than 1 Myr. Only orbits with large amplitudes can survive. Four large amplitude asteroids were still alive when the integration was stopped at 421 Myr. However, not all large amplitude orbits are stable, and thus this is not a sufficient condition for stability. The variations of the eccentricity play a fundamental role in the development of the instability. The orbits become unstable when e grows above 0.1. In some cases this happens only after a long time (in an example after 60 Myr), but when this limit is reached, instabilities occur.

De la Barre *et al* (1996) developed a Hamiltonian perturbation theory for the Saturnian Trojans, considering the perturbing effect of Jupiter. They determined the GI separatrices in the plane of the (rotated) h, k variables ($h = e \sin \varpi$, $k = e \cos \varpi$). The positions of the separatrices and of the saddle points strongly depend on the size of the libration orbits. For small libration orbits the separatrices are close to the origin and an asteroid with eccentricity 0.1 gets close to them. On the other hand, for large libration orbits the separatrices move outward from the centre and affect the asteroid only at large eccentricities ($e > 0.2$). Thus the GI has stronger influence on small tadpole orbits. Asteroids can keep their small eccentricities on orbits with large libration amplitude and with librating perihelion (with respect to Saturn's perihelion). The four long-lived Saturnian Trojans satisfied these conditions.

Mikkola and Innanen (1992) studied the evolution of orbits near the triangular Lagrangian points of *Uranus* and *Neptune* for 20 Myr. Changing the initial semi-major axis in the range $\Delta a/a = 0 - 0.025$ with stepsize 0.0025, they found stable tadpole orbits existing for the whole integration time-span. The libration takes place in the interval $30^o - 120^o$. The horseshoe orbits become unstable in a few million years. Some general conclusions can be drawn from the evolution of the unstable orbits. There are two possible ways of evolution. The asteroids either are ejected out of the Solar System, most often due to close encounters with Jupiter, or tend to populate orbits beyond Neptune's orbit. Their perihelion remains close to Neptune's orbit, but their semi-major axis and inclination tend to values resembling those of Pluto's orbit. They are often trapped into resonances (3:2, 3:1, 4:3 with Neptune) for millions of years. The asteroids ejected outside Neptune's orbit can serve as possible sources of comets.

Holman and Wisdom (1993) also found stability regions around the L_4, L_5 points of Uranus and Neptune. They used their symplectic mapping method (Wisdom and Holman, 1991) to integrate the orbits for 20 Myr considering the perturbations of the outer major planets. They integrated about 1000 orbits for each planet and obtained stability regions in the two-dimensional space of the initial semi-major axis and mean longitude, centered at the Lagrangian points. An interesting result is the asymmetry of the stability regions at the L_4 and L_5 points of Neptune.

Acknowledgements

I would like to express my thanks to Prof. R. Dvorak for sending me the results of his numerical integration on which Figures 4-6 are based. I thank also Prof K. Innanen and Dr. S. Mikkola for their suggestions on the paper. The support from the Hungarian Foundation for Scientific Research through grant OTKA T019409 is acknowledged.

References

Bien, R.: 1978, *Astron. Astrophys.* **68**, 295.
Bien, R.: 1980a, *Astron. Astrophys.* **81**, 255.
Bien, R.: 1980b, *Moon and Planets* **22**, 163.
Bien, R. and Schubart, J.: 1983, in V. V. Markellos, Y. Kozai, ed(s)., *Dynamical Trapping and Evolution in the Solar System*, D. Reidel Publ. Co., p. 153.
Bien, R. and Schubart, J.: 1984, *Celest. Mech.* **34**, 425.
Bien, R. and Schubart, J.: 1986, in C.-I. Lagerkvist, H. Rickman, ed(s)., *Asteroids, Comets, Meteors II*, Uppsala University, p. 157.
Bien, R. and Schubart, J.: 1987, *Astron. Astrophys.* **175**, 292.
de la Barre and Kaula, W.: 1995, in A. E. Roy, B. A. Steves, ed(s)., *From Newton to Chaos: Modern Techniques for Understanding and Coping with Chaos in N-body Dynamical Systems*, Plenum Press, New York, p. 379.
de la Barre, C. M., Kaula, W. M., and Varadi, F.: 1996, *Icarus* **121**, 88.
Chebotarev, G. A., Belyaev, N. A., and Eremenko, R. P.: 1974, in Y. Kozai, ed(s)., *The Stability of the Solar System and of Small Stellar Systems*, D. Reidel Publ. Co. Dordrecht, Holland, p. 63.
Érdi, B.: 1977, *Celest. Mech.* **15**, 367.
Érdi, B.: 1978, *Celest. Mech.* **18**, 141.
Érdi, B.: 1979, *Celest. Mech.* **20**, 59.
Érdi, B.: 1981, *Celest. Mech.* **24**, 377.
Érdi, B.: 1983a, *Celest. Mech.* **30**, 3.
Érdi, B.: 1983b, in Y. Kozai, V. V. Markellos, ed(s)., *Dynamical Trapping and Evolution in the Solar System*, D. Reidel Publ. Co., Dordrecht, Holland, p. 165.
Érdi, B.: 1984, *Celest. Mech.* **34**, 435.
Érdi, B.: 1988, *Celest. Mech.* **43**, 303.
Érdi, B.: 1995, in S. Ferraz-Mello, B. Morando and J. E. Arlot, ed(s)., *Dynamics, Ephemerides and Astrometry in the Solar System*, Kluwer Academic Publishers, Dordrecht, Holland, p. 171.
Érdi, B. and Presler, W. H.: 1980, *Astron. J.* **85**, 1670.
Érdi, B. and Váradi, F.: 1983, in C.-I. Lagerkvist, H. Rickman, ed(s)., *Asteroids, Comets, and Meteors*, Uppsala Universitet, p. 155.
Garfinkel, B.: 1977, *Astron. J.* **82**, 368.
Garfinkel, B.: 1978, *Celest. Mech.* **18**, 259.
Garfinkel, B.: 1980, *Celest. Mech.* **22**, 267.
Garfinkel, B.: 1983, *Celest. Mech.* **30**, 373.
Garfinkel, B.: 1985, *Celest. Mech.* **36**, 19.
Holman, M. J.and Wisdom, J.: 1993, *Astron. J.* **105**, 1987.
Holt, H. E. and Levy, D.: 1990, *IAU Circ.*, No. 5045.
Innanen, K. A. and Mikkola, S.: 1989, *Astron. J.* **97**, 900.
Kevorkian, J.: 1970, in G. E. O. Giacaglia, ed(s)., *Periodic Orbits, Stability and Resonances*, D. Reidel Publ. Co. Dordrecht, Holland, p. 286.
Levison, H. F., Shoemaker, E. M., and Wolfe, R. F.: 1991, *Lunar and Planetary Science* **XXII**, 803.
Mikkola, S. and Innanen, K.: 1990, *Astron. J.* **100**, 290.
Mikkola, S. and Innanen, K.: 1992, *Astron. J.* **104**, 1641.
Mikkola, S. and Innanen, K.: 1994, *Astron. J.* **107**, 1879.
Mikkola, S., Innanen, K., Muinonen, K., and Bowell, E.: 1994, *Celest. Mech.* **58**, 53.

Milani, A.: 1993, *Celest. Mech.* **57**, 59.
Milani, A.: 1994, in A. Milani et al., ed(s)., *Asteroids, Comets, Meteors 1993*, Kluwer Academic Publ., p. 159.
Nakai, H. and Kinoshita, H.: 1985, *Celest. Mech.* **36**, 391.
Nobili, A. M., Milani, A., and Carpino, A.: 1989, *Astron. Astrophys.* **210**, 313.
Schubart, J.: 1964, *SAO Spec. Rep.* **No. 149**, 1.
Schubart, J. and Bien, R.: 1984, *Celest. Mech.* **34**, 443.
Schubart, J. and Bien, R.: 1986, in C-I. Lagerkvist, H. Rickman, ed(s)., *Asteroids, Comets, Meteors II*, Uppsala Univ. Publ., p. 153.
Schubart, J. and Bien, R.: 1987, *Astron. Astrophys.* **175**, 299.
Shoemaker, E. M., Shoemaker, C. S., and Wolfe, R. F.: 1989, in R. P. Binzel, T. Gehrels, M. S. Mattews, ed(s)., *Asteroids II*, Arizona University Press, p. 487.
Szebehely, V.: 1967, *Theory of Orbits*, Academic Press, New York and London.
Varadi, F., de la Barre, C. M., Kaula, W. M., and Ghil, M.: 1995, *Celest.Mech.* **62**, 23.
Wisdom, J. and Holman, M.: 1991, *Astron. J.* **102**, 1528.
Zhang, S. P. and Innanen, K. A.: 1988a, *Astron. J.* **96**, 1983.
Zhang, S. P. and Innanen, K. A.: 1988b, *Astron. J.* **96**, 1989.
Zhang, S. P. and Innanen, K. A.: 1988c, *Astron. J.* **96**, 1995.
Zagretdinov, R. V.: 1986, *Kinematika i Fizika Nebesnih Tel* **2**, N 3, 68.

DYNAMICAL TRANSPORT TO PLANET-CROSSING ORBITS

CHRISTIANE FROESCHLÉ
Observatoire de la Côte d'Azur, B.P. 229
F-06304 Nice Cedex 4, France

1. Introduction

The study of small bodies on planet–crossing orbits brings out two main issues i.e. the identification of their birth place and how they got their current orbit.

Every day interplanetary bodies fall into the Earth atmosphere. In effect the total influx of interplanetary material for the entire Earth's surface is about 1.7×10^8 kg/year, over a wide mass range lying between 10^{-21} to 10^{15} kg (Ceplecha 1992). As shown by Ceplecha, the fluxes across this vast range cannot be expressed by a simple power law and this suggests many different intermixed populations of different origins.

In the following we will briefly recall what we know presently on the origins of the different populations of interplanetary bodies which fall on the Earth, considering size classes, then we will present the new results obtained in the last few years on the dynamical transport of NEAs.

2. The Different Populations Falling on the Earth's Surface

We may classify the interplanetary bodies according to their size:

The dust (formed of particles smaller than 1 mm) is the most important interplanetary component and is mainly of cometary origin, though asteroids release also dust through collisional processes. The dynamical evolution of such particles is very complex due to various dissipative effects (Poyinting Robertson drag, Yarkovsky effect etc...)

Another type of bodies falling on the surface of the Earth is the micrometeorites and meteorites which sizes are smaller than 10 m. Meteorites are rocks of diverse compositions and their collisions with the Earth are frequent enough (the impact scale lies between 1 day to 1 year) to allow human collection. Today there is an evidence that a tiny fraction of meteorites comes from Mars and the Moon; however in the last few years a lot of work has been done to compare asteroids and meteorites properties (see the review of Farinella et al. 1994a) concluding that most of meteorites are of asteroidal origin. Until now no recovered small meteorites have been identified to be of cometary origin. If it is very rare that the fall of such bodies hurts humans, many damages on animals and cars have been registered.

The fall of larger interplanetary bodies also occurs with a frequency F - according to the typical size distribution of interplanetary complex - roughly proportional to the inverse of the square of the chosen threshold diameter D. There are first the

Celestial Mechanics and Dynamical Astronomy **65**: 165–173, 1997.

SEAs (Small Earth–Asteroids) and bodies of sizes between 10 to 100 m which may also impact our planet, with a frequency - according to the typical size distributions of interplanetary complex - roughly proportional to the -2 power of the chosen threshold diameter. Following the relation $F \approx D^{-2}$ the typical interval between two impacts for a 50 meter–size body, like the one which felt in Tunguska in 1908 producing an explosion equivalent to 10 Mton of TNT, is of the order of a few centuries. Until the recent discoveries of several such bodies by the spacewatch Survey (Rabinowitz 1991) the statistics of such bodies were not known. Rabinowitz (1993) showed that there is an excess of Earth–approaching asteroids with diameter less than 50 m, relative to the population inferred from the distribution of larger objects. Furthermore Rabinowitz et al. (1993) argued that the SEAs characterized by low eccentricities ($e < 0.35$) and unusual spectral properties, form a previously undetected asteroid belt concentrated near the Earth. Their composition is not known and many theories on their origin have been proposed: they could be the fragments of lunar and planetary impact ejecta (Gladman et al. 1995, Bottke et al. 1996), they may also be the resulting fragments of cometary or asteroidal break up (Rabinowitz 1993, Rabinowitz and Wetherill 1995).

Finally the largest bodies included in the interplanetary complex are the NEAs (Near Earth Asteroids) of diameter $D \geq 100$ m. Fortunately for civilization and life, their impact time scale is large, since - as for the SEAs - it is proportional to D^{-2}. Thus, it turns out that for a body of 10 km–size the impact time scale is of the order of 10^8 years; such an impact would cause a global climatic disaster such as that recorded at the K/T boundary.

At the present time, about 250 NEAs orbits are known, classified into three groups: the Apollos which perihelion distance q is smaller than 1.017 AU and semi–major axis a is larger than 1 AU; the Amors characterized by $1.017 < q \leq 1.3$ AU and the Atens which have orbits interior to that of the Earth (a<1 AU and Q >0.983 AU). The physical properties of the AAAO (Apollos–Amor–Aten Objects) population show that most of them come from the main asteroid belt (Macfadden et al. 1989), since almost all the taxonomic types found in the asteroid belt are present in the NEA population. However it is not excluded that some NEAs may be of cometary origin. Indeed recently the asteroid (4015) 1979 VA has been shown to be the same object as comet Wilson Harrington, and the association of (3200) Phaeton with the orbit of the Gemenid meteor shower raises the question that this asteroid is an extinct cometary nucleus.

3. Dynamical Transport Mechanisms.

Since many meteorites and NEAs have been recognized to originate in the main belt, we have now to find how they have been transported on planet–crossing orbits. When collisions occur between asteroids, fragments are usually ejected at velocities of the order of 100 m / s, far too small for causing the drastic orbital changes needed to reach planet–crossing orbits. Also, the interplay between collisions and subtle dynamical effects is necessary to transport material from the asteroid belt to the Earth. In the recent years, significant work has been done on these problems and we will describe the main dynamical transports which all involve resonances.

During the 70s and the 80s, Williams (1973), Wetherill and Williams (1979) suggested that secular resonances may be efficient to transport material near the Earth. Wisdom (1983), in the frame of the elliptic restricted 3–body problem, pointed out the existence of a chaotic zone in the 3/1 mean motion resonance with Jupiter which causes random jumps of the eccentricity up to 0.35 causing Mars–grazing orbits. Using a similar hypothesis ($e_J \neq 0$), Yoshikawa (1989, 1990, 1991) showed that in some other mean motion resonances like for instance the 5/2 and 4/1 mean motion commensurabilities with Jupiter, the eccentricity of small bodies may be pumped up to a Mars–crossing value. Then close approaches due to Mars may change the semimajor axis of those small bodies, taking them away from the resonances and even transporting them to the Earth. Thus, it was shown that mean motion resonances could be a mechanism for transporting material to the inner planets. However, the time scales are very long (about 100 Myrs) since the probability to have an effective encounter with Mars is very small.

In the following we will describe in more details the transport mechanisms related to each type of resonances i.e. *secular resonances* and *mean motion resonances*. We will mainly present numerical integrations in order to illustrate the main features of dynamical transport.

3.1. THE SECULAR RESONANCES

In the asteroid belt, among the three strongest secular resonances: ν_5, ν_6 (which occur when the asteroid's perihelion longitude is precessing at the same rate as that of Jupiter and Saturn, respectively) and ν_{16} (which involves the precession of the nodes), the ν_6 is the most effective as far as material transport is concerned, since it can pump up the eccentricity almost to unity, while the resonance ν_5 increases the eccentricity only to values producing Mars–grazing or Mars–crossing orbits and ν_{16} may increase only the inclination.

The dynamics of fictitious fragments of real asteroids injected into the ν_6 resonance have been studied both by purely numerical experiments (Farinella et al. 1993a, b) and semi–analytical theory (Morbidelli 1993, Morbidelli et al. 1994). It has been shown that the ν_6 resonance is a potential effective channel for fragment collection and delivery of meteorites, not only near the inner edge of the belt but also for several asteroids on orbits with moderate inclinations (15° to 20°) at semimajor axes of about 2.4 to 2.7 AU. The time scale for the fragments to be transported to a planet–crossing orbit is of the order of 1 Myr.

Since the ν_6 secular resonance is an efficient mechanism of production of planet–crossers, it can be expected that a fraction of NEAs are in/near the ν_6 resonance. Froeschlé et al. (1995) found that 19 over a sample of 181 NEAs are presently connected to the resonance. These 19 bodies were integrated numerically (taking

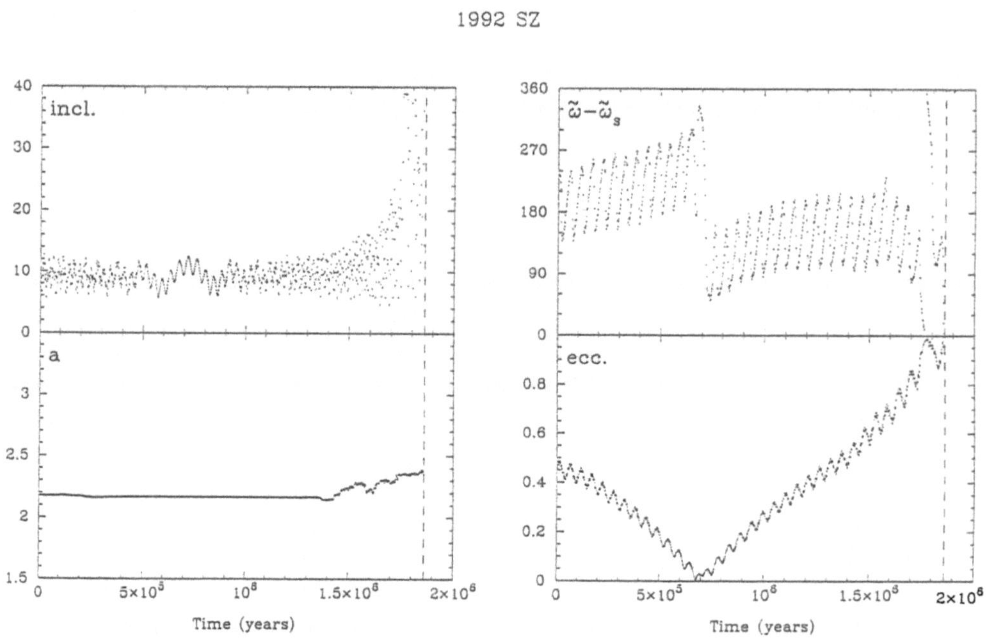

Fig. 1. Orbital evolution in the time span $0 \leq t \leq 2$ Myr of the asteroid 1992 SZ.The figure shows the semimajor axis in AU, eccentricity e, inclination i (in degrees) and the critical argument $\varpi - \varpi_S$ of the ν_6 secular resonance (in degrees) versus time. The eccentricity e is pumped up regularly almost to 1., due to the ν_6 resonance and the asteroid falls into the Sun at $t = 1.86$ Myr (vertical dashed line, and the integration has been interrupted.)

into account the perturbations of the planets from Venus to Neptune) over at least 2 Myrs. They found that 16 of them have minimal eccentricity of 0.2, which is a typical value of the main belt population. This result confirms that these bodies are fragments of main belt asteroids which fall in the ν_6 transport track. Moreover 5 over the 19 bodies fall into the Sun due to the increase of their eccentricity almost to unity, so that their perihelion distance became smaller than the Solar radius (Fig. 1). Valsecchi et al. (1995) have pointed out the important role of the ν_6 resonance on the dynamical behaviour of asteroids having orbits resembling that of the Taurid meteoroid complex and of the comet P/Encke. On one hand the ν_6 resonance depletes the population in the Taurid region by pumping up the eccentricity almost to unity: 12 over 21 bodies (including P/Encke) finally fall into the Sun in a few millions of years. On the other hand the ν_6 resonance refills the Taurid region by transporting both main belt asteroids and comets from Jupiter family.

Jopek et al. (1995) analyzed the long term dynamics of 17 brightest bolids. The orbits were integrated backwards in time over a time span of at least 1 Myr. The results show that the ν_6 secular resonance affects also the dynamics of meter–size

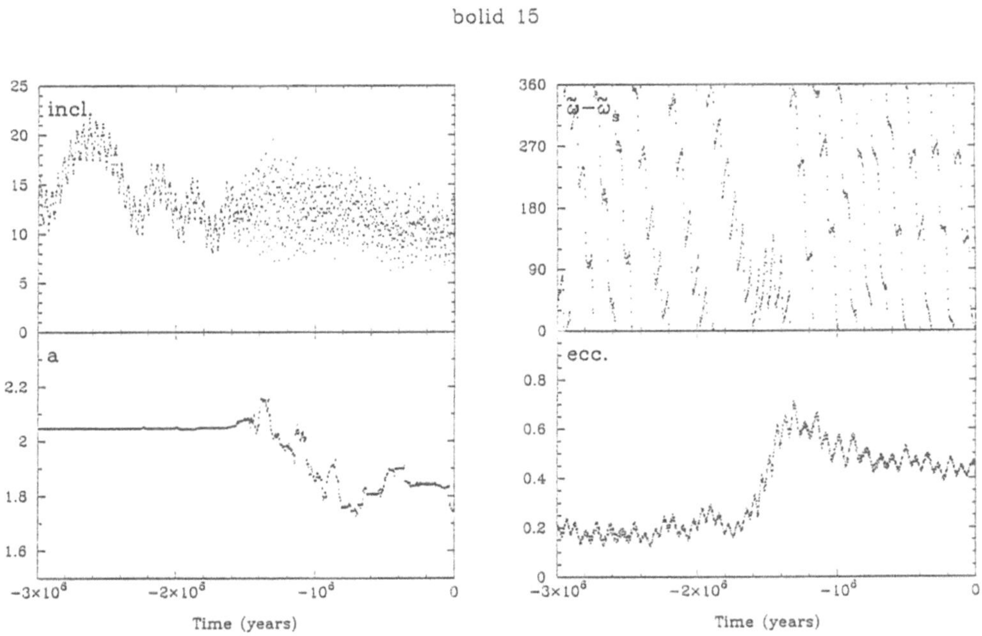

Fig. 2. Same as Fig. 1 but for bolid 15 (associated to the Lost City meteorite fall), in the time span -3 Myr $\leq t \leq 0$. The body is injected in the ν_6 resonance due to close approaches with Mars. In the interval $-1.6 \leq t \leq -1.3$ Myr the resonance ν_6 pumps up the eccentricity to about 0.6 and it becomes Earth–crosser.

bodies. For example the bolid associated to the Lost City meteor fall (Fig. 2) was an object located at the edge of the inner belt 3 Myrs ago with an eccentricity e around 0.2. Then due to close approaches with Mars it entered the ν_6 resonance. Consequently e was pumped up to 0.6 and the bolid became Earth–Crosser.

3.2. The Mean Motion Resonances

As already pointed out, the mean motion resonances are also efficient mechanisms to produce planet–crossing orbits, (see Moons this volume). While in the frame of the 3–body restricted elliptic problem it has been shown (Wisdom 1983, Yoshikawa 1989, 1990, 1991) that the eccentricity of bodies located in mean motion resonances with Jupiter is pumped up to only Mars–crossing orbits, the situation changes drastically when one considers more realistic models. Numerical integrations taking into account the four giant planets along with their orbital precession rates (Wisdom 1986) show that the maximal value of the orbital eccentricity of the small body is increased and its orbit becomes strongly chaotic. In effect Morbidelli and Moons (1993), Moons and Morbidelli (1995), Moons (this volume) have shown by semi–analytical methods, that large chaotic regions are present in almost all the mean motions resonances. These chaotic zones are due to the overlapping of secular

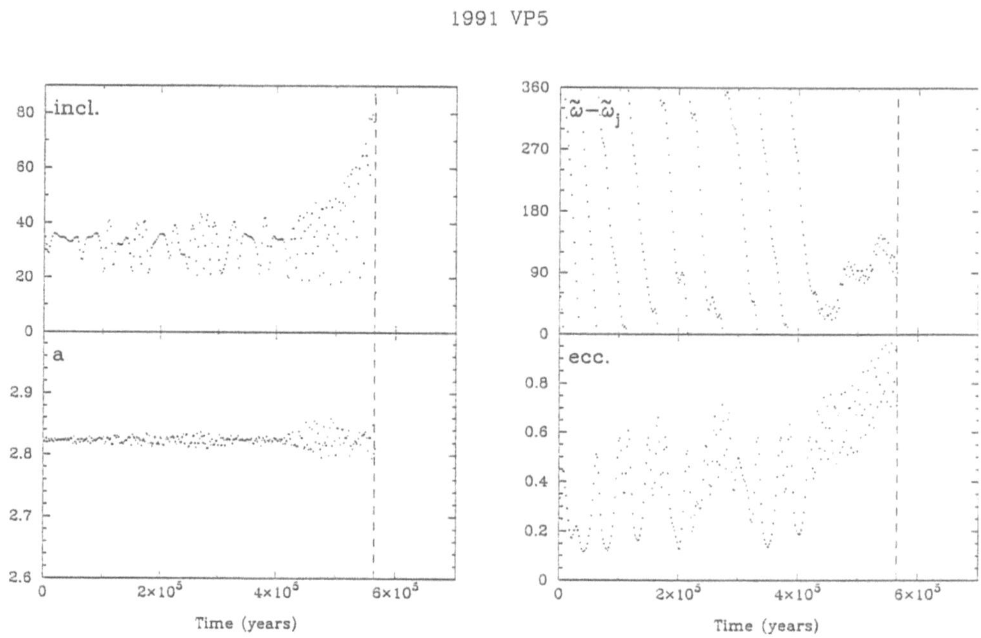

Fig. 3. Orbital evolution of the asteroid 1991 VP5. The figure shows the three osculating elements: semimajor axis a (in AU), the eccentricity e, inclination i (in degrees), and the critical argument $\varpi - \varpi_j$ (in degrees) of the secular resonance ν_5 versus time. When the body is in both resonances 5:2 and ν_5 the eccentricity is pumped up almost to unity, the integration has been interrupted when the perihelion distance reaches 1 solar radius (dashed vertical line)

resonances inside the mean motion ones; the chaotic regions may extend to all eccentricities so that e may jump in a very irregular manner to Earth–crossing values and may even increase up to almost unity. Farinella et al. (1993) pointed out that the 3/1 resonance is an efficient mechanism of fragment generation and of transport of meteorites as is the ν_6 resonance. Morbidelli et al. (1994) showed that a large fraction of fictitious fragments of asteroids located near the 3/1 and 5/2 mean motion resonances and injected in these resonances becomes planet–crossers. .

An example of transport mechanism due to a mean motion resonance is given in Fig. 3 for the asteroid 1991 VP5. This asteroid (integrated by Farinella et al. 1994) collides with the Sun near $t = 5.7 \times 10^5$ years. Until $t = 4 \times 10^5$ years, the asteroid is only locked in the 5/2 resonance: the semimajor axis a oscillates around the value $a = 2.82$ AU of the resonance and the eccentricity increases slowly. But when the asteroid enters also in the ν_5 secular resonance (the critical argument of this resonançe, $\varpi - \varpi_j$, librates), the eccentricity is pumped up rapidly almost to unity. Other asteroids affected by mean motion and secular resonances were found to hit the Sun (see Farinella et al. 1994).

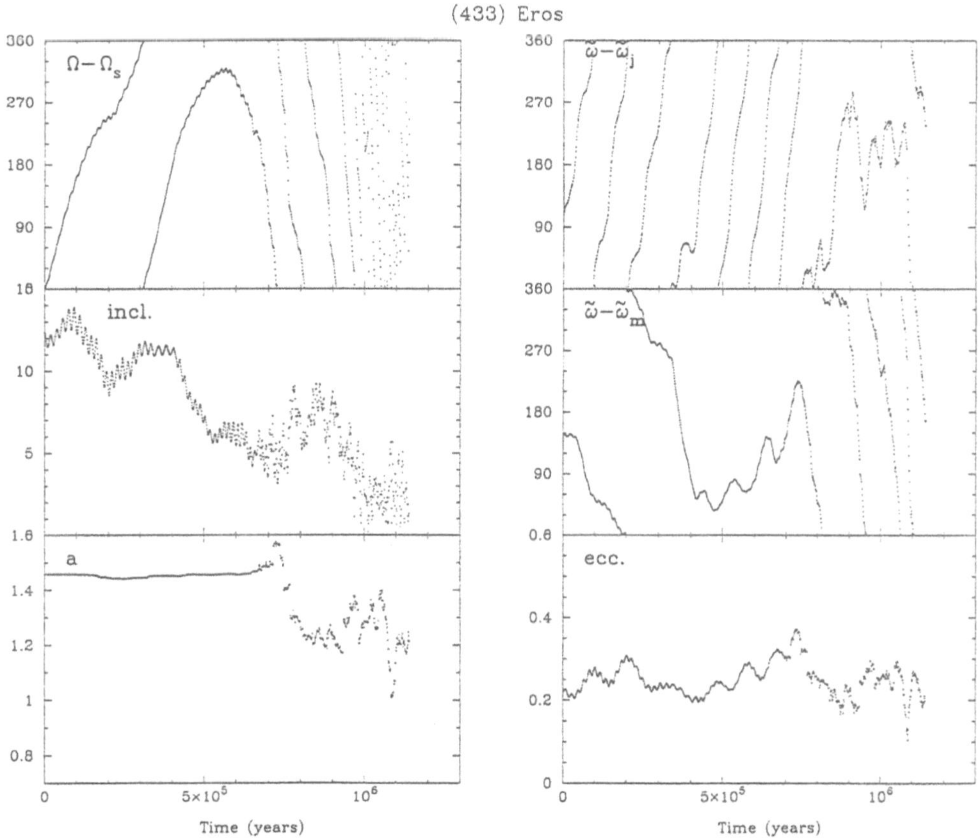

Fig. 4. Orbital evolution of the asteroid (433) Eros. The figure shows the three osculating elements: semimajor axis a (in AU), eccentricity e, and inclination i (in degrees), and the critical arguments: $\Omega - \Omega_s$, $\varpi - \varpi_m$, $\varpi - \varpi_j$ of respectively the secular resonances ν_{16}, ν_4, ν_5 versus time. When the asteroid is located in the ν_{16} resonance the inclination decreases, increasing the probability of close encounters. Then the body is injected in the ν_4 resonance and the eccentricity increases up to 0.33, the body gets into the region strongly dominated by Earth close encounters and finally hits the Earth at $t = 1.14$ Myr.

4. Conclusions.

The large amount of numerical and theoretical works done in the recent years has shown that mainly the secular resonance ν_6 and the mean motion resonances with Jupiter (*not only the 3/1 one*) are efficient mechanisms of transport of material from the main belt to the inner planets. The results of Morbidelli and Moons (1993), Moons and Morbidelli (1995), have shown that secular resonances exist also inside the mean motion ones and that such overlapping of resonances produces large chaotic zones and then explains almost all the Kirkwood gaps except the one associated to 2/1 resonance. Nevertheless, until now, no systematic dynamical study has been performed in the region of the inner planets i.e. for $a \leq 2$ AU, where most NEAs evolve. A new subject of research is the role of secular resonances not only with the giant planets but with the inner planets on the dynamical evolution of the NEAs. Indeed, in this region, the close encounters with the planets are numerous and important. However, numerical studies performed by Froeschlé et al. (1995) have shown that the ν_5 and ν_{16} resonances are also effective in the region of $a < 2$ AU. Moreover Michel et al. (1996a, 1996b) have found by numerical integrations that the combined effects of the secular resonances ν_{16} and ν_4 (secular perihelion resonance with Mars) may produce a collision of the asteroid Eros with the Earth (Fig. 4). Recently, Michel and Froeschlé (1996) developed a semi–analytical theory (taking into account the possible close encounters with the planets) allowing to locate the secular resonances in the region of $a < 2$ AU. The results show the location of the perihelion secular resonances with Venus ν_2, the Earth ν_3 and Mars ν_4 and of the nodal secular resonances with the same planets namely ν_{12}, ν_{13} and ν_{14}. All these resonances are present in this region. Beside, as the frequencies, on one hand g_3 and g_4 and on the other hand s_3 and s_4 have values which differ very little, the secular resonances ν_3 and ν_4, ν_{13} and ν_{14} are very close to each other. The results show also that the main secular resonances ν_5, ν_6 and ν_{16} appear in the region $a < 2$ AU and that some overlappings of resonances occur which could lead to chaotic regions. Further studies have to be done concerning the region $a < 2$ AU in order to estimate the effects of the secular resonances on the dynamics of small bodies, and also to detect chaotic regions. Knowing more precisely the dynamical mechanisms at work in this region will give a better understanding of the dynamical evolution of NEAs and especially of the SEAs for which the origin is still not known.

References

Bottke, W. F., Nolan, M. C., Melosh, H. J., Vickery, A. M., and Greenberg, R.: 1996, *Icarus*, **122**, 406–427..

Ceplecha, Z.: 1992, *Astron.& Astrophys* **263**, 313–366.

Farinella, P., Gonczi, R., Froeschlé, Ch. and Froeschlé, C. : 1993a, *Icarus* **101**, 174–187.

Farinella, P., Gonczi, R. and Froeschlé, Ch.: 1993b, *Cel. mech & Dyn. Astron.* **56**, 287–305.

Farinella, P., Froeschlé, C. and Gonczi, R.: 1994a. *ACM 1993*, Kluwer Academic Publishers.

Farinella, P., Froeschlé, Ch., Froeschlé, C., Gonczi, R., Hahn, G., Morbidelli, A. and Valsecchi, G. B.: 1994b, *Nature* **371**, 314–317.

Froeschlé, Ch., Hahn, G., Gonczi, R., Morbidelli, A. and Farinella, P.: 1995, *Icarus* **117**, 45–61.

Gladman, B., Burns, J. A., Duncan, M. J., and Levison, H.: 1995, *Icarus* **118**, 302–321.

Jopek, T. J., Farinella, P., Froeschlé, Ch. and Gonczi, R.: 1995, *Astron & Astrophys*. **302**, 290–300.

MacFadden, L. A., Tholen, D. J. and Veeder, G. J.: 1989, in *Asteroids II*, the university of Arizona Press, Tucson.

Michel, P., Farinella, P. and Froeschlé, Ch.: 1996a, *Nature* **380**, 689–691.

Michel, P., Froeschlé, Ch. and Farinella, P.:1996b, *Astron. & Astrophys*., **313**, 993–1007..

Michel, P. and Froeschlé, Ch.: 1996, *Icarus*, in press.

Moons, M. and Morbidelli, A.: 1995, *Icarus* **114**, 33–50.

Morbidelli, A.: 1993, *Icarus* **105**, 48–50.

Morbidelli, A. and Moons, M.: 1993, *Icarus* **102**, 316–332.

Morbidelli, A., Gonczi, R., Froeschlé, Ch. and Farinella, P.: 1994, *Astron. & Astrophys*. **282**, 955–979.

Rabinowitz, D. L.: 1991, *Astron. J.* **101**, 1518–1529.
Rabinowitz, D. L.: 1993, *Astrophys. J.* **407**, 412–427.
Rabinowitz, D. L., Gerhels, T., Scotti, J. V., McMillan, R. S., Perry, M. L., Wisniewski, W., Larson, S. M., Howell, E. S., and Muller, B. E. A.: 1993, *Nature* **363**, 704–706.
Rabinowitz, D. L., and Wetherill, G. W.: 1995, (preprint).
Valsecchi, G. B., Morbidelli, A., Gonczi, G., Farinella, P., Froeschlé, Ch. and Froeschlé, C.: 1995, *Icarus* **118**, 169–180.
Wetherill, G. W. and Williams, J. G.: 1979, in *Origin and distribution of Elements*, Pergamon Press, Oxford.
Williams, J. G.: 1973, *Eos* **54**, 233.
Wisdom, J.: 1983, *Icarus* **56**, 51–74.
Wisdom, J.: 1985, *Nature* **315**, 731–733.
Yoshikawa, M.: 1989, *Astron. & Astrophys.* **213**, 436–458.
Yoshikawa, M.: 1990, *Icarus* **87**, 78–102.
Yoshikawa, M.: 1991, *Icarus* **92**, 94–117.

REVIEW OF THE DYNAMICS IN THE KIRKWOOD GAPS

MICHÈLE MOONS
Département de mathématique FUNDP
8, Rempart de la Vierge, B-5000 Namur, Belgique

Abstract. The study of mean motion resonance dynamics was motivated by the search for an explanation for the puzzling problem of the Kirkwood gaps. The most important contributions in this field within the last 32 years are reviewed here. At the beginning of that period, which coïncides with the first long-term numerical investigations of resonant motion, different hypotheses (collisional, gravitational, statistical and cosmological) to explain the origin of the gaps were still competing with each other. At present, a general theory, based on gravitational mechanisms only, is capable of explaining in a uniform way all the Kirkwood gaps except the 2/1 one. Indeed, in the 4/1, 3/1, 5/2 and 7/3 mean motion commensurabilities, the overlap of secular resonances leads to almost overall chaos where asteroids undergo large and wild variations in their orbital elements. Such asteroids, if not thrown directly into the Sun, are sooner or later subject to strong close encounters with the largest inner planets, the typical time scale of the whole process being of the order of a few million years. Unfortunately, this mechanism is not capable of explaining the 2/1 gap where the strong chaos produced by the overlapping secular resonances does not attain orbits with moderate eccentricity, of low inclination and with low to moderate amplitude of libration. In the light of the most recent studies, it appears that the 2/1 gap is the global consequence of slow diffusive processes. At present, the origin of these processes remains under study.

Key words: Asteroids, resonances, Kirkwood gaps

1. Introduction

The problem of the Kirkwood gaps is a 130 year-old riddle in astronomy. Its origin lies in the discovery by Kirkwood (1867) of gaps in the asteroidal distribution, located at certain values of the semi-major axis. The histogram of the number of asteroids as function of the semi-major axis value shows clearly a lack of asteroids in concordance with the 4/1, 3/1, 5/2, 7/3 and 2/1 mean motion commensurabilities with Jupiter (Figure 1). It is worth mentioning that, in contrast, the 3/2 and 1/1 mean motion commensurabilities host many asteroids.

Since the discovery of this peculiarity of the asteroidal distribution, many researchers in astronomy have, at some time or another, tried to understand the mechanisms at work. In the early searches for a description of the asteroidal motion in resonances, famous names can be found like that of Hill(1902a, 1902b), Poincaré (1902a, 1902b) or Andoyer (1903). A detailed list of the pioneer works in this field was provided by Hagihara (1961) and an early attempt to explain the gaps on a statistical basis was provided by Brouwer (1963).

In this review paper, attention is focused on the advances in the understanding of the Kirkwood gaps problem in the last 32 years, namely since 1964. Due to the important number of papers published on this subject within that period, it will be impossible to provide an extensive review citing all contributions. For that reason, the emphasis will be put on the works which can be considered, more or less objectively, as the most important ones in the light of the actual state of the art in the Kirkwood gaps problem. Readers who are interested in a complete

Celestial Mechanics and Dynamical Astronomy **65**: 175–204, 1997.

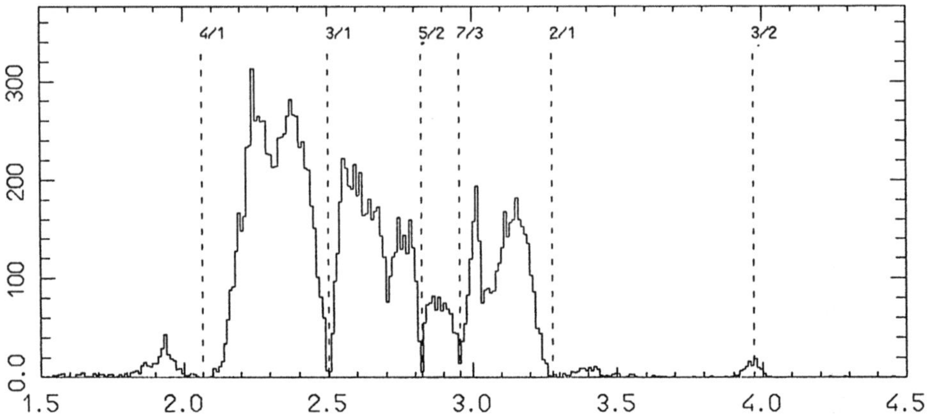

Fig. 1. Histogram of the number of asteroids as function of the semi-major axis (in AU).

coverage of the subject can find more detailed reviews in (Greenberg and Scholl, 1979; Henrard, 1988; Froeschlé and Greenberg, 1989; Ferraz–Mello, 1994b). The different hypotheses (collisional, gravitational, statistical and cosmogonical) will not be presented here; they are discussed in detail in (Scholl, 1979). Moreover, although a (at least partial) cosmogonical explanation of the gaps cannot be rejected, attention will be focused on works which do not consider the early stages of the Solar System formation. Finally, most of the results presented here were obtained within the framework of averaged models. The validity of the averaging principle will thus be taken as a starting hypothesis; a discussion and a list of references dealing with this subject can be found in (Wisdom, 1982).

The paper is organized as follows. Section 2 is devoted to the period ranging from 1964 to 1981. This period is dominated by works based on the integration of numerically averaged equations of motion (without expansion of the disturbing function) in the framework of the elliptic restricted three-body problem, mainly the work of Giffen (1973) and that of Scholl and Froeschlé (1974, 1975) or Froeschlé and Scholl (1976). Section 3 is devoted to the period ranging from 1982 to 1987, which will be called "Wisdom's era". In 1982, Wisdom constructed, for the 3/1 resonance, a mapping of the elliptic restricted three-body problem which allowed the integration time scale to be significantly extended. With his mapping, he showed that, when the third dimension is taken into account, most of the trajectories of fictitious asteroids in the 3/1 gap spend at least 50% of the time being Mars crossers and eventually also become Earth crossers, providing a simple gravitational explanation for the formation of the 3/1 gap. Although it has now been proved (see Section 4) that much more powerful mechanisms can explain, in a uniform way, not only the 3/1 gap but also the 4/1, 5/2 and 7/3 ones, the name of Wisdom remains associated with the Kirkwood gaps explanation as he definitely stated the relevance of the gravitational hypothesis. Besides the work by Wisdom (1982, 1983, 1985b,

1987), this period is also marked by the works of Murray (1986) and of Henrard and Lemaitre (1986, 1987). Unfortunately, all of them used truncated expansions of the disturbing function, which drastically limited the validity of the results to low eccentricities. Finally, Section 4 is devoted to the period ranging from 1988 to 1996. This period, and especially the end of it, is marked by a renewed interest in the Kirkwood gaps problem. Additional effects have been taken into account, such as the secular resonances which occur when the frequency of the longitude of perihelion (resp. of the node) of the asteroid is commensurable with the frequency of the longitude of perihelion (resp. of the node) of a planet. Moreover, the theories again avoided the usual expansion of the disturbing function in order to give accurate coverage of the full range of eccentricity. To select the most interesting works in that period is a very hard task as the problem is not yet completely understood. However, we would draw attention to the works of Yoshikawa (1989), Ferraz–Mello and Klafke (1991), Morbidelli and Moons (1993), Franklin (1994), Ferraz–Mello (1994c), Moons and Morbidelli (1995), Henrard, Watanabe and Moons (1995) and of Morbidelli (1996).

2. Before Wisdom

In 1964, Schubart presented a series of typical phase space portraits of several mean motion resonances in the framework of the averaged planar-circular restricted three-body problem (R3BP) Sun–Jupiter–asteroid. His work was based on an extension of Poincaré's formalism (Poincaré, 1902b) and used a numerical averaging procedure (Schwarzschild, 1903) for the computation of the secular and long-period resonant effects of the disturbing function. Schubart's procedure (Schubart, 1964) has a double advantage: the use of Poincaré's variables avoids the singularity at $e = 0$ and the numerical averaging procedure avoids the expansion of the averaged disturbing function in power series of the eccentricity. His method is thus valid and accurate in the whole eccentricity range of the small particle and the only limitation lies in the fact that the averaging procedure breaks down when the asteroid comes too close to Jupiter (typically, when its distance to Jupiter becomes less than 1 AU). Schubart's study of the circular averaged R3BP confirmed that asteroids in mean motion resonances present large variations of their averaged semi-major axis (or, equivalently, of their averaged mean motion), but contributed to invalidate the hypothesis that these variations lead to strong close encounters with Jupiter. The resonant mechanism acts as a protection from close encounters with Jupiter for all asteroids which are in a stable configuration, since the configuration is preserved due to the resonance. Schubart's results were confirmed by Message (1966) who made a detailed study of the 2/1 resonance by a completely different approach.

Before proceeding further in the review of the different mechanisms acting to deplete the gaps and their chronological discovery, a quick look at the mathematical description of the problem is necessary.

The behaviour of particles in mean motion resonances can be studied with the Hamiltonian of the restricted three-body problem

$$\mathcal{H} = L' - \frac{1-\mu}{2a} - \mu\left(\frac{1}{|\mathbf{r}-\mathbf{r}'|} - \frac{\mathbf{r}|\mathbf{r}'}{r'^3}\right)$$

where $\mathbf{r}$ is the heliocentric position vector of the asteroid, $\mathbf{r}'$ that of Jupiter, μ the mass of Jupiter and L' the conjugate momentum to the mean longitude of Jupiter (the universal gravitational constant, the semi-major axis of Jupiter and the total mass of the Sun-Jupiter system are chosen as units). The number of degrees of freedom of the above Hamiltonian varies, according to the adopted representation of Jupiter's orbit and the number of dimensions taken into account. However, before any simplification, it is at least two in the simplest model (planar-circular problem). In the study of the long-term behaviour of asteroids, the first step consists generally in the elimination of the short-period terms associated with the orbital motion of Jupiter. After this elimination, and assuming Jupiter to be on a fixed elliptic orbit, the geometry of the averaged R3BP in the case of a $(p+q)/p$ mean motion resonance can be described with the use of the canonical variables

$$\begin{aligned} \sigma &= \frac{p+q}{q}\lambda' - \frac{p}{q}\lambda - \varpi \quad , & S &= L - G \quad , \\ \sigma_z &= \frac{p+q}{q}\lambda' - \frac{p}{q}\lambda - \Omega \quad , & S_z &= G - H \quad , \\ -\nu &= \frac{p+q}{q}\lambda' - \frac{p}{q}\lambda - \varpi' \quad , & N &= \frac{p+q}{p}L - H \ , \end{aligned} \tag{1}$$

where $\lambda, \varpi = \omega + \Omega, \Omega, \lambda'$ and ϖ' are the usual averaged orbital elements (primed quantities refer to Jupiter) and L, G, H are Delaunay's momenta. Schubart (1964) and Message (1966) studied the behaviour of asteroids within the framework of the planar-circular problem. In this case, the averaged Hamiltonian, denoted here by $\mathcal{H}_0(\sigma, S, N)$, is integrable. As it is independent of σ_z, S_z and ν, the particle evolves on one of the lines $N = C^t$, which are almost horizontal in the (a, e) representative plane (Figure 2): in the planar-circular approximation, the semi-major axis of resonant asteroids can present large variations, while the variations in eccentricity are always relatively small. In the planar-elliptic problem, the long-term behaviour is described by a two-degree-of-freedom Hamiltonian $\mathcal{H}(\sigma, S, \nu, N)$ and the action N is no longer constant. Most of the time, the time scales associated with the libration in σ and with the averaged motion of -ν (or, equivalently, the motion of $\varpi - \varpi'$) are well separated, the first being typically a few hundred years while the second is at least several thousand years. Therefore, a description of the long-period dynamics can be obtained by further averaging over the shortest time scale (i.e. the period of libration of σ), and the Hamiltonian of the secular problem, here denoted by $\mathcal{K}(J, \langle\nu\rangle, N)$, is again reduced to one degree-of-freedom. The action J, the area enclosed by the trajectories of the integrable approximation $\mathcal{H}_0(\sigma, S, N)$,

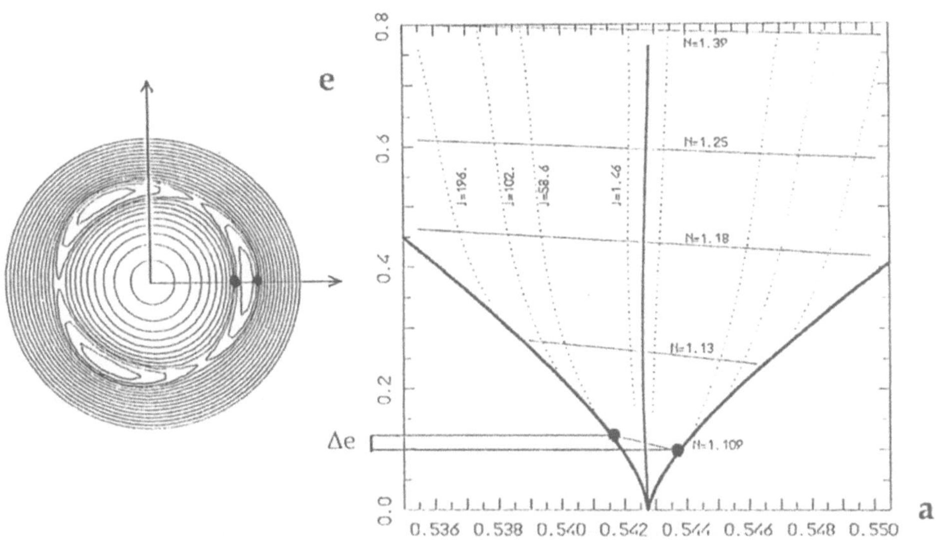

Fig. 2. Dynamics within the framework of the planar-circular and planar-elliptic problems, 5/2 mean motion commensurability. On the left: Poincaré diagram of the circular problem for a givem value of N. The coordinates are $(e\cos\sigma, e\sin\sigma)$. On the right: Location of the curves $N = C^t$ (circular problem) and $J = C^t$ (elliptic problem) in the (a, e) plane. The semi-major axis unit is a'. For a given value of N, each resonant trajectory of $\mathcal{H}_0(\sigma, S, N)$ is identified by its two intersections with the axis $\sigma = 0$ in the Poincaré diagram, and is represented by two points, located on each side of the stable equilibrium point (central bold line), in the right diagram. The bold dots correspond to the separatrices of the circular problem and indicate the maximum variation in semi-major axis and eccentricity on the corresponding N-level. Within the framework of the elliptic problem, each trajectory evolves from one line $N = C^t$ to another one, but the area J enclosed by the trajectory is a quasi-invariant of motion. Trajectories are thus identified by two lines, one on each side of the central bold line. The limits of variation on these lines will be determined by the study of the secular Hamiltonian.

is adiabatically conserved during the long-period evolution. As shown in Figure 2, the level curves of the quasi-integral J are almost vertical lines in the (a, e) representative plane. The maximum variation in eccentricity along lines $J = C^t$ will be determined by the study of $\mathcal{K}(J, \langle\nu\rangle, N)$.

The first exploration of asteroidal motion within the framework of the planar-elliptic problem was performed by Schubart (1968). With a view to studying the Hilda family, located at the 3/2 mean motion commensurability, Schubart extended his procedure to take account of Jupiter's eccentricity and constructed an integrator of the averaged planar-elliptic R3BP. Schubart's integrations of Hilda family members revealed large and regular variations of the eccentricity (from 0.1 to 0.25 in the case of Hilda) forced by the eccentricity of Jupiter.

Due to the removal of short periodic terms in the computation of the averaged derivatives of the disturbing function, the integration step used in the Schubart integrator can be as long as 1.5 to 2 years. With respect to classical integrations at that epoch, this saved a considerable amount of time, allowing the study of secular variations of asteroids over time scales of the order of 10^5 years.

The Schubart integrator was used by Giffen (1973) who performed a systematic search for periodic orbits corresponding to commensurabilities between the period of libration and that of ϖ within the 2/1 and 3/2 mean motion resonances. Giffen was the first to point out a major difference between these two cases: in the 3/2, the existence of invariant curves allows regular resonant motion over a large range of eccentricities while, in the 2/1, these curves exist only for large eccentricities ($\sim e \geq 0.3$) while a very complex motion appears at small eccentricities.

Scholl and Froeschlé extensively used the Schubart integrator to explore the dynamics within the Kirkwood gaps. In the 3/1 (Scholl and Froeschlé, 1974), they found a chaotic region at small eccentricity and, in most of the cases, large increases in the eccentricity of the test particles (up to 0.3 and more). Their study of the 5/2, 7/3 and 2/1 (Scholl and Froeschlé, 1975) confirmed the general tendency to eccentricity increase within the gaps, with the exception of initially circular orbits, borders of the resonances and of the 7/3 mean motion commensurability. Unfortunately, the study was made within the framework of the collisional hypothesis and Scholl and Froeschlé did not relate the increases in eccentricity to possible close encounters with Mars or other inner planets. Although they did so later on (Scholl and Froeschlé, 1977), the explanation of the gaps by purely gravitational effects remained definitely associated with Wisdom's name.

In the 2/1 resonance, within the framework of the averaged planar-elliptic R3BP, Froeschlé and Scholl (1976) proved that invariant curves bound the chaotic region found by Giffen to low eccentricities ($\sim e \leq 0.2$). In other words, the 2/1 gap cannot be explained within the framework of the planar-elliptic R3BP. A subsequent systematic three-dimensional exploration of the 2/1 mean motion commensurability (Froeschlé and Scholl, 1982) with the aid of a new version of Schubart integrator (Schubart, 1978) did not change the results: asteroids did not escape the resonance zone. Moreover, when the third dimension is taken into account, a supplementary protection mechanism can exist: the resonance in argument of perihelion, later on called "Kozai resonance" as it was first studied by Kozai for non-resonant asteroids (Kozai,1962). The detailed location of the Kozai resonance for orbits deep in mean motion commensurability can be found in (Jefferys and Standish, 1972).

3. Wisdom's Era

In 1982, Wisdom created a surprise in the Kirkwood gaps problem by showing that a test particle in the 3/1 can spend as much as a hundred thousand years or longer with $e < 0.1$ and then jump suddenly to $e > 0.3$. That result was obtained with the help of a mapping of the planar-elliptic R3BP (Wisdom, 1982) based on Chirikov's procedure (Chirikov, 1979). In Wisdom's mapping, the long-period terms come from the averaged 3/1 resonant Hamiltonian, truncated at order two in eccentricities, and the usual short-period terms are removed and simulated by means of delta functions. Wisdom claimed that his mapping was 1000 times faster than the Schubart integrator. Even if this factor seems somewhat overestimated (Murray and Fox, 1984), Wisdom's mapping allowed the "integration" time scale to be significantly increased, putting for the first time millions of years within reach. With his mapping, Wisdom studied the evolution over 2 Myr of 300 asteroids with initial conditions near the 3/1 commensurability. Starting from a relatively uniform distribution, Wisdom created a gap at the 3/1 by removing the test asteroids which became Mars crossers (i.e. such that $e > 0.3$). However, the gap was too narrow, the real Kirkwood gap being approximately twice as large. After that, Wisdom performed the task all over again with a more sophisticated mapping extended to the third dimension (3D) and taking the secular perturbations of Jupiter's orbit into account. Unfortunately, the gap was still too narrow and, on the basis of a particular case which showed an increase in eccentricity after 18 Myr, Wisdom concluded that 2 Myr is probably too short a time to see the full width of the gap to open. Wisdom's 1982 paper ends with an account of a series of problems raised by this experience:

- The actual probability of collision with Mars must be estimated. The typical lifetime of a Mars crosser is of the order of 200 Myr (Wetherill, 1975). However, as it was shown that test asteroids spend only a fraction of their time ($\simeq 10\%$) with a large enough eccentricity, their lifetime could be very long.
- The generalization of that explanation to other gaps is to be proved: these gaps are farther from Mars and higher eccentricities are thus required to cross the orbit of Mars.
- The mapping is particularized to the 3/1 commensurability. It should be adapted to other resonances.
- The validity of the mapping itself should be confirmed by accurate numerical integration or approximate analytic theory. Indeed, the sudden large jumps in eccentricity are attributed to the high-frequency terms which have been artificially created.

One year later, Wisdom (1983) definitely proved the adequacy of his mapping, by comparisons with numerical integrations and comparative computations of Lyapounov characteristic exponents (LCEs). Then, using his mapping not only to find quasi-periodic Mars crossers but also to determine, by computation of LCEs, the extent of the chaotic zones associated with the 3/1 resonance (in the 3/1 case, the chaotic trajectories are not confined to eccentricities below the Mars crossing limit), he succeeded in reproducing fairly well the 3/1 gap. The remaining doubts about the actual efficiency of Mars to deplete the 3/1 gap were ruled out by taking the third dimension into account. Indeed, in 3D "integrations", test particles reach eccentricities large enough to be Earth crossers ($e > 0.6$) and spend roughly 50% of the time above the Mars crossing limit. This led Wisdom to conclude that "No extra hypotheses are needed beyond the dynamics of the three-dimensional elliptic-restricted three-body problem and the presence of Mars to obtain the 3/1 Kirkwood gap's precise size and shape". The role of Mars was then clarified (Wisdom, 1985a): in 3D, trajectories are most of the time Mars crossers and successive minor encounters with Mars can cause a shift from one chaotic zone to another until the trajectory is injected into Earth's control. However, even in the planar case, the validity of the mapping seemed to be limited to $e \simeq 0.4$ and eccentricities as large as 0.6 could be considered as suspect. A few real numerical integrations, taking account of the four major planets, allowed Wisdom to prove the validity of his scenario (Wisdom, 1985a).

Murray and Fox (1984) studied the 3/1 commensurability in the planar-elliptic R3BP case using three numerical methods: (a) integrating the full equations of motion, (b) integrating averaged equations of motion truncated in eccentricities, and (c) using Wisdom's mapping. All three methods adequately reflect the qualitative behaviour of chaotic trajectories and determine the bounds of chaotic motion. In the regular regions of the phase space, and for truncature at order two in eccentricities in method (b), Murray and Fox found excellent agreement between methods (b) and (c), which shows that the high-frequency terms in Wisdom's mapping were relatively unimportant. Comparison with method (a) was not so good, even for eccentricities of the order of 0.4, and the inclusion of higher-order terms in eccentricity in model (b) did not improve the results. According to Murray and Fox, it was due to the poor convergence of the expansion of the disturbing function for eccentricities greater than 0.3.

The problem of convergence increases with decreasing distance with respect to Jupiter (Sundman, 1916; Ferraz–Mello, 1994a) and will result, for the other gaps, in a drastic limitation to low eccentricities of the validity of the results obtained with truncated models.

In order to study in more detail the chaotic zones as well as the zones of quasi-periodic librations in the 3/1, Wisdom (1985b) performed, within the framework of the planar-elliptic R3BP, an analytic study of the long-term dynamics associated with the 3/1. Taking account of the separation between the time scales of σ and ν (see Section 2), Wisdom represented the dynamics inside the 3/1 by means of a pendulum (the motion in σ) with slowly varying coefficients. Wisdom's perturbative treatment of the 3/1 confirmed that the eccentricity in the resonance zone can grow to values as high as 0.35 and highlighted a "zone of uncertainty" corresponding to the region in between the libration and the circulation of σ (Figure 3). In that zone, the perturbative treatment breaks down, as the quasi-integral J is no longer defined. In order to see the extent of the chaotic zone associated with the slow separatrix crossing, Wisdom displayed several surfaces of section of the full two-degree-of-freedom Hamiltonian $\mathcal{H}(\sigma, S, \nu, N)$. These surfaces of section confirm that large-scale chaos is associated with separatrix crossing (Figure 3): chaos at the borders of the 3/1 can be explained within the framework of long-term resonant dynamics and has very little to do with the high-frequency terms in the disturbing function.

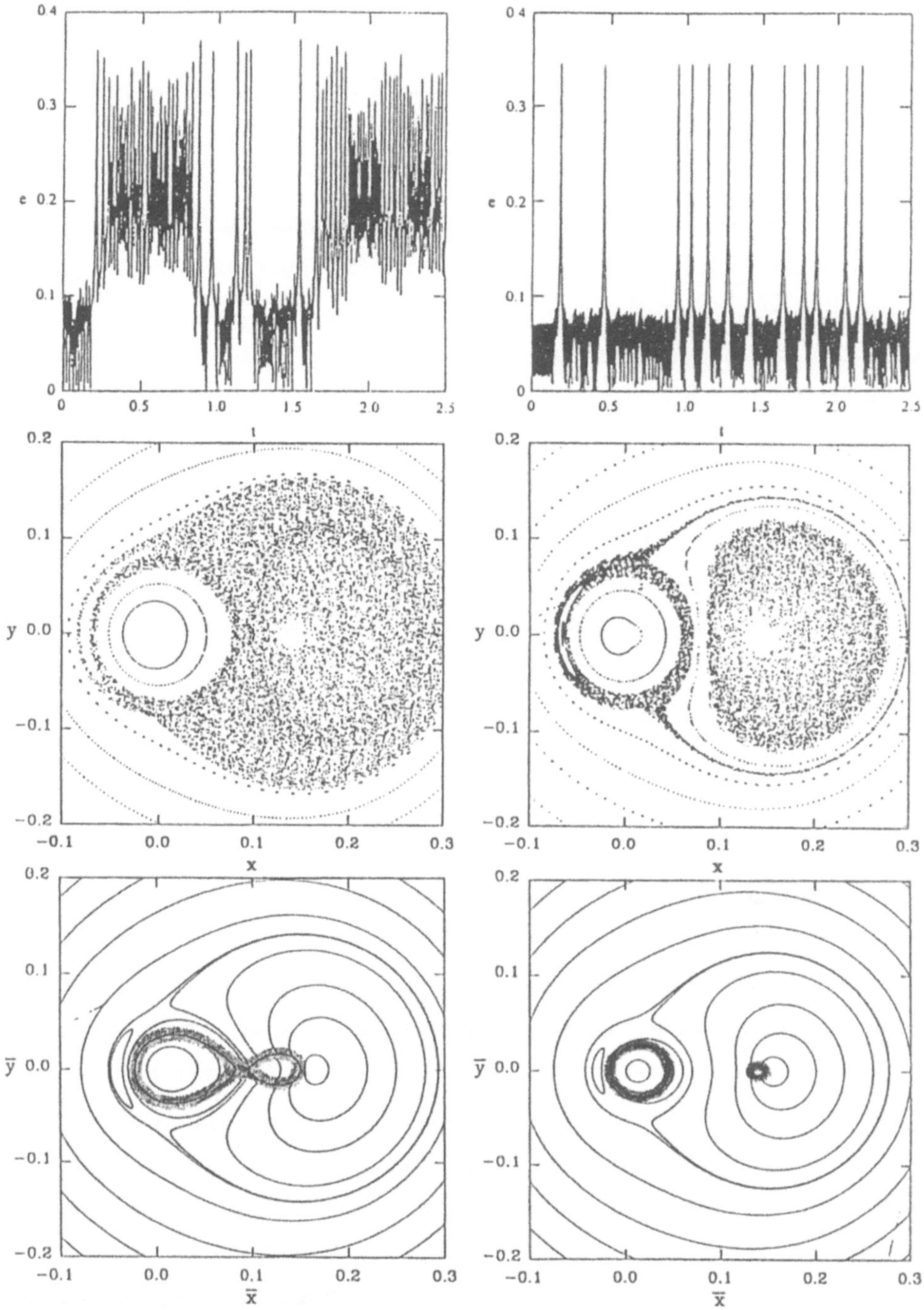

Fig. 3. Chaotic behaviour in the 3/1 mean motion commensurability. On the left: typical behaviour. On the right: relatively rare behaviour. From top to bottom: Evolution of the eccentricity as function of the time (expressed in millions of years), surface of section of the averaged planar-elliptic R3BP for a trajectory similar to the trajectory shown in the top picture, guiding trajectories of the long-period motion for the same energy level (the "zone of uncertainty" is shaded). The coordinates are: $\bar{x} \simeq x = e\cos(\varpi - \varpi')$, $\bar{y} \simeq y = e\sin(\varpi - \varpi')$. From Wisdom (1983, 1985).

Wisdom's mapping was adapted to the 5/2 resonance by Šidlichovský and Melendo (1986). The planar-elliptic version, truncated at order 3 in eccentricities, was used to recompute Scholl and Froeschlé's 1975 results on the 5/2. Šidlichovský and Melendo found more cases with $e > 0.3$ for low eccentricities and at the borders of the resonance. Surprisingly, at that time they still worked within the framework of the collisional hypothesis and did not push their computations to the Mars crossing limit ($e \simeq 0.4$). They also proposed extensions of their mapping to 3D, including secular variations of Jupiter's orbit and higher order terms in eccentricity. One year later, the mapping was applied to compute LCEs throughout the 5/2 resonance (Šidlichovský, 1987a) and an analytic theory, very similar to Wisdom's analytical approach of the 3/1, was also constructed (Šidlichovský, 1987b). Results indicated that the existence of the 5/2 gap can be explained by the same mechanisms as those acting in the 3/1 gap.

Murray (1986) extended the planar-elliptic R3BP Wisdom's mapping to general first order resonances of the form $(p+1)/p$ and used it to compute the maximum LCEs throughout the 2/1 and 3/2 mean motion resonances. Results were reported on a (a, e) diagram of initial conditions taken at $\sigma = \nu = 0$ and showed, for the first time, the extent of the chaotic zone found by Giffen in the 2/1 resonance. Murray computed percentages of errors in the second order expansion of the disturbing function and, on that basis, suggested that his results for the 2/1 and 3/2 can be believed up to approximately $e = 0.2$ and $e = 0.1$ respectively. However, although the percentages of error were 21.3% in the 2/1 and 22.9% in the 3/2, part of Murray's results were completely distorted: he found a large chaotic region at the center of the 2/1 for $e \geq 0.1$ and more chaos in the 3/2 than in the 2/1. By comparison, the value he found for $e = 0.4$ in the 3/1 is 17.1%, which questions the validity of Wisdom's mapping for such an eccentricity.

The early analytical studies of the 2/1 resonance within the framework of the planar-elliptic R3BP were due to Henrard and Lemaitre, at first with a pendulum-like formalism (Henrard and Lemaitre, 1987) and then with a semi-numerical method (Henrard and Lemaitre, 1986). In both cases, they used the reducing transformation (Wisdom, 1986; Henrard *et al.*, 1986) which allows the inclusion of the term in $e' \cos \nu$ in the integrable part of the Hamiltonian (Sessin and Ferraz–Mello, 1984). By the way, the inversion of dates between both Henrard and Lemaitre's papers is not an error but a simple demonstration that the refereeing process can sometimes be faster in Celestial Mechanics than in Icarus!

In their 1987 paper, Henrard and Lemaitre highlighted the perverse effects of early truncature of the disturbing function in the 2/1 case: the description of the long-term dynamics appeared order-dependent, especially at the center of the resonance. Henrard and Lemaitre's results, in good agreement with those of Murray for a disturbing function truncated at order two in e, became completely different when the disturbing function was truncated at higher orders. Indeed, higher order models agreed up to $e \simeq 0.3$ and did not show large variations in eccentricity for $0.1 \leq e \leq 0.2$ at the center of the resonance.

In their 1986 paper, Henrard and Lemaitre used a much more refined procedure to describe the dynamics near the 2/1 mean motion commensurability. There, instead of using a pendulum approximation for the leading terms in the Hamiltonian, they introduced action-angle variables, taking account of the whole integrable part of the Hamiltonian. These action-angle variables describe very accurately the three topological zones associated with resonant dynamics: internal circulation, libration and external circulation of the resonant angle. The price to pay for such accuracy is the unavailability of explicit analytic formulas. However, this is not a problem as the computation can be done numerically at any given point of the phase space. For that reason, the method was characterized as "semi-numerical" by its author (Henrard, 1990). The semi-numerical method of Henrard is a general procedure allowing the description of the dynamics associated with a two-or-more-degree-of-freedom separable Hamiltonian. It allows the computation of the basic frequencies of the integrable part of the Hamiltonian, the coefficients of the averaged perturbation with respect to the fastest degree-of-freedom, and the value of the resulting quasi-integral of motion on given surfaces of section. In Henrard and Lemaitre's 1986 paper, it was applied to the two-degree-of-freedom averaged Hamiltonian $\mathcal{H}(\sigma, S, \nu, N)$, truncated at order two in eccentricities. Results are thus second-order in the eccentricity of the asteroid and more or less second-order also in e', as the leading term of the perturbation ($e' \cos \nu$) has been incorporated in the integrable part of the Hamiltonian. Due to the early truncature in e, the description of the long-term dynamics in regions where σ is much faster than ν was not trusted and only served as a check for the method. On the contrary, at low eccentricity in the resonance zone, Henrard and Lemaitre located the 2/1 secondary resonance (at that time called "super resonance") between the period of libration of σ and that of circulation of $\varpi - \varpi'$ (Figure 4a). They also showed that the 3/1 secondary resonance crosses the outside critical curve of the 2/1 secondary resonance, identifying a new source of chaotic behaviour inside mean motion commensurabilities and providing an explanation for the peculiar behaviour found by Giffen in that region. The slow Hamiltonian chaos associated with separatrix crossing and detected by Wisdom in the 3/1 case is relatively marginal in the 2/1 case, where the dominant features of chaotic dynamics are associated with secondary resonances.

Wisdom's investigations of the 2/1 and the 3/2 (Wisdom, 1987) were performed by means of direct numerical integrations of the full equations of motion. Within the framework of the planar-elliptic R3BP, he integrated, over a 100000-year time span, a series of orbits and reported on the (a, e) plane the initial conditions leading to chaotic behaviour. The results he obtained for the 2/1 were in good agreement with those of Murray up to $e = 0.1$, which appeared to be the limit of validity of the planar mapping in the 2/1 case. As expected, the other results completely disagreed with Murray's conclusions: Wisdom found no chaos near the 3/2 exact commensurability and did not recover the central chaotic zone at $e \geq 0.1$ in the 2/1 resonance. In the 2/1, there was qualitatively good agreement

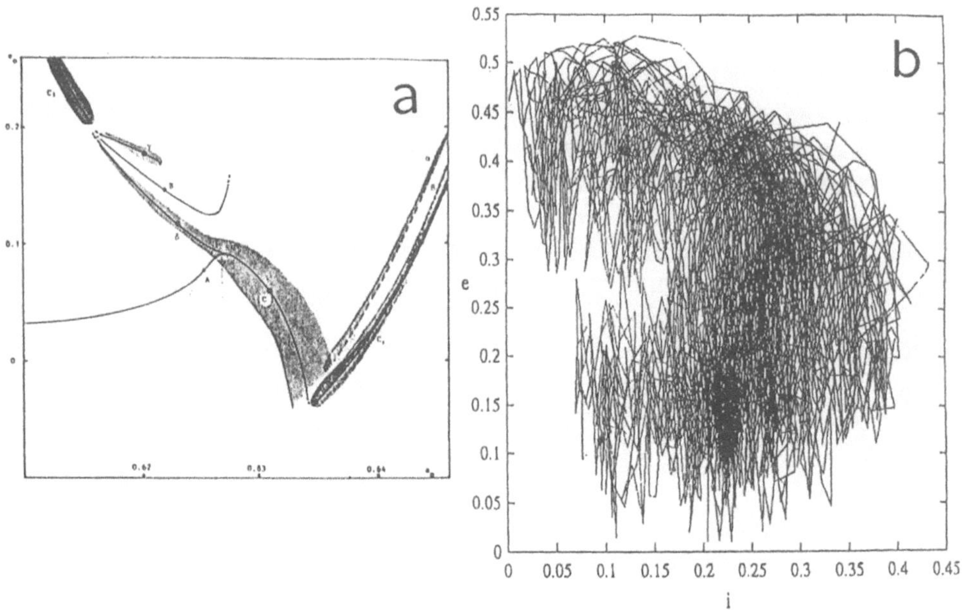

Fig. 4. **(a)** Location of the 2/1 secondary resonance between the period of libration of σ and that of circulation of $\varpi - \varpi'$ in the 2/1 representative plane. The semi-major axis unit is a'. The thick lines on the sides are the separatrices of the 2/1 mean motion resonance. The detailed features of the 2/1 secondary resonance are explained in the caption of Figure 5. The chaotic zone was shaded according to Murray's results. From Henrard and Lemaitre (1986). **(b)** Chaotic trajectory in the 2/1 mean motion commensurability. In the beginning of the integration, the inclination is low and the eccentricity is limited to values below about 0.25. Over the span of the integration the trajectory seems to trace out a pathway to high eccentricity which temporarily takes it through inclinations as high as 0.44 radians. From Wisdom (1987).

between the V-shaped borders of the chaotic zone and the borders of the 2/1 gap. Wisdom did not explore much further the 2/1 commensurability. Few numerical integrations, performed over 5 Myr and within the framework of the full 3D problem taking account of the perturbating effects of the four major planets, resulted in the discovery of a very peculiar behaviour (Figure 4b). In the light of the present knowledge of the Kirkwood gaps dynamics, it appears that this kind of behaviour, which, according to Wisdom, was "at the border of believability", has a reasonable dynamical explanation (see Section 4) and is not an artifact of the computation.

Although they were published at a later date, three more papers are to be mentioned here as they can be considered as a direct continuation of the two above-quoted papers by Henrard and Lemaitre. All of them consist of "analytical" studies of mean motion resonances performed with the aid of the semi-numerical method of Henrard. The first two were written by Lemaitre and Henrard and concern the 3/2 (1988) and the 2/1 (1990) cases. The third paper was by Henrard and Caranicolas (1990) and concerns the 3/1 case.

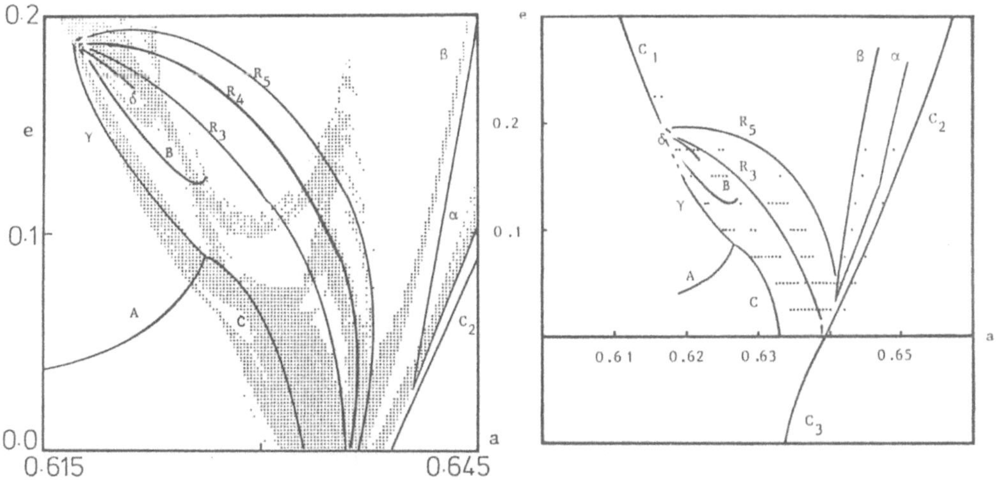

Fig. 5. Secondary resonances in the 2/1 mean motion resonance. The semi-major axis unit is a'. The main features of the 2/1 secondary resonance are described by the curves (B, C, γ, δ): B locates the stable equilibrium, C locates the unstable equilibrium, γ and δ locate the separatrices. The curves (R_3, R_4, R_5) denote the unperturbed location of the (3/1, 4/1, 5/1) secondary resonances. The lines (C_1, C_2) are the separatrices of the mean motion resonance. On the left: the shaded area corresponds to initial conditions leading to chaos, according to Murray's results. On the right: the dots correspond to initial conditions leading to chaos, according to Wisdom's results. From Lemaitre and Henrard (1990).

In the study of the 2/1 and 3/2, the semi-numerical method was applied to Hamiltonians truncated at different orders in eccentricity (up to order eight). According to Lemaitre and Henrard, the more accurate model allows correct qualitative description of long-term behaviour up to $e = 0.35$ in the 2/1 case and to $e = 0.15$ in the 3/2 case. These two papers also contain the location of the most important secondary resonances between the period of libration of σ and the period of circulation of $\varpi - \varpi'$. In the 2/1 case (Lemaitre and Henrard, 1990), results were in very good agreement with both the location of Giffen's periodic orbits and the shape of the chaotic zones found by Murray and Wisdom (Figure 5). Due to the overlap of secondary resonances and the limitation in order of the semi-numerical method, the exact zone of influence of each secondary resonance cannot be predicted by the theory. A detailed study of this region, performed with the aid of surfaces of section of the averaged planar-elliptic R3BP, can be found in (Moons and Morbidelli, 1993). A detailed study of the 3/2 secondary resonance zone is still awaited, but little chaos is present in this case as overlap does not occur (Michtchenko and Ferraz-Mello, 1995).

The study of the 3/1 mean motion commensurability (Henrard and Caranicolas, 1990) was performed with a Hamiltonian truncated at the order two in eccentricity. A detailed analysis of the long-term behaviour inside the 3/1 was presented, with many comparisons with respect to Wisdom's perturbative treatment of the same

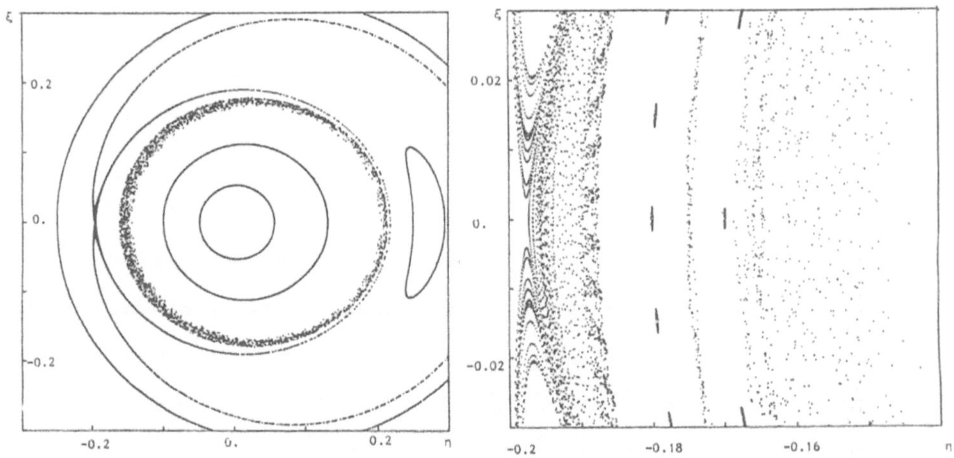

Fig. 6. Surface of section of the averaged planar-elliptic R3BP with $e' = 0.026$ in the 3/1 case. The coordinates are similar to $\left(e\cos(\varpi - \varpi'), e\sin(\varpi - \varpi')\right)$. In the large chaotic domain associated with separatrix crossing (center of the left picture), the eccentricity stays almost constant. On the contrary, the very thin chaotic layer associated with the corotation resonance produces an increase in eccentricity from $e = 0.2$ to about $e = 0.4$. The two chaotic domains, which are disconnected for $e' = 0.026$, enter in connection for higher values of e', which explains the random jumps in eccentricity found by Wisdom (Figure 3). The right picture shows an enlargment of the unstable equilibrium zone. The two chaotic domains can be seen on each side of the picture. From Henrard and Caranicolas (1990).

resonance (Wisdom, 1985b). Inside the 3/1 resonance, Henrard and Caranicolas detected the presence of the "secondary resonance" $\dot{\varpi} = \dot{\varpi}'$, which occurs when the asteroidal longitude of perihelion moves at the same rate as Jupiter's longitude of perihelion and, for that reason, was subsequently called the "corotation resonance". Henrard and Caranicolas showed that, although the Chirikov separatrix layer associated with that resonance is very thin, it is responsible for the random jumps between low-eccentricity and medium-eccentricity modes observed by Wisdom, the slow chaotic layer corresponding to separatrix crossing remaining in the low-eccentricity mode (Figure 6).

4. After Wisdom

Even with the most optimistic limits of validity, the classical expansion of the disturbing function does not allow either the study of the Hilda family in the 3/2 or that of the depletion of the 2/1 Hecuba gap using Wisdom's scenario. In order to get rid of the limitation in asteroidal eccentricity, subsequent works generally came back to (at least partial) numerical procedures to compute the averaged disturbing function or the coefficients of its expansion about a high-eccentricity center. Digital filtering techniques were also applied to the results of direct numerical integration.

Mappings, on the other hand, were adapted to high-eccentricity computations by the introduction of selected artificial terms, chosen in order to ensure the correct location and stability of the fixed points. A detailed analysis of mappings is beyond the scope of the present review and requires a complete review by itself. Adequate mapping construction is directly related to knowledge of periodic orbits of the real model. Both of them were extensively studied by Hadjidemetriou and thorough references to these subjects can be found in (Hadjidemetriou, 1995, 1996).

Using an expansion about the libration center, Ferraz–Mello (1988) constructed an analytical theory to study the behaviour of asteroids deep in the 3/2 resonance within the framework of the planar problem. The use of an asymmetric expansion about the libration center (Ferraz–Mello, 1987) allowed the study of resonant motion in a much larger eccentricity range. However, in that case, the gain in eccentricity was compensated by a loss in amplitude of libration. For that reason, the phase space was later on divided into a series of cells and an asymmetric expansion, extended to the full eccentricity range (Ferraz–Mello and Sato, 1989), was performed about the center of each cell, providing a complete coverage of the zone of interest.

Combining exploration by semi-analytical models and numerical integration, Yoshikawa (1989, 1990, 1991) analysed in a uniform way many mean motion commensurabilities and, in particular, the 3/1, 5/2, 7/3 and 2/1 gaps. Yoshikawa's semi-analytical treatments have in common two major advantages: they can be applied in a uniform way to any mean motion commensurability and they are valid in the whole eccentricity range.

At first, Yoshikawa (1989) studied the motion of test particles near 28 mean motion commensurabilities, with the aid of a semi-analytical method (Nakai and Kinoshita, 1985) and with direct numerical integration of the Sun–Jupiter–Saturn–asteroid system. The semi-analytical method described the long-term behaviour of asteroids by means of level curves of the one-degree-of-freedom Hamiltonian obtained after a series of numerical quadratures. Starting from the Hamiltonian of the full R3BP with Jupiter on a precessing ellipse, Yoshikawa performed a first numerical quadrature (Schwarzschild, 1903) to remove the short periodic part of the Hamiltonian, followed by a second numerical quadrature with respect to the argument of perihelion, which was supposed to move uniformly and much faster than the node. Moreover, the semi-major axis was fixed at the exact commensurability and σ was fixed at the stable equilibrium point of the resonance, which gives a good approximation for orbits deep in mean motion commensurability. Models for orbits with higher amplitude of libration were obtained by computing the mean value of models based on different values of σ. Yoshikawa's 1989 semi-analytical results were displayed for a mean inclination equal to 5° and provided a good description of long-term dynamics for orbits deep in resonance (Figure 7). These diagrams show the presence of a corotation resonance at high eccentricity at the center of all the Kirkwood gaps and of the 3/2. Due to that, orbits deep in mean motion commensurability can undergo very large variations in eccentricity in the

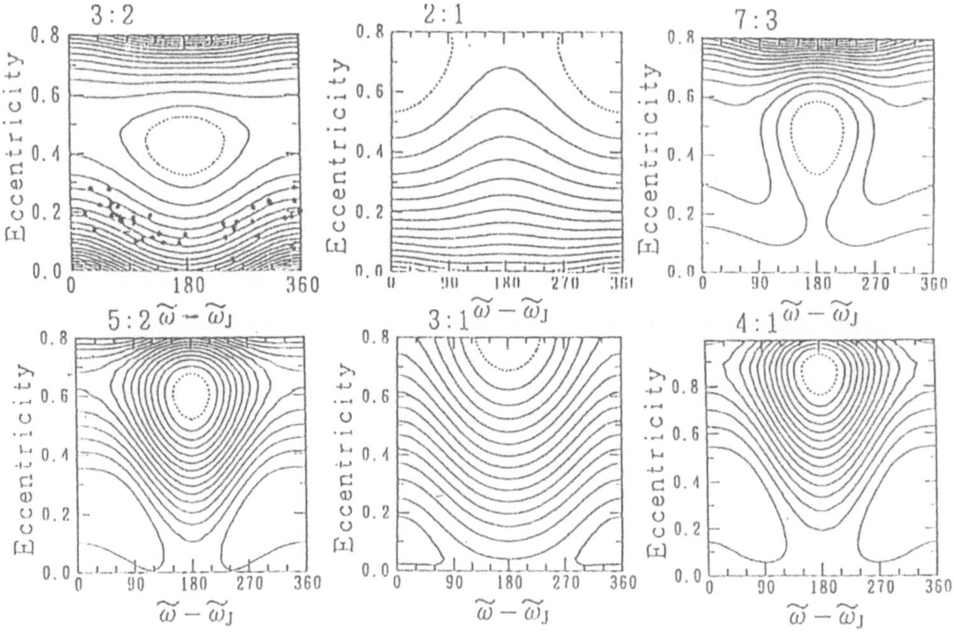

Fig. 7. Level curves of the secular problem for nearly planar orbits deep in mean motion commensurability ($\varpi_J \equiv \varpi'$). The dots in the 3/2 case denote the location of the Hilda asteroids. From Yoshikawa (1989).

4/1, 5/2 and 7/3 cases. Numerical integrations of the four-body problem confirmed these large variations. In general, the results of numerical integrations did not follow the curves of the semi-analytical model, but drift from one curve to another. In addition, they showed that chaotic behaviour occurs in large regions of the phase space. In the 4/1 case, an increase from $e = 0.1$ to $e \simeq 1$ is not rare, and can occur in a time scale much shorter than 1 Myr.

Yoshikawa's global survey of asteroids near mean motion commensurabilities was followed by specific studies dedicated to the Kirkwood gaps: the 3/1 (Yoshikawa, 1990), 5/2, 7/3 and 2/1 (Yoshikawa, 1991). In these studies, two semi-analytical models were used. The first gives the averaged motion of (a, σ) for fixed averaged values of e, i, ϖ ,and ϖ'. The second represents the long-term behaviour of asteroids deep in resonance. However, due to the quadrature in argument of perihelion, results for non-negligible inclination are to be taken with caution, as Kozai resonance can generally be found inside mean motion commensurabilities (Jefferys and Standish, 1972). Results of numerical integrations, performed within the framework of the planar-elliptic R3BP, were also presented. It was shown that, in the 3/1 and in the 5/2, large eccentricity variations generally occur while, for the 7/3, they occur only at the center of the resonance. For the 7/3 and 2/1 resonances, Wisdom's scenario is not supported.

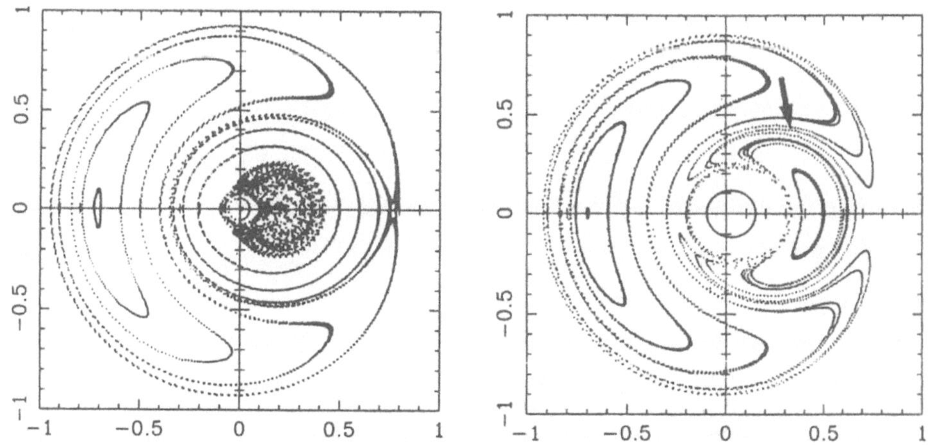

Fig. 8. Surfaces of section of the averaged planar-elliptic R3BP in the 3/1 case for two different energy levels. The coordinates are $(e\cos(\varpi - \varpi'), e\sin(\varpi - \varpi'))$. On the left, the chaotic domain found by Wisdom (Figure 3) is seen in the inner part of the figure. Regular curves separate this chaotic domain from the regular corotation region at high eccentricity. On the right, for another energy level, a heteroclinic bridge (denoted by the arrow) connects the medium-eccentricity and high-eccentricity corotation regions. From Ferraz–Mello and Klafke (1991).

Ferraz–Mello and Klafke (1991) and Klafke, Ferraz–Mello and Michtchenko (1992) confirmed and extended Yoshikawa's 1989 results. Using the very-high-eccentricity asymmetric expansion of the disturbing function near resonances (Ferraz–Mello and Sato, 1989) to map the phase space, these authors integrated averaged equations of the planar-elliptic R3BP and displayed surfaces of section of the two-degree-of-freedom Hamiltonian $\mathcal{H}(\sigma, S, \nu, N)$ for the 3/1, 4/1 and 5/2 cases. In all these mean motion commensurabilities, they found the corotation resonance at high eccentricity. In the corotation region, the two time scales of $\mathcal{H}(\sigma, S, \nu, N)$ are different and, apart from chaos associated with the corotation separatrix, the motion presents very large but very regular oscillations in eccentricity. The particles behave like in a one-degree-of-freedom system and seem to follow invariant curves like those of Yoshikawa's 1989 semi-analytical model. The small chaotic zones may be very important, though. In the the 4/1 case (Klafke, Ferraz–Mello and Michtchenko, 1992), an orbit starting at $e \simeq 0.15$ in the chaotic zone can suffer an increase in eccentricity up to about 1. In the 3/1 case, Ferraz–Mello and Klafke (1991) pointed out the existence of a small chaotic bridge connecting the separatrices of the medium-eccentricity and the high-eccentricity corotation regions (Figure 8), thus providing a continuation of the small chaotic bridge between low-eccentricity and medium-eccentricity regions already found by Wisdom (1983, 1985b) and explained by Henrard and Caranicolas (1990). Orbits starting at relatively low eccentricities ($e \simeq 0.25$) in the 3/1 can thus reach eccentricities as high as 0.9. In both cases, the time scale of the process is of the order of a few times 10^5 years. However, these phenomena occur only in very small portions of the phase space.

The regularity of the high-eccentricity region completely disappears when Saturn's indirect effects are taken into account. The disturbing action of Saturn was already pointed out in many cases, and several mappings or numerical integrations included its secular perturbations on Jupiter's orbit. The theoretical explanation of the resulting chaotic behaviour of resonant asteroids, the overlap of secular resonances inside mean motion commensurabilities, was provided by Morbidelli and Moons (1993).

At this stage, a return to the mathematical description of the subject is needed. As mentioned above, Yoshikawa's semi-analytical theories took the precession of Jupiter's orbit into account. However, this effect does not change the location and size of the corotation resonance, as Jupiter's orbit precesses very slowly and $e' = 0.048$. When Saturn is included in the model, a supplementary oscillation of e' and ϖ' is to be considered. At a first approximation, the secular variations of Jupiter's orbit due to Saturn's action can be expressed as quasi-periodic functions of the time:

$$\begin{aligned}
e' \cos \varpi' &= m_{5,5} \cos(g_5 t + \lambda_5^0) + m_{5,6} \cos(g_6 t + \lambda_6^0) \,, \\
e' \sin \varpi' &= m_{5,5} \sin(g_5 t + \lambda_5^0) + m_{5,6} \sin(g_6 t + \lambda_6^0) \,, \\
\sin \tfrac{i'}{2} \cos \Omega' &= n_{5,6} \cos(s_6 t + \mu_6^0) \,, \\
\sin \tfrac{i'}{2} \sin \Omega' &= n_{5,6} \sin(s_6 t + \mu_6^0) \,,
\end{aligned} \tag{2}$$

where $m_{5,5}, m_{5,6}, n_{5,6}, g_5, g_6, s_6, \lambda_5^0, \lambda_6^0$ and μ_6^0 are constants whose values can, for example, be found in (Nobili *et al.*, 1989). The inclusion of Saturn's indirect effects necessarily involves a slight modification of the definition of the canonical variables (1) presented in Section 2. Keeping the same definition for σ, S, σ_z and S_z, the other variables are:

$$\begin{aligned}
-\nu &= \frac{p+q}{q}\lambda' - \frac{p}{q}\lambda \,, & N &= \frac{p+q}{p} L - H \,, \\
\varpi_J &\equiv g_5 t \,, & &\Lambda_5' \,, \\
\varpi_S &\equiv g_6 t \,, & &\Lambda_6' \,, \\
\Omega' &\equiv s_6 t \,, & &\Lambda_{16}' \,,
\end{aligned} \tag{3}$$

and the full averaged problem is now five-degree-of-freedom as the Hamiltonian depends only on ν through the combinations $\nu + \varpi_J, \nu + \varpi_S$ and $\nu + \sigma_z + \Omega'$.

The inclusion of Saturn's effects in the disturbing function introduces three new secular resonances (Williams, 1969):

- the ν_5, which corresponds to $\dot{\varpi} = \dot{\varpi}_J$ (already present in the R3BP);
- the ν_6, which corresponds to $\dot{\varpi} = \dot{\varpi}_S$;
- the ν_{16}, which corresponds to $\dot{\Omega} = \dot{\Omega}'$.

Secular resonances were primarily studied outside of mean motion resonances (Williams, 1969), and their location in that case lead, for instance, many authors to think that the ν_6 could have importance at low inclination only for asteroids near 2AU. However, the geography of secular resonances depends on the mean motion resonance under study, as was already pointed out long ago in the case of the Kozai resonance (Jefferys and Standish, 1972). Mathematically, this can be explained by the degeneracy of the Hamiltonian of the restricted problem. In the two-body problem, the secular angles (perihelion and node) are stationary. Therefore, the behaviour of these angles is completely dependent on the form of the perturbation. Outside of mean motion resonances, the perturbation can be averaged over λ and λ', while inside mean motion resonances, it can be averaged over λ' only and the result depends on the resonant angle σ. The form of the averaged perturbation and, hence, the behaviour of the secular angles, is thus completely different in the resonant and non-resonant cases, and changes according to the mean motion resonance under consideration. Consequently, the location of secular resonances *outside* of mean motion resonances has nothing in common with their location *inside* mean motion resonances. Moreover, the latter is resonance-dependent.

A systematic planar search for the ν_5 and ν_6 secular resonances inside mean motion resonances was performed by Moons and Morbidelli (1993). Using the semi-numerical method of Henrard (1990), they computed the basic frequencies of the integrable part of the Hamiltonian by numerically integrating $\mathcal{H}_0(\sigma, S, N)$ and located the secular resonances ν_5 and ν_6 in the 4/1, 3/1, 5/2, 7/3, 2/1 and 3/1 cases. The derivatives needed for such an integration were computed by numerically averaging (Schwarzschild, 1903) the derivatives of the disturbing function expressed in terms of non-singular variables (Poincaré, 1902b). The procedure is thus very accurate at any point of the phase space. Moons and Morbidelli showed that the ν_6 secular resonance is generally present at high eccentricity in the vicinity of the ν_5 one (Figure 9). The location of the ν_5 secular resonance for orbits deep in resonance agreed with Yoshikawa's 1989 results. Moreover, the ν_6 secular resonance was found in the domain of influence of the ν_5 secular resonance: a large overlap was expected.

The overlap of ν_5 and ν_6 secular resonances in the planar problem was first studied in the 2/1 and 3/2 cases (Morbidelli and Moons, 1993) and later in the 4/1, 3/1, 5/2 and 7/3 cases (Moons and Morbidelli, 1995). At first, the one-degree-of-freedom Hamiltonians describing the isolate secular effects of both resonances were computed independently and their separatrices were located, confirming a large overlap for any amplitude of libration. Later, the two-degree-of-freedom secular Hamiltonian describing the cumulative secular effects of both resonances

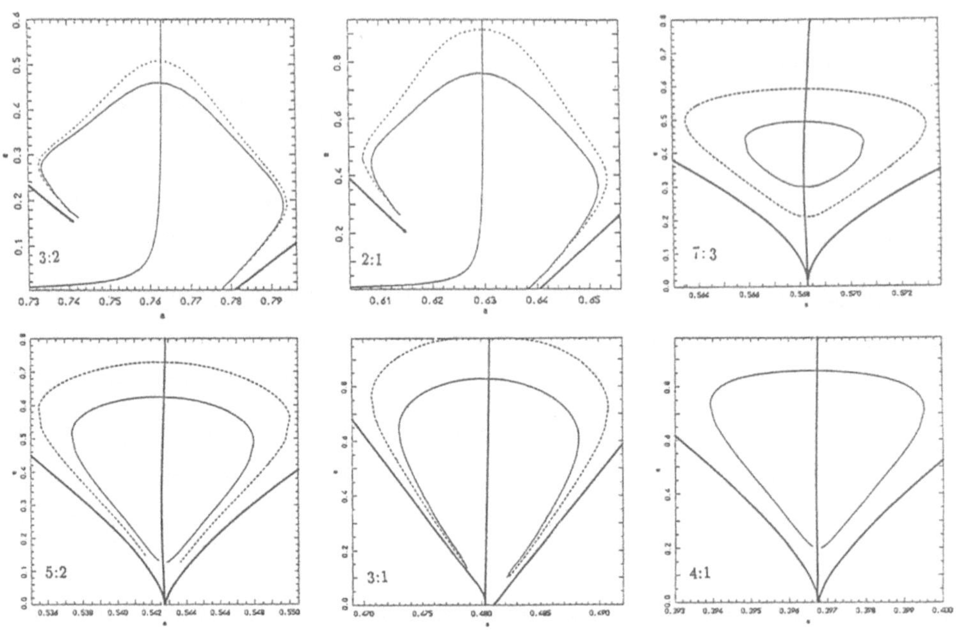

Fig. 9. Location of the secular resonances in the (a, e) representative plane. The solid lines denote ν_5 and the dashed lines denote ν_6. The central vertical line corresponds to the stable equilibrium point of $\mathcal{H}_0(\sigma, S, N)$ and the bold lines on the sides are the separatrices. The semi-major axis unit is a'. From Moons and Morbidelli (1993).

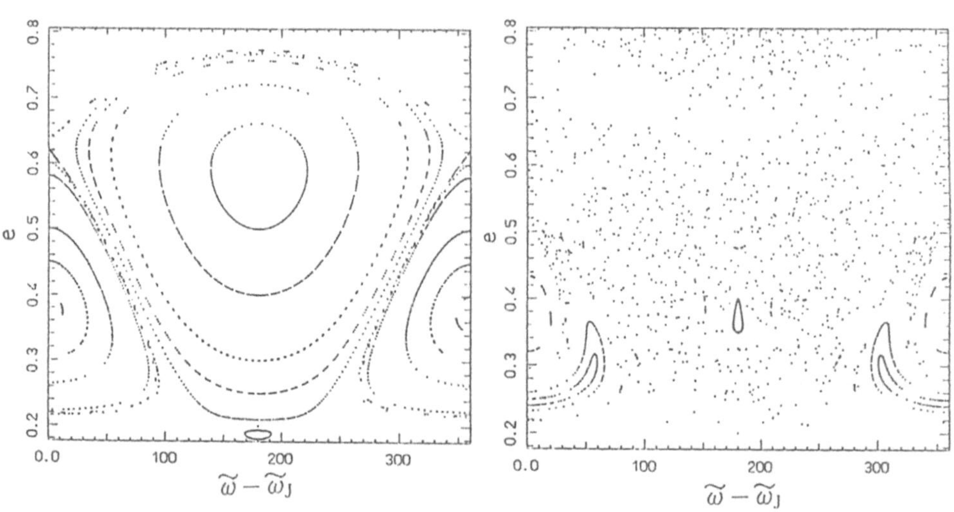

Fig. 10. Secular dynamics on the level $J = 58.6$ (see Figure 2) in the 5/2 mean motion resonance. On the left: results of the numerical integration of the secular Hamiltonian associated with the ν_5 secular resonance. Very large but very regular eccentricity variations can take place (the apparent chaos is due to a lack of points in the computed trajectories). On the right: surface of section of the secular Hamiltonian associated with both the ν_5 and the ν_6 secular resonances. From Moons and Morbidelli (1995).

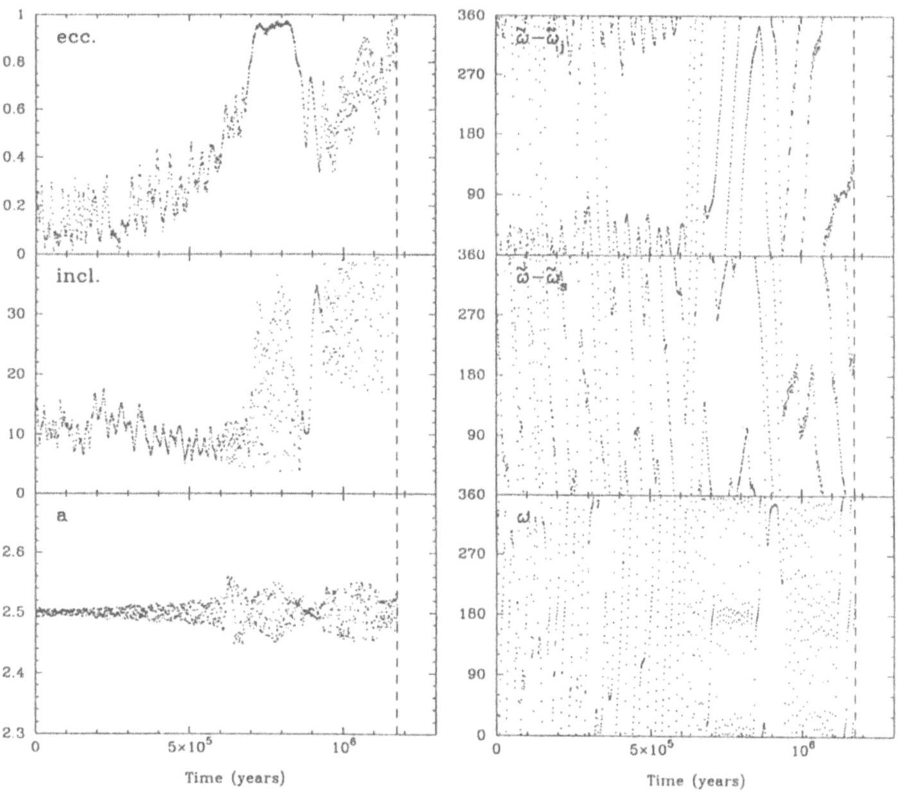

Fig. 11. The evolution of the fictitious fragment n° 319 of the asteroid 6 Hebe. From Farinella *et al.* (1993). The first 6×10^5 year are dominated by the ν_5 resonance, with some passages through the ν_6 resonance or the Kozai resonance. The eccentricity grows to almost 0.6. The time from 6×10^5 year to 8.5×10^5 year is dominated by the ν_6 resonance. The eccentricity grows to 0.95 and then decreases to 0.5. Afterward, the Kozai resonance is responsible for the anticorreletad oscillations of e and i. Then, since 9.5×10^5 year, the ν_6 resonance first and the ν_5 resonance later pump up the eccentricity to 1, with the consequent fall of the asteroid into the Sun.

was integrated and results were displayed on surfaces of section for fixed values of the adiabatic invariant J (see Section 2). These results (Figure 10) are striking, to say the least: the large regular excursions in eccentricity are replaced by a huge chaotic zone when both ν_5 and ν_6 are taken into account.

In the 4/1, 3/1, 5/2 and 7/3 gaps (Moons and Morbidelli, 1995), the chaotic zone extends to almost all the resonance region, providing a nice explanation for the lack of asteroids observed in these gaps. Indeed, asteroids in this big chaotic sea undergo strong and wild increases in eccentricity (up to 0.8 and more), and they become deep Earth crossers in a time scale of the order of 1 Myr. The behaviour described in Moons and Morbidelli's 1995 paper was in good agreement with the behaviour observed in realistic numerical integrations (Figure 11), and provided an explanation for the frequent occurrences of collisions with the Sun in the 3/1 case (Farinella *et al.*, 1993, 1994). The numerical integration of hundreds of resonant fictitious particles has been recently performed by Migliorini *et al.* (1997) and Gladman *et al.* (1997). Such integrations show, in a statistical reliable way, what are the dominating phenomena associated with the 3/1, 5/2, 7/3 and other mean motion resonances. It comes that, in the 3/1 and 5/2 resonances, the eccentricity of the majority of resonant particles is pumped to Earth and Venus crossing values on 1 Myr time scale. In the 3/1 case, only 5% of the particles are extracted from the resonance by Mars, while 66% are extracted by the Earth and Venus. The remaining 29% of the particles reach $e = 1$ within the mean motion resonance and collide with the Sun. Most of the particles extracted by the Earth or Venus are reinjected into some resonance, and subsequently increase their eccentricity to 1 (colliding with the Sun), or are delivered to Jupiter's control and are subsequently ejected from the Solar System. The resulting "mortality" of initially resonant particles is impressive: 50% of the bodies die within 2 Myr, while the population of survivors is reduced to 10% in 7 Myr. These dramatic phenomena are well explained by the mechanism of secular resonances overlap inside mean motion commensurabilities and no other mechanism can account for such strong disturbing action. As a check, A. Morbidelli integrated the same sample of fictitious particles taking only account of Jupiter on very eccentric orbit ($e' = 0.0596$). The results are eloquent: all particles were still alive after 2 Myr and only 14% of the particles (instead of 55%) became Earth crosser on the same time span. This shows the importance of secular resonances.

In the 2/1 and 3/2 cases (Morbidelli and Moons, 1993), chaos due to overlapping secular resonances does not cover the whole phase space. In both cases, a large regular zone was found below the overlapping region in the (a, e) plane (Figure 12). This zone is in good agreement with the location of the Hilda family in the 3/2 case. In the 2/1, it is located at medium eccentricity and concerns orbits with small to medium amplitude of libration. The observed lack of asteroids in that region cannot be explained within the framework of a planar theory, even when the long-term indirect disturbing action of Saturn is taken into account.

Morbidelli and Moons (1993) also constructed a 3D semi-analytic model describing the Kozai and ν_{16} secular resonances for asteroids deep in mean motion commensurability. They showed that the location of the ν_{16} secular resonance in the 3/2 coïncides with the border of the Hilda family. In the 2/1, the ν_{16} is at lower eccentricity and would cross a family like that of the Hildas (Figure 12). However, a priori it could not be considered as an explanation for the 2/1 gap as the dynamical effect of the ν_{16} is an increase in inclination and not in eccentricity. Moreover, the ν_{16} secular resonance appeared isolated from the Kozai and other secular resonances. Even within the framework of the 3D model, non-negligible regular regions would persist in the 2/1.

Henrard, Watanabe and Moons (1995) extended the 3D analysis of the secular perturbations in the 2/1 resonance to larger amplitudes of libration, at first with

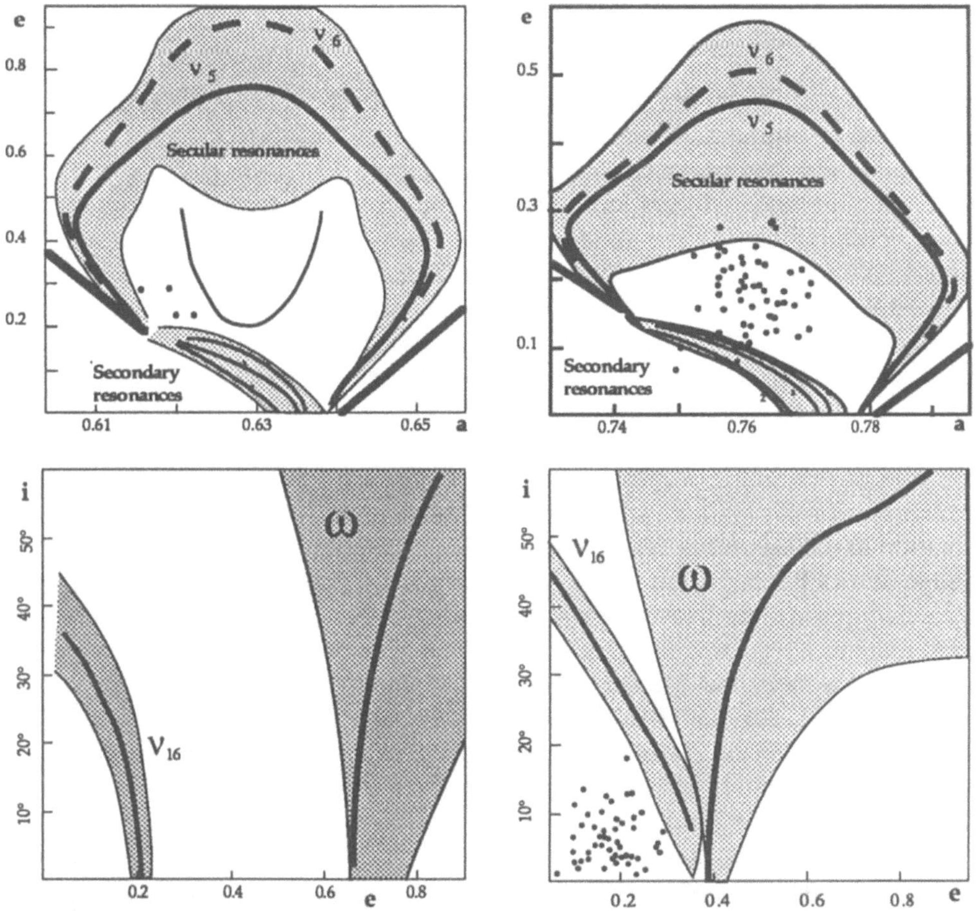

Fig. 12. Comparative study of the dynamics inside the 2/1 and 3/2 mean motion commensurabilities. The left pictures correspond to the 2/1 case and the right pictures correspond to the 3/2 case. The top pictures show the result of the planar study and the bottom pictures are dedicated to the 3D exploration of asteroids deep in mean motion commensurability. The study of the planar case shows the extent of the chaotic zones associated with the overlap of the ν_5 and the ν_6 secular resonances. The secondary resonances zones (Lemaitre and Henrard, 1990) are also indicated. In both resonances, a large zone separates the two chaotic regions. In the 3/2 mean motion resonance, this zone hosts a lot of asteroids while, in the 2/1, it is almost empty. The dots inside the right pictures indicate the Hilda asteroids (osculating elements). The separatrices of the $\nu 5$ secular resonance being computed at $\varpi - \varpi_J = \pi$ in the 3/2 case, some Hilda members seem inside the chaotic zone in the top picture although they are phase-protected (see Figure 7). The 3D study of asteroids deep in mean motion commensurability shows the location of the Kozai and the ν_{16} secular resonances. In the 2/1 mean motion resonance, the ν_{16} secular resonance is located at a lower eccentricity, but it appears disconnected from the Kozai resonance. In the 3/2 mean motion resonance, the ν_{16} is located at a higher eccentricity and seem to bound the Hilda family. From Morbidelli and Moons (1993).

an approximate semi-analytical theory and later on using an extension of the Schubart integrator taking account of the secular perturbations of Jupiter's orbit (Moons, 1994). With the semi-analytical theory, Henrard, Watanabe and Moons located the most important secondary resonances in the (e, i) plane and, for low amplitudes of libration, found a connection between these secondary resonances and the ν_{16} secular resonance (left picture of Figure 13). They also showed that, when the amplitude of libration increases, the Kozai resonance moves towards lower eccentricities. As the semi-analytical theory was not able to account for large amplitudes of libration, Henrard, Watanabe and Moons performed a series of numerical integrations with the extended Schubart integrator. These integrations showed an additional displacement of the ν_{16} secular resonance towards higher eccentricities, which produces a shrinking of the regular zone in between the ν_{16} and what they called "the complex of pericentric resonances" ($=\nu_5 + \nu_6+$ Kozai) (Figures 13abc). At large amplitudes of libration, a bridge was found between the complex of pericentric resonances and the secondary resonance zone, providing an explanation for Wisdom's 2/1 peculiar orbit (Figure 13d). As the amplitude of libration in the secondary resonance zone is not a quasi-invariant of motion, this bridge allows the depletion of the low-eccentricity chaotic zone associated with secondary resonances. Nevertheless, for low to moderate amplitudes of libration, there remains a regular zone free of secular perturbations. This region corresponds to the place were Ferraz–Mello (1994c) and Franklin (1994) computed the smallest LCEs in the 2/1. Henrard (1996) showed that a similar bridge exists in the 3/2. The difference between the 2/1 and the 3/2 cases lies in the time scales: for comparative amplitudes of libration, orbits in the secondary resonance zone took less time to find the bridge in the 2/1 than in the 3/2.

Even if the description of the dynamics inside the 2/1 and the 3/2 appeared very similar, comparative computations of LCEs highlighted a major difference between both resonances. Indeed, when the disturbing action of Saturn is taken into account, the lower limit of LCEs in the 2/1 seems superior to their upper limit in the 3/2. Franklin, Lecar and Murison (1993) integrated, within the framework of the planar problem, the orbits of 25 Hilda family members, selecting those with a larger amplitude of libration. For these asteroids, they found LCEs of less than $10^{-5.5}$/year. Similar values in the 3/2 case were obtained by Ferraz–Mello, Dvorak and Michtchenko (1994). On the other hand, Ferraz–Mello (1994c) computed LCEs within the Hecuba gap and, apart from a stable zone at the center of the resonance, generally found values greater than 10^{-5}/year for orbits with an initial inclination of 5°. Similar values in the 2/1 case were obtained by Franklin (1994), but the most regular regions he found correspond to a moderate amplitude of libration. As the models are different, a direct comparison of these values is meaningless. Nevertheless, results indicated that, if diffusive phenomena exist in both resonances, they would be faster in the 2/1 than in the 3/2. This led some of these authors to conclude that the age of the solar system is probably too short a time to be able to see the Hilda asteroids leave the 3/2 "gap". In other words,

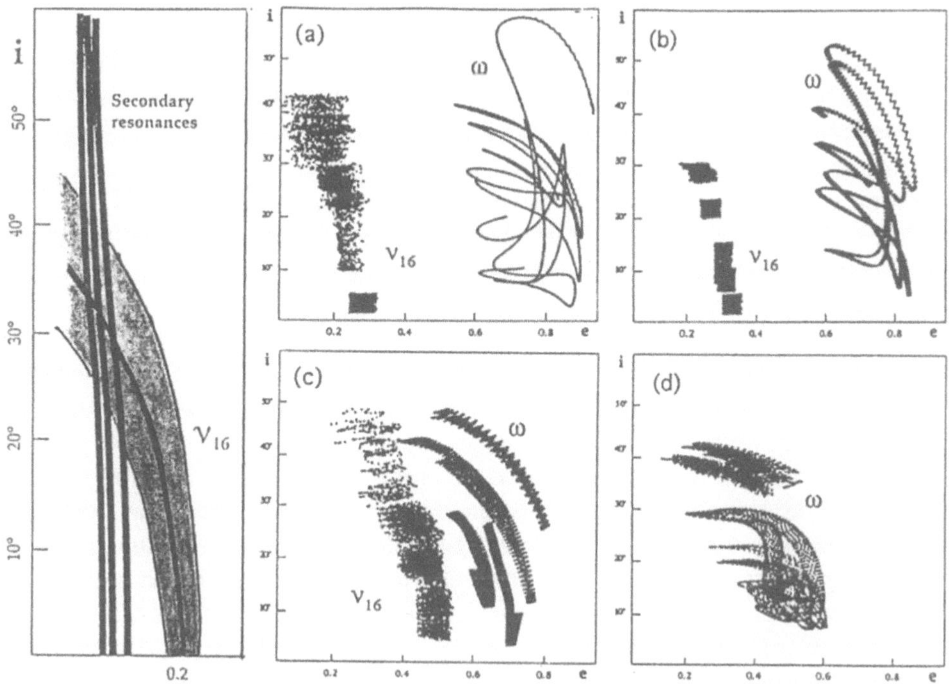

Fig. 13. Three-dimensional exploration of the 2/1 mean motion resonance. On the left: result of a semi-analytical theory. Other pictures: projection on the (e, i) plane of portions of orbits corresponding to different amplitudes of libration: 10° **(a)**, 45° **(b)**, 100° **(c)** and 130° **(d)**. See text for comments. From Henrard, Watanabe and Moons (1995).

we are faced with a mechanism which is more or less completed in the 2/1 while it is still at work in the 3/2. However, the demonstration of such a conjecture by numerical integration requires integration time spans which are beyond the actual possibilities.

With the use of numerical integration, on-line digital filtering (Carpino *et al.*, 1987) and Fourier analysis of the results, Michtchenko and Ferraz–Mello (1996) performed a comparative study of the 2/1 and 3/2 mean motion resonances within the framework of the 3D restricted three-body problem. The secondary resonances between the three basic frequencies of the 3D model were analysed, as well as their association with chaos. Michtchenko and Ferraz–Mello also showed that, at variance with the 3/2 case, the regular asteroidal motions in the 2/1 are not nearly periodic, the analysis of the power spectra of the orbital elements revealing an accumulation of many frequency lines of comparable intensities. The inclusion of Saturn in the model (Ferraz–Mello *et al.*, 1996) introduces many additional spectral lines, mainly associated with g_6, creating multiplets of linear combinations with other independent modes and transitions between neighbouring secondary resonances can occur.

Ferraz–Mello (1996) studied the 2/1 over time spans of the order of 1Gyr with the use of a planar symplectic mapping based on an asymmetric expansion of the disturbing function (Ferraz–Mello and Sato,1989). This study indicated an escape time scale in the range $10^7 - 10^9$ years and revealed the disturbing role played by the short-period perturbations of Jupiter's orbit associated with the great inequality.

Morbidelli (1996) numerically integrated over 1 Gyr the orbits of 24 fictitious particles with initial eccentricities below 0.35 in the 2/1 resonance. Results showed that all orbits diffuse chaotically and 20 of the 24 fictitious particles escaped from the resonance within the integration time span. However, in some parts of the resonance, the diffusion speed is very slow and particles can stay safely at least 1 Gyr in the resonance. Morbidelli also integrated over a 38 Myr time span the orbits of several real asteroids near the 2/1 mean motion commensurability. Amongst the seven which were actually in libration, four asteroids showed a very regular behaviour. These asteroids are located in the region where the lowest diffusion rate has been found and were Franklin computed the smallest LCEs. Their resonant proper elements (a minimum, e maximum) are plotted in the (a, e) plane in Figure 12. Morbidelli identified three main channels for large increase in the inclination, which seemed related to secondary resonances, and Kozai resonances, but also to the ν_{16} resonance, which progressively enters in contact with the Kozai resonance when the libration amplitude increases, as was hinted by Henrard *et al.* (1995). Recently, a study of the ν_{16} secular resonance was performed by Moons and Morbidelli (unpublished results) with an accurate 3D semi-analytical theory based on the algorithm of successive elimination of perturbation harmonics (Morbidelli, 1993). The location of the ν_{16} secular resonance in the plane is indicated by the parabolic curve seen at the center of the "regular zone" in the 2/1 mean motion resonance (Figure 12). It divides this regular zone in two parts: the central part (above the ν_{16}) where Ferraz–Mello computed the smallest LCEs and the peripheric part (below the ν_{16}) where Franklin computed the smallest LCEs. The seemingly regular zone in the 2/1 is thus crossed by a diffusive channel directly connected to high-eccentricity chaotic regions.

5. Conclusions

Substantial progress has been made in the understanding of the Kirkwood gaps problem within the last 32 years. Several sources of chaotic behaviour inside the gaps were identified: separatrix crossing (Wisdom, 1985b), secondary resonances (Henrard and Lemaitre, 1986) and overlap of secular resonances (Morbidelli and Moons, 1993). A scenario was proposed to deplete the 3/1 gap by combined effects of chaoticity within the 3D restricted problem and gravitational scattering by the inner planets (Wisdom, 1983). However, the time scales needed are very long and the scenario is not applicable to all gaps. A similar scenario, based on the disturbing action of overlapping secular resonances, was proved to be much faster and applicable to all the inner gaps (Moons and Morbidelli, 1995). Moreover, it could explain the frequent occurrence of collisions with the Sun in the 3/1 and the 5/2 cases. The study of the 2/1 Hecuba gap and the 3/2 Hilda group revealed many similarities and some differences (location of the ν_{16}, time scales). Further studies are needed to understand the nature and precise the time scales of the diffusive phenomena acting in these mean motion resonances.

Acknowledgements

The author wishes to thank C. Froeschlé, J. Henrard and A. Morbidelli for a careful reading of the manuscript. Special additional thanks to C. Froeschlé for his encouragements to write this review paper, to J. Henrard for his help with the graphic software and to A. Morbidelli for a lot of very helpful comments and suggestions. Many thanks also to P. Kelly, for his precious help in improving the quality of the text.

References

Andoyer, H.: 1903, "Contribution à la théorie des petites planètes dont le moyen mouvement est sensiblement double de celui de Jupiter", *Bull. Astron.*, **20**, 321–356.

Brouwer, D.: 1963, "The Problem of the Kirkwood Gaps in the Asteroidal Belt", *Astron. J.*, **68**, 152–159.

Carpino, M., Milani, A. and Nobili, A.M.: 1987, "Long-Term numerical integrations and synthetic theories for the motion of the outer planets", *Astron. Astrophys.*, **181**, 182–194.

Chirikov, B.V.: 1979, "A universal instability of many-dimensional oscillator systems", *Phys. Rept.*, **52**, 263–379.

Farinella, P., Froeschlé, Ch. and Gonczi, R.: 1993, "Meteorites from the asteroid 6 Hebe", *Celest. Mech.*, **56**, 287–305.

Farinella, P., Froeschlé, Ch., Froeschlé, C., Gonczi, R., Hahn, G., Morbidelli, A. and Valsecchi, G.B.: 1994, "Asteroids falling into the Sun", *Nature*, **371**, 314–317.

Ferraz–Mello, S.: 1987, "Expansion of the disturbing force–function for the study of high-eccentricity librations", *Astron. Astrophys.*, **183**, 397–402.

Ferraz–Mello, S.: 1988, "The high-eccentricity libration of the Hildas", *Astron. J.*, **96**, 400–408.

Ferraz–Mello, S.: 1994a, "The convergence domain of the Laplacian expansion of the disturbing function", *Celest. Mech.*, **58**, 37–52.

Ferraz–Mello, S.: 1994b, "Kirkwood gaps and resonant groups", in *Asteroids, Comets, Meteors 1993* (A. Milani, M. Di Martino and A. Cellino, eds.), Kluwer, Dordrecht, 175–188.

Ferraz–Mello, S.: 1994c, "Dynamics of the asteroidal 2/1 resonance", *Astron. J.*, **108**, 2330–2337.

Ferraz–Mello, S.: 1996, "On the Hecuba gap", in *Dynamics, Ephemerides and Astrometry of the Solar System* (S. Ferraz–Mello, B. Morando and J.-E. Arlot, eds.), Kluwer, Dordrecht, 177–182.

Ferraz–Mello, S. and Klafke, J.C.: 1991, "A model for the study of very-high-eccentricity asteroidal motion: the 3:1 resonance", in *Predictability, Stability, and Chaos in N–body Dynamical Systems* (A.E. Roy, ed.), Plenum Press, New–York, 177–184.

Ferraz–Mello, S. and Sato, M.: 1989, "The very-high-eccentricity asymmetric expansion of the disturbing function near resonances of any order", *Astron. Astrophys.*, **225**, 541–547.

Ferraz–Mello, S., Dvorak, R. and Michtchenko, T.A.: 1994, "Depletion of the asteroidal belt at resonances", in *From Newton to Chaos* (A.E. Roy and B.A. Steves, eds.), Plenum Press, New–York, 157–169.

Franklin, F.: 1994, "An examination of the relation between chaotic orbits and the Kirkwood gap at the 2:1 resonance. I., *Astron, J*,, **107**, 1890–1899.

Franklin, F., Lecar, M. and Murison, M.: 1993, "Chaotic orbits and long term stability: an example from asteroids of the Hilda group", *Astron. J.*, **105**, 2336–2343.

Froeschlé, C. and Greenberg, R.: 1989, "Mean motion resonances", in *Asteroids II* (R. P. Binzel, T. Gehrels and M. S. Matthews, eds.), University of Arizona Press, Tucson. 827–844.

Froeschlé, C. and Scholl, H.: 1976, "On the Dynamical Topology of the Kirkwood Gaps", *Astron. Astrophys.*, **48**, 389–393.

Froeschlé, C. and Scholl, H.: 1982, "A Systematic Exploration of Three-dimensional Asteroidal Motion at the 2/1 Resonance", *Astron. Astrophys.*, **111**, 346–356.

Giffen, R.: 1973, "A study of Commensurable Motion in the Asteroid Belt", *Astron. Astrophys.*, **23**, 387–403.

Gladman B.J., Migliorini, F., Morbidelli, A., Zappalà, V., Michel, P., Cellino, A., Levison, H., Froeschlé, Ch. and Bailey, M.E.: 1997, to be submitted to *Science*.

Greenberg, R. and Scholl, H.: 1979, "Resonances in the asteroidal belt", in *Asteroids* (T. Gehrels, ed.), The University of Arizona Press, Tucson, 310–333.

Hadjidemetriou, J.D.: 1995, "Mechanisms of Generation of Chaos in the Solar System", in *From Newton to Chaos* (A.E. Roy and B.A. Steves, eds.), Plenum Press, New–York, 79–96.

Hadjidemetriou, J.D.: 1996, "Symplectic mappings", in *Dynamics, Ephemerides and Astrometry of the Solar System* (S. Ferraz–Mello, B. Morando and J.-E. Arlot, eds.), Kluwer, Dordrecht, 255–266.

Hagihara, Y.: 1961, "Gaps in the distribution of asteroids", *Smithsonian Contr. to astrophysics*, **5,6**, 59–67.

Henrard, J.: 1988, "Resonances in the Planar Elliptic Restricted Problem", in *Long-Term Dynamical Behaviour of Natural and Artificial N-Body Systems* (A.E. Roy, ed.), Kluwer, Dordrecht, 405–425.

Henrard, J.: 1990, "A semi-numerical perturbation method for separable Hamiltonian systems", *Celest. Mech.*, **49**, 43–67.

Henrard, J.: 1996, "A note concerning the 2:1 and the 3:2 resonances in the asteroid belt", *Celest. Mech.*, **64**, 107–114.

Henrard, J. and Caranicolas, N.D.: 1990, "Motion near the 3/1 resonance of the planar elliptic restricted three body problem", *Celest. Mech.*, **47**, 99–121.

Henrard, J. and Lemaitre, A.: 1986, "A perturbation method for problems with two critical arguments", *Celest. Mech.*, **39**, 213–238.

Henrard, J. and Lemaitre, A.: 1987, "A Perturbative Treatment of the 2/1 Jovian Resonance", *Icarus*, **69**, 266–279.

Henrard, J., Watanabe, N. and Moons, M.: 1995, "A Bridge between Secondary and Secular Resonances inside the Hecuba Gap", *Icarus*, **115**, 336–346.

Henrard, J., Lemaitre, A., Milani, A. and Murray, C.D.: 1986, "The Reducing Transformation and Apocentric Librators", *Celest. Mech.*, **38**, 335–344.

Hill, G.W.: 1902a, "Illustration of periodic solutions in the problem of three bodies. I", *Astron. J.*, **22**, 93–97.

Hill, G.W.: 1902b, "Illustration of periodic solutions in the problem of three bodies. II", *Astron. J.*, **22**, 117–121.

Jefferys, W.H. and Standish, E.M.: 1972, "Further Periodic Solutions of the Three-Dimensional Restricted Problem. II", *Astron. J.*, **77**, 394–400.

Kirkwood, D.: 1867, Meteoric Astronomy, Lippincott, Philadelphia, ch. 13.

Klafke, J.C., Ferraz–Mello, S. and Michtchenko, T.A.: 1992, "Very-high-eccentricity librations at some higher order resonances", in *Chaos, Resonance and Collective Dynamical Phenomena in the Solar System* (S. Ferraz–Mello, ed.), Kluwer, Dordrecht, 153–158.

Kozai, Y.: 1962, "Secular Perturbations of Asteroids with High Inclination and Eccentricity", *Astron. J.*, **67**, 591–598.

Lemaitre, A. and Henrard, J.: 1988, "The 3/2 resonance", *Celest. Mech.*, **43**, 91–98.

Lemaitre, A. and Henrard, J.: 1990, "On the Origin of Chaotic Behavior in the 2/1 Kirkwood Gap", *Icarus*, **83**, 391–409.

Message, P.J.: 1966, "On nearly–commensurable periods in the restricted problem of three bodies, with calculation of the long-period variations in the interior 2:1 case", in *The Theory of Orbits in the Solar System and in Stellar Systems* (G. Contopoulos, ed.), Academic, London, 197–222.

Michtchenko, T.A. and Ferraz–Mello, S.: 1995, "Comparative study of the asteroidal motion in the 3:2 and 2:1 resonances with Jupiter. I. Planar model", *Astron. Astrophys.*, **303**, 945–963.

Michtchenko, T.A. and Ferraz–Mello, S.: 1996, "Comparative study of the asteroidal motion in the 3:2 and 2:1 resonances with Jupiter. II. Three-dimensional model", *Astron. Astrophys.*, **310**, 1021–1035.

Migliorini F., Morbidelli, A., Zappalà, V., Gladman, B.J., Bailey, M.E. and Cellino, A.: 1997, "Vesta Fragments for ν_6 and 3:1 resonances: implications for V-type NEAs and HED meteorites", submitted to *Meteoritics*.

Moons, M.: 1994, "Extended Schubart averaging", *Celest. Mech.*, **60**, 173–186.

Moons, M. and Morbidelli, A.: 1993, "The main mean motion commensurabilities in the planar circular and elliptic problem", *Celest. Mech.*, **57**, 99–108.

Moons, M. and Morbidelli, A.: 1995, "Secular Resonances in Mean Motion Commensurabilities: The 4/1, 3/1, 5/2 and 7/3 Cases", *Icarus*, **114**, 33–50.

Morbidelli, A.: 1993, "On the Successive Elimination of Perturbation Harmonics", *Celest. Mech.*, **55**, 101–130.

Morbidelli, A.: 1996, "The Kirkwood gap at the 2/1 commensurability with Jupiter: new numerical results", *Astron. J.*, **111**, 2453–2461.

Morbidelli, A. and Moons, M.: 1993, "Secular resonances in mean motion commensurabilities: The 2/1 and 3/2 Cases", *Icarus*, **102**, 316–332.

Murray, C.D.: 1986, "Structure of the 2/1 and 3/2 Jovian Resonances", *Icarus*, **65**, 70–82.

Murray, C.D. and Fox, K.: 1984, "Structure of the 3:1 Jovian Resonance: A Comparison of Numerical Methods", *Icarus*, **59**, 221–233.

Nakai, H. and Kinoshita, H.: 1985, "Secular Perturbations of Asteroids in Secular Resonance", *Celest. Mech.*, **36**, 391–407.

Nobili, A., Milani, A. and Carpino, M.: 1989, "Fundamental frequencies and small divisors in the orbits of the outer planets", *Astron. Astrophys.*, **210**, 313–336.

Poincaré, M.H.: 1902a, "Les solutions périodiques et les planètes du type d'Hécube", *Bull. Astron.*, **19**, 177–198.

Poincaré, M.H.: 1902b, "Sur les planètes du type d'Hécube", *Bull. Astron.*, **19**, 289–310.

Šidlichovský, M.: 1987a, "Application of the method of Wisdom to resonances, 5/2 resonance of asteroids", in Figure and Dynamics of the Earth, Moon, and Planets (M. Šidlichovský, ed.), Praha, Czechoslovakia, 571–580.

Šidlichovský, M.: 1987b, "Comparison of perturbation and Wisdom methods for 5/2 resonance", in *Dynamics of the Solar System*, (M. Šidlichovský, ed.), Praha, Czechoslovakia,

33–37.
Šidlichovský, M. and Melendo, B.: 1986, "Mapping for 5/2 asteroidal commensurability", *Bull. Astron. Inst. Czechosl.*, **37**, 65–80.
Scholl, H.: 1979, "Recent work on the origin of the Kirkwood gaps", in *Dynamics of the Solar System* (R.L. Duncombe, ed.), 217—222.
Scholl, H. and Froeschlé, C.: 1974, "Asteroidal motion at the 3/1 commensurability", *Astron. Astrophys.*, **33**, 455–458.
Scholl, H. and Froeschlé, C.: 1975, "Asteroidal motion at the 5/2, 7/3 and 2/1 resonances", *Astron. Astrophys.*, **42**, 457–463.
Scholl, H. and Froeschlé, C.: 1977, "The Kirkwood gaps as an asteroidal source of meteorites", in *Comets, asteroids, meteorites* (E.H. Delsemme, ed.), 293–295.
Schubart, J.: 1964, "Long-Period effects in nearly commensurable cases of the restricted three–body problem", *Smithsonian Astrophys. Obs.* Spec. Rept. **149.**
Schubart, J.: 1968, "Long-Period effects in the motion of Hilda–type planets", *Astron. J.*, **73**, 99–103.
Schubart, J.: 1978, "New results on the commensurability cases of the problem Sun–Jupiter–Asteroid", in *Dynamics of Planets and Satellites and Theories of Their Motion* (V. Szebehely, ed.), D. Reidel, Dordrecht, 137–143.
Schwarzschild, K.: 1903, "Über die periodischen Bahnen vom Hecubatypus", *Astron. Nachr.*, **160**, 385–400.
Sessin, W. and Ferraz–Mello, S.: 1984, "Motion of two planets with periods commensurable in the ratio 2:1. Solutions of the Hori auxiliary system", *Celest. Mech.*, **32**, 307–332.
Sundman, K.F.: 1916, "Sur les conditions nécessaires et suffisantes pour la convergence du développement de la fonction perturbatrice dans le mouvement plan", *Öfversigt Finska Vetenskaps-Soc. Föhr.*, **58 A**, (24).
Wetherill, G.W.: 1975, "Late heavy bombardment of the Moon and terrestrial planets", in *Proceedings of the Sixth Lunar Science Conference*, Pergamon, Oxford, 1539–1561.
Williams, J.G.: 1969, "Secular Perturbations in the Solar System", Ph. D. Dissertation, Univ. of California, Los Angeles.
Wisdom, J.: 1982, "The origin of the Kirkwood gaps: a mapping for asteroidal motion near the 3/1 commensurability", *Astron. J.*, **87**, 577–593.
Wisdom, J.: 1983, "Chaotic behavior and the origin of the 3/1 Kirkwood gap", *Icarus*, **63**, 272–289.
Wisdom, J.: 1985a, "Meteorites may follow a chaotic route to Earth", *Nature*, **315**, 731–733.
Wisdom, J.: 1985b, "A perturbative treatment of the motion near the 3/1 commensurability", *Icarus*, **63**, 279–282.
Wisdom, J.: 1986, "Canonical Solution of the Two Critical Argument Problem", *Celest. Mech.*, **38**, 175–180.
Wisdom, J.: 1987, "Urey Price Lecture: Chaotic Dynamics in the Solar System", *Icarus*, **72**, 241–275.
Woltjer, J.: 1928, "The motion of Hyperion", *Ann. Sterrew. Leiden*, **16**,3.
Yoshikawa, M.: 1989, "A survey on the motion of asteroids in commensurabilities with Jupiter", *Astron. Astrophys.*, **213**, 436–458.
Yoshikawa, M.: 1990, "Motions of asteroids at the Kirkwood gaps. I. On the 3:1 resonance with Jupiter", *Icarus*, **87**, 78–102.
Yoshikawa, M.: 1991, "Motions of asteroids at the Kirkwood gaps. II. On the 5:2, 7:3 and 2:1 resonances with Jupiter", *Icarus*, **92**, 94–117.

OPEN PROBLEMS ON THE EVE OF THE NEXT MILLENNIUM

VICTOR SZEBEHELY
University of Texas, Austin, Texas, USA

Abstract. To predict the future of Celestial Mechanics and to investigate unsolved problems for the next millennium contradicts the nature of our field of science which is based on non-integrable and transcendental equations with short time predictability. The reliability and predictability in our field depend on several factors, most of them unpredictable. Numerical predictions depend on the development of computer technology, which is itself not predictable. New analytical approaches call for the genius of Newton. We must realize that between Ptolemaeus (100-178) and Copernicus (1473-1543) our field received little attention. Between 1700 and 1800 the explosion occurred (Newton, Euler, Laplace, Lagrange and Gauss) followed by the period of 1800-1900 associated with Hamilton, Hill, Jacobi, Poincaré and Tisserand. Can we expect anyone to make similar contributions in the next 1000 years?

The entrance and emphasis of probability in our field certainly will continue. Speaking about bounds of trajectories instead of individual orbits gains significant popularity. Theory of stability and chaos are expected to receive more and more attention together with Poincare's conjectures concerning the denseness of periodic orbits in the set of bounded orbits.

Planetary computations for long time have shown considerable advances, partly due to computer advances as well as the selection of the proper formulation of the problems. The preparation of "exact" short time ephemerides and qualitative long time predictions require initial conditions with small observational errors, the use of proper models (including relativity effects and non-conservative forces) and the use of correct values of astronomical constants. The results of such calculations in the future should include approximate estimations of errors in the outcome.

The observations and orbital calculations of Earth-crossing asteroids should receive special attention along with development of various techniques of space dynamics to avoid collisions. Interplanetary orbit calculations of space probes with optimizations included, shall receive additional attention since interplanetary flights expect to become regular travels in the next millennium. Here, again, the use of proper models, constants and planetary ephemerides will be essential.

This paper was prepared with the help of many of my colleagues whose suggestions and ideas were incorporated with my many thanks.

1. Introduction

To list unsolved problems in celestial mechanics might be considered an easy task, compared to other fields of science. To predict for the next millennium the emergence of new problems and the expected progress concerning the present problems is much more difficult.

This paper attempts to list new and old unsolved problems in our field, hoping that it will inspire the readers to join our club of dynamicists, astronomers, mathematicians - all of us following in Newton's footsteps.

The subjects of selected problems should be considered very flexible and many other (not listed) problems must exist in our field. My selections were strongly influenced by the suggestions and advices I received from my colleagues as well as by my limited areas of contributions (and not solutions) to celestial mechanics.

We shall start with a short description of the history of our field to show when and how some of the major problems emerged (Newton, Leibnitz, Laplace, Lagrange, etc.). This will be followed by the treatment of the concept of non-integrability,

Celestial Mechanics and Dynamical Astronomy **65**: 205–211, 1997.

since all problems of today's celestial mechanics belong to this category; integrable systems might be considered as approximations.

Stability and chaos might be the most important unsolved problems. The reduction of the order of the system of our differential equations is of primary importance along with elimination of time. Using proper transformations and variables, regularization techniques can be of significant importance. Formulation of these equations, the terms of importance, the physical effects, the accuracy of constants and observations (giving initial conditions) follows. As one of the practical problems of the detection and possible deflection of Earth-crossing asteroids will be mentioned. In order to establish accurate physical descriptions and accurate constants, we face the inverse problem of dynamics.

The unpredictable and important role of computer development must be emphasized. The basic non-integrable dynamical system in our field is known as the problem of three bodies which might be advanced further by numerical integration of the equations offering ranges and densities of chaoticity, periodicity, escapes, etc.

Once again, it should be emphasized that our problems are basically non-integrable and no techniques (analytical or numerical) can change this fact.

2. Historical Remark

If predictions are made on a time scale one might review the past and base the predictions on the history, without attaching too much importance to such findings. In our field of science progress started in the middle of the fifteenth century with Copernicus, Galilei and Kepler (pervious efforts by the philosopher Aristotle and the geometer Ptolemaeus might be mentioned; both of them using the concept of the geocentric solar system, in fact of the geocentric universe).

The next step is very significant in our history and I do not believe that we can overemphasize the importance of Sir Isaac Newton (1642-1727), in fact I prefer to refer to him as a singularity in our history. The next group consists of Euler, Lagrange, Laplace and Gauss contributing important developments during the 18th century. The 19th century names are Hamilton, Jacobi, Hill, Poincaré, Tisserand, Moulton, Birkhoff and Whittaker. This is followed by the 20th century contributors, most of whom you know.

The reason for the above short historical review is to show you that not much happened between Aristotle (384-322 BC) and the Copernicus group. The singularity (Newton) is around 1700, i.e. about 200 years after the Copernicus group and about 1800 years after Aristotle. If you wish to predict the entrance of another singularity you can extrapolate based on Aristotle-Newton or based between Newton and 1900. For the next singularity, according these 2 extrapolations, we might have to wait 2 millennia or 2 centuries. Such predictions, of course, might have little significance but I could not come up with a better evaluation of our history.

3. Open Problem

3.1. ANALYTICAL TECIINIQUES

If one of the important contributions in our field is to produce accurate predictions we must have the proper inputs and the proper techniques to handle our equations of motion. The inputs might be arbitrary or are obtained from previous observations. To have information we do not only mean initial conditions but we need reliable values of the physical constants entering our equations. In addition the formulation of our equations can be of great importance since neglecting terms besides gravitational force can result in an erroneous model.

One of the greatest advances during the next millennium could be to formulate simpler equations (without neglecting terms!) describing our dynamical system. Such simplifications are often called reductions of the order and usually are performed in the framework of Hamiltonian dynamics. I wish to mention another reduction technique which has not received enough attention up to now and (if it is applicable) might result in useful reformulation. This reduction uses the energy conservation and the technique of eliminating the time. It is emphasized that our interest is concentrated on reduction of the order of our differential equations and not necessary substituting geometry for dynamics. I wish to present a very simple 4th order system and show how the above-mentioned process can reduce the system to 2nd order. Note that our system is still not-integrable, but such simplifications sometimes help to understand some qualitative features of the system. Our example is

$$\ddot{x} = \frac{\partial \phi}{\partial x}, \quad \ddot{y} = \frac{\partial \phi}{\partial y}$$

where dots indicate time derivatives, x and y are Cartesian coordinates and the force function (= negative potential) is $\phi = \phi(x, y)$. (Note that the potential does not depend explicitly on time.) The energy integral is

$$\dot{x}^2 + \dot{y}^2 = 2\phi + c$$

and we can eliminate $\dot{y}$ by using the relation

$$\dot{y} = \frac{dy}{dx}\dot{x} .$$

The derivative of this equation with respect to time gives

$$\ddot{y} = y'\ddot{x} + \dot{x}^2 y'' ,$$

where primes denote derivatives with respect to x.

Substituting from the original equations of motion $\ddot{x}$ and $\ddot{y}$ and using the energy integral in the form

$$\dot{x}^2 = \frac{c + 2\phi}{1 + y'^2}$$

we obtain

$$y''(c+2\phi) = (\phi_y - y'\phi_x) \ .$$

This second order differential equation relates the dependent variables x and y and it represents the geometry of our dynamical problem. If the solution $y = y(x,c)$ can be obtained in closed form from the second order equation (which in our case cannot be done because of the non-integrability principle) we can reintroduce the time using our equation for $\dot{x}^2$ as

$$t = t_0 + \int_{x_0}^{x} \sqrt{\frac{1+y'^2}{c+2\phi}} dx \ .$$

It is mentioned that elimination of time will not make our system integrable but it might suggest the application of further transformations or it might lead us to some qualitative information.

The above second order differential equation (in y) can lead to the determination of the potential for a given family of orbits. In fact this is a linear partial differential equation for $\phi(x,y)$ when $y = y(x,c)$ is given and it is often referred to as the inverse problem of dynamics. These ideas and equations have been generalized to higher order systems and their applications to celestial mechanics present still open problems. Note that solution for the potential (ϕ) does not have to be a Legendre polynomial or other, generally not convergent series. Consequently, the idea of elimination of time presented above can be useful for two open problems: to improve our knowledge concerning the geometrical properties of the solutions of the system and to establish the potential function resulting in observed orbits.

In addition, regarding the group of problems mentioned in item (1) we suggest work in the future to represent potential functions of interest as the solutions of the above mentioned partial differential equation. Note that with the selection of proper variables, the two-dimensional restricted problem of three bodies can also be reduced to a linear second order partial differential equation, which, for given family of orbits will determine the potential.

At this point we should emphasize the importance of the determination of the Earth's (and other bodies') potential. At present the Legendre polynomials are most frequently used to describe the Earth's potential. This does not seem to be a very satisfactory approach since the Legendre series can show divergences in the solutions. To find the distribution of point masses representing the Earth's potential was initiated by Vinti in 1959. To represent the Earth's gravitational potential by a finite number of point masses should result in a simpler solution than the Legendre polynomials and no divergence in the solutions is expected. It is suggested that this new approach to find the Earth's potential should not use presently existing Legendre coefficients but it should be based on the results of actual satellite orbit observations.

3.2. Numerical Techniques

Unless a completely new analytical approach is presented to handle gravitational dynamics, we must rely on numerical techniques and computers. The developments and advances in computer technology are not predictable. Consider, for instance, our Humbolt colloquium in 1984 where Dr. Kinoshita described results concerning the behavior of the Solar System for 5×10^6 years. Today we accept the validity of results for 10^9 years. The prediction of computer development is one of the difficult, if not impossible subjects. As Dr. Marchal describes this "anybody is now able to do in a single day more computations than all mankind before 1950."

Further improvements concerning numerical integration techniques, fitting various dynamical problems can be expected. Related to this we could discuss further developments of families of orbits obtained by numerical integration. The use of such results can improve our perturbation techniques. While two-body solutions are useful to start perturbation approaches, it has been suggested that we establish numerically some better approximations and base our perturbation theory on such orbits.

Along the numerical work we should mention that three and n-body problems will benefit significantly by the improvement of computer technology. It can be expected that qualitative results will be established regarding the dynamics of these non-integrable problems. To find the fuzzy regions and denseness of escape and of bounded orbits in the general problem of three bodies is in progress trying to establish Poincaré's conjectures. But even simpler problems such as modifications of Sitnikov's problem require large number of computer results.

To put emphasis on numerically established periodic solutions is a classical approach in celestial mechanics, since long-time predictions become valid for periodic orbits. It can be expected that families of periodic orbits will be available for not only the restricted problem of three bodies but for other problems.

3.3. Statistical Mechanics

One of the recent results emerging in celestial mechanics as well as in dynamics is related to stability and chaos. The many different definitions of stability are useful since different dynamical systems will show various behaviors. Liapunov in 1892 suggested an approach now known as the evaluation of Liapunov coefficients for isochronous correspondence. Some results concerning chaos are based on the computation of these coefficients, but significant improvements are needed in order to clarify this subject. While we know that chaotic motions exist in deterministic dynamical system, we can ask the inverse question, regarding the existence of potential functions resulting in a given chaotic system of orbits.

In connection with stability we might expect some new statistical results in celestial mechanics. First it is important to realize that Newton did not believe in perfect predictability of dynamical systems, while Leibnitz and later Laplace felt very strongly about predictability. Today we know that, once again, Newton was correct!

One future aspects of dynamics is the statistical approach, which is the consequence of limited predictability. Instead of computing single trajectories for given sets of initial conditions we take the initial conditions x_0^i and change them to $(x_0^i + \Delta x^i)$ and compute the set of trajectories close to the originally given initial conditions. In this way we obtain a set of trajectories which will show the sensitivity to initial conditions. Since the accuracy of the integration is limited, we shall obtain a bundle of orbits due to changed initial conditions and numerical error-propagation. This set has definitely physical meaning, while a single trajectory can be quite questionable. Note that the initial conditions often come from observations which process also introduces limited predictability. To obtain a bundle of trajectories, therefore, is more realistic than the information obtained from a single one.

It can be expected that results of statistical mechanics will be applied to celestial mechanics in the future.

3.4. Theory and Observations

The list of open problems is concluded with mentioning Lagrange who established analytically the possible existence of asteroids around the Sun-Jupiter stable triangular equilibrium points. It took observational astronomers 134 years after Lagrange's theoretical announcement to find the Trojan asteroids.

The message in this historical remark is that theoretical results should not be ignored by observational projects and collaboration between these two approaches should be encouraged. Another example is the existence of stable periodic orbits in the systems of binary stars. It is well accepted that most of the stars of our galaxy form binaries and we are looking for planets in this system for the existence (based on stability) of planetary systems. The stable regions of planets in binary systems have been in principle established theoretically by Hill (1878) and the recent observational results (1995) concerning the possible existence of other planetary systems took again some time.

4. Summary

In summary we may conclude that the new problems are approached either analytically, numerically or by observations. Those three fields will experience large forward steps during the next millennium and from a scientific point of view, should be available through supporting agencies, even if the actual applications are not presently known.

As an example consider that observations of the dynamic behavior of bodies, or groups of bodies, in other galaxies might lead to non-Newtonian potentials. This is one of the future possibilities, at least in principle, using the idea of inverse dynamics.

Acknowledgements

Without the help of many of my colleagues this paper could not have been prepared. They are all very well known in our field and I will give only their last names, Bond, Contopoulos, Dvorak, Eichhorn, Erdi, Kovalevsky, Liu, Marchal, Mark, Prigogine, Seidelmann, Vicenti, Waldvogel, Zare and others.

Research supports obtained from my Richard B. Curran Centennial Chair at The University of Texas, from USAF Office of Scientific Research and from NATO are gratefully acknowledged.

References

Hill, G. W.: 1878, Researches in the Lunar Theory, *American Journal of Mathematics*, **1**, 129–245.
Lagrange, J.: 1867-1892, Oeuvres, Gauthier-Villars, Paris.
Liapunov, A.A.: 1892, The General Problem of the Stability of Motion, *Commun. Math. Soc. Krakow*, **2**, 1.
Poincaré, H.: 1892-1899, Les Méthodes Nouvelles de la Mécanique Céleste, Gauthier-Villars, Paris.
Sitnikov, K.: 1960, The Existence of Oscillatory Motions in the Three-body Problem, *Dokl. Akad. Nauk USSR*, **133**, 303.
Strömgren, E.: 1922, Forms of Periodic Motion in the Restricted and in the General Problem of Three Bodies, *Copenhagen Obs. Publ.*, **39**.
Szebehely, V.: 1989, Adventures in Celestial Mechanics, University of Texas Press.
Szebehely, V.: 1967, Theory of Orbits, Academic Press.
Vinti, J.P.: 1959, A New Method of Solution of Unretarded Satellite Orbits, National Bureau of Standards.

LIMITS OF COMPREHENSION, REFERENCE FRAMES AND GENERAL RELATIVITY

HEINRICH EICHHORN*
Universitätssternwarte Wien
Türkenschanzstraße 17, A-1180 Wien, Austria

Abstract. The paper hypothesizes that human comprehension is limited because only those human traits and abilities emerged in the course of evolution which are important for the preservation of the species. For this reason, man can probably become aware of only a part of the entire reality. We are probably forever prevented from intuitively understanding the intrinsic nature of spacetime. Our notions about space and time developed and evolved in an environment where the maximum speeds consciously experienced are far smaller than that of light. As a consequence, the notions of Newtonian mechanics, of which the concept of inertial frames of reference is a part, attained the status of intuitively self-evident laws of nature. With the development of general relativity a (counterintuitive) insight emerged, *viz.* that gravitation is only a manifestation of the fact that particles in spacetime describe geodesics, and that the geometry of spacetime, described by the line element, is determined by the distribution of mass and energy in space or alternatively, that masses are a manifestation of the curvature of spacetime. This effectively obviates gravitational force. The only space accessible to immediate intuition is flat Euclidean space, where the geodesics are straight lines—arguably of privileged shape. There, inertial reference frames which can be realized in practice by a number of different and mutually independent procedures, are also mathematically privileged because in them, the coordinates of (mass) points moving freely (*i.e.*, under the influence of no force) are linear functions of time. Gravitational force appears there as something extraneous that keeps the points from moving uniformly on the geodesics of flat space, *i.e.*, on straight lines. In the curved spacetime of general relativity, the geodesics may have any shape, are thus no longer privileged and for this reason general relativity, in contrast to Newtonian mechanics and special relativity, no longer produces any privileged frames of reference.

Key words: philosophy of science – inertial reference frames – general relativity

1. Introduction: The Unavoidable Limits of Human Comprehension

We know The World through our sensory organs and our brains. For this reason, we can never be sure that Nature, the entirety of the existing—in some metaphysical sense that defies a clear definition in terms of previously well and unambiguously defined entities—can be perceived and fully comprehended by human beings. I rather think, as first apparently suggested by Parmenides and later discussed in much detail by Platon, that the limitations our senses and the structure of our brain impose on our ability to reason and on our perception, restricts us to perceiving and understanding only a subset of Nature. These restrictions probably keep the Faustian Dream "daß ich erkenne, was die Welt im Innersten zusammenhält" (so I may understand what keeps the world together in the innermost) forever from becoming true for mankind. All we can do is to attempt to bring, in a rational fashion, order into that subset of existing phenomena accessible to us.

Our mental and sensory abilities as well as limitations impose certain categories on us to which our perception of Nature is restricted. Among them are space and

* On leave from Department of Astronomy, University of Florida

Celestial Mechanics and Dynamical Astronomy **65**: 213–222, 1997.

time, and our unenlightened raw intuition compels us to hold certain notions for self-evident: Anything existing must exist in space and time. Our intuitive perception forces us to think of any event as happening in a definite and uniquely specified location, and this event must occur at a definite and uniquely specified time. This is to say that two otherwise identical events occuring even simultaneously (however this concept may be defined) at two distinctly different locations cannot be regarded as the same event, and events occurring at the same location and with all circumstances identical must be regarded as different unless they also occur at the same time. In addition, our perception dictates that we regard both space and time as continuous and measurable, time as a one-dimensional and space as a three-dimensional manifold. This translates into the statement that each instant ("point in time") is uniquely determined by exactly one real number, and each location ("point in space") is determined unambiguously by exactly one triple of real numbers.

It is remarkable that there is no structural *a priori* reason for some of our seemingly self-evident perceptions. Time is regarded as "obviously" one-dimensional because this is how people perceive time and think in terms of time, not because time can intrinsically and essentially not be anything but one-dimensional in nature. Anything different would be regarded as counterintuitive. Yet, theories with even three-dimensional time have been developed.

One of the most intriguing properties of time is, that *no procedure is known by which time can be measured directly*. While the necessities of our thinking patterns—if I may extrapolate from my own perception to that of others—dictate that we accept the "existence" of time as a oneparametric entity, we can experimentally determine the value of this parameter, as it is associated with a specified event, only by observing the state of a physical system which changes, as described by theory, as a function of time. Never in the history of science was time itself ever measured. It was alway so, that time was postulated as the parameter that characterized different spacial (or other) arrangements of a physical system. Astronomers originally "measured time" by counting oscillations of a pendulum or in terms of one of the Eulerian angles of the Earth, until it became clear (in the 1930ies) that the Earth's angular speed is not proportional to the time expressed in terms of the reading of a quartz clock. Later, they determined the numerical value of the time-parameter—defined as the independent variable in the differential equations of motion in celestial mechanics—by measuring the Moon's position and currently, time is measured by the output of atomic clocks. No one has yet suggested what time, qua "Ding an sich," "is" when it is regarded totally separated from its entanglement with other physical entities.

Does it then make sense at all to postulate the existence of time as an entity that can meaningfully exist apart from *any* other physical entity? This is a question in the same group as whether it makes sense to postulate the existence of space—likewise plausible enough to our intuitive thinking—independently of masses (or fields) that occupy it.

Relativity and quantum theory have taught us to be suspicious of common sense statements that are intuitively self-evident. In somehow existing complete reality, which is inaccessible to full human understanding, an instant of time might require more than one number for a complete and unambiguous specification. Similarly, our primitive intuition suggests the existence of an *arrow of time*, and some authors have suggested that there are physical reasons why such an arrow must actually exist. A simple thought-experiment will show, however, that the arrow of time may in fact only be a necessity in the world of human thinking and perception. Imagine a fictitious sentient bacterium (without spatial extension but, like human beings, gifted with the same intuitive sensation of time and an understanding of its existence) attached to a log floating down a river. Since the bacterium cannot move on its own but has its (one spatial) coordinate forced upon it by the flow of the river, the notion of one-dimensional space may be impressed upon the bacterium which may also be compelled to think that there is necessarily an "arrow of space," because this is so in the bacterium's world, and all the physics the bacterium develops may reflect this, while the human world certainly does not support the notion of an arrow of space. By analogy, however, may not the complete "real" world relate to the human world as the human world relates to that of the bacterium?[1]

Whether space must intrinsically be at least four-dimensional to accommodate curved three-dimensional space is an open question: For human intuition, a one-(two-)dimensional space seems to require at least a second (third) dimension in which to be curved, even though this is not necessary from the standpoint of differential (or generally any non-Euclidean) geometry, but then mathematics, as a human construct, need not have a strict analogue in physical reality.

2. A Hypothesis on the Nature of Human Comprehension

The question why the intellectual and sensory apparatus with which humans are equipped keeps them from comprehending reality fully cannot definitively be answered, but one may speculate about it. The statements of even special relativity are counterintuitive. The fact that the speed of incoming light in a vacuum is always measured to be the same regardless of how fast and in which direction emitter and receiver change their mutual distance flies in the face of the intuitive certainty that all velocities add like vectors. This certainty originates—so one may speculate—from the empirical fact that it really is so in observable nature within the

[1] For the purposes of this thought-experiment, we had to assume the bacterium to be a dimensionless but sentient entity with an intuitive appreciation of the concept of time. Otherwise one could have argued that the bacterium itself must perceive space as having the same dimensions it has itself. On the other hand, we humans might be of a dimension higher than three without realizing it and we take the three-dimensionality of space for granted not because physical space is undoubtedly and manifestly three-dimensional, but because the sensory and intellectual limitations of our human nature unavoidably subject us to perceive space as three-dimensional. (There have been speculations that space was originally eleven-dimensional but that only three dimensions have remained important in its macroscopic manifestations.)

available measuring precision as long as the observations are restricted to speeds very small compared to that of light, and the everyday speeds we encounter in daily life on Earth are indeed extremely small compared to that of light. If, as we have good reason to believe, intuitive certainties are the manifestations of evolved and inherited thinking patterns which are necessary consequences of the structure of our brain and that they have evolved and become established in response to stimuli by the physical world, it is not difficult to appreciate how the notion of vectorial addition of velocities attained (almost) the stature of a fundamental law of nature: It had the appearance (again, within the framework of achievable accuracy and precision) of always being rigorously verified for all cases involving speeds extremely small compared to that of light. I venture to say that the laws of (at least special) relativity would appear self-evident to us if our (distant) ancestors would in everyday life routinely have experienced speeds comparable to that of light.

Another consideration which may explain some of the shortcomings of the human intellectual and sensory apparatus is that in the course of evolution, those abilities to perceive and comprehend that were the most useful, important and necessary for survival are the ones which emerged and developed preferentially. For example, it is most likely no accident that the sensitivity function of the typical human eye emulates the spectral intensity function of the Sun quite closely, thus having provided us with an efficient sensory organ without which our emergence as the dominant species on the Earth would probably not have happened. X-rays exist, but a (natural) sensory organ for it has not evolved. It would indeed be quite useless, because the Earth's atmosphere blocks (for all practical purposes) all X-rays from space, so that there was no stimulus that could have triggered the evolution of an appropriate sensory organ. The view that we developed preferentially those (and only those) abilities and thinking propensities that helped us deal with the most important, but by no means all, among those facets of our natural environment that were indispensable for the emergence of ours as the dominant species seems very reasonable.

This process may have kept certain insights (namely those without survival value) from us as well as kept us from realizing certain existing relationships, especially as space and time are concerned. Space appears to us as an entity in itself which "exists" independently of the matter (or energy) contained in it. Objects seem to have shapes that depend in no other way on the "surrounding" space except that they occupy it, and we take it for intuitively evident that the reference system organizing the space which the object occupies cannot influence the object's "shape," which is intuitively perceived as an intrinsic attribute. And yet, there are models in which mass is just another word for the structure of space, and special relativity teaches the change a "shape" undergoes through the Lorentz contraction when it moves with respect to a certain reference system, while the shape of the same object is different at the same time when the object moves in a different way with respect to another reference system.

3. Spacetime and Reference Frames

We perceive space as that physical entity which relates to objects such that the object's location in it at a given time is unambiguously fixed as soon as space is *organized*, which means that a device—a coordinate system—is introduced that connects each point in space unambiguously with a triple of numbers and *vice versa*. Such a device has traditionally been called a *coordinate system* or a *system of coordinates*.[2] There is always an infinite variety of coordinate systems that can organize a reference frame, which is a physical entity when its motion (rotation and shift of the origin) with respect to a sufficiently numerous set of objects is known. Note that an *extended* rigid object—and all objects are intrinsically extended (even though not intrinsically rigid)—can serve as an object rigidly connected to (*i.e.*, motionless within) a certain reference frame which they thus define. (One might then call this the object's "proper" reference frame, as is usual in investigations involving the theory of relativity; the time in this frame is the "proper time".)

In order to conduct investigations in physics, that is establish equations that relate measurable properties of physical entities to each other, we must define and implement frames of reference whenever the measurable properties involve spacetime, as it is the case with motion- or space-related properties such as *e.g.*, volume. However, a reference frame is as a rule much more easily defined than implemented; note that it is only possible to realize (more or less accurate) approximations, but not the frames themselves.

Suppose we restrict ourselves for the purpose of this train of thought (without thereby restricting the generality of the salient arguments) to the motion of particles within the conceptual framework of Newtonian mechanics. Newton's first so-called law (it is actually a postulate) could be rephrased as follows: *There are frames of reference with respect to which particles not subjected to a force always move uniformly*. *Prima facie*, this cannot be experimentally verified directly because nowhere in the universe will there be a region in which the attraction of all masses of the universe is neutralized. The only possible verification is to find (guess at?) the best available model for the force (*e.g.*, Newton's law of gravitation), set up the equations of motion by using Newton's Second Postulate of Motion and his Third Law of Motion, integrate these equations and then try to find a frame of reference such that the theoretically found locations coincide with those actually observed, keeping in mind Karl Popper's convincing argument that definitive verification of any law of nature is intrinsically unattainable. This is, of course, possible only within the framework of available accuracy and precision of the observations.

Consider a frame of reference and a coordinate system that organizes it. We could, for example, *define* a reference frame as inertial and co-moving with the

[2] To insist on introducing a new name or to assigning a new meaning to the concept of "coordinate system" would isolate the community of users of these neologisms from the community of users at large and thus create confusion. Unfortunately, such a situation has emerged due to some astronomers' assigning a meaning to the concept "frame of reference" or "reference frame" that differs from what these terms mean in the (much larger!) community of physicists.

barycenter of the solar system. For this purpose, we need to assume the solar system's barycenter to move uniformly (which is not quite true because of its revolution round the center of the Galaxy) and we also make the (not entirely correct) assumption that the only forces in the solar system are the gravitational attractions of its constituent bodies upon each other. We can then set up the equations of motion. Even though we cannot write down any closed solutions, the existence-and-uniqueness theorem for differential equations assures us that a solution exists and that it is unique for a given set of initial conditions. We can then (in principle) find arbitrarily accurate approximations by *e.g.*, numerical integration. Next, we can define a system of coordinates (for example the system Q of kinematic astronomy in which right ascension and declination are the angular polar coordinates) and measure the positions (locations) of the constituent bodies of the solar system with respect to it. Therefore we could arrive, at least in principle, at (an approximation to) an inertial frame of reference by finding that transformation (it would ideally be a time dependent rotation) which will make the coordinates predicted from the theory of the solar system bodies agree with the measured ones.

We presume that the axes have been defined by relation to physical principle, either dynamical (*e.g.*, the invariable plane) or kinematic (*e.g.*, the axis of rotation of the solid model Earth). Unfortunately, the implementation (meaning the practical realization of the directions of the axes) is frought with difficulties.

The model of the solar system as isolated from the rest of the universe is patently unrealistic and thus in principle false, even though the forces originating from outside the solar system may not produce any effects that would be noticeable within the framework of contemporary accuracy and precision. We know (as remarked above) that closed uniform solutions of the equations do not exist (as Poincaré has shown). Whatever solutions we obtain will therefore *eo ipso* be approximate ones; this is not a very serious objection as the approximations can—in principle though not realistically in practice—be made more accurate and precise than the measurements carried out to check them. (We must not forget the relativistic "corrections" which can also be calculated with sufficient accuracy.)

The principal handicap is the necessity to rely on measurements. Measurements are intrinsically of limited precision; all a measurement can yield is an approximate value of the target parameter. It is therefore obvious that every result derived by a process in which measurements play an essential and unavoidable rôle will deviate from what would have been obtained had the measurements been totally free of errors.

4. Procedures for the Practical Realization of an Inertial Reference Frame

As a rule, there are several different procedures available for the estimation of the same parameter. We have just described how one can establish an approximation to an inertial frame of reference by investigating the various (theoretical and practical) aspects of the motion of the constituent bodies of the solar system. Another principle available for the empirical establishment of an approximation to the (an) inertial system draws on a totally different source: the stars in the Galaxy. In this context, it is no longer possible to trace the orbit (motion) of each individual star. The best we can expect from dynamics is to find the theoretical probability distribution function of the velocities of the individual stars at each point (in practice: in each region) of the Galaxy. When we know what sample of stars we will incorporate in our investigations in each direction (in practice: region of the sky), we can—again in principle—derive the probability distribution function of the observable parameters, *i.e.*, proper motion components and radial velocities in this region. If we also have a model for the group motion of the stars in the Galaxy (such as galactic rotation and shearing), we can analyze the observed stellar motions for the estimation of the parameters of the group motions (*e.g.*, Oort's constants, the Sun's velocity components) and those (time dependent ones) of rotation which transform the system in which the measurements are made to an inertial reference frame. This has been done: The late W. Fricke corrected the precession parameters in this way. In comparison with finding an approximation to an inertial frame of reference by involving only solar system bodies, this (so to speak) galactic-body-based approximation has an intrinsic upper limit on the accuracy of the relevant parameters in that the velocities of the stars (in contrast to those of the solar system bodies) cannot be individually determined but must be regarded as statistical variates.

This disadvantage could be avoided by picking stars (or other objects) for this investigation which are known not to move. Such objects do not exist, of course, but very distant galaxies (with good observational characteristics, meaning that they produce star-like nearly circular images) come close to fulfilling this requirement: Their proper motions are so small that they do not change their inertial directions perceptively and thus may be regarded as 0 over a sufficiently long time span.

It lies in the nature of statistical investigations that each of the abovementioned three independent methods (and there are more) will yield an approximation to an inertial system and that therefore each of these approximations will rotate with respect to the others. The IAU recommends that the system based on the investigation of some 200 distant galaxies *define* "the" inertial reference frame. It would at first appear that there are in principle no objections to this procedure because *the* target inertial reference frame is not available, yet one expects that the combination of the results obtained by this approach with those obtained by other methods by way of a simultaneous adjustment of the available observations would have produced a more accurate approximation.[3] Also, those who have much experience in statis-

[3] Before the solution of overdetermined systems of equations by the method of least squares became commonplace, investigators who had measured a quantitiy several times and obtained (of course) a different value each time selected what was in their judgement the "best" measurement (or observation) and discarded the others instead of taking the mean of all measurements. Still prevailing practices such as using only one of several independent methods for the realization of an intertial reference frame make me wonder if we have completely moved away from the renaissance and into

tical adjustments will object to the identification of any *approximation*, no matter how accurate, with the target entity itself—in the theory of statistical adjustment it is often important to distinguish beween an entity and its approximation (or estimate). Another reason not to hinge a fundamental definition to an approximation is that some approximations are intrinsically sample-dependent, meaning that the same method, applied to different samples from the same parent population of observations, will always produce different results.

The difference between "catalogue equinox" and "dynamical equinox," with which those who work in fundamental astrometry trying to fix coordinate systems are familiar, is a striking example of the confusion created by not distinguishing between an entity and its approximation. *There can only be one equinox*, and its definition is anchored in physical privilege. Two independent approximations to the same entity *must* disagree (and may even give the impression of being incompatible within their formal errors), and one may well argue that the combination of several kinds of material for the unique estimation of a parameter would give an estimation more accurate than each of the individual ones; provided of course, that the reduction models are realistic. Incompatibility of different independent estimates of the same parameter within the formal errors is an indicator (provided that the formal errors were correctly computed) that at least one of the reduction models was unrealistic, and the central limit theorem would then suggest the combination of even quite heterogeneous materials if they all can make contributions to the estimation of a set of common parameters.

5. Inertial Reference-Frames and General Relativity

Above, we discussed inertial frames in Newtonian mechanics. Their definition (in terms of Newton's postulates and laws) makes them privileged in the sense that *Newton's first postulate and third law of motion are true only in inertial frames of reference.* This does not say that the motion of a particle cannot be correctly described with respect to a noninertial frame of reference. Consider an arbitrary reference frame A organized by a cartesian coordinate system K. Let the particle move freely, i.e., not influenced by any force, and describe its motion with respect to K by $\boldsymbol{x}^A(t)$. Since A is not inertial, the curve $\boldsymbol{x}^A(t)$ will be neither a straight line nor will the particle move uniformly along it. Now assume an inertial frame of reference I, also organized by a cartesian system of coordinates and let A and I be connected by the matrix $\mathbf{M}(A, I)$ so that the curve $\boldsymbol{x}^I(t) = \mathbf{M}(A, I)\boldsymbol{x}^A(t)$ which the particle describes with respect to I is (as a consequence of Newton's First Postulate) a straight line, which will, in addition, be traversed uniformly if t is the time. *This shows that the path even of a freely moving particle depends on the choice of the reference frame and is a uniformly traversed straight line only with respect to an inertial system and if the parameter t is the time.* The laws of Newtonian mechanics are invariant with respect to Galilei transformations and

the modern age.

those of special relativity with respect to Lorentz transformations. To obtain any kind of nonuniform motion in an inertial frame of reference, there must be forces acting on the particles.

The view of general relativity is different. There, in the absence of nongravitational effects, *all* particles move along the geodesics of curved spacetime which are the solutions of the well-known differential equations

$$\ddot{u}^{\mu} + \Gamma^{\mu}_{\kappa\lambda}\dot{u}^{\kappa}\dot{u}^{\lambda} = 0.$$

The geometry of spacetime is determined by the field equations

$$R^{\mu\nu} - \frac{1}{2}Rg^{\mu\nu} = \kappa T^{\mu\nu}$$

which connect the contravariant Ricci-tensor $R^{\mu\nu}$ (a symmetrical contraction of the curvature tensor, which therefore has ten independent components in four-dimensional spacetime) with the stress tensor $T^{\mu\nu}$, which contains all forms of energy—therefore also mass—and momentum. The field equations are essentially a set of partial differential equations for the components of the metric tensor and therefore allow us to calculate the Ricci tensor and the (symmetrical) metric tensor $g_{\mu\nu}$. No king's way is known for this process. The connection coefficients $\Gamma^{\mu}_{\kappa\lambda}$ necessary for setting up the differential equations of the geodesics can be calculated routinely from the components of the metric tensor. The geodesics are curved and their shapes do not depend on the coordinate system employed to organize spacetime. In view of the variety of the possible shapes of the geodesics, all reflecting the stress tensor (with respect to the reference frame and not the specific coordinate system used to organize it), there is (except for freak cases, e.g., when $g_{\mu\nu} \rightarrow \eta_{\mu\nu}$ for "$r \rightarrow \infty$") no global physically privileged reference frame, although one can define a reference frame in the neighborhood of each point of spacetime in which events may be accurately described in terms of special relativity. *The globally privileged inertial reference frames disappear together with flat space and gravity as a force.* One may thus argue that flat inertial reference frames as well as gravity are artifacts that emerge when the curvature of space vanishes and the geodesics become straight lines; inertial frames of reference are thus a device that produces the laws of motion and gravitation in the simple form that was first stated by Newton.

This is different from what usually happens when certain laws of physics are subsumed under and derived from more general ones as physical theory is generalized, *e.g.*, when Kepler's laws are derived from the general principles of Newtonian mechanics when these are applied to investigate the motion of two isolated mass points, or when the laws of classical mechanics emerge from those of special relativity as $v/c \rightarrow 0$. The transition from special to general relativity is less a process of generalization than a genuine change of paradigm because in general relativity, the concept of gravitational force no longer exists. One will therefore not find a concept in general relativity that specializes to an inertial reference frame when one goes to special relativity or Newtonian mechanics; inertial reference frames rather have to be newly introduced to make sure that Newton's first postulate of motion can be pronounced and is correct. Space, in Newtonian mechanics and in special relativity is implicitly Euclidean; a theory which describes correctly the actual curved paths of moving bodies *must* therefore introduce gravitational forces. Particles not moving under the influence of forces will move uniformly that is, on geodesics of Euclidean space. In special relativity and Newtonian mechanics there emerge accordingly those reference frames as privileged in which the geodesics are straight lines, *and these are exactly the inertial ones.*

Acknowledgements

I profited from enlightening conversations with my colleagues Henry Kandrup and Reva Kay Williams. George Contopoulos, who refereed the originally submitted version of this paper, made very perceptive and cogent comments. Remaining misconceptions and errors are, however, my sole responsibility.

STABLE CHAOS: A FIRST MODEL

ANNE LEMAITRE
FUNDP, Département de Mathématiques
Rempart de la Vierge, 8
B-5000 Namur, Belgium
Tél : (32-81) 72 49 08 Fax : (32-81) 72 49 14
E-mail: alemaitr@math.fundp.ac.be

Abstract. This paper presents a very simple analytical model, so to test the apparition of stable chaos in the asteroidal belt; the idea, introduced by Milani et al (1996), is that the superposing of all the critical terms with different multiples of the longitude of the pericentres, in the neighborhood of a high-order mean-motion resonance, could explain important jumps in the semi-major axis, even in stable regions, and explain the very short Lyapounov times calculated for test asteroids. The cases of 490 Veritas and 564 Dudu are chosen as test examples.

Key words: Chaotic motion – stability – asteroids

1. Introduction

This paper is a test concerning a possible explanation for "stable chaos" in the main asteroid belt (following the terminology introduced by Pisa's team, see for example Milani and Nobili (1992) or Milani and Farinella (1994-1995)). It has been shown (Milani et al.1996) that several asteroid orbits are characterized by very stable proper elements (variations of proper eccentricities and proper inclinations less than 10^{-3}) but are associated with short Lyapunov times ($\leq$ 10000 years). Chaos is (without any doubt) present in those orbits but it is confined in very thin regions of the phase space. We reproduce here in table I parts of tables I and II taken from the above-mentioned paper to provide examples of those cases.

The three authors try to explain such a strange behavior, especially the very large excursions of about 10^{-3} observed in the mean semi-major axis a. Their main observation is the presence in all the treated cases of one (or even two) high order mean motion resonance with Jupiter; these resonances (denoted by $\frac{p+q}{p}$) are probably responsible for the observed chaos. Jumps in semi-major axis are visible on filtered numerical integrations and occur randomly, always on the same side of the main mean-motion resonance. As an example, the group around the asteroid 490 Veritas is influenced by the 21/10 mean motion resonance with Jupiter (also by the 44/21) and the group around 564 Dudu is close to the 13/5 resonance. We reproduce (see figure 1) one of these figures (case of 490 Veritas, figure 2 in the original paper). The mean motion resonance introduces (in the perturbing function) a summation (over m) of terms like $A_m(a)\, e'^{q-m}\, cos(p\lambda-(p+q)\lambda'+m\varpi+(q-m)\,\varpi')$ where a is the semi-major axis of the asteroid, λ and λ' are the mean longitudes of the asteroid and of Jupiter and ϖ (respectively ϖ') is the longitude of the pericentre of the asteroid (respectively of Jupiter). A coefficient $A_m(a)$ alone is certainly not large enough to explain the excursions of a, but the summation over m and the proximity of the resonance could significantly increase the size of the oscillations

R. Dvorak and J. Henrard (eds.), The Dynamical Behaviour of our Planetary System, 223.

ASTEROID	LCE		Resonances
490 Veritas	9.9	10^{-5}	21/10 44/21
3542 1964TN2	11	10^{-5}	21/10 44/21
564 Dudu	25.4	10^{-5}	13/5
485 Genua	15.5	10^{-5}	13/5
795 Fini	9.9	10^{-5}	13/5

TABLE I
Results from Milani, Nobili and Kneževič 1996, for the group of 490 Veritas and the group of 564 Dudu

of a. This was one of the conjectures of the three authors, and our purpose was to check it, using a minimal analytical model.

We present the context of our analysis in the first section, defining Delaunay's variables and then action-angle variables. We average over the fastest variables and we obtain the averaged oscillations of the semi-major axis. To keep a complete analytical expression, we use Bessel functions expansions.

In the second section we apply our model to the case of 490 Veritas and associated asteroids and to the case of 564 Dudu and similar cases.

The results are briefly commented in the conclusion (third section); they are not in favor of a global explanation of stable chaos by superpimposition of various terms of the same mean motion resonance; this explanation works qualitatively (the jumps in the semi-major axis are much more important using the summation formula than expected by using only the main resonant term) but not quantitatively, in the sense that for the case of Veritas, even the summation is not large enough to reproduce the observed jumps. It seems that local individual explanations have to be added in each case.

2. The Model

2.1. THE CONTEXT

To isolate as much as possible the effect of one particular resonance in the averaged motion of an asteroid, we introduce a simplified dynamical model, based upon the restricted 3-body problem, Sun-Jupiter-asteroid, planar and averaged over the periods of the planets around the Sun, and close to a $\frac{p+q}{p}$ mean motion resonance (with rather large q). Let us give the expression of the Hamiltonian of the problem expressed in usual canonical variables :

$$
\begin{aligned}
H &= -\frac{\mu^2}{2L^2} + n'\Lambda' + CS + D\sqrt{2S}\cos\theta \\
&+ \sum_{m=0}^{q} \sqrt{2S}^{m} A_m e'^{q-m} \cos(p\lambda - (p+q)\lambda' - m\theta + (m-q)\,\theta')
\end{aligned} \qquad (1)
$$

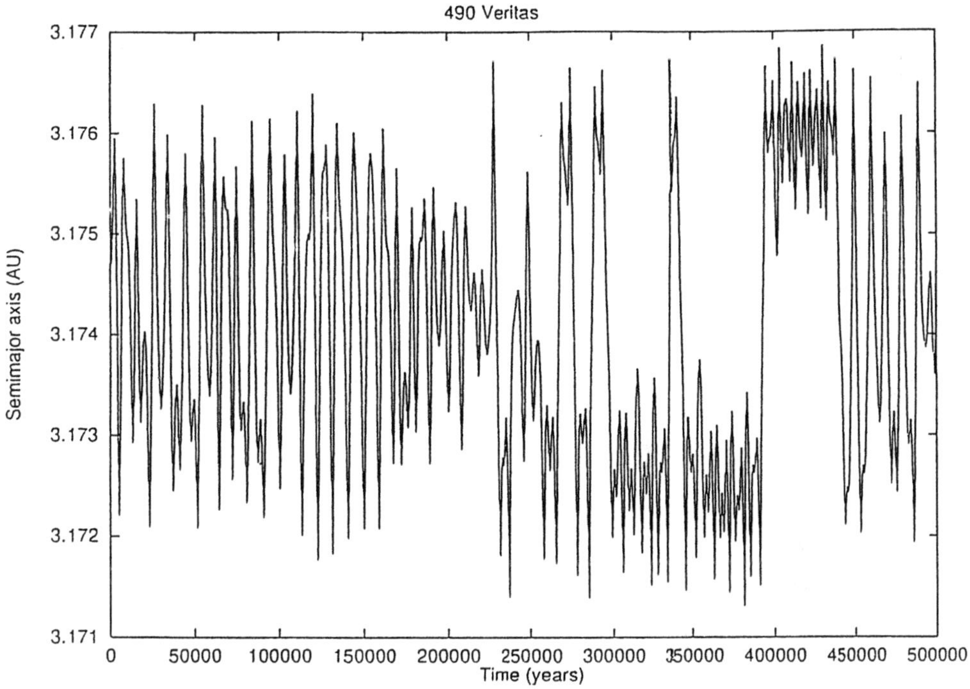

Fig. 1. variations of the filtered semi-major axis of 490 Veritas with respect to time for 500000 years, taken from Milani, Nobili and Knežević 1996 (Figure 2)

with

$L = \sqrt{\mu a}$		λ	= the mean longitude
$S = L\,(1-\sqrt{1-e^2})$		θ	$= -\varpi$
Λ'		λ'	= the mean longitude of Jupiter
e'	= the eccentricity of Jupiter	$-\theta'$	= the longitude of the pericentre of Jupiter

The coefficients C, D or A_m are functions of the momentum L and are evaluated by numerical Fourier transform, for given values of the semi-major axis.

Introducing the resonant angle ϕ as a canonical variable, we obtain :

$$H = -\frac{\mu^2 p^2}{2q^2 L_1^2} - n'\frac{p+q}{q}L_1 + n'\Lambda + CS + D\sqrt{2S}\cos\theta \tag{2}$$

$$+ \sum_{m=0}^{q} \sqrt{2S}^{m} e'^{q-m}\, A_m \cos(q\phi - m\theta + (m-q)\,\theta') \tag{3}$$

where

$$
\begin{array}{rclcrcl}
L_1 & = & \frac{q}{p}L & & \phi & = & \frac{p}{q}\lambda - \frac{p+q}{q}\lambda' \\
S & = & \sqrt{\mu a}\,(1-\sqrt{1-e^2}) & & \theta & = & -\varpi \\
\Lambda & = & \Lambda' - \frac{p+q}{p}L & & \lambda' & &
\end{array}
$$

We decide to choose $\theta' = 0$ so to simplify the trigonometric argument.

2.2. Action - Angle variables

The angle θ is present in the first part of the Hamiltonian, which we shall indicate by H_0, before the summation over the resonant terms, which is indicated by H_1. For the purpose of eliminating this angular dependence from H_0, we introduce action-angle variables J and ψ :

$$
\begin{aligned}
H &= H_0 + H_1 \\
H_0(L_1, S, \theta) &= H_0(M, J) \\
H_1(L_1, \phi, S, \theta) &= H_1(M, \sigma, J, \psi)
\end{aligned} \tag{4}
$$

The transformation to action-angle variables corresponds to a simple translation, the result of which is a set of linear proper elements :

$$
\begin{array}{rclcrclcl}
x & = & \sqrt{2S}\cos\theta & \rightarrow & \xi & = & x - \alpha(L_1) & = & \sqrt{2J}\cos\psi \\
y & = & -\sqrt{2S}\sin\theta & & \eta & = & y & = & -\sqrt{2J}\sin\psi
\end{array}
$$

In order to preserve the canonical structure (see for example Henrard 1990), we also modify the second degree of freedom (L_1, ϕ) which is now called (M, σ) with the relations :

$$
\begin{aligned}
\sigma &= \phi + y\beta(L_1) \\
M &= L_1
\end{aligned} \tag{5}
$$

The third degree of freedom (λ' and Λ') is not concerned by the canonical transform and stays unchanged. The coefficients α and β are functions of L_1, i.e. of the semi major axis :

$$
\alpha(L_1) = -\frac{D}{C} \qquad \beta(L_1) = \frac{\partial\alpha(L_1)}{\partial L_1}
$$

In the new canonical set of variables (ξ, η, M, σ) or (J, ψ, M, σ), H is given by :

$$
\begin{aligned}
H = -\frac{\mu^2 p^2}{2q^2 M^2} & - n'\frac{p+q}{p}M + n'\Lambda' + CJ + \frac{D}{C^2} \\
& + \sum_{l=0}^{q} \sqrt{2J}^{\,l} E_l \cos(q\sigma + q\beta\eta - l\psi)
\end{aligned}
$$

or

$$H = -\frac{\mu^2 p^2}{2q^2 M^2} - n'\frac{p+q}{p} M + n'\Lambda' + CJ + \frac{D}{C^2} + \sum_{l=0}^{q} \sqrt{2J}^l E_l \cos(q\sigma - q\beta\sqrt{2J}\sin\psi - l\psi) \tag{6}$$

The new coefficients E_l are linear combinations of the initial coefficients A_m and of powers of α and e':

$$E_l = \sum_{m=l}^{q} A_m(M)\, e'^{q-m}\, \alpha^{m-l} \binom{m}{l} \tag{7}$$

2.3. Bessel Functions

The last summation (6) is a combination of cosine of sine; the classical way to describe it analytically is a Bessel functions expansion :

$$\begin{aligned} \cos(\epsilon \sin\psi) &= J_0(\epsilon) + 2\sum_{k=1}^{\infty} J_{2k}(\epsilon)\cos 2k\psi \\ \sin(\epsilon\sin\psi) &= 2\sum_{k=0}^{\infty} J_{2k+1}(\epsilon)\cos(2k+1)\psi \end{aligned} \tag{8}$$

Including these developments into the expression of H_1, we get the following double sum :

$$\begin{aligned} H_1 &= \sum_{l=0}^{q} \sqrt{2J}^l E_l\, J_0(\epsilon)\cos(q\sigma - l\psi) \\ &+ \sum_{l=0}^{q} \sqrt{2J}^l E_l \sum_{k=-N,k\neq 0}^{N} J_{|2k|}(\epsilon)\cos(q\sigma - l\psi + 2k\psi) \\ &+ \sum_{l=0}^{q} \sqrt{2J}^l E_l \sum_{k=-N}^{N} J_{|2k+1|}(\epsilon)\cos(q\sigma - l\psi + (2k+1)\psi) \end{aligned} \tag{9}$$

where : $\epsilon = q\beta\sqrt{2J}$, $\sqrt{2J}$ being proportional to the linear proper eccentricity.

The Bessel functions are all computed for the same value of the parameter ϵ, which mainly depends on the parameter β. We computed the variation of ϵ for the two regions of interest (Veritas and Dudu regions) and this parameter does not exceed 0.05. Let us remind that a Bessel function $J_n(x)$ behaves as x^n (apart from the J_0 case). This means that the summation over k from $-N$ to N can be truncated very rapidly.

2.4. AVERAGE OVER THE RESONANT ANGLE

Our Hamiltonian function is now ready to apply a first order perturbation method, the first part being only function of the two momenta J and M and the second part, the perturbation, given as a Fourier series with respect to the two angles ψ and σ. The idea is to average over the fastest angle, which is, for high order resonances, the resonant angle σ. The averaged canonical variables are denoted by $(\bar{M}, \bar{\sigma}, \bar{J}, \bar{\psi})$ and the Hamiltonian is transformed into $\bar{H}$ which reduces to $\bar{H}_0(\bar{M}, \bar{J})$ neglecting all terms of order two or more. The generating function for this transformation can be given analytically, again as a sum of Bessel functions :

$$\begin{aligned} W &= \sum_{l=0}^{q} \sqrt{2J}^{l} E_l\, J_0(\epsilon) \frac{\sin(q\sigma - l\psi)}{q\omega_1 - l\omega_2} \\ &+ \sum_{l=0}^{q} \sqrt{2J}^{l} E_l \sum_{k=-N}^{N} J_{|2k|}(\epsilon) \frac{\sin(q\sigma - l\psi + 2k\psi)}{q\omega_1 - (l-2k)\omega_2} \\ &+ \sum_{l=0}^{q} \sqrt{2J}^{l} E_l \sum_{k=-N}^{N} J_{|2k+1|}(\epsilon) \frac{\sin(q\sigma - l\psi + (2k+1)\psi}{q\omega_1 - (l-2k-1)\omega_2} \end{aligned} \tag{10}$$

where ω_1, ω_2 are the derivatives of H_0 with respect to M, J respectively, and can be computed from (6).

2.5. AVERAGED DYNAMICS

The averaged dynamics is obvious : the two momenta are constant and the two angles are linear functions of time, associated with the proper frequencies ω_1 and ω_2.

$$\begin{aligned} \dot{\bar{J}} &= 0 \quad \dot{\bar{\psi}} = \frac{\partial \bar{H}_0}{\partial \bar{J}} = \omega_1 \\ \dot{\bar{M}} &= 0 \quad \dot{\bar{\sigma}} = \frac{\partial \bar{H}_0}{\partial \bar{M}} = \omega_2 \end{aligned}$$

The connection between the doubly averaged (over the short periods and over the resonant angle) variables $(\bar{M}, \bar{\sigma}, \bar{J}, \bar{\psi})$ and the usual once averaged variables (M, σ, J, ψ) is given through the generating function W :

$$\begin{aligned} J &= \bar{J} - \frac{\partial \bar{W}}{\partial \bar{\psi}} \qquad \psi = \bar{\psi} + \frac{\partial \bar{W}}{\partial \bar{J}} \\ M &= \bar{M} - \frac{\partial \bar{W}}{\partial \bar{\sigma}} \qquad \sigma = \bar{\sigma} + \frac{\partial \bar{W}}{\partial \bar{M}} \end{aligned}$$

where

$$\bar{W} = W(\bar{M}, \bar{\sigma}, \bar{J}, \bar{\psi})$$

Let us recall that we are interested in measuring the order of magnitude of the oscillations of the semi-major axis induced by the mean motion resonance $\frac{p+q}{p}$ and all the associated multiples of ϖ. Thanks to equations (2.5), the motion of the once averaged semi-major axis can be written as a function F of time :

$$a = \bar{a} + F(t, \bar{M}, \bar{J}, \omega_1, \omega_2)$$

3. Results

3.1. 490 VERITAS & 3542 1964TN2

Let us give the characteristics of the planets in that area, very close to the $21/10$ mean motion resonance. We do not analyze exactly planet 490 or planet 3542 but we are interested in following the behavior of the semi-major axis for similar planets. The situation of such planets in the asteroid belt is : $\bar{a} \simeq 3.173$ AU; the eccentricities are about 0.05 and the frequency $\omega_1 \simeq -92$ arcsec/yr (directly given by the calculation of C).

$\bar{a}$	$\bar{e}$	frequency ω_1	$\bar{a}_{max} - \bar{a}_{min}$
3.174 AU	0.06	-91.8 arcsec/yr	0.003 10^{-4} AU
3.173 AU	0.10	-91.8	0.1
	0.20	-95.1	1

TABLE II
Examples of maximal jumps in semi-major axis measured in the Veritas region, for various values of the averaged semi-major axis $\bar{a}$, of the averaged eccentricity $\bar{e}$ and of ω_1 for 2 500 000 years

The variation of the semi-major axis calculated by the above method is less than 10^{-5} AU for eccentricities $\bar{e} \leq 0.1$ (see table II). The optimal value for $\bar{a}$ is of course dependent on the proximity of the resonance $21/10$, in the framework of this model. To get by this model the characteristic value of 10^{-3} AU which was the main unexpected feature about Veritas motion, we have to calculate all coefficients at $\bar{a} = 3.173$ and to increase the eccentricity to 0.2, which has nothing anymore to do with the real planet Veritas.

3.2. 564 DUDU - 795 FINI - 485 GENUA

Only partial results were given in the paper of Milani, Nobili, Knežević (1996) about "Dudu-type" planets. We have first integrated them again numerically, over 2.5 million years, using Milani's ORBIT9 software, adapted for variable step integration, as it is necessary for an accurate integration at rather high eccentricities. As a check we have found similar values for the LCE's and for the proper elements. In the three cases (and even in other ones not reproduced here) we have noticed, as for Veritas group, variations of the semi-major axis (filtered to eliminate all periods between 2 and 200 years) of the order of magnitude of 2 or 3 10^{-3} AU i.e. much larger than expected for asteroids in "stable" regions (see figure 2).

Here also, we are interested in cases like Dudu, Fini or Genua, to observe the effect of the resonance $13/5$ over the semi-major axis and not in the exact motion of these three planets. The similarities in their behavior give us reasons to analyze the general behavior in that area, characterized by $\bar{a} \simeq 2.75$ AU; the

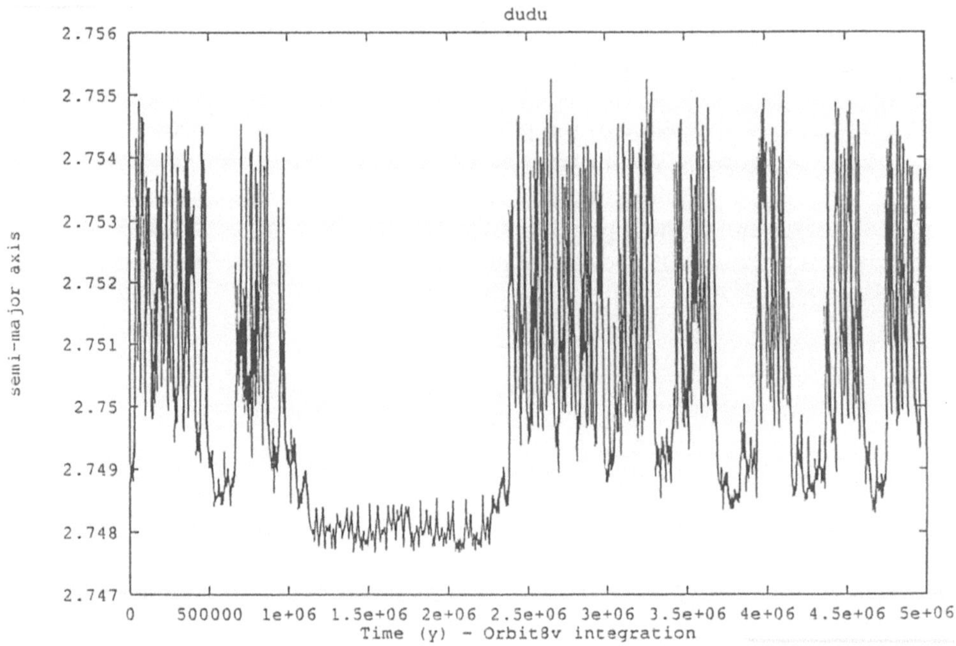

Fig. 2. Variations of the filtered semi-major axis of 564 Dudu with respect to time for more than 4500000 years

proper eccentricities are much larger than in the previous region (≥ 0.18) and the frequency $\omega_1 \simeq -58$ arcsec/yr.

In this case, the variation of the semi-major axis is always close to 10^{-3} AU at least for eccentricities greater than 0.2 and a close enough to the exact resonance $13/5$. Let us give a few examples in table III:

$\bar{a}$	$\bar{e}$	frequency ω_1	$\bar{a}_{max} - \bar{a}_{min}$
2.7520 AU	0.25	-58.4 arcsec/yr	58 10^{-4} AU
2.7525 AU	0.18	-57.5	6
	0.20	-57.7	11
	0.25	-58.4	37
2.7528 AU	0.20	-57.8	7
2.7530 AU	0.20	-57.8	6
	0.15	-58.4	2

TABLE III
Examples of maximal jumps in semi-major axis measured in the Dudu region, for various values of the averaged semi-major axis $\bar{a}$, of the averaged eccentricity $\bar{e}$ and of ω_1 for 2 500 000 years

4. Conclusions

Thc model built up for Veritas case i.e. truncated in e and associated with a high order mean motion resonance (order 11 for Veritas) does not give the expected variation of a. The resonance gives impulses to the semi-major axis, but the order of magnitude is not correct. To explain the phenomenon, we can mention the proximity of a very important mean motion resonance, the $2/1$, the effect of which is very large; of course the short periodic effects are eliminated by the filtering of the numerical integration output, but the second order effects are by no means negligible.

Another possible improvement would be to take into account many more terms in the perturbing function, which is in reality a double summation aver the powers of e and e' and not a single summation, as presented here; for this first analysis I have only chosen the terms selected by Milani, Nobili and Knežević (formula (1) in their paper) but it would be interesting to complete the model by other resonant terms and to check their importance.

The case in which it seems to work (the Dudu case) is such that it is not possible to generalize to most other cases of stable chaos : the eccentricity is much larger, and the order of the resonance is smaller (8 instead of 11). The result depends on two parameters : the eccentricity and the proximity to the resonance. Changing one of them is enough to get a much smaller variation of a. On the other hand, at any position in the asteroid belt we are close to a high order mean motion resonance. Escaping from one means being under the influence of another one. So the final question should be : is it possible to explain the stable chaos behavior, everywhere in the asteroid main belt, by the same arguments ? Our first answer is dubitative, on the base of this first simple analysis. This might indicate that a detailed theory of the interaction of secular perturbations and high order mean motion resonances needs to be done, case by case, with different approximations and analytical methods, a very annoying conclusion.

Acknowledgements

I would like to thank Andrea Milani for his collaboration to this paper; I have appreciated his comments and suggestions for future improvements of the present model.

References

Henrard, J.: 1990, A semi-numerical Perturbation Method for Separable Hamiltonian Systems, *Celest. Mech.*,**49**,43–67

Milani, A., Nobili, A. and Knežević, Z.: 1996, Stable chaos in the asteroid belt, *Icarus*, accepted.

Milani, A. and Nobili, A.: 1992, An example of stable chaos in the Solar System, *Nature*,**357**,569-571

Milani, A. and Farinella P.: 1994, The age of the Veritas Family deduced by chaotic chronology, *Nature*,**370**, 40-42

Milani, A. and Farinella P.: 1995, An Asteroid on the brink, *Icarus*,**104**,209-212

TRANSPORT IN PERTURBED INTEGRABLE HAMILTONIAN SYSTEMS AND THE FRACTALITY OF PHASE SPACE

H. VARVOGLIS, CH. VOZIKIS AND B. BARBANIS
Section of Astrophysics, Astronomy and Mechanics,
Department of Physics, Aristotle University of Thessaloniki
GR-54006 THESSALONIKI, GREECE

Abstract. We study transport in a model perturbed integrable Hamiltonian system by calculating the volume, $V(t)$, of elementary phase space cells visited by a trajectory, as a function of time. We use this function in order to "measure" the fractality of phase space. We argue that the "degree" of fractality is related to the well known difficulties in assigning unambiguously Lyapunov Characteristic Numbers (LCN's) to trajectories. Moreover we show that transport in phase space regions with pronounced fractality cannot be described as "normal diffusion", since the self-similar properties of $V(t)$ imply that it is governed by Levy statistics, while the correlation dimension of dV/dt implies that, in some cases, the process is strongly non-Markovian.

Key words: Hamiltonian systems – diffusion – fractality – Lyapunov numbers

1. Introduction

Transport in phase space is an interesting theoretical aspect of chaotic behaviour in perturbed integrable Hamiltonian systems, since it is encountered in many problems of physical interest (e.g. see Kaneko and Konishi 1989, Meiss 1992, Shlesinger *et al.* 1993, Benkadda *et al.* 1994 and references therein). Recently, in a series of papers by Lecar *et al.* (e.g. see Murison *et al.* 1994 and references therein), a correlation was reported between transport in action space of the Elliptical Restricted Three Body Problem (ERTBP) and the Lyapunov Characteristic Numbers (LCN's). In particular the authors, using as a model the ERTBP, presented numerical evidence for the existence of a power law, relating the exit times of asteroidal trajectories from action space regions with the corresponding LCN's of the trajectories. A theoretical interpretation of this correlation, in cases of "strong" perturbation, where the motion can be considered as a random walk in action space, has been attempted by Varvoglis and Anastasiadis (1996) and Morbidelli and Froeschlé (1996).

However, in attempting a statistical description of the process, in order to relate the local rate of trajectory divergence to transport, there are two important problems that have to be solved. The first refers to the difficulty in calculating reliably LCN's in cases where a trajectory is continuously migrating to new regions of phase space, so that the usual method of calculating LCN's (as the limit of $\chi(t) = ln(\xi/\xi_0)/t, t \to \infty$) does not show signs of convergence. The second refers to the fact that, in the case of the main asteroidal belt, the perturbation to the motion of asteroids by Jupiter cannot be considered as "strong". Both problems originate from the fact that the invariant tori in the regions of interest are far from being completely destroyed, so that the phase space is characterised by a pronounced fractality. This, in turn, implies that transport (i) obeys Levy rather than classical Brownian kinetics (e.g. Shlesinger et al. 1993, Benkadda et al. 1994) and (ii) cannot be considered as a random walk, since it is not a pure Markovian process (there exist long tails in the autocorrelation function, e.g. see Meiss et al. 1983, Meiss 1992 and references therein). Therefore, if one wants to recover theoretically a law analogous to that of Lecar *et al.* in cases where the perturbation is not strong, he has: (a) to circumvent the problems in the convergence of the function $\chi(t)$ and (b) to take into account the non-Brownian and, possibly, non-Markovian nature of transport in the statistical description of the process.

In this paper we discuss both problems. We use a method for the study of transport in Hamiltonian systems proposed by Varvoglis *et al.* (1995) in order to calculate the fractality of phase space of the model Hamiltonian

$$H = \frac{1}{2}p_x^2 + \frac{1}{2}p_y^2 + \frac{1}{2}p_z^2 + \frac{1}{2}Ax^2 + \frac{1}{2}By^2 + \frac{1}{2}Cz^2 - \epsilon xz^2 - \eta yz^2 \tag{1}$$

for the particular case $A = 0.9$, $B = 0.4$, $C = 0.225$, $\epsilon = .56$, $\eta = 0.20$ and $h = 0.00765$ which has been studied extensively by Contopoulos and Barbanis (1989, 1995) and Barbanis *et al.* (1997). We conjecture that the fractality may be correlated to the evolution of the function $\chi(t)$ and we discuss the possible usefulness of a "Local Lyapunov time" of a trajectory, in accordance to the ideas of Benkadda et al. (1994) and Morbidelli (1997).

In order to study transport in the phase space of the Hamiltonian (1) we work as follows. We construct a six-dimensional grid in phase space and we calculate the phase space volume, $V(t)$, of the elementary cells explored by a trajectory up to time t. All numerically integrated trajectories start on the $z = 0$ plane with an initial velocity perpendicular to it ($p_x = p_y = 0$ and p_z calculated from the integral of energy). A typical example of the evolution of $V(t)$ for three trajectories, a "central" one and two adjacent with x_0 differing by $\Delta x = \pm 10^{-7}$, is given in Fig. 1. We see that, in accordance to its definition, $V(t)$ is the superposition of "step functions" and we note that these "steps" seem to have sizes of widely different values. We interpret the form of $V(t)$ as follows. Consecutive small steps indicate that the trajectory is diffusing slowly, according to the "brownian" model of random walk, in a phase space region characterised by common properties and, in particular, by a common rate of separation, implying a well defined LCN. A "large" step indicates that the trajectory has moved to an adjacent phase space region with different properties.

2. Results

We test this interpretation by plotting simultaneously the functions $V(t)$ and $\chi(t)$. A typical example of our results is presented in Fig. 2, where we have plotted the above functions for the "central" trajectory of Fig. 1. We see that, up to $t_1 \approx 1 \cdot 10^5$, the trajectory moves in a phase space region whose volume increases slowly with

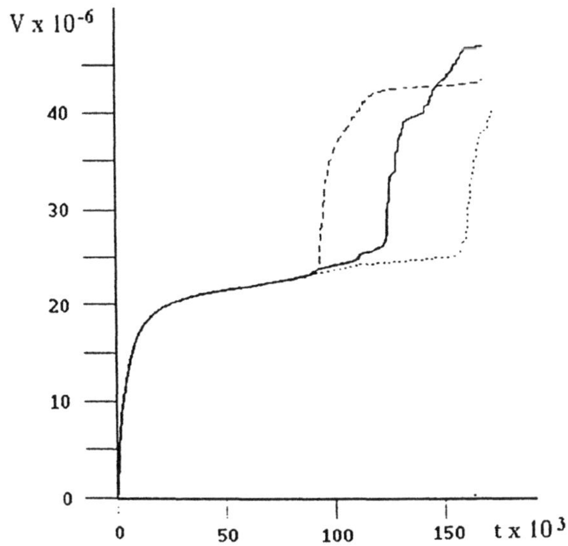

Fig. 1. The function $V(t)$ for three trajectories with initial conditions $y_0 = 0.032$ and $x_0 = 0.01725$ (dotted curve), $x_0 = 0.0172499$ (solid curve) and $x_0 = 0.0172501$ (dashed curve).

time. From this segment one would infer that the LCN is close to zero, since the corresponding function $\chi(t)$ decays almost exponentially. After t_1, however, the trajectory migrates suddenly to a region "outside" the initial one and the function $\chi(t)$ does not show signs of levelling off up to the end of our numerical calculations, at $t_{fin} = 3 \cdot 10^5$. Aside from the fact that this picture corroborates our conjecture, concerning the relation between the transport process and the evolution of $\chi(t)$, Fig. 2 is an example of the practical difficulties arising in the numerical calculation of LCN's: the limiting behaviour of $\chi(t)$ is governed by the properties of the "outside" regions of phase space and, therefore, it does not reflect the properties of the initial region where the trajectory started. In the next section we discuss how these difficulties can be circumvented by the definition of a suitable "Local Lyapunov time".

Proceeding now to the second problem mentioned, we note that, if the phase space has a fractal structure, then the function $V(t)$ should show some sort of self-similarity. However the time series describing numerically $V(t)$ cannot be used in a fractal analysis algorithm, since it is not *stationary* (i.e. it shows long-range changes in the mean level). Therefore we decided to use instead its derivative, dV/dt, which has the above property. We see that, indeed, dV/dt consists of "spikes" of all orders, each one corresponding to a "jump" in $V(t)$ (Figs. 3). After a transient of large spikes, corresponding to the "spreading" of the trajectory in the initially available phase space region, the value of the function drops until $t \approx 1.5 \cdot 10^5$, where from the "forest" of high "spikes" it is obvious that the trajectory migrates to a

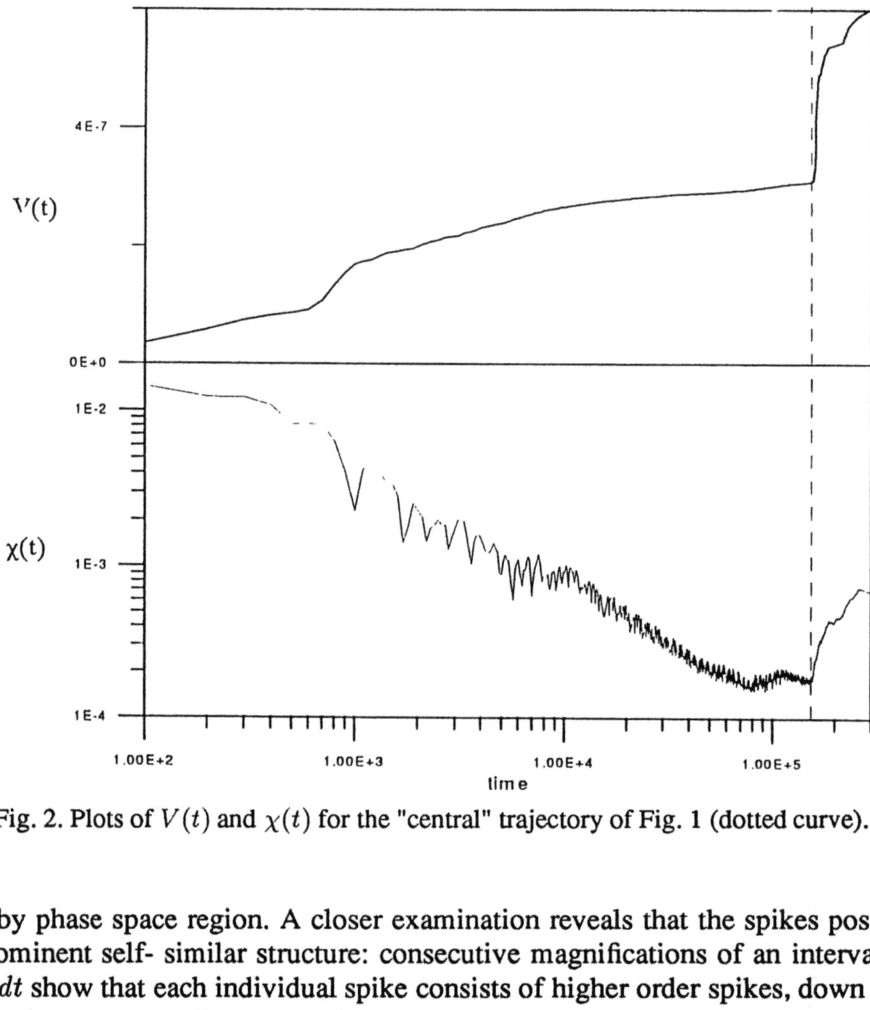

Fig. 2. Plots of $V(t)$ and $\chi(t)$ for the "central" trajectory of Fig. 1 (dotted curve).

nearby phase space region. A closer examination reveals that the spikes possess a prominent self- similar structure: consecutive magnifications of an interval of dV/dt show that each individual spike consists of higher order spikes, down to a scale size corresponding to the 6-D volume of an elementary cell (Figs. 3b and 3c). However, apart from this qualitative approach, we can obtain a quantitative measure of the "fractality" of the function dV/dt through the calculation of its generalised dimensions, $D(q)$, q = 0, 1, 2, ... (e.g. see Schuster, 1988; McCauley, 1995). In particular $D(2)$ is the *correlation dimension.* A value of $D(2) = 1$ implies total randomness, while a value of $D(2) = 0$ complete predictability. It should be noted that the canonical variables $q_i(t), p_i(t)$ (or any function of them) are not useful for the study of phase space fractality, since the corresponding sets are "fat fractals" (Benettin *et al.*, 1986).

From the calculation of the $D(q)$ spectrum we find that $0 < D(2) < 1$ (Fig. 5), which corroborates the results of the qualitative analysis, showing that dV/dt is a *multifractal* (for a definition see the Appendix). Moreover it shows that the transport process is not completely random. From the simultaneous plotting of

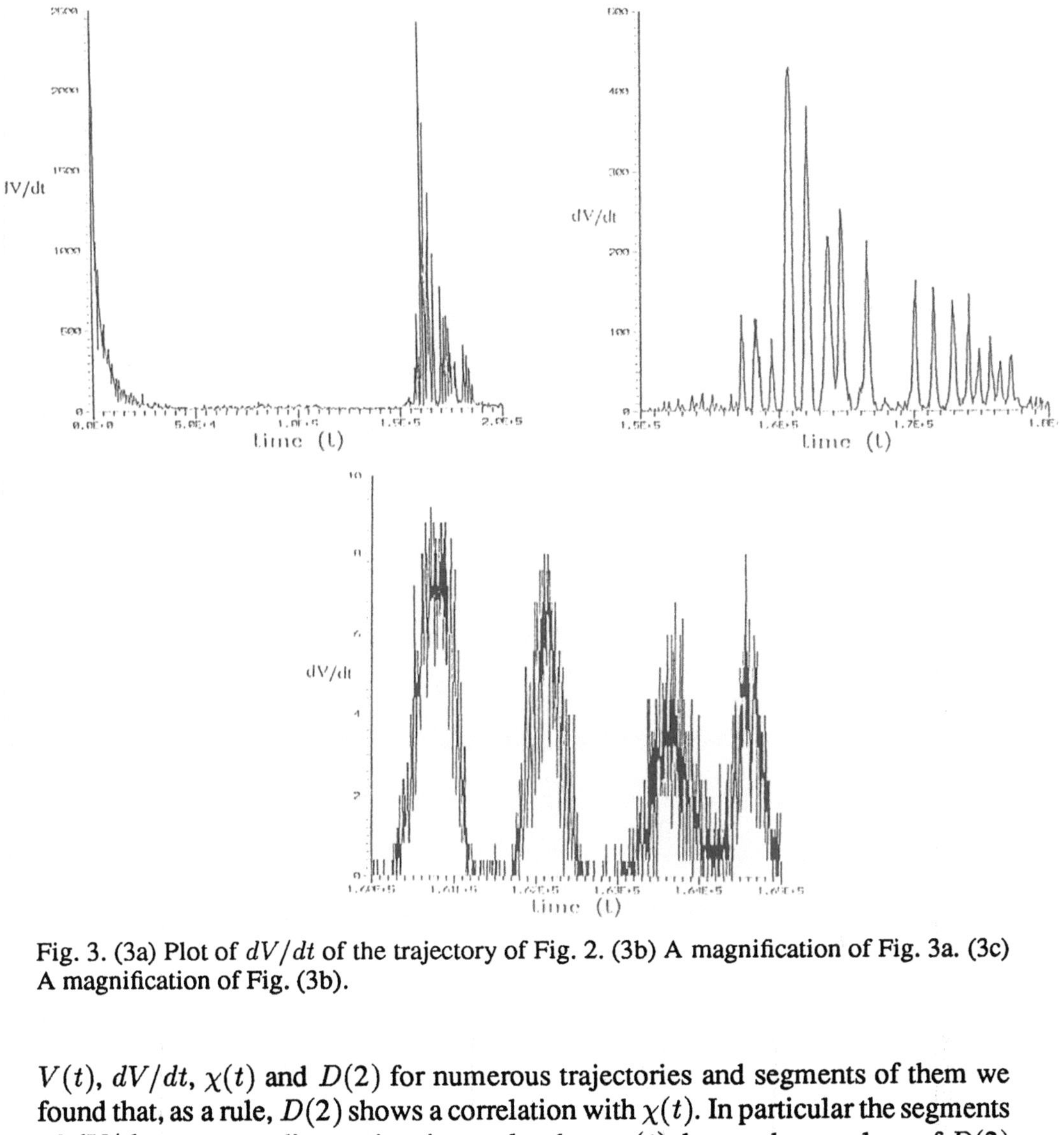

Fig. 3. (3a) Plot of dV/dt of the trajectory of Fig. 2. (3b) A magnification of Fig. 3a. (3c) A magnification of Fig. (3b).

$V(t)$, dV/dt, $\chi(t)$ and $D(2)$ for numerous trajectories and segments of them we found that, as a rule, $D(2)$ shows a correlation with $\chi(t)$. In particular the segments of dV/dt, corresponding to time intervals where $\chi(t)$ decays, have values of $D(2)$ lower than those of the segments of dV/dt, corresponding to time intervals where $\chi(t)$ increases. On the other hand, if the trajectory is treated as a whole, the resulting value of $D(2)$ takes an intermediate value. This fact is in agreement with our interpretation, since, in places where the divergence of trajectories is almost linear with time, we have weak chaos and, therefore, a low correlation dimension. In contrast, in regions where the divergence of trajectories in exponential, we have strong chaos and, therefore, a high correlation dimension.

However this is not always the case. An interesting example is given in Figs. 4 and 5. In Fig. 4 we give the function $\chi(t)$ for a trajectory with initial conditions x_0

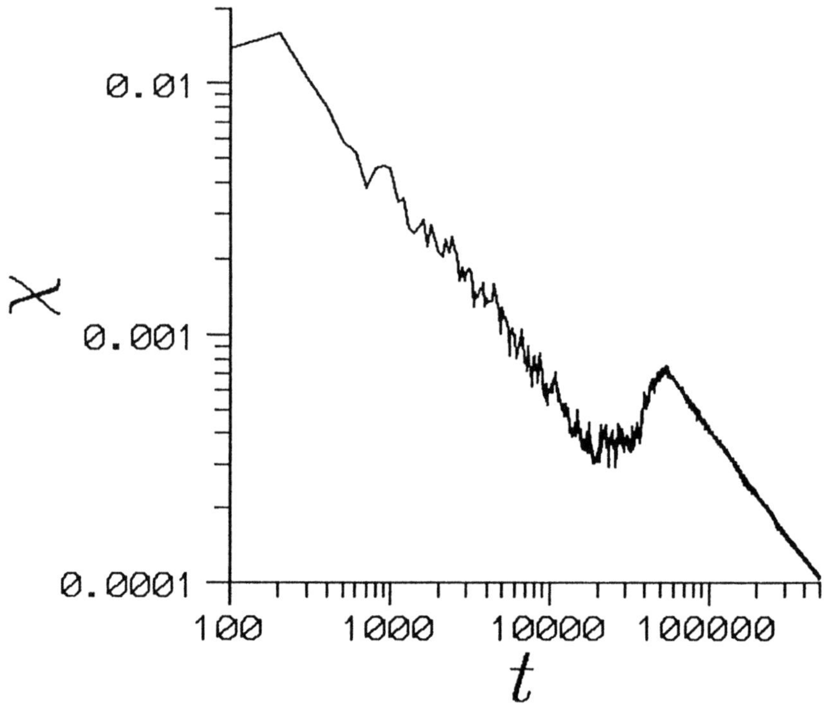

Fig. 4. Plot of $\chi(t)$ for a trajectory with initial conditions x_0 = 0.0095 and y_0 = 0.0415.

= 0.0095 and y_0 = 0.0415. We see that three different trajectory segments may be defined: one for the time interval $0 < t < 2 \cdot 10^4$, one for $2 \cdot 10^4 < t < 6 \cdot 10^4$ and one for $6 \cdot 10^4 < t < 5 \cdot 10^5$. In the first and third segments $\chi(t)$ shows an almost exponential decay, a typical behaviour of a weakly chaotic trajectory, while in the second segment it shows a steep increase, characteristic of strong chaos. From what has been said above, the trajectory has migrated from one region characterised by an almost linear divergence of trajectories to another one with the same property, passing briefly through a region characterised by an exponential divergence. In Fig. 5 we give the $D(q)$-spectra of the three segments and of the trajectory as a whole. We see that, considering only the three segments, the "typical" relation, discussed above, between $\chi(t)$ and $D(2)$ still holds. However in this case the $D(2)$ of the trajectory *as a whole* is considerably lower. This implies that, even while the trajectory has drifted from a phase space region of weak chaos to another, going through one with prominent chaotic behaviour, the third segment of the trajectory "keeps a memory" of the first, which shows that the process in the second segment is not Markovian!

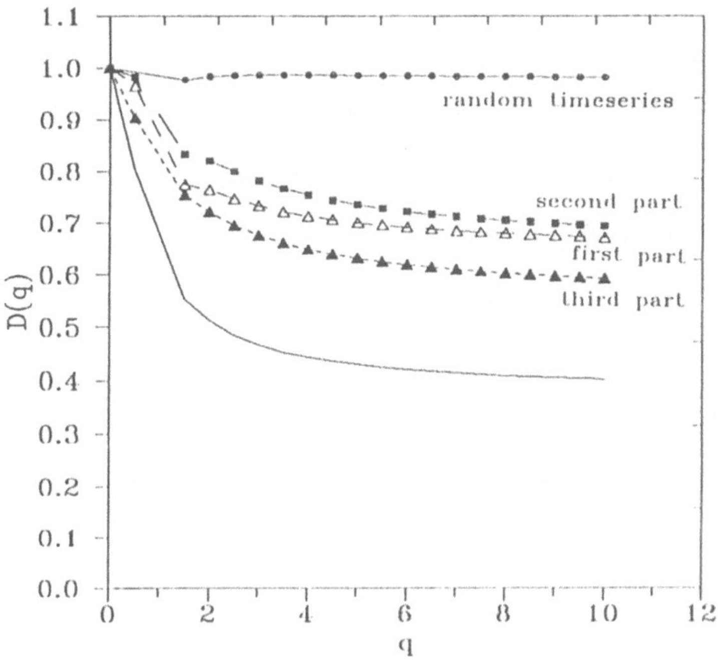

Fig. 5. The $D(q)$-spectrum of the trajectory of Fig. 4 (solid curve) as well as that of the three segments described in the text.

3. Discussion

In a perturbed integrable Hamiltonian system we can distinguish three regimes, according to the "amplitude" of the perturbation: those of "small", "medium" or "large" perturbations, where the "amplitude" of the perturbation is inferred by the dominant properties of the transport mechanism. In the regime of small perturbation, transport is expected to be governed by Arnold diffusion, as discussed recently by Morbidelli and Froeschlé (1996). In the regime of "strong" perturbation the following analytic result may be obtained, according to Varvoglis and Anastasiadis (1996). Assuming that transport in a Hamiltonian system can be described as a *normal* diffusion process, it can be shown that the exit time, T_E, of a trajectory from a "compact" region of action space depends on its Lyapunov time, $T_L = 1/\lambda$ (where λ stands for the maximal Lyapunov Characteristic Number, LCN), through a power law. Since transport in perturbed integrable Hamiltonian systems can be modeled as normal diffusion only in regions where most of the KAM tori are destroyed, the power law dependence appears when the perturbation is strong.

In this paper we showed that the function $V(t)$ may be used to obtain a quantitative measure of the fractality of phase space of a Hamiltonian dynamical system. We discussed how the fractality of phase space is related to the evolution of the function $\chi(t)$, in the regime of "medium" perturbations, and the ensuing difficulties in the calculation of LCN's. A way to circumvent these difficulties might be the definition of a "new" quantity, related to the local (rather than the usual average) properties of the phase space. Following the ideas of Benkadda *et al.* (1994) and Morbidelli (1997), this quantity might be a "Local Lyapunov time", LLT, describing strictly the "autocorrelation time", i.e. the time interval after which a trajectory "looses memory" of its initial conditions. This quantity may be estimated by the evolution of $V(t)$ of nearby trajectories. E.g. from Fig. 1 we can estimate that LLT $\approx 100 \cdot 10^3$. Note that its inverse, the "local LCN" (LLCN), turns out to be of the order 10^{-5}, an order of magnitude lower than the value of LCN inferred from Fig. 2.

In order to establish a statistical relation between local properties (i.e. rate of trajectory divergence) and global properties (transport) in the regime of "medium" perturbations, the fractality of phase space has to be taken into account, since it affects not only the calculation of LCN's, but the process of transport as well. In particular the fact that dV/dt is a multifractal implies that transport in phase space follows Levy rather than normal diffusion kinetics, while the results of the $D(q)$ analysis show that in several cases the non-Markovian nature of the transport process is pronounced. Therefore it is not possible to describe transport through a classical, Fokker-Planck type, diffusion equation and any attempt to use the "regular" random walk theory in order to infer the statistical behaviour of trajectories in a "medium" perturbation regime, as it is the case with the Sun-Jupiter-asteroid problem in the ERTBP approximation, might lead to erroneous results. It should be emphasized that, while the Fokker-Planck equation can be appropriately generalised, in order to take into account the Levy statistics of the random walk process (Zaslavsky, 1994), the problem of the long-tail correlations has not been addressed in a satisfactory way up to now. The use of the function $V(t)$, presented here, may constitute a useful tool in the study of transport in Hamiltonian systems, in particular when the approximation of random walk (Brownian or Levy) may not be adequate.

Appendix

Here we give, for completeness, some basic definitions on the concepts of fractal analysis used in the paper. The main property of a fractal set is *self-similarity*, i.e. the fact that the set looks the same if one scales appropriately the x_i axes $(i = 1, N)$ of the space where this set is embedded. In generalising the definition of the "dimension" of a geometrical object, a fractal set has a *non-integer* dimension, which can be understood in the following way. If for a set of points in d dimensions the number $N(l)$ of d-spheres of diameter l needed to cover the set increases like

$$N(l) \propto l^{-D} \quad \text{for} \quad l \to 0 \tag{2}$$

then D is the *fractal dimension* of the set. For a "regular" geometrical object D turns out to be an integer, which shows that the above definition agrees with the geometrical dimension of a non-fractal set. In the case where the scaling depends on the region of the set, the geometrical object is a multifractal. "Naturally" occurring fractals fall in this category.

A multifractal is characterised by more than one "fractal dimensions" and, in this way, it contains a lot more information than a pure fractal. These so-called *generalised dimensions* or *$D(q)$ spectrum* can be calculated as follows. If we divide the space in cells of linear dimension l and we denote by p_i the probability that a point of the set lies in the $i-th$ box, then the generalised dimensions $D(q)$ are given by the formula

$$D(q) = -\lim_{l\to 0} \frac{1}{q-1}\frac{1}{\log l}\log(\sum_i p_i^q) \tag{3}$$

All generalised dimensions $D(q)$ of a pure fractal set have the same value, that of the *capacity dimension*, which corresponds to $q = 0$, i.e. $D(0)$.

It is obvious that comparing two multifractals is much more difficult than comparing two pure fractals, since it involves comparing an infinite sequence of numbers, rather than only two (the two capacity dimensions). In applications one usually compares only the values of $D(2)$, since in experimental results this is the number that can be more easily calculated. $D(2)$ has the name *correlation dimension*, since it is related to the *correlation integral*, $C(l)$, of the time-series. It can be proved that, in general, $D(0) > D(1) > D(2)... > D(n)$.

Acknowledgements

The authors would like to thank M. Georgoulis for his help in the calculation of the $D(q)$ spectra.

References

Barbanis, B., Varvoglis, H. and Vozikis, Ch.: 1997, in preparation

Benettin, G., Casati, D., Galgani, L., Giorgilli, A. and Sironi, L.: 1986, *Phys. Lett.*, **118A**, 325

Benkadda, S., Elskens, Y., Ragot, B. and Mendonca, J.T.: 1994, *Phys. Rev. Lett.*, **72**, 2859

Contopoulos, G. and Barbanis, B.: 1989, *Astron. Astrophys.*, **222**, 329

Contopoulos, G. and Barbanis, B.: 1995, *Astron. Astrophys.*, **294**, 33

Kaneko, K., and Konishi, T.: 1989, *Phys. Rev. A*, **40**, 613

McCauley, J.L.: 1995, *Chaos, Dynamics and Fractals, an algorithmic approach to deterministic chaos*, Cambridge University Press, Cambridge

Meiss, J.D., Cary, J.R., Grebogi, C., Crawford, J.D., Kaufman, A.N. and Abarbanel, H.D.I.: 1983, *Physica* **6D**, 375

Meiss, J.D.: 1992, *Rev. Mod. Phys.*, **64**, 795 **294**, 33

Morbidelli, A.: 1997, this conference

Morbidelli, A., and Froeschlé, C.: 1997, *Cel. Mech.*, submitted

Murison, M.A., Lecar, M. and Franklin, F.A.: 1994, *Astron. J.*, **108**, 2323

Schuster, H.G.: 1988, *Deterministic Chaos*, VCH Verlagsgesellschaft, Weinheim

Shlesinger, M.F., Zaslavsky, G.M., and Klafter, J.: 1993, *Nature* **363**, 31
Varvoglis, H., Vozikis, Ch. and Barbanis, B.: 1995, *Proceedings of the 2nd Hellenic Astronomical Conference*, Hellenic Astronomical Society, Athens, in press
Varvoglis, H. and Anastasiadis, A.: 1996, *Astron. J.*, **111**, 1718
Zaslavsky, G.M.: 1994, *Physica* **D76**, 110

EQUILIBRIA BIFURCATIONS OF SATELLITE ORBITS

VIKTOR W. KUDIELKA
Peter Jordan Str. 165, A-1180 Vienna, Austria
e-mail: oe1vkw@oe1xtu.ampr.org

Abstract. Equilibria of circular satellite orbits around an oblate central body exist in three planes, as long as the perturbing influence of a third body is comparatively small. In terms of inclination and longitude of node, two of these equilibria are stable and one unstable; for small eccentricities all three equilibria are stable. If the influence of a third body is in the range of 1/4 to 1/2 of the perturbations caused by the oblateness, all three equilibria become unstable in eccentricity. At all these boundary lines of stability a bifurcation into two equilibria of elliptic orbits originates. Even more complex configurations of these equilibria can be observed in the Laplacian plane for inclinations of the third body between 68.87° and 111.13°, and in the orthogonal Laplacian plane for inclinations of the third body between 84.09° and 95.91°. The lines of apsides of these "balanced" elliptic orbits lie either in the plane of the disturbing third body or in the plane orthogonal to the line of intersection of the plane of the disturbing third body and the equatorial plane of the oblate central body.

Key words: Satellite theory – third body perturbations – highly inclined orbits – orbit stability

1. Introduction

P.S. Laplace (1805) was the first to recognize the interplay of the non-central gravitational field of a planet and the disturbing fields of third bodies, and the resulting plane of stable equilibrium for satellite orbits. With the advent of artificial satellites in the late fifties, many papers on long-term satellite orbit evolution appeared, for example, M.L. Lidov (5), R.R. Allan and G.E. Cook (1), and J. Lorell (7). Equilibria of circular geosynchronous orbits in three planes were described in a paper by N. Sekiguchi (8). M.L. Lidov and M.V. Yarskaya (6) offered a summary of the directly integrable cases, including the case of stable elliptic orbits in the plane orthogonal to the line of intersection of the orbital plane of the perturbing body and the equatorial plane of the planet (orthogonal to the direction of equinox). M.A. Vashkovyak (9) investigated the stability of circular orbits by linearizing the perturbation equations at the equilibria in the three planes. H. Kinoshita and H. Nakai (3) observed "complicated" and possibly chaotic motions of satellites in the region where the two major perturbing forces were approximately of equal size.

This paper intends to show the application of another approach – proposed by G.E. Cook (2) and completed by the author (4) – to the analysis of the stability of circular orbits at libration points and the systematic search for equilibria of elliptic orbits.

2. Long-Term Gravitational Perturbations of Satellite Orbits

We consider satellite orbits under the influence of two gravitational perturbations, one due to the oblateness of the central body, characterized by the second zonal harmonic, and the second due to a remote third body, circulating also around the central body. Two simplifying assumptions are made. First, the movement of the

R. Dvorak and J. Henrard (eds.), The Dynamical Behaviour of our Planetary System, 243.

third body during one revolution of the satellite is negligible, that is, the mean motion of the third body is small compared to the mean motion of the satellite. Second, the distance of the satellite from the central body is small compared to the distance of the third body. These assumptions allow the integration of the perturbations over one revolution of the satellite. Another averaging can be done by deleting terms from the perturbation equations which are dependent upon the position of the third body. The remaining terms represent long-term perturbations. When we describe the satellite orbit by the classical equatorial Kepler elements, the semi-major axis remains constant due to the first assumption mentioned above, and the mean anomaly is no longer significant due to the integration. The long-term perturbations are therefore described by a system of four first-order differential equations, with eccentricity, inclination, longitude of node, and argument of perigee as variables, see equations 21.1 through 21.4 in the Appendix (2)(4). The notation used in this paper is summarized at the beginning of the Appendix.

3. Circular Orbits around an Oblate Central Body with Third Body Perturbations

For circular orbits ($e = 0$) the long-term perturbations (equations 21.1 through 21.4 in the Appendix) reduce to

$$\dot{i} = c_3 \left(\sin 2i_d \cos i \sin h + \sin^2 i_d \sin i \sin 2h \right) \tag{1.1}$$

$$\dot{h} = c_3 \left(\cos i \left(-2(1+\gamma) + (3 + \cos 2h) \sin^2 i_d \right) + \sin 2i_d \cos h (\csc i - 2 \sin i) \right). \tag{1.2}$$

Figure 1 shows the trajectories of the system represented by two non-linear differential equations (1.1 and 1.2) for $\gamma = 1$ and two values for i_d, corresponding to a simplified Earth-Moon-Sun system and a Uranus-Sun system, respectively. Equilibria can be seen at $h = 0$ in the Laplacian plane as well as in the orthogonal Laplacian plane, and at $h = \pi/2$ in the plane orthogonal to the direction of equinox ($i = \pi/2$).

3.1. Circular orbits in the orthogonal equinox plane

3.1.1. Equilibria

Substituting $i = \pi/2$ and $h = \pi/2$ in equations 1.1 and 1.2, we obtain immediately the confirmation for the equilibrium of circular orbits.

3.1.2. Stability

When analyzing orbits with small eccentricities, we cannot immediately use the Kepler elements as variables. As usual, two new variables, $l = e \cos g$ and $k =$

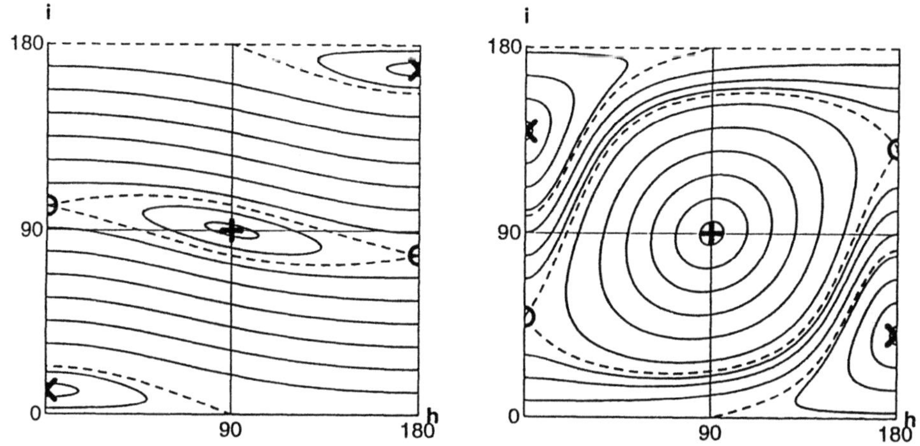

Fig. 1. i–h diagrams ($e = 0,\ \gamma = 1$) for $i_d = 23.44°$ (Earth – left diagram) and $i_d = 97.9°$ (Uranus – right diagram) with the following equilibria: (1) Laplacian plane (cross signs) (2) Orthogonal equinox plane (plus signs) (3) Orthogonal Laplacian plane (circles)

$e \sin g$, are introduced. With

$$\dot{l} = \dot{e} \cos g - \dot{g}\, e \sin g, \qquad \dot{k} = \dot{e} \sin g + \dot{g}\, e \cos g,$$

and the substitutions

$$\cos g \to l/\sqrt{l^2 + k^2}, \qquad \sin g \to k/\sqrt{l^2 + k^2}, \qquad e \to \sqrt{l^2 + k^2},$$

we obtain from equations 21.1 through 21.4 a system of differential equations with the variables i, h, l, and k. Linearizing the equations at the equilibrium point $i = \pi/2, h = \pi/2, l = 0, k = 0$, we obtain the Jacobi matrix (ignoring a factor c_3)

$$\begin{pmatrix} -\sin 2i_d & -1 + \cos 2i_d & 0 & 0 \\ 1 + 2\gamma + \cos 2i_d & \sin 2i_d & 0 & 0 \\ 0 & 0 & -5 \sin 2i_d & 1 + \gamma + 5 \cos 2i_d \\ 0 & 0 & -1 - \gamma + 5 \cos 2i_d & 5 \sin 2i_d \end{pmatrix} \quad (2)$$

The eigenvalues of the Jacobi matrix are

$$\pm\sqrt{-4\,\gamma \sin^2 i_d} \quad \text{and} \quad \pm\sqrt{(4 - \gamma)(6 + \gamma)}.$$

For $\gamma > 4$ all four eigenvalues are imaginary and pairwise conjugated complex, describing a center point. For $\gamma < 4$ the eigenvalues of the second pair are real and of opposite sign, describing a saddle point. This change of stability at $\gamma = 4$ is independent of i_d. For retrograde orbits $(i = \pi/2, h = 3\pi/2)$ we obtain identical eigenvalues.

3.2. Circular orbits in planes aligned with the direction of equinox

3.2.1. *Equilibria*

Substituting $h = 0$ in equations 1.1 and 1.2, we obtain

$$\dot{i} = 0 \tag{3.1}$$
$$\dot{h} = -c_3 \csc i\Big(\gamma \sin 2i + \sin 2(i - i_d)\Big). \tag{3.2}$$

We find equilibria ($\dot{h} = 0$) immediately from equation 3.2 at

$$\gamma_c = \sin 2(i_d - i)/\sin 2i. \tag{4}$$

Equation 4 can be solved easily for i, and we get the inclinations for the Laplacian and the orthogonal Laplacian planes

$$i_L = 1/2 \ \arctan\Big(\sin 2i_d/(\gamma + \cos 2i_d)\Big) \tag{5}$$
$$= 1/2 \ \arctan\Big(\sin i_d/(\cos i_d + (\gamma - 1)/(2\cos i_d))\Big), \tag{6}$$
$$i_{oL} = \pi/2 + 1/2 \ \arctan\Big(\sin 2i_d/(\gamma + \cos 2i_d)\Big). \tag{7}$$

Equation 5 is instructive for the case $\gamma = 0$, while the equivalent equation 6 is more instructive for the case $\gamma = 1$. Please note that in this case (circular orbits) we can describe the equilibria explicitly either by $\gamma_c(i, i_d)$ or by $i_L(\gamma, i_d)$.

3.2.2. *Stability of circular orbits in the Laplacian plane*

We can use here again the same system of differential equations with the variables $i, h, l = e\cos g$, and $k = e\sin g$ (see 3.1.2), linearize the equations at $i = i_L, h = 0, l = 0$, and $k = 0$, and obtain the following Jacobi matrix (ignoring a factor c_3):

$$\begin{pmatrix} 0 & L_{12} & 0 & 0 \\ L_{21} & 0 & 0 & 0 \\ 0 & 0 & 0 & L_{34} \\ 0 & 0 & L_{43} & 0 \end{pmatrix} \tag{8}$$

with

$$L_{12} = 2\cos(i_d - i_L)\sin i_d \tag{9.1}$$
$$L_{21} = 1/2 \ \csc^2 i_L\Big(\sin(2i_d - 3i_L) - 3\sin(2i_d - i_L)\Big) + 2\gamma \sin i_L \tag{9.2}$$
$$L_{34} = 1/4 \csc i_L\Big(-\gamma(\sin i_L - 5\sin 3i_L) + 10\sin(2i_d - 3i_L) - 6\sin(2i_d - i_L) + 8\sin i_L\Big) \tag{9.3}$$
$$L_{43} = 1/4\,\gamma(1 + 5\csc i_L \sin 3i_L) + 3 - \csc i_L \sin(2i_d - i_L). \tag{9.4}$$

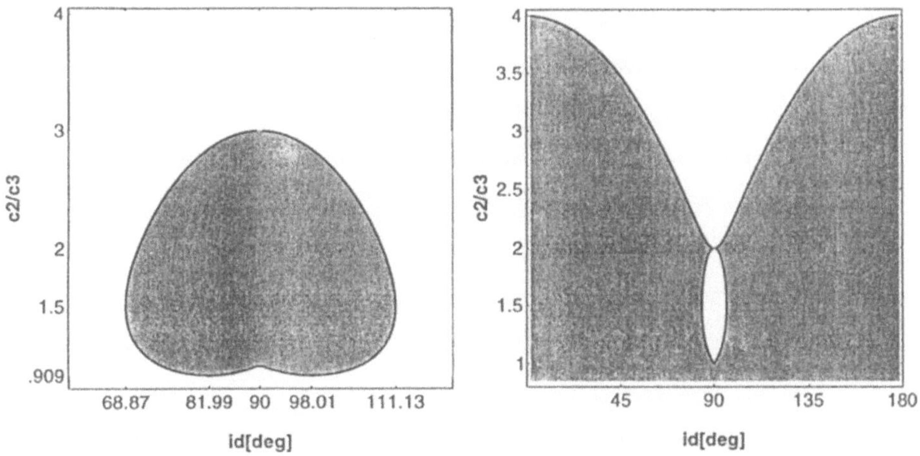

Fig. 2. Areas of unstable circular orbits (shaded) in the Laplacian plane (left part) and in the orthogonal Laplacian plane (right part) ($c_2/c_3 = \gamma$)

The four eigenvalues are in the form of $\pm\sqrt{f_1(\gamma, i_d) \pm \sqrt{f_2^2(\gamma, i_d)}}$. We can then evaluate the boundary lines of stability from $f_1(\gamma, i_d) \pm |f_2(\gamma, i_d)| = 0$, and obtain an unstable area at $68.87° \leq i_d \leq 111.13°$ and $3 \geq \gamma \geq 0.909$ (see left part of figure 2).

In our planetary system this will concern satellites in the Laplacian plane of Uranus ($i_d = 97.9°$) only (see section 4.3.2).

3.2.3. Stability of circular orbits in the orthogonal Laplacian plane

The Jacobi matrix for the orthogonal Laplacian plane is similar to that for the Laplacian plane and also the structure of the eigenvalues is the same as for the Laplacian plane.

Evaluating the boundary lines of stability, we obtain a fairly complex structure (see right part of figure 2). The major boundary line of stability is at $4 \geq \gamma \geq 2$, depending on i_d. There is another small stable area at $84.09° \leq i_d \leq 95.91°$ and $2 \geq \gamma \geq 1$.

4. Elliptic Orbits

4.1. THIRD BODY PERTURBATIONS IN ORBITAL KEPLER ELEMENTS

Using the plane of the disturbing body as reference plane ($i_d = 0$) and setting $\gamma = 0$, we get another special case. The variables for this case are designated by an asterisk (i^*, g^*, l^*, k^*). For $i_d = 0$ the equations 21.1 and 21.2 reduce to

$$\dot{e} = 5\, c_3\, e\,(1-e^2)^{1/2} \sin 2g^* \sin^2 i^* \qquad (10.1)$$
$$\dot{i}^* = -5/2\, c_3\, e^2\,(1-e^2)^{-1/2} \sin 2g^* \sin 2i^*. \qquad (10.2)$$

By separation of the variables and integration we derive the well-known integral of motion: $(1-e^2)\cos^2 i^* = C = \cos^2 i_c^*$. Using the integration constant C and by defining $e_0 = \sin i_c^* = \sqrt{1-C}$, we get

$$\dot{e} = 5\, c_3\, e\,(1-e^2)^{-1/2} \sin^2 g^* (e_0^2 - e^2) \qquad (11.1)$$
$$\dot{g}^* = 2\, c_3\,(1-e^2)^{-3/2}\Big(2(1-e^2)^2 + 5\sin^2 g^*\big(e^2(2-e^2) - e_0^2\big)\Big). \qquad (11.2)$$

Together with the integral of motion, the two equations are sufficient to describe the whole case, since the position of the node remains undefined. Equilibria exist for $g^* = \pi/2$ and $3\pi/2$ at $\eta^4 = 5/3\ C$, where $\eta = \sqrt{1-e^2}$. For $C \geq 3/5$, the equilibrium remains at $\eta = 1$.

Trajectories in the $l^* - k^*$ plane are shown in figure 3. We can see the stable equilibrium for circular orbits for $C \geq 3/5$ ($i_c^* \leq 39.23°$) and the bifurcation into two stable equilibria for $C < 3/5$ ($i_c^* > 39.23°$). Please note that this bifurcation is caused by the third body perturbation only and the two forks lie in a plane perpendicular to the plane of the disturbing body.

4.2. Elliptic Orbits in the Orthogonal Equinox Plane

4.2.1. *Equilibria*

Substituting $h = \pi/2$, and $i = \pi/2$ in equations 21.1 through 21.4, we obtain

$$\dot{e} = 5\, c_3\, e\,(1-e^2)^{1/2} \sin 2\,(g - i_d) \qquad (12.1)$$
$$\dot{i} = 0 \qquad (12.2)$$
$$\dot{h} = 0 \qquad (12.3)$$
$$\dot{g} = c_3\Big((1-e^2)^{1/2}\big(5\cos 2\,(g-i_d) - 1\big) - \gamma(1-e^2)^{-2}\Big). \qquad (12.4)$$

For $\gamma < 4$, we find the equilibria ($\dot{g} = 0$) at $g = i_d, \pi + i_d$ and $e_s = \sqrt{1-(\gamma/4)^{2/5}}$. With $\eta = \sqrt{1-e^2}$, the equilibria can be described somewhat shorter by $\eta^5 = \gamma/4$.

4.2.2. *Stability*

For analyzing the stability we can use here the equations 21.1 through 21.4 immediately, with the variables i, h, e, g (in this order) and linearize the equations at the equilibrium $i = \pi/2, h = \pi/2, e = \sqrt{1-(\gamma/4)^{2/5}}$, and $g = i_d$. Alternatively we can, for example, rotate and shift the coordinates to the equilibrium, convert the equations for variables $\bar{\imath} = i - \pi/2, \bar{h} = h - \pi/2, \bar{l} = e\cos(g - i_d) -$

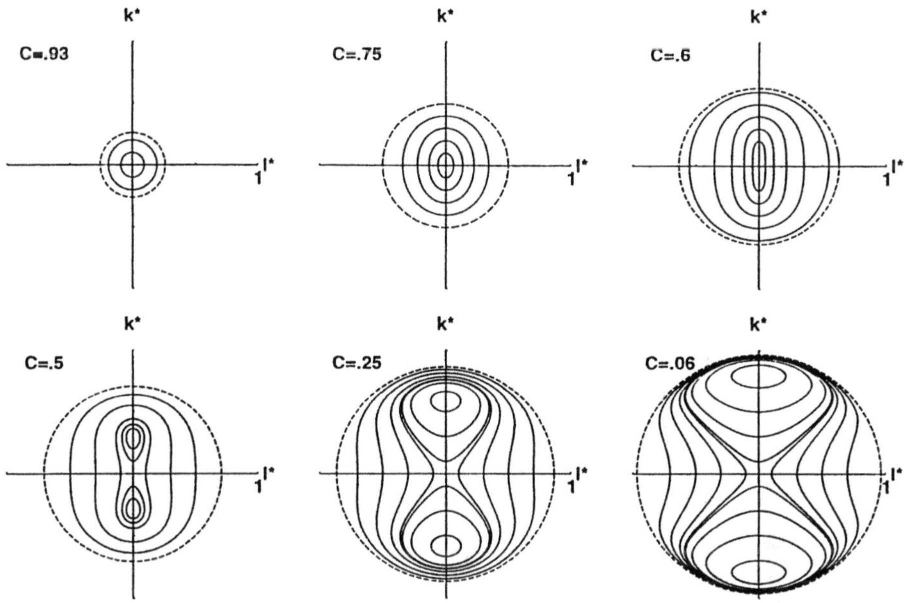

Fig. 3. l^*–k^* diagrams for $\gamma = 0, i_d = 0$, and $i_c^* = 15°, 30°, 39.23°, 45°, 60°$, and $75°$. The dotted lines represent circular trajectories where $\dot{e} = 0$.

$\sqrt{1-(\gamma/4)^{2/5}}, \bar{k} = e\sin(g - i_d)$ and linearize at $\bar{\imath} = 0, \bar{h} = 0, \bar{l} = 0$, and $\bar{k} = 0$. In both cases we obtain the same Jacobi matrix for $\gamma < 4$ (ignoring again a factor c_3):

$$(\gamma/4)^{1/5}\begin{pmatrix} -\sin 2i_d & -1+\cos 2i_d & 0 & 0 \\ 9+\cos 2i_d & \sin 2i_d & 0 & 0 \\ 0 & 0 & 0 & 0 \\ 0 & 0 & 0 & 0 \end{pmatrix}. \tag{13}$$

One pair of eigenvalues of the Jacobi matrix is $\pm 4\,(\gamma/4)^{1/5}\sqrt{-\sin^2 i_d}$, and the second pair is zero. We have therefore a weak stability independent of i_d. The case $i_d = 0$ will need a different analysis.

For retrograde orbits ($i = \pi/2, h = 3\pi/2$), the arguments of the trigonometric functions in equations 12.1 and 12.4 are $2\,(g + i_d)$, otherwise the formulae are identical. The equilibria are in the same plane as for the prograde orbits.

For Earth satellite orbits ($i_d \simeq \pi/8$) figure 4 shows l–k diagrams for different values of γ. For $\gamma \geq 4$, where the oblateness perturbation dominates, we see the stable circular orbit, while for $\gamma < 4$ the two forks for stable elliptic orbits occur. The apsides of stable elliptic orbits in the orthogonal equinox plane lie in the plane of the perturbing third body.

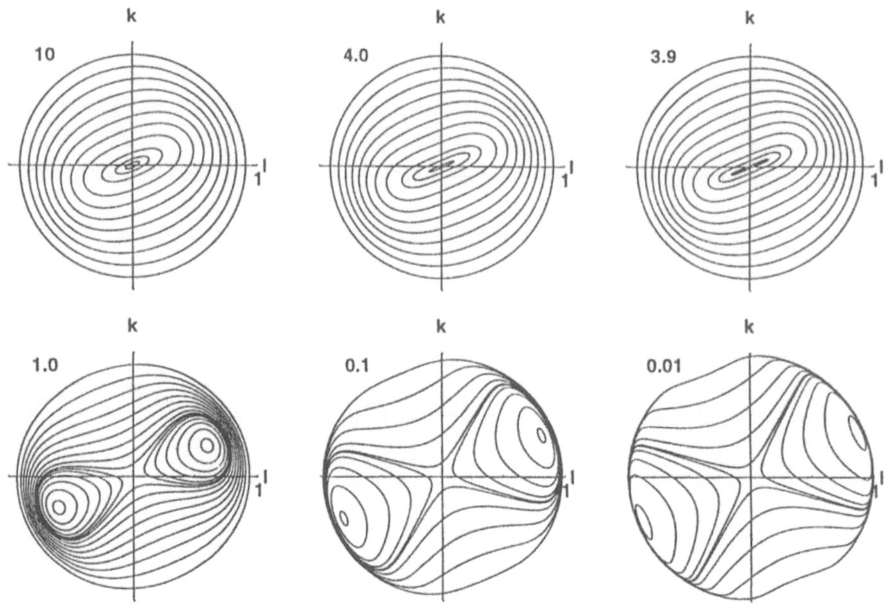

Fig. 4. l–k diagrams for $h = \pi/2, i = \pi/2, i_d = \pi/8$ and $\gamma = 10, 4, 3.9, 1, 0.1,$ and 0.01

4.3. Elliptic orbits in highly inclined planes aligned with the direction of equinox

Inserting $h = 0$ in equations 21.1 and 21.2 we obtain

$$\dot{e} = 5\, c_3\, e\,(1 - e^2)^{1/2} \sin 2g\, \sin^2(i - i_d) \tag{14.1}$$
$$\dot{i} = -5/2\, c_3\, e^2\,(1 - e^2)^{-1/2} \sin 2g\, \sin 2\,(i - i_d). \tag{14.2}$$

Inspecting these equations for potential equilibria, we find $i = i_d$ as one alternative. This case has been covered in section 4.1. Other possibilities are $g = \nu\pi/2$ with $\nu = 0, 1, \ldots$, which will be analyzed in the following sections.

4.3.1. Line of apsides in the direction of equinox

Inserting $h = 0$ and $g = 0$ in equations 21.3 and 21.4 we obtain

$$\dot{h} = c_3\Big(-2\gamma\eta^{-4}\cos i - \eta \sin 2(i - i_d)/\sin i\Big) \tag{15.1}$$
$$\dot{g} = c_3\Big(\gamma\eta^{-4}(5\cos^2 i - 1) + \eta(3 - \cos 2(i - i_d) - \cot i \sin 2(i - i_d))\Big). \tag{15.2}$$

For elliptic orbits – independently of their eccentricity – we find from equations 15.1 and 15.2 equilibria ($\dot{g} = \dot{h} = 0$) at certain inclinations $i = i_b$, which can be defined implicitly (4) by

$$(5\cos^2 i_b - 1)\sin 2(i_d - i_b) + 3\sin 2i_b + 2\cos i_b\, \sin(i_b - 2i_d) = 0 \tag{16}$$

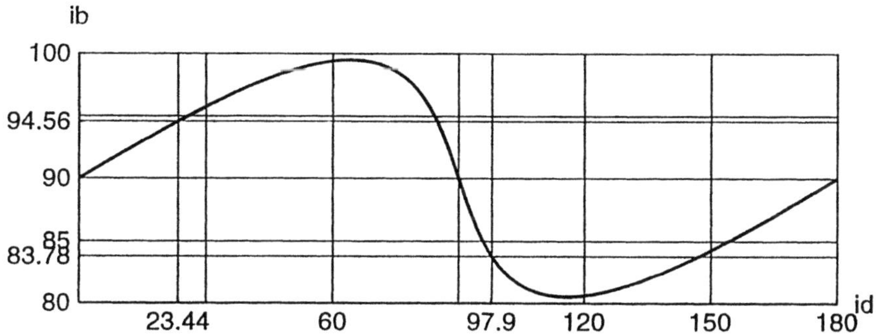

Fig. 5. Inclination i_b of equilibria of elliptic orbits ("balanced" orbits) as function of i_d, the inclination of the perturbing third body ($h = 0$, $g = 0$ or $g = \pi$)

and

$$\gamma_b = \eta^5 \sin 2(i_d - i_b)/\sin 2i_b. \tag{17}$$

In figure 5 i_b is shown as a function of i_d.

The maximum of i_b ($= \arctan(-\sqrt{2}(2+\sqrt{5})) = 99.4767\ldots°$) occurs at an inclination of the perturbing third body of $i_d = \arctan((5+3\sqrt{5})/(4\sqrt{2})) = 64.2123\ldots°$.

Inserting the numerical values of i_b into equation 4 (or equation 17 with $\eta = 1$), we get γ_c as a function of i_d, identical to the boundary line of stability of circular orbits, as shown in the right part of figure 2.

For high values of γ stable circular orbits follow the inclination of the orthogonal Laplacian plane. When, with decreasing γ ($4 \geq \gamma \geq 2$, dependent upon i_d) i_b is reached, the bifurcation occurs and the inclination for stable elliptic orbits remains constant, while the eccentricity increases with decreasing γ.

4.3.2. *Line of apsides in the plane orthogonal to the direction of equinox*

To cover the other alternative for equilibria of elliptic orbits, we insert $h = 0$ and $g = \pi/2$ in equations 21.3 and 21.4 and obtain

$$\dot{h} = c_3\Big(-2\gamma\eta^{-4}\cos i - (5-4\eta^2)/\eta \sin 2(i - i_d)/\sin i\Big) \tag{18.1}$$

$$\dot{g} = c_3\Big(\gamma\eta^{-4}(5\cos^2 i - 1) + 2\eta(-1 + 2\cos 2(i - i_d)) + (5-4\eta^2)/\eta \cot i \sin 2(i - i_d)\Big). \tag{18.2}$$

From $\dot{h} = 0$ and $\dot{g} = 0$ we derive $\eta(i, i_d)$ and $\gamma(\eta, i, i_d)$:

$$\eta^2 = 5/4\Big((3\cos^2 i - 1)\sin 2(i_d - i)\Big)/\Big((5\cos^2 i - 1)\sin 2(i_d - i) + \cos i(\sin i + 2\sin 2(i_d - i))\Big) \tag{19.1}$$

$$\gamma = \eta^3(5 - 4\eta^2)\sin 2(i_d - i)/\sin 2i. \tag{19.2}$$

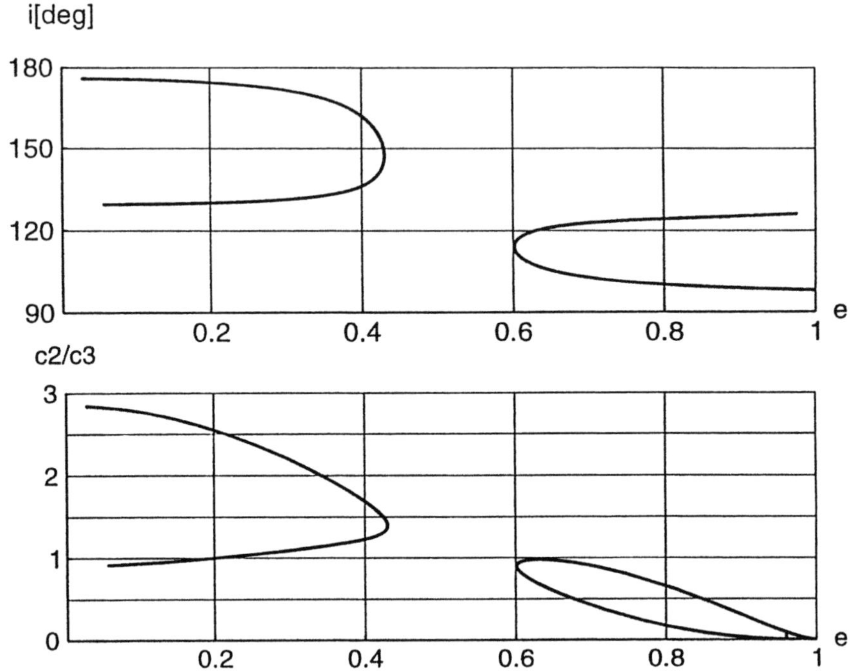

Fig. 6. "Balanced" elliptic orbits at $h = 0$ with $g = \pi/2$ or $g = 3\pi/2$ for Uranus ($i_d = 97.9°$) ($c_2/c_3 = \gamma$)

With $\eta = 1$ we get the boundary line of stability for circular orbits:

$$\gamma_c = \sin 2(i_d - i)/\sin 2i = 2(2\sin(2i_d - i)/\sin i - 1)/(5\cos^2 i + 1). \qquad (20)$$

The two equations 20 describe exactly the boundary lines of stability of circular orbits in the Laplacian plane at $3 \geq \gamma \geq .909$ (left part of figure 2) and in the orthogonal Laplacian plane at $2 \geq \gamma \geq 1$ (right part of figure 2).

As one example of the equilibria of elliptic orbits the case of Uranus ($i_d = 97.9°$) is shown in figure 6.

5. Summary

The results of the analytic model proposed by G.E. Cook, which are presented in this paper, prove the usefulness of the model. The results on stability of circular orbits match those obtained by M.A. Vashkovyak. The boundary lines of stability of circular orbits can be described by two short equations. Equilibria of elliptic orbits have been found, in addition to those in the orthogonal equinox plane (described by M.V. Yarskaya and M.L. Lidov), in planes aligned with the direction of equinox. Some cases, applicable only to restricted ranges of inclinations of the third body, like that for the Laplacian plane of Uranus or for the orthogonal Laplacian plane for $84.09° \leq i_d \leq 95.91°$, will need more detailed analysis.

Acknowledgements

The author would like to thank Prof. R. Dvorak of the University of Vienna and Prof. J. Waldvogel of the ETH Zurich for their advice and help. Prof. J. Waldvogel provided the exact solution for the maximum of i_b (equation 16 and figure 5).

Special thanks go to two anonymous referees, who were successful in finding weak points and errors in the original manuscript.

Appendix

NOTATION AND ABBREVIATIONS

Orbital elements of the satellite:

a	semi-major axis
e	eccentricity
i	inclination
h	longitude of ascending node
g	argument of perigee
n	angular mean motion
n_0	angular mean motion ($a = R$)

Elements of the disturbing body:

r_d	distance from central body
i_d	inclination
$h_d = 0$	longitude of ascending node
m_d	mass
n_d	angular mean motion

Other:

m_c	mass of central body
R	radius of central body
J_2	2nd zonal harmonic (central body)
c_2	factor of oblateness perturbation
c_3	factor of third body perturbation
$\gamma = c_2/c_3$	ratio of oblateness and third body perturbations

$$
\begin{aligned}
n &= (Gm_c/a^3)^{1/2} \qquad n_0 = (Gm_c/R^3)^{1/2} \qquad n_d = (Gm_d/r_d^3)^{1/2} \\
c_2 &= 3/4\, J_2\, n_0 (n/n_0)^{7/3} = 3/4\, J_2 (Gm_c/R^3)^{1/2} (R/a)^{7/2} \\
c_3 &= 3/8\, n_d^2/n = 3/8\, (G\, m_c/a^3)^{-1/2}\, G\, m_d/r_d^3 \\
&= 3/8\, (G\, m_c/a^3)^{1/2} (m_d/m_c)(a/r_d)^3 \\
\gamma &= c_2/c_3 = 2\, J_2\, (n\, n_0/n_d^2)(n/n_0)^{7/3} = 2\, J_2\, (m_c/m_d)(R/a)^5 (r_d/R)^3
\end{aligned}
$$

A. Long-term perturbations in (equatorial) Kepler elements

$$
\begin{aligned}
\dot{e} = {} & 5\, c_3\, e\, (1 - e^2)^{1/2} \Big((1 + 3\cos 2i_d)/4 \sin^2 i \sin 2g + \\
& \sin^2 i_d \Big(\cos^4 \left(\frac{i}{2}\right) \sin 2(h + g) - \sin^4 \left(\frac{i}{2}\right) \sin 2(h - g) \Big) - \\
& \sin 2i_d \sin i \Big(\cos^2 \left(\frac{i}{2}\right) \sin(h + 2g) + \sin^2 \left(\frac{i}{2}\right) \sin(h - 2g) \Big) \Big)
\end{aligned} \qquad (21.1)
$$

$$
\begin{aligned}
\dot{i} = {} & -5/2\, c_3\, (1-e^2)^{-1/2} \\
& \Big(-\left(2+3e^2\right)/5 \left(\sin 2i_d \cos i \sin h + \sin^2 i_d \sin i \sin 2h \right) + \\
& \quad e^2 \Big(\left(1 + 3 \cos 2i_d\right)/4 \sin 2i \sin 2g - \\
& \qquad \sin^2 i_d \sin i \left(\cos^2 \left(\frac{i}{2}\right) \sin 2(h+g) + \sin^2 \left(\frac{i}{2}\right) \sin 2(h-g) \right) - \\
& \qquad 1/2 \sin 2i_d \Big(\left(\cos i + \cos 2i \right) \sin(h+2g) + \\
& \qquad\qquad \left(\cos i - \cos 2i \right) \sin(h-2g) \Big) \Big) \Big)
\end{aligned} \tag{21.2}
$$

$$
\begin{aligned}
\dot{h} = {} & -c_3\, (1-e^2)^{-1/2} \\
& \Big(\left(1 + 3 \cos 2i_d\right)/4 \cos i \left(2 + e^2 (3 - 5 \cos 2g)\right) - \\
& \quad \left(1 + 3/2\, e^2\right) \left(\sin 2i_d \cos 2i \csc i \cos h + \sin^2 i_d \cos i \cos 2h \right) + \\
& \quad 5/2\, e^2 \Big(\sin^2 i_d \left(\cos^2 \left(\frac{i}{2}\right) \cos 2(h+g) - \sin^2 \left(\frac{i}{2}\right) \cos 2(h-g) \right) + \\
& \qquad 1/2 \sin 2i_d \csc i \Big(\left(\cos i + \cos 2i \right) \cos(h+2g) - \\
& \qquad\qquad \left(\cos i - \cos 2i \right) \cos(h-2g) \Big) \Big) \Big) - \\
& 2\, c_2\, (1-e^2)^{-2} \cos i
\end{aligned} \tag{21.3}
$$

$$
\begin{aligned}
\dot{g} = {} & c_3 (1-e^2)^{-1/2} \\
& \Big(\big(1 + 3 \cos 2i_d\big)/4 \cos^2 i \big(2 + e^2(3 - 5 \cos 2g)\big) - \\
& \quad \big(1 + 3/2\, e^2\big) \Big(\sin 2i_d \cos 2i \cot i \cos h + \sin^2 i_d \cos^2 i \cos 2h \Big) + \\
& \quad 5/2\, e^2 \cos i \\
& \qquad \Big(\sin^2 i_d \Big(\cos^2 \Big(\frac{i}{2}\Big) \cos 2(h+g) - \sin^2 \Big(\frac{i}{2}\Big) \cos 2(h-g)\Big) + \\
& \qquad\quad 1/2 \sin 2i_d \csc i \Big(\big(\cos i + \cos 2i\big) \cos(h+2g) - \\
& \qquad\qquad \big(\cos i - \cos 2i\big) \cos(h-2g)\Big)\Big)\Big) + \\
& c_3 (1-e^2)^{1/2} \\
& \Big((1 + 3 \cos 2i_d)/4 \big(2 - \sin^2 i\, (3 - 5 \cos 2g)\big) + \\
& \quad 3/2 \big(\sin 2i_d \sin 2i \cos h + \sin^2 i_d \sin^2 i \cos 2h \big) + \\
& \quad 5 \sin^2 i_d \Big(\cos^4 \Big(\frac{i}{2}\Big) \cos 2(h+g) + \sin^4 \Big(\frac{i}{2}\Big) \cos 2(h-g)\Big) + \\
& \quad 5 \sin 2i_d \sin i \Big(-\cos^2 \Big(\frac{i}{2}\Big) \cos(h+2g) + \sin^2 \Big(\frac{i}{2}\Big) \cos(h-2g)\Big)\Big) + \\
& c_2 (1-e^2)^{-2} \big(5 \cos^2 i - 1\big)
\end{aligned} \tag{21.4}
$$

References

[1] Allan, R.R. and Cook, G.E.: 1964,"The long-period motion of the plane of a distant circular orbit", *Proc. R. Soc. Lond.* A **280**, 97–109.

[2] Cook, G.E.: 1962,"Luni-Solar Perturbations of the Orbit of an Earth Satellite", *GeophysJ. R.Astr.Soc.* **6**, 271–291.

[3] Kinoshita, H. and Nakai, H.: 1991, "Secular perturbations of fictitious satellites of Uranus", *Celest. Mech.* **52**, 293–303.

[4] Kudielka, V.: 1994,"Balanced Earth Satellite Orbits", *Celest. Mech.* **60**, 455–470.

[5] Lidov, M.L.: 1961, "The evolution of orbits of artificial satellites of planets under the action of gravitational perturbations of external bodies", *translated in Planet. Space Sci.* **9**, 1962, 719–759.

[6] Lidov, M.L. and Yarskaya, M.V.: 1974,"Integrable cases in the problem of the evolution of a satellite orbit under the joint effect of an outside body and of the noncentrality of the planetary field", *Kosm. Issled.* **12**, *No.* 2, 155–170.

[7] Lorell, J.: 1965, "Long Term Behaviour of Artificial Satellite Orbits Due to Third-Body Perturbations", *Journ. of the Astronaut. Sci.* **XII**, No.4, 142–152.

[8] Sekiguchi, N.: 1961,"On the libration of the orbital plane of a stationary artificial satellite with a circular orbit", *Publ. Astron. Soc. Jpn.* **13**, *No.* 2, 207–211.

[9] Vashkovyak, M.A.: 1974,"The stability of circular satellite orbits for combined action of perturbations from an external body and from the noncentrality of the planetary gravitational field", *Kosm. Issled.* **12**, *No. 6*, 834–847.

FAST LYAPUNOV INDICATORS
COMPARISON WITH OTHER CHAOS INDICATORS
APPLICATION TO TWO AND FOUR DIMENSIONAL MAPS.

E. LEGA
CNRS LATAPSES, 250 Rue A.Einstein,
06560 Valbonne
Observatoire de Nice B.P.4229,
06304 Nice cedex 4

and

C. FROESCHLÉ
Observatoire de Nice B.P.4229,
06304 Nice cedex 4

Abstract. Using two and four dimensional symplectic maps as model problem we test a simple and fast method to separate chaotic from regular orbits and we compare with 3 other chaos indicators: the frequency, the sup-map and the Lyapunov characteristic exponent.

Key words: Chaotic motion – Lyapunov numbers – symplectic maps

1. Introduction

In these last years there has been a renewal of interest, at least in the astronomic scientific world, about the chaos in the solar system. The most important results are those of Laskar (1990) on the chaotic motion of inner planets and on the movement of the rotation axis of Mars (Laskar and Robutel 1993). Moreover, the question of stable chaos raised by Milani and Nobili (1992) and the eventual connection with the global stability of some regions of the asteroidal belt have lead Morbidelli and Froeschlé (1996) to make the conjecture that the asteroid belt is in the Nekhoroshev regime, i.e. the majority of asteroids have zero or very small Lyapunov characteristic indicators (hereafter LCI).

The computation of small LCIs takes a very long time and therefore the implementation of indicators of detections and measure of weak chaos is a starting point for such a study. Moreover, recent theoretical developments in the line of the famous KAM and Nekhoroshev theorem have brought new lights on the stability and diffusion processes for Hamiltonian systems (Morbidelli and Giorgilli 1994, Morbidelli and Guzzo in this volume). Of course as usual the theoretical conditions on the magnitude of the perturbation are very stringent and far to be realistic as far as real problems are concerned. Therefore it is very useful to have a set of numerical methods in order to explore the complicated structure of invariant tori within tiny chaotic zones.

In Section 2 we revise the methods already used to detect and measure weak chaos and present a new fast indicator which goes back to the spirit of the LCEs. In Section 3 we test the sensitivity of the new indicator in comparison with the

R. Dvorak and J. Henrard (eds.), The Dynamical Behaviour of our Planetary System, 257.

frequency analysis (Laskar et al. 1992) and the sup-map method (Froeschlé and Lega 1996) . In Section 4 we revisit the Froeschlé 4 dimensional map (Froeschlé 1971,1972) using the 4 methods of chaos detection.

2. Methods of Detection of Weak Chaos. Definition and Test of a New Fast Lyapunov Indicator.

The analysis of the chaotic behaviour of a dynamical system has been done since the 70's (Froeschlé 1970, Froeschlé and Scheidecker 1973) by means of the Lyapunov characteristic indicators. As far as strong chaos is concerned this remains a simple and fast method. The cost to pay in computation time becomes too high when looking at the detection of weak chaos, in particular if we want to analyze a set of orbits of a given system. The frequency map analysis (hereafter FMA) introduced by Laskar et al. (1992) resulted to be a very well suited method for weakly chaotic motion. It allows to detect chaos over a smaller time span than with the LCIs and to give an estimate of the size occupied by the chaotic zones, by islands and by invariant curves (Laskar 1992, Lega and Froeschlé 1996). Another method, which gives results of the same quality of the FMA is that of the sup-map analysis (Froeschlé and Lega 1996). This method simply consists in the detection of the maximum of the action for a given orbit. Despite a strong dependence on the vicinity to a rational frequency (Froeschlé and Lega 1996) this method has the advantage to be very easily extended to multi-dimensional systems.

Coming back to the LCIs, it is well known that when computing them a renormalization procedure is applied at given intervals of time on the tangential vectors evolving with the flow. This procedure is applied to avoid overflow of the lengths of the vectors or underflow of the angle between the vectors in the case of a chaotic orbit. Curiously a systematic study and use of the time at which such renormalization is necessary seems not to have been done yet.

The time necessary to reach a given value, either of the length of any vector or of the angle between vectors is taken as an indicator of stochasticity.

Following this general idea we define three different fast Lyapunov indicators. The first is simply the time necessary for the norm of a given vector to become greater than a threshold L. It is well known that any vector will evolve under the action of the largest characteristic exponent. However, a transitory regime could occur if the initial vector is almost perpendicular to the expanding manifold. Such a drawback is smoothed when considering as an indicator the evolution of the angle between the vectors starting with a p-dimensional basis embedded in the n-dimensional phase space ($p \leq n$). This second indicator is in fact closely related to the first one, given the conservation of the volume in the phase space. In average it is of the order of the inverse of the first indicator to the power p.

Studying the variation with time of these two indicators for specific orbits of the standard map we have found a third indicator which eliminates almost completely the problem, evoked above, of the dependence on the initial vector basis. Starting again with a $p-$dimensional basis in the phase space, we take at each iteration the largest among the vectors of the evolving basis. Given a $p-$dimensional basis $V_p(0) = (v_1(0), v_2(0), ...v_p(0))$ and an initial condition $P(0) = (x_1(0), x_2(0), ...x_n(0))$, in order to have homogeneous quantities we define the three indicators as functions of t as follows:

$$\begin{cases} \psi_1(t) = 1/||v_1(t)||^p \\ \psi_2(t) = 1/(\prod_{j=1}^{p} ||v_j(t)||) \\ \psi_3(t) = 1/(\sup_j ||v_j(t))||^p) \end{cases} \tag{1}$$

Of course the first method requires only to compute the evolution of a single vector but for Hamiltonian-like systems with more than one degree of freedom it is interesting to compute all the LCEs and therefore the evolution of the $n-$dimensional basis.

As mentioned above the FLIs may depend on the direction of the initial vectors of the V_p basis. The third indicator has been shown to be (Froeschlé et al. 1996) the less dependent on the direction of the initial vectors. In the following we will take ψ_3 as fast Lyapunov indicator.

2.1. Test on the Standard Map

We have taken as a model problem the bi-dimensional standard map, for which we recall the set of equations:

$$\begin{cases} x' = x + k\sin(x+y) \mod (2\pi) \\ y' = x + y \qquad\qquad \mod (2\pi) \end{cases} \tag{2}$$

Fig. 1 a displays orbits of this area preserving mapping for $k = 0.3$. The usual features of invariant curves appear. The origin corresponds to a hyperbolic invariant point surrounded by a small chaotic zone shown on Fig. 1 b.

Fig. 2 a shows the variation of ψ_3 as a function of time for a chaotic orbit, with initial conditions $x_0 = 0.001$, $y_0 = 0.001$, belonging to the small chaotic zone associated with the hyperbolic invariant point. We clearly see a sharp drop down to 10^{-20} of ψ_3 on only 200 iterations. Fig. 2 b shows the evolution of the largest Lyapunov characteristic indicator for the same orbit. The LCI (whose limit, as t goes to infinity, is nothing but the LCE) stops to decrease after about 10 000 iterations. Of course the computation of the LCE needs a much larger number of iterations and the indication of stochasticity given by the LCI is obtained quicker with the FLI. On the other hand, when looking at a regular orbit (the torus on the right of Fig. 1 a) the variation with time of the FLI (Fig. 3a) is quite similar to that of the LCI (Fig. 3b). Actually this is related to the fact that for such very regular orbit it would take a long time to have problems of overflow or underflow. Let us remark however, that for $t = 200$ the value of ψ_3 for the regular orbit is about 10^{-5}, that is 15 order of magnitude greater than for the chaotic orbit. This result is confirmed for any initial conditions on the orbits (Froeschlé et al. 1996).

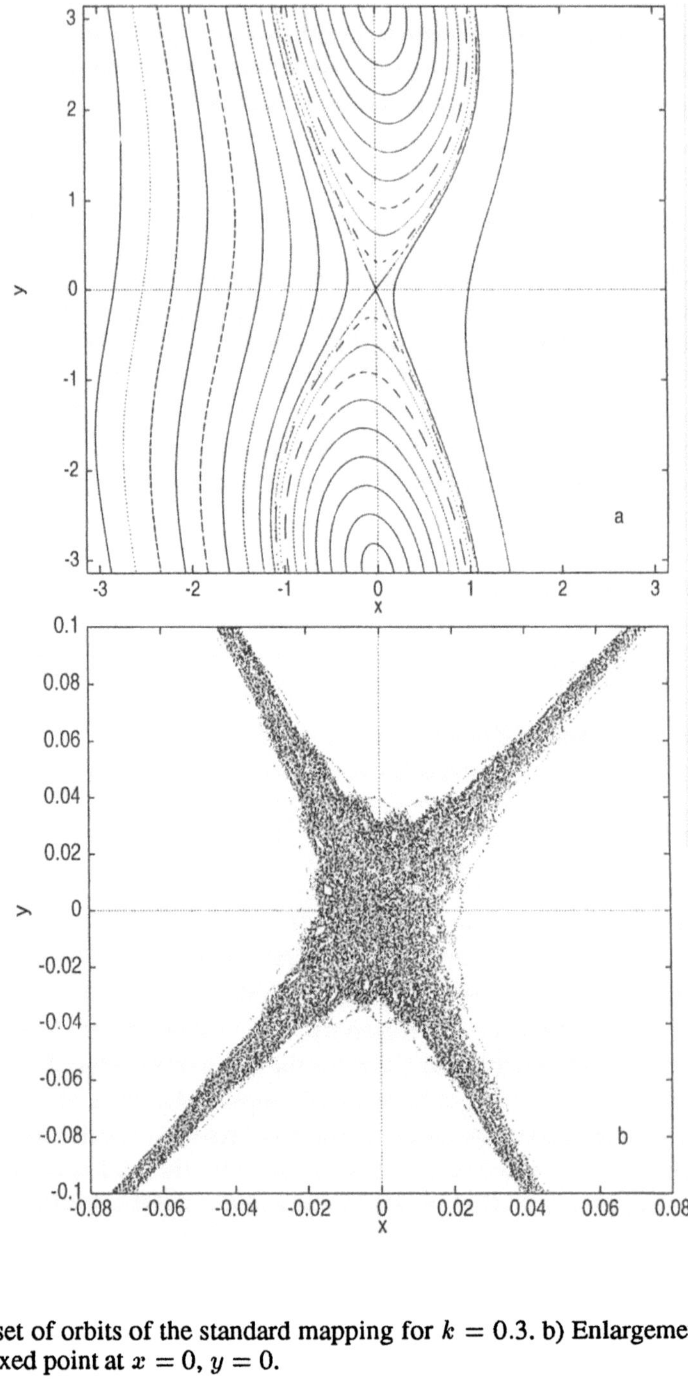

Fig. 1. a) A set of orbits of the standard mapping for $k = 0.3$. b) Enlargement around the hyperbolic fixed point at $x = 0$, $y = 0$.

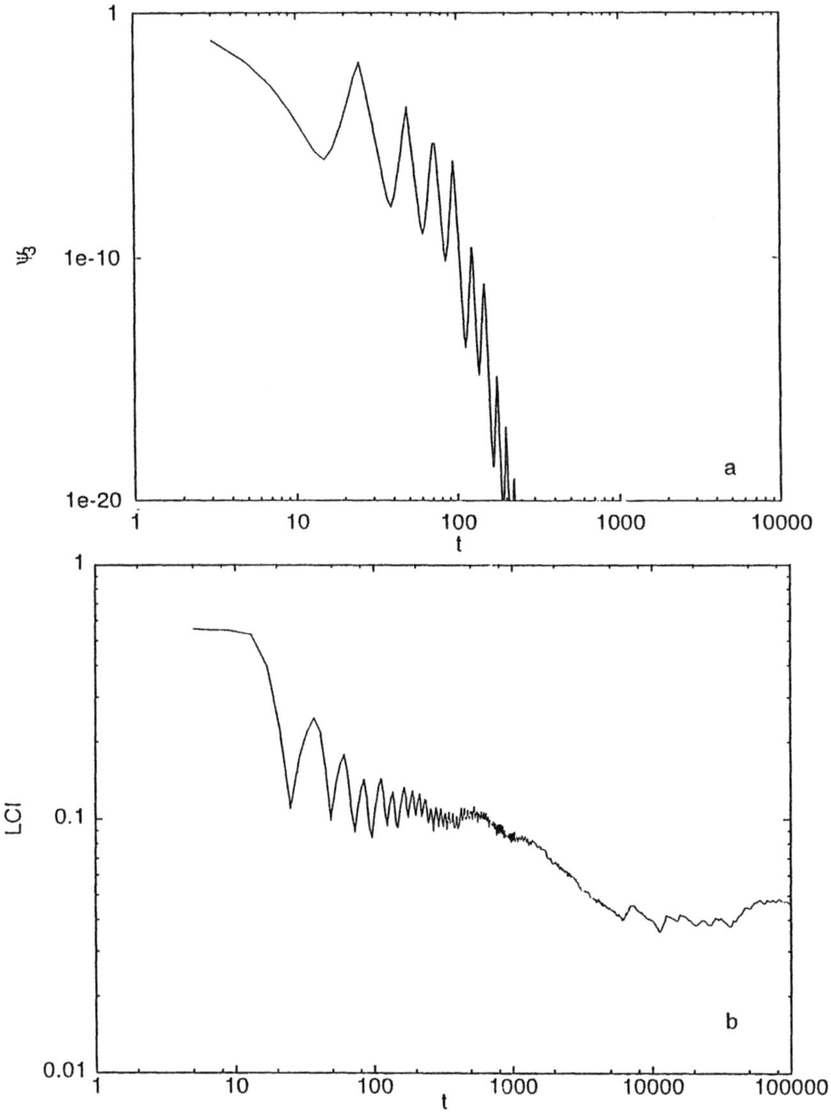

Fig. 2. a) Variation of ψ_3 as a function of time for a chaotic orbit of the standard map with perturbation parameter $k = 0.3$ and initial conditions: $x_0 = y_0 = 0.001$. b)Variation of the largest Lyapunov Characteristic Indicator for the same orbit.

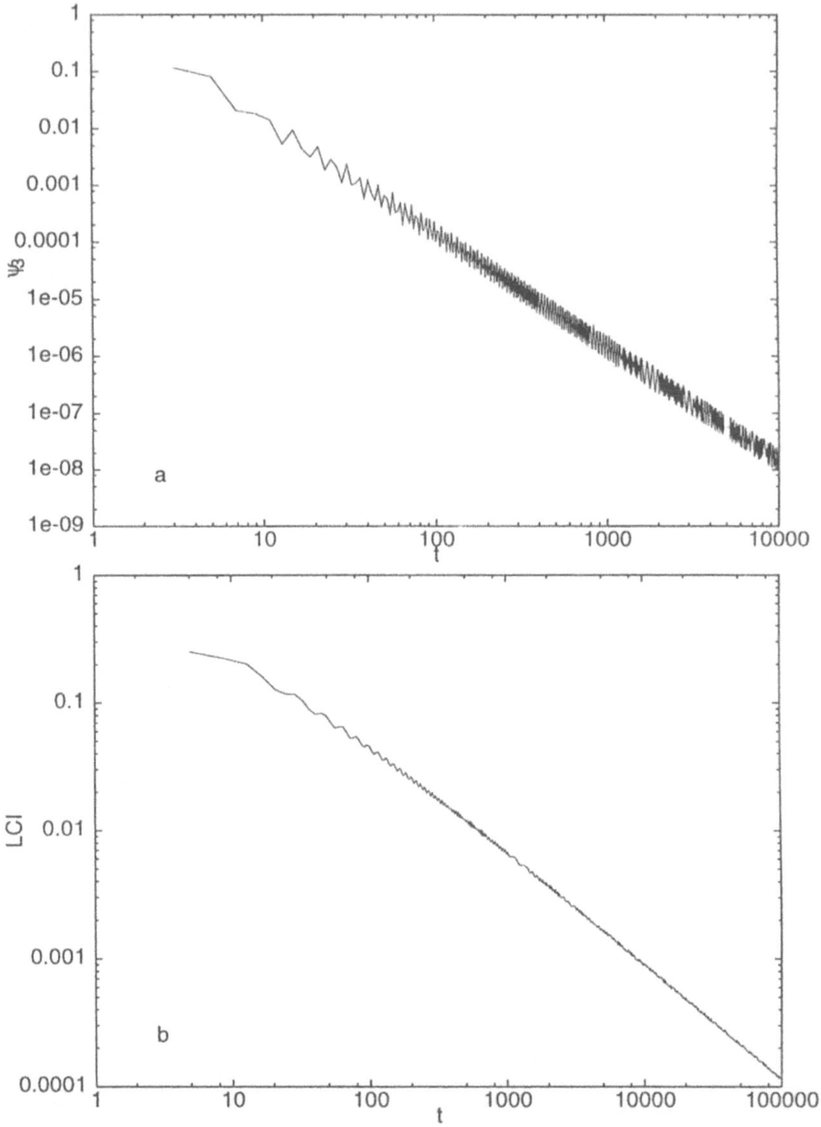

Fig. 3. a) Variation of ψ_3 as a function of time for an invariant torus of the standard map with perturbation parameter $k = 0.3$ and initial conditions: $x_0 = 1$, $y_0 = 0$. b) Variation of the largest Lyapunov Characteristic Indicator for the same orbit.

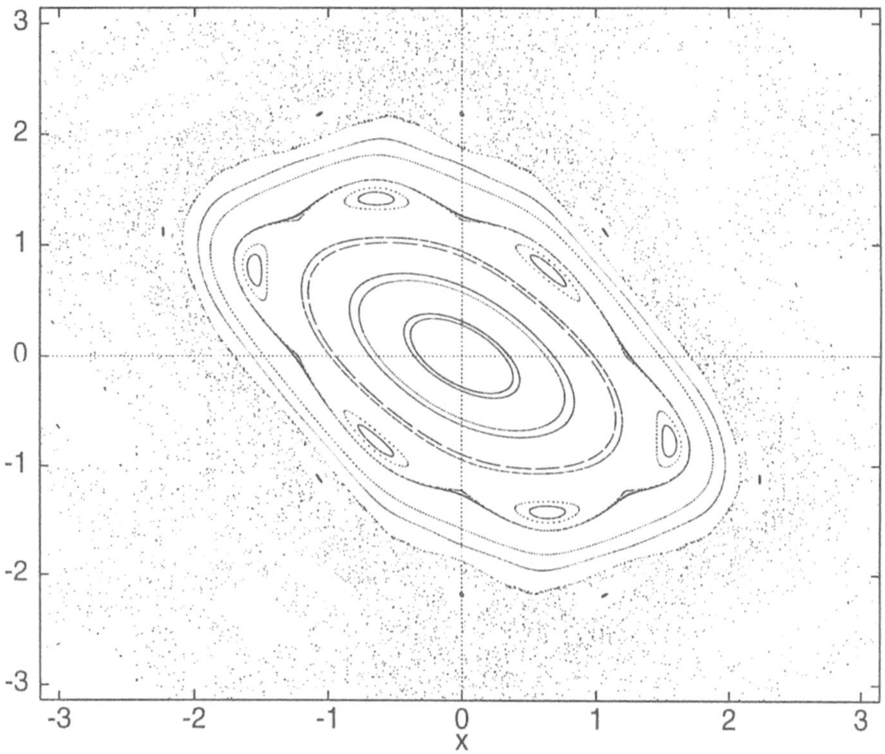

Fig. 4. A set of orbits of the standard mapping for $k = -1.3$

3. Test of the Sensitivity of the Fast Indicator

It is a relatively easy exercise to distinguish between regular and chaotic orbits, as far as strong chaos is concerned. An indicator like the first twenty values of the LCIs, in the case of a mapping, (Contopoulos, private communication) is already a sensitive one. What happens in the case of very thin chaotic layers? The problem was already raised by Laskar et al. (1992) who detected chaotic orbits looking at the variation of their frequency with time. To test our method we have explored the same case as Laskar et al. (1992), i.e., again for the standard map but with $k = -1.3$, the vicinity of an hyperbolic point connected with a chain of 6 islands embedded in the big island around the origin as shown on Fig. 4. For such an orbit Laskar et al. claim that using the LCI: "5×10^6 iterations were necessary to clearly detect the chaotic motion, while it was already visible with the frequency analysis with 20 000 iterations". Fig. 5 shows for the same orbit the computation of ψ_3 and only 2 000 iterations are necessary to reach a value of 10^{-20}.

To be completely sure of the sensitivity of the result we have to compare this behaviour to that of nearby invariant orbits. Therefore we make a cross-section (Fig. 6a) at $x = 0$ in the interval $-1.223 < y < -1.220$ taking as an indicator

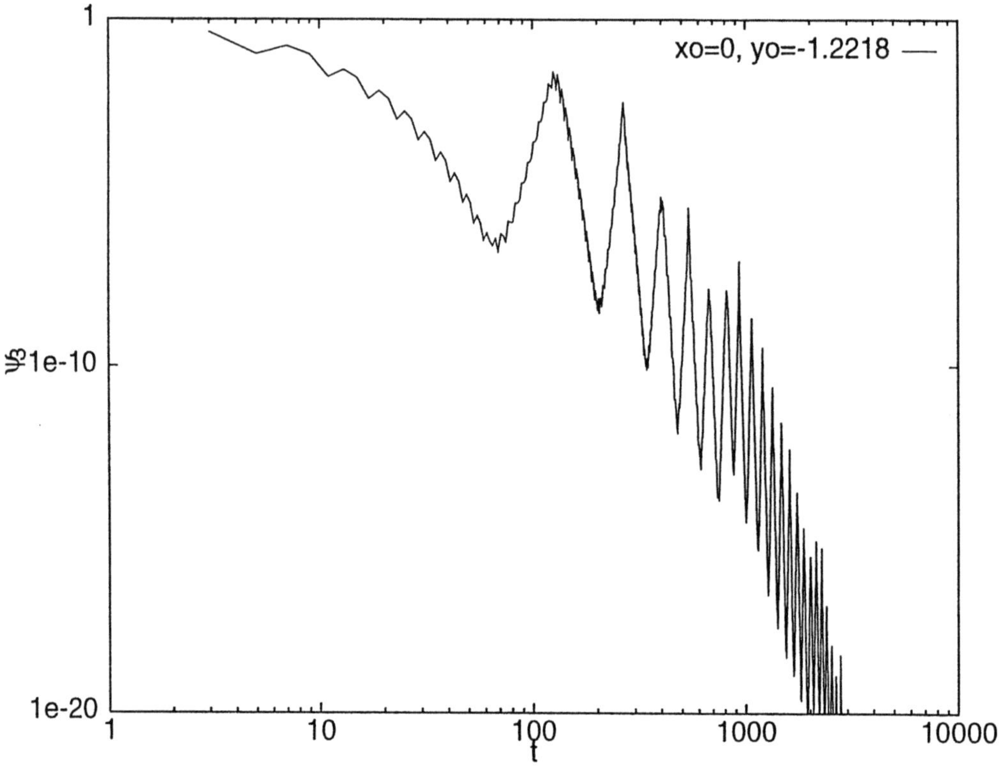

Fig. 5. Variation of ψ_3 as a function of time for a chaotic orbit of the standard map with $k = -1.3$. The initial condition is $x_0 = 0$, $y_0 = -1.2218$.

the time $N(\epsilon)$ necessary for ψ_3 to reach a threshold ϵ with $\epsilon = 10^{-10}$. We made the computation over a total number of $N_m = 2\,000$ iterations. Of course, if the threshold is not reached in less than 2 000 iterations, we take $N(\epsilon) = N_m$. Fig. 6b shows the same cross-section for the frequency analysis. The frequency has been computed with the Hénon method (Hénon, private communication, Celletti and Froeschlé 1995) using 20 000 iterations for each orbit. All the features appearing in the frequency analysis also appear when looking to the variation of $N(\epsilon)$.

To test even more the sensitivity of our indicator we made an enlargement of the zone $-1.223 < y < -1.222$ just before the crossing of the hyperbolic point. On the same figure (Fig.7) we show both, the variation of the frequency and of the number $N(\epsilon)$ rescaled by a factor $N(\epsilon)2\,10^{-7} + 0.165$. Again all the very small chaotic layers indicated by a jump in the frequency are clearly shown when looking at $N(\epsilon)$. Moreover at this resolution ($\Delta y = 10^{-6}$) the variation of $N(\epsilon)$ indicates chaotic layers which are not clearly detected with the frequency analysis. Of course our method is not able to distinguish between tori and islands (conversely to the frequency map) which are both invariant curves with zero LCE. We emphasize that this very sensitive variation of $N(\epsilon)$ was obtained using only 2 000 iterations.

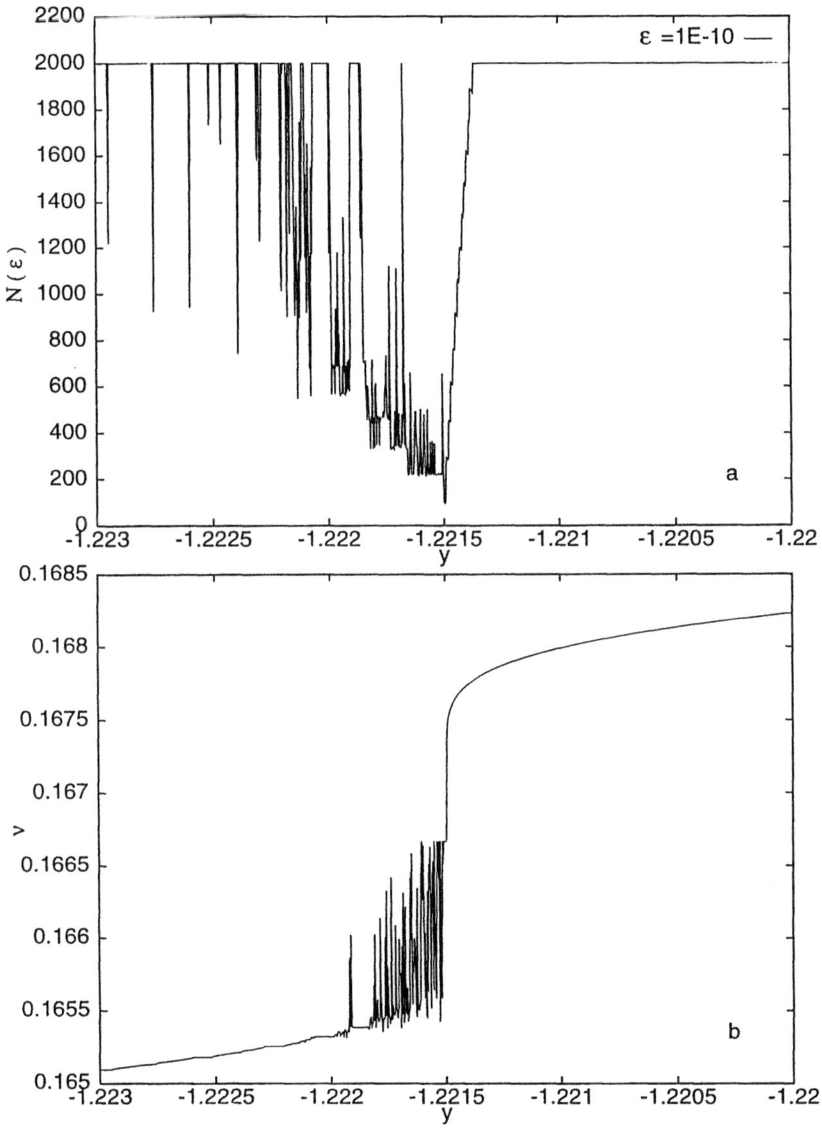

Fig. 6. Cross section in the vicinity of the hyperbolic point associated to the 1/6 resonance of the standard map with perturbation parameter $k = -1.3$. a) Number of iterations necessary for ψ_3 to reach a threshold value of 10^{-10} versus the initial action. The total number of iterations for each orbit is $N = 2\,000$. b) Frequency value for the same set of orbits computed over 20 000 iterations.

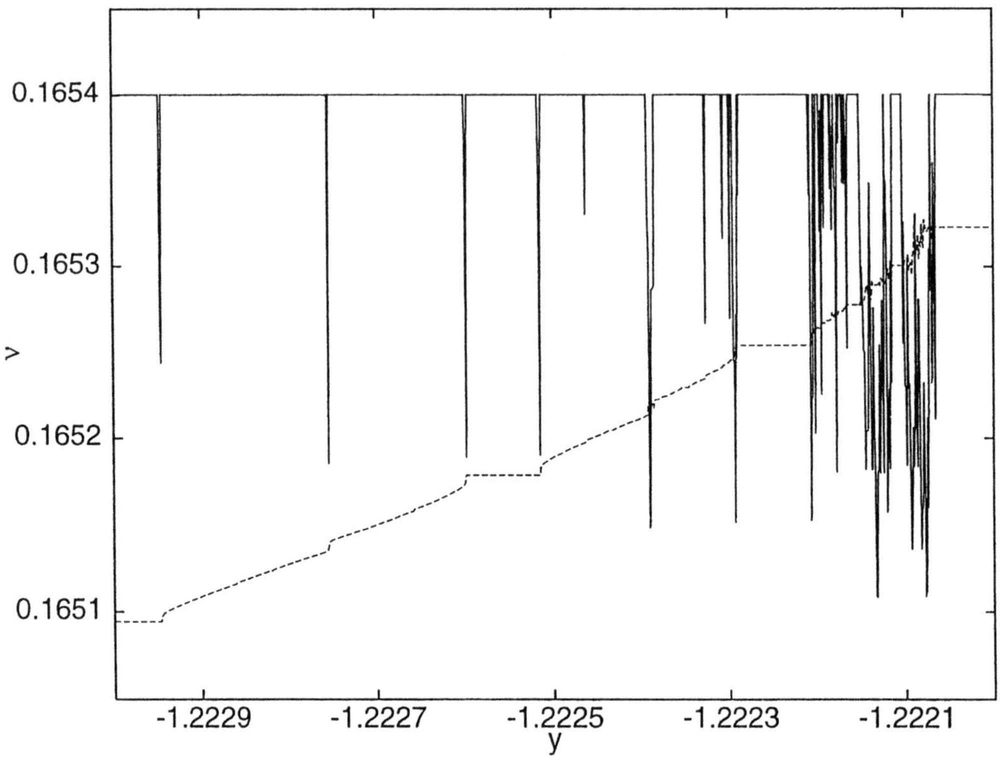

Fig. 7. Same as Fig. 6 for an enlargement before the hyperbolic point. Frequency map (dashed line), $N(\epsilon)$ (continuous line). $N(\epsilon)$ has been rescaled in order to have the two plots on the same picture.

Fig.8a shows the same cross-section for the sup-map analysis using 20 000 iterations for each orbit. As already stated by Froeschlé and Lega (1996) a monotonic variation of the maximum of the action corresponds to a set of invariant KAM tori, a jump followed by a plateau indicates the crossing of a chaotic zone and the crossing of nested islands gives rise to v-shaped structures. These structures appear clearly on Fig.8a and all the feature obtained with the frequency analysis can be found with the sup-map analysis. The larger islands, i.e. the larger v-shaped structures have been cut at a level of the sup of 0.15816 in order to have a sensitive magnification of the smaller structures. This analysis has been repeated for 100 000 iterations (Fig.8b). Some of the smaller v-shaped peaks have disappeared. The corresponding orbits are, in fact, invariant tori with frequencies very close to rational.

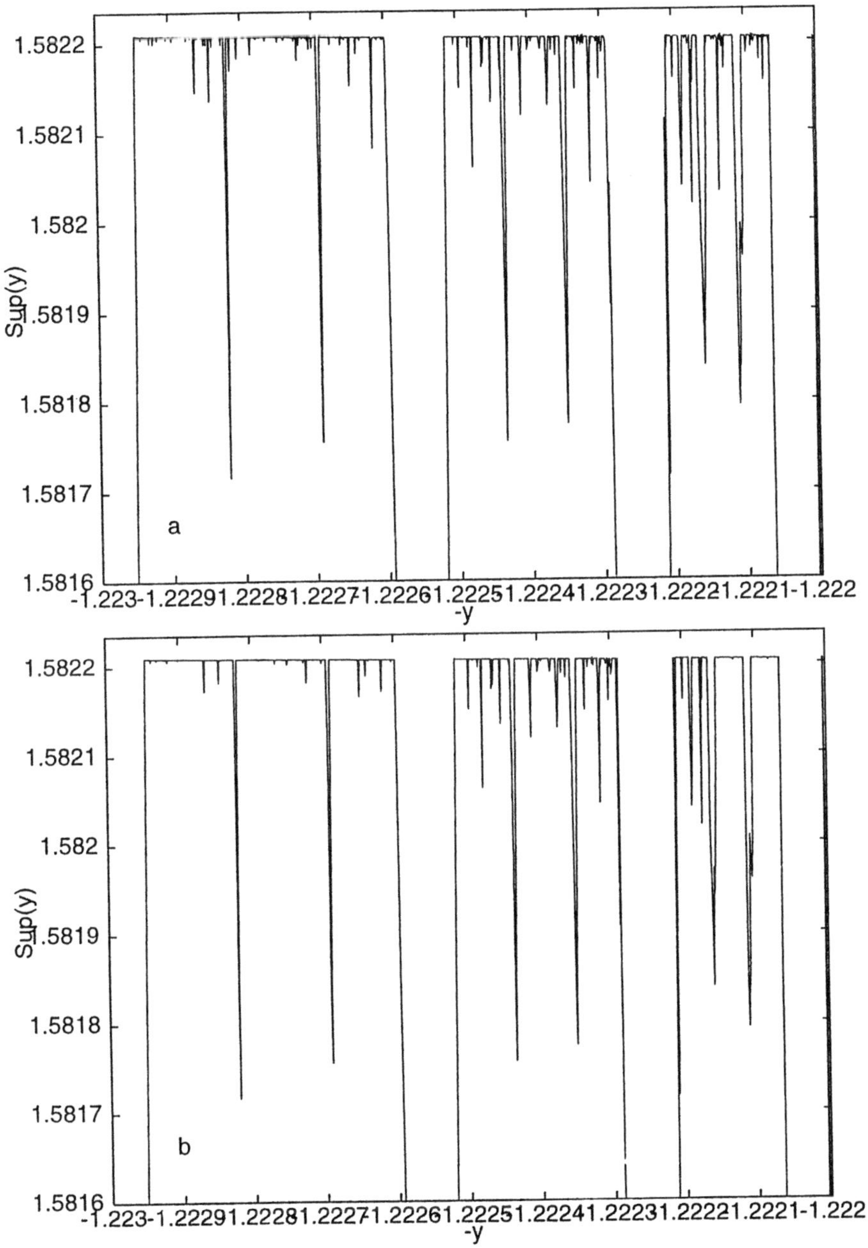

Fig. 8. Variation of the sup y versus y for the same enlargement of Fig. 7. The larger v-shaped structures have been cut at a level of the sup of 0.15816 in order to have a sensitive magnification of the smaller structures. The analysis has been done using respectively 20,000 (Fig.8a) and 100 000 (Fig.8b) iterations.

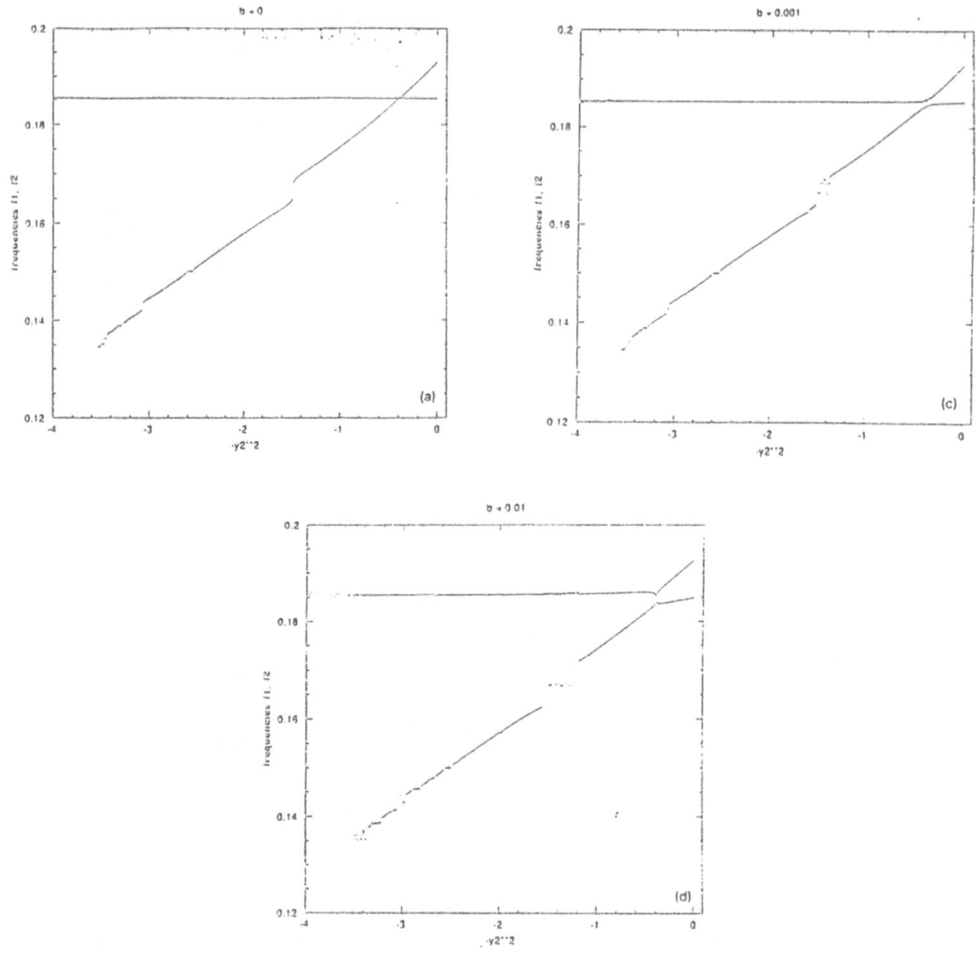

Fig. 9. Frequency f_1 (nearly horizontal curve) and f_2 (diagonal curve) with respect to y_2 for the 4D standard map with $k_1 = k_2 = -1.3$ and various values of the coupling parameter b (From Laskar 1993).

4. Application to 4 Dimensional Maps

In this fourth part we will use all the methods already introduced to study the global dynamics of the 4 dimensional symplectic map already used by several authors (Froeschlé 1971,1972; Laskar et al. 1992). We recall the set of equations:

$$\left\{ \begin{array}{lll} x'_1 & = x_1 + k_1 \sin(x_1 + y_1) + b \sin(0.5(x_1 + y_1 + x_2 + y_2)) & \mathrm{mod}\,(4\pi) \\ y'_2 & = x_1 + y_1 & \mathrm{mod}\,(4\pi) \\ x'_2 & = x_2 + k_2 \sin(x_2 + y_2) + b \sin(0.5(x_1 + y_1 + x_2 + y_2)) & \mathrm{mod}\,(4\pi) \\ y'_2 & = x_2 + y_2 & \mathrm{mod}\,(4\pi) \end{array} \right. \qquad (3)$$

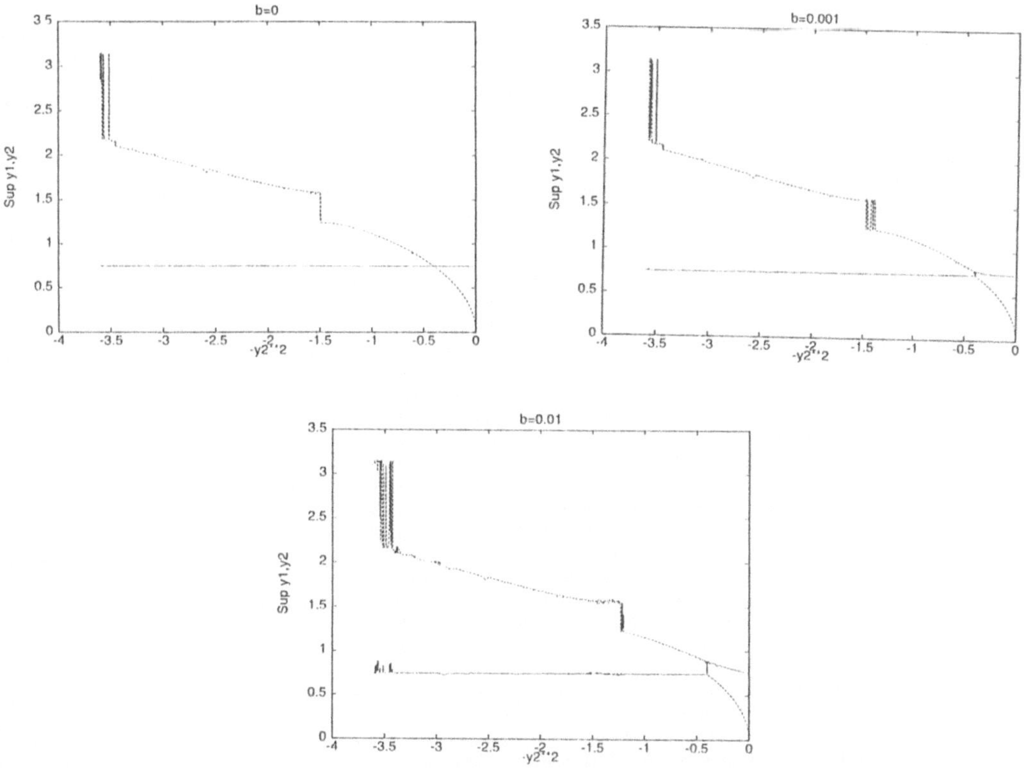

Fig. 10. Maximum of y_1 (nearly horizontal curve) and of y_2 (diagonal curve) with respect to y_2 for the 4D standard map with $k_1 = k_2 = -1.3$ and various values of the coupling parameter b.

In the following we take for the perturbing parameters the values: $k_1 = k_2 = -1.3$. Let us remark than when introducing a modulo we have to be cautious that the functions involved have to be periodic with respect to the argument even if the variables y_1 and y_2 are thought to be angle variables. We have tested the sup-map and the FLI methods for the same set of experiments than those performed by Laskar (1993) using the Frequency Map Analysis method.

Fig.9 (taken from Laskar, 1993) corresponds to the variation of the two fundamental frequencies f_1 and f_2 for three different values of the coupling parameter b: $b = 0, 0.001, 0.01$ for a cross section along y_2, y_2 ranging from -1.9 to 0. The other initial conditions are $x_1 = x_2 = 0$, and $y_1 = -0.63$. Fig.10 and Fig.11 show the same analysis made respectively for the sup-map method and for the FLI. More precisely in Fig.11 we plot the time $N(\epsilon)$ necessary for ψ_3 to reach the threshold value of 10^{-12}. In all cases the total number of iterations is 516. All the features detected by Laskar using the FMA analysis are also detected with the two other methods. The appearance of the 1:1 resonance at the intersection of the two curves

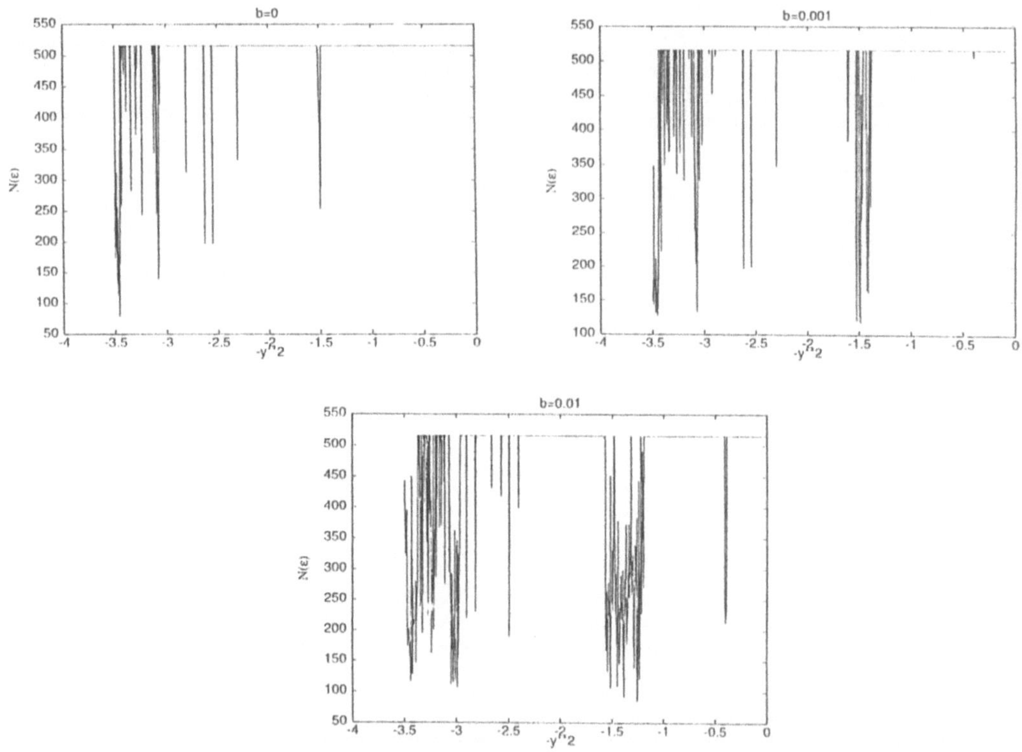

Fig. 11. Time $N(\epsilon)$ necessary for ψ_3 to reach the threshold value of 10^{-12} with respect to y_2 for the 4D standard map with $k_1 = k_2 = -1.3$ and various values of the coupling parameter b.

f_1 and f_2 induces a distortion of both the frequencies curves and the sup curves. The amplitude of the distortion is of about the same magnitude. This can be easily understood since the distortion of the frequency is mainly due to the variation of the action variables. The variation of ψ_3 shows also clearly the small chaotic zone associated with this 1:1 resonance. We notice also with the three methods a dramatic increase in the size in action of the chaotic zones, particularly visible around the 1:6 resonance as the coupling parameter goes from 0 to 0.01.

The very small islands which appear on the f_2 frequency curves appear also on the sup-map curve and the tiny chaotic zones associated to this set of islands are clearly detected with the FLI. Using the same number of iterations, i.e. for the same computational cost the three methods show the same sensitivity but bring different pieces of information and therefore they are strongly complementary. Despite the fact that the FLI does not allow to distinguish between invariant tori and islands it shows a better sensitivity than the other methods to detect very tiny chaotic zones.

To visualize the complete mapping we map the plane defined by $x_1 = x_2 = 0$

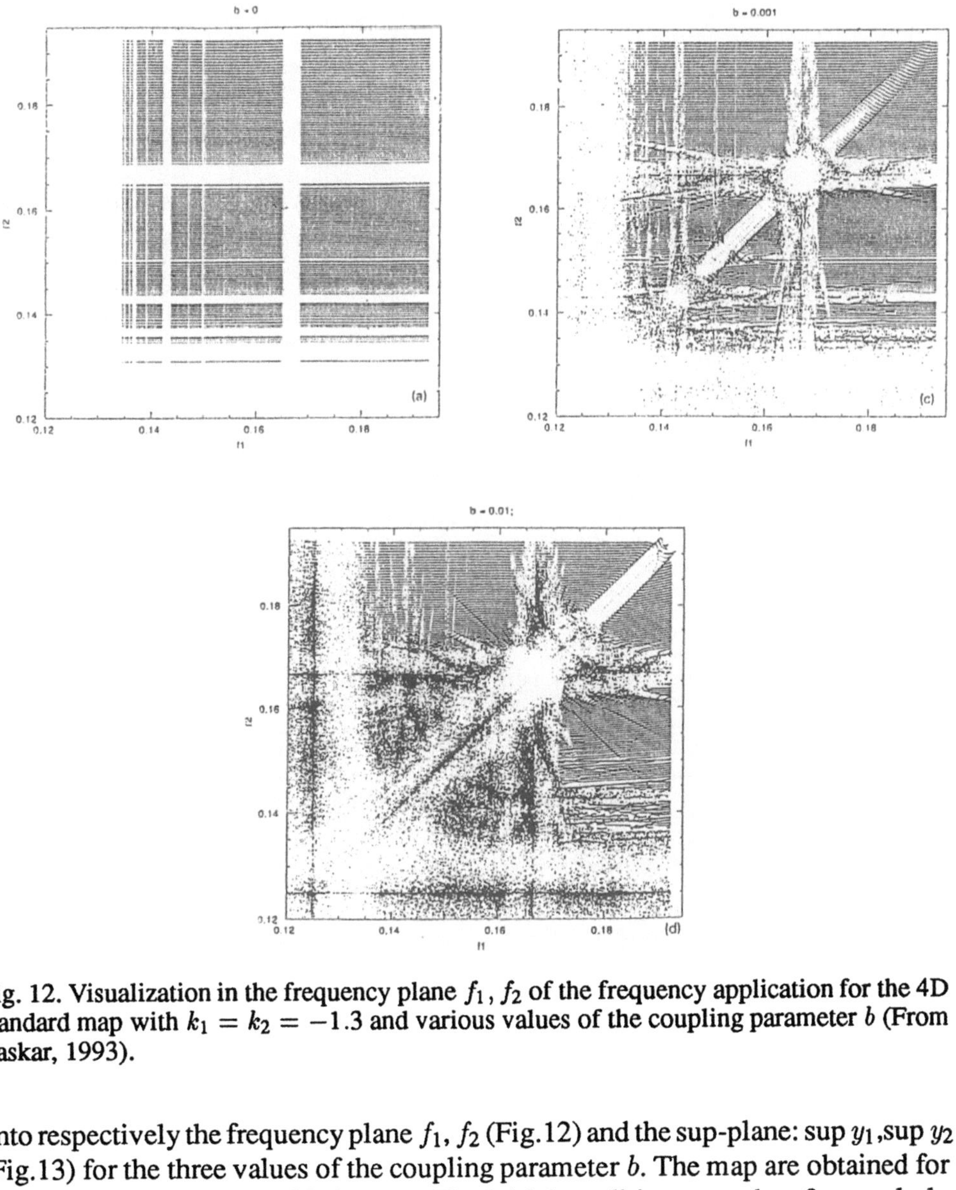

Fig. 12. Visualization in the frequency plane f_1, f_2 of the frequency application for the 4D standard map with $k_1 = k_2 = -1.3$ and various values of the coupling parameter b (From Laskar, 1993).

onto respectively the frequency plane f_1, f_2 (Fig.12) and the sup-plane: sup y_1,sup y_2 (Fig.13) for the three values of the coupling parameter b. The map are obtained for 120 orbits along y_1 and 1200 along y_2. The initial conditions are taken for regularly spaced values of y^2 instead of y. We have also considered using a representation on a grey-scale the estimated values of the largest LCE for the usual 516 iterations (Fig.14 left column) and for 10 000 iterations (Fig.14 right column) as well as the FLI (Fig.15 left column) and the time $N(\epsilon)$ for $\epsilon = 10^{-12}$ (Fig.15 right column). The images, for the three values of the coupling parameter, are obtained on a grid of 256×256 pixels regularly spaced on y^2. All these figures show clearly the interaction between the resonances trough the Arnold web structure. Of course,

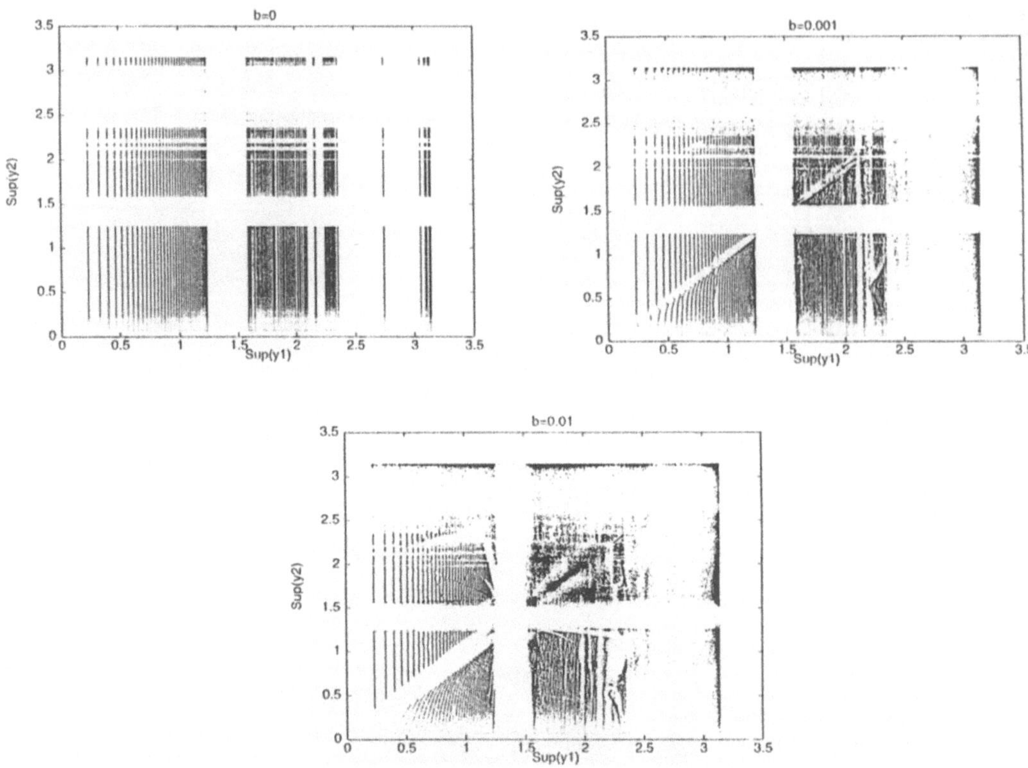

Fig. 13. Visualization in the sup-plane $\sup y_1, \sup y_2$ of the sup-application for the 4D standard map with $k_1 = k_2 = -1.3$ and various values of the coupling parameter b.

as expected, when looking to the chart of the largest LCI the structure reveals himself only for the 10 000 iterations. The first three approaches are really fast. We emphasize that for such a qualitative approach the sup-map method and the FLI-chart are really easy to handle.

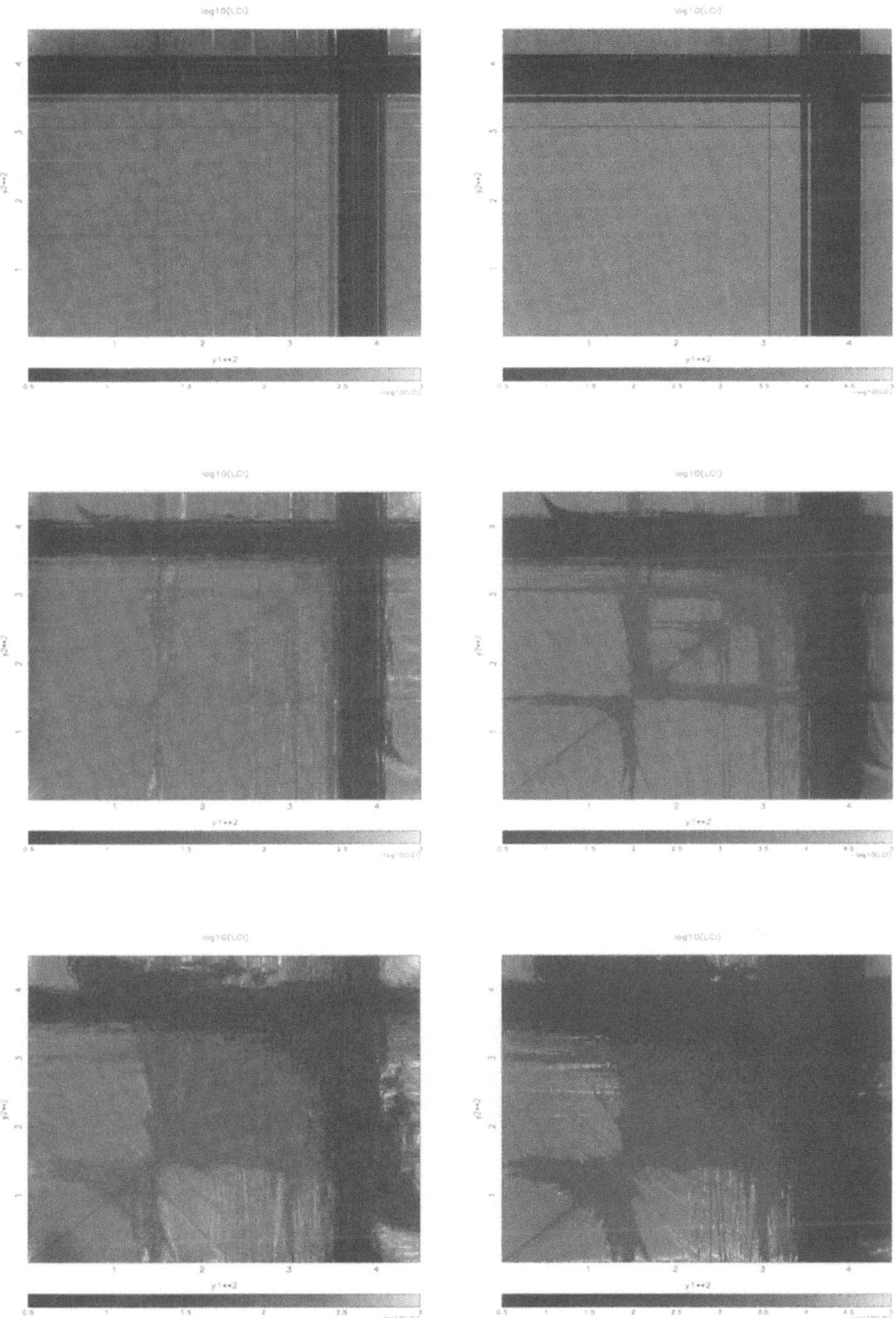

Fig. 14. The logarithm of the LCI for the same zone of the phase space explored for the frequency plane and for the three different values of the coupling parameter b: $b = 0, 0.001, 0.01$ (from top to bottom). The darker points correspond to the chaotic zones. Left column: the computation is done using 516 iterations. Right column: the computation is done using 10 000 iterations.

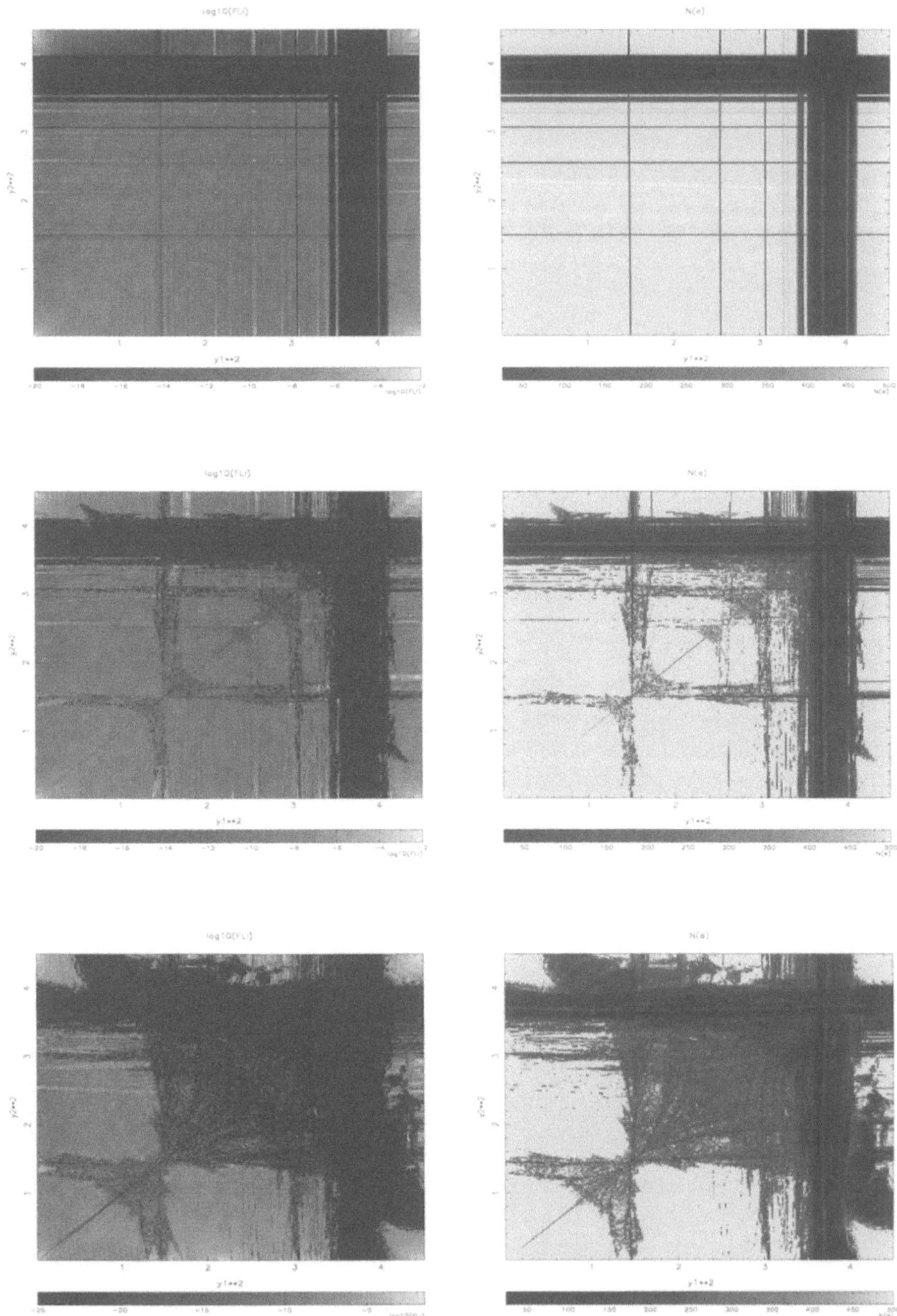

Fig. 15. The FLI for the same zone of the phase space explored for the frequency plane and for the three different values of the coupling parameter b: $b = 0, 0.001, 0.01$ (from top to bottom). The darker points correspond to the chaotic zones. Left column: the values of $\psi(3)$ after 516 iterations. Right column: the time $N(\epsilon)$ necessary for ψ_3 to reach a threshold value of $\epsilon = 10^{-12}$ on a maximum number of 516 iterations.

5. Conclusion

The four methods tested in this paper, namely the frequency map, the sup-map, the fast Lyapunov and the Lyapunov characteristic exponent appear to be complementary tools for the numerical exploration of the reach structure of a Hamiltonian dynamical system. Indeed the FLI is very fast and very promising for a first systematic exploration of the LCI, but fails to separate islands from original invariant tori which is not the case for both the FMA and the sup-map methods.

References

Celletti, A. and Froeschlé, C.: 1995, On the determination of the stochasticity threshold of invariant curves. *Int. J. of Bifurcation and Chaos*, 5, n.6.

Froeschlé, C.: 1970, A numerical study of the stochasticity of dynamical systems with two degrees of freedom. *Astron. Astrophys.*, **9**, 15–23.

Froeschlé, C.: 1971, On the number of isolating integrals in systems with three degrees of freedom. *Astrophys. and Space Sciences*, **14**, 110.

Froeschlé, C.: 1972, Numerical study of a four-dimensional mapping. *Astron. Astrophys.*, **16**, 172.

Froeschlé, C. and Lega, E.: 1996, On the measure of the structure around the last KAM torus before and after its break-up. *Celest. Mech. and Dynamical Astron.*, (in press).

Froeschlé, C. Lega, E., and Gonczi, R.: 1996, Fast Lyapunov indicators. Application to asteroidal motion. *Celest. Mech. and Dynamical Astron.*, submitted.

Froeschlé, C. and Scheidecker, J.P.: 1973, Numerical study of a four dimensional mapping II. *Astron. Astrophys.*, **22**, 431–436.

Froeschlé, C. and Scheidecker, J.P.: 1973, On the disappearance of isolating integrals in systems with more than two degrees of freedom. *Astrophys. and Space Sc.*, **25**:373–386.

Laskar, J.: 1990, The chaotic motion of the solar system. A numerical estimate of the size of the chaotic zones. *Icarus*, **88**, 266.

Laskar, J., Froeschlé, C., and Celletti, A.: 1992, The measure of chaos by the numerical analysis of the fundamental frequencies. Application to the standard mapping. *Physica D*, **56**, 253.

Laskar, J. and Robutel, Ph.: 1993, The chaotic obliquity of the planets. *Nature*, **361**, 608.

Lega, E. and Froeschlé, C.: 1996, Numerical investigations of the structure around an invariant KAM torus using the frequency map analysis. *Physica D*, **1345**, 1.

Milani, A. and Nobili, A.: 1992, *Nature*, **357**, 569.

Morbidelli, A. and Froeschlé, C.: 1996, On the relationship between Lyapunov times and macroscopic instability times. *Celest. Mech. and Dynam. Astr.*, In press.

Morbidelli, A. and Giorgilli, A.: 1994, Superexponential stability of KAM tori. *J. Stat. Phys.*, **78**, 1607.

Morbidelli, A. and Guzzo, M.: 1996, The Nekhoroshev theorem and the asteroid belt dynamical system. *published in this volume.*

ASTEROID MOTION NEAR THE 2:1 RESONANCE: A SYMPLECTIC MAPPING APPROACH

JOHN D. HADJIDEMETRIOU
University of Thessaloniki, Department of Physics
GR-540 06 Thessaloniki, Greece

and

ANNE LEMAITRE
FUNDP, Département de Mathématiques
Rempart de la Vierge, B-5000 Namur, Belgium

Abstract. We present a symplectic mapping model that is valid close to the 2:1 resonance in a Sun-Jupiter-asteroid system, for planar motion. The mapping is based on the averaged Hamiltonian close to this resonance, where an additional correction term has been introduced in order to restore the correct position and stability of the fixed points. The topology of the mapping is similar to that of the Poincaré map of the real system. The evolution of the eccentricity of the asteroid is presented and it is shown that for a large region of phase space there is a diffusion to very large values, leading to escape of the asteroid from the 2:1 resonance. There are however regions in phase space where the eccentricity remains bounded, in the framework of this model.

Key words: Symplectic maps – Jovian resonances

1. Introduction

The problem of resonant asteroid motion is an old problem of dynamical astronomy and is related to the distribution of the asteroids and the formation of the Kirkwood gaps at several resonances with Jupiter. This problem has an interest in its own as a problem of dynamics and many of its features that were long unsolved, were made clear recently by applying methods of modern chaotic dynamics. This lead to the explanation of the gap at the 3:1 resonance by the work of Wisdom (1982,1983,1985,1987), who used a mapping model to study asteroid motion at this resonance. He showed that chaotic motion can appear at the 3:1 resonance, resulting to sudden jumps of the eccentricity of the asteroid. In Wisdom's work the model that was used (based on the averaged Hamiltonian) was not valid for values of the eccentricity of the asteroid larger than 0.3, and consequently the jumps did not exceed this value. It can be shown however that by including to the model the high eccentricity resonances (that were missing in Wisdom's model) we can have jumps of the eccentricity to very high values, so that the asteroid can become an Earth crosser or even fall on the Sun. This was shown by Hadjidemetriou (1991,1993) who used a symplectic mapping approach, based also on the averaged Hamiltonian, which was corrected in order to include the high eccentricity resonances that are present to the real problem. Similar results were obtained by Klafke, Ferraz-Mello and Michtchenko (1992) and by Moons and Morbidelli (1995).

The mapping technique of Wisdom has been applied to the 5:2 resonance by Šidlichovský (1988,1992) and similar results were obtained. Despite however the

R. Dvorak and J. Henrard (eds.), The Dynamical Behaviour of our Planetary System, 277.

success of these methods to the 3:1 and the 5:2 resonances, many features of the problem of resonant asteroid motion are not yet clear. In particular, the study of the 2:1 resonance appears more complex and work is still in progress (Henrard (1988), Henrard and Lemaitre (1987), Lemaitre and Henrard (1990), Henrard et al. (1995), Ferraz-Mello (1994a, 1996a,b), Morbidelli (1996), Morbidelli and Moons (1993), Morbidelli and Giorgilli (1990a,b)). In the present study we will approach this problem by constructing a symplectic mapping model, using the method presented in Hadjidemetriou (1991,1993). The mapping is based on the averaged Hamiltonian of the three body problem (we shall use the planar case in this study) near the 2:1 resonance, given by Lemaitre and Henrard (1987), which is corrected in order to include all the resonances, both at low and high values of the eccentricity of the asteroid, by introducing suitable *correction terms.*

The dynamics of the asteroid motion can be structured in a hierarchy of perturbations: the integrable problem is the *two-body problem* "Sun-asteroid" and the simplest nonintegrable model is the system "Sun-Jupiter-asteroid", assuming Jupiter revolving around the Sun in a circular orbit (*restricted circular three body problem*). Although this model is not realistic, as we shall see in the following, it is useful in our study, because it provides the basic framework upon which all other perturbations are applied. A more realistic model is obtained by considering Jupiter to revolve around the Sun in a fixed elliptic orbit, the *elliptic restricted three body problem.*

Mapping models are very useful because they are very fast, compared to numerical integrations, provided they are realistic, i.e. they have the same topology as the Poincaré map of the actual system. A necessary condition for this is that the mapping model must has the same fixed points, with the correct stability, as the real system. The comparison of the fixed points of the mapping model with the periodic orbits/fixed points of the Poincaré map of the real system is a good test of the validity of the model. Thus, it is essential to know some basic facts, for example the basic families of periodic orbits of the real model and some Poincaré surfaces of section.

2. The Basic Model: the Restricted Three Body Problem

As we mentioned in the introduction, the knowledge of the basic facts of the dynamics of the elliptic restricted three body problem Sun-Jupiter-asteroid is essential in constructing the mapping model. For this reason we shall give in the following the basic families of periodic orbits, in order to compare them with the fixed points of the mapping model and test how realistic it is. We will start with the circular case and then proceed to the elliptic case. We will use normalized units, so that the sum of the masses of the Sun and Jupiter is equal to one, the gravitational constant is equal to one and the radius of the orbit of Jupiter is also equal to one. In these units, the value of the mass of Jupiter is equal to $\mu = 0.00095387535$ and its period is equal to $T = 2\pi$.

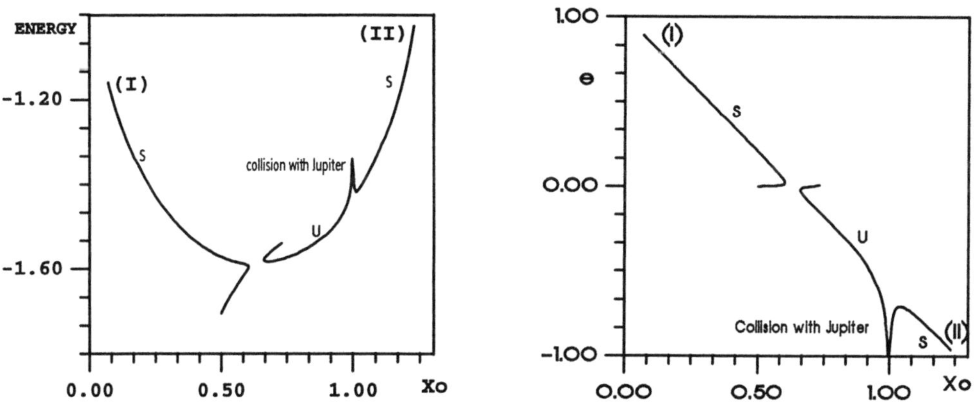

Fig. 1. The initial conditions of the two families of resonant 2:1 periodic orbits in the planar circular three body problem, for $\mu = 0.00095387535$. (a) Variables: *Energy* - x_0, (b) variables: *eccentricity* - x_0N. S denotes stability and U instability.

2.1. THE CIRCULAR CASE

In this model Jupiter revolves around the Sun in a circular orbit and the asteroid moves in the orbital plane of these two bodies. There are two basic families of resonant periodic orbits at the 2:1 resonance (in the rotating frame of reference), along which the eccentricity of the asteroid increases, starting with values close to zero. The initial conditions of a periodic orbit are x_0, $y = 0$, $dx/dt = 0$ and $(dy/dt)_0$. Thus, the initial conditions of a periodic orbit can be presented in a diagram x_0 - $(dy/dt)_0$, or in a diagram $x_0 - energy\ \ constant$. The above two families differ in the initial phase at $t = 0$: in family I the asteroid is at perihelion and in family II at aphelion. In Figure 1 we present the initial conditions of these two families. In Figure 1a we give the energy constant vs. x_0 and in Figure 1b the eccentricity e vs. x_0. Family I corresponds to $\sigma = 0$ and family II to $\sigma = \pi$. Family I is stable and family II starts as unstable until a value of the eccentricity equal to $e = 0.6$, where we have collision with Jupiter. Beyond this point there is a complicated exchange between stability and instability in a small region close to the collision and after that the family becomes stable.

We present also, in Figure 2, Poincaré surfaces of section at several energy levels. The surface of section is defined by $y = 0$ for a fixed energy, and we consider the points x and dx/dt. For reasons of comparison with the mapping in Figure 4, we present these points in the axes $(e \sin\sigma,\ \ e\cos\sigma)$, where e is the osculating eccentricity and σ the resonance angle defined by Equation (1). The knowledge of the Poincaré surfaces of section is essential in checking the applicability of the method of averaging. By the method of averaging the system is reduced to an integrable dynamical system of one degree of freedom and this is acceptable only if the original system behaves as integrable, i.e. if it has smooth invariant curves. We note from Figure 2 that beyond a certain energy level, corresponding to a value of the eccentricity of the resonant periodic orbit equal to $e = 0.34$, chaotic motion appears. We will comment on this in the following.

2.2. THE ELLIPTIC CASE

The case of a circular orbit of Jupiter is the simplest nonintegrable dynamical system for the study of the perturbations on the Keplerian orbit of the asteroid. Although it is not realistic for the study of actual asteroid motion, it is essential in understanding the dynamics of the complete system. The actual dynamical system describing the motion of an asteroid can be structured in a hierarchy of perturbations: (1) circular orbit of Jupiter, (2) fixed elliptic orbit of Jupiter, (3) variable orbit of Jupiter due to perturbations by Saturn and the other planets. On going from one approximation to the next in this hierarchy of models, small additional perturbing terms are involved and consequently the dynamics of the new model can be understood better if it is looked at as a modification of the previous model. In the present paper we shall consider a fixed orbit for Jupiter, with eccentricity equal to $e_j = 0.048$.

The topology of the phase space of the elliptic restricted three body problem depends critically on the fixed points and their stability character. These appear as fixed points in a Poincaré map (in this case a four dimensional stroboscopic map at integral multiples of the period $T = 2\pi$ of Jupiter's orbit) and a mapping model must have the same fixed points with the correct stability, in order to be realistic.

For a fixed value of the eccentricity of Jupiter, the periodic orbits of the elliptic problem are isolated. We can define in this case a family of periodic orbits if we consider the eccentricity of Jupiter, e_j, as a parameter. These families bifurcate from the families of periodic orbits of the circular problem (Figure 1) at the points where the period is exactly equal to 2π. From the numerical computations it turns out that there exists only one such point, on the stable family I, and the value of the eccentricity of the asteroid at this periodic orbit is equal to $e = 0.72$. We remark that this bifurcation point is very sensitive to the approximations involved in the construction of an approximate model (averaged Hamiltonian or mapping model). This is so, because the period of all resonant periodic orbits along the two families of the circular problem is very close to 2π (it is exactly equal to 2π in the unperturbed problem $\mu = 0$) and the approximations involved in constructing a model affect the position of this point.

There are two families of periodic orbits of the elliptic restricted three body problem that bifurcate from the above mentioned point, one stable and the other unstable. The initial conditions are given by x_0, $y = 0$, $dx/dt = 0$ and $(dy/dt)_0$ (in the rotating frame) for each value of e_j. In Figure 3 we present the families of periodic orbits in the 3-D space x_0,$(dy/dt)_0$,e_j. The curves on the x_0,$(dy/dt)_0$ space are the two families of the circular problem, that are given in Figure 1 (in different axes).

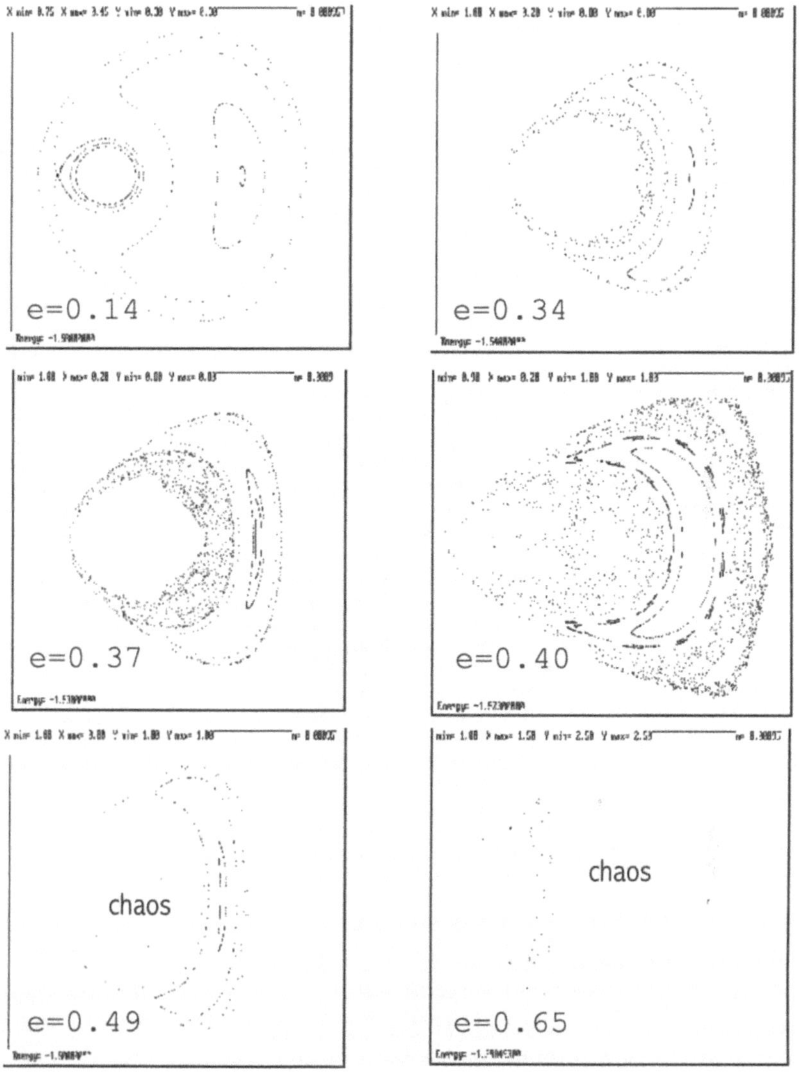

Fig. 2. Poincaré surfaces of section of the circular planar restricted three body problem at the 2:1 resonance. The value of the eccentricity of the stable resonant periodic orbit/fixed point is indicated. The axes are $e \cos \sigma$ and $e \sin \sigma$.

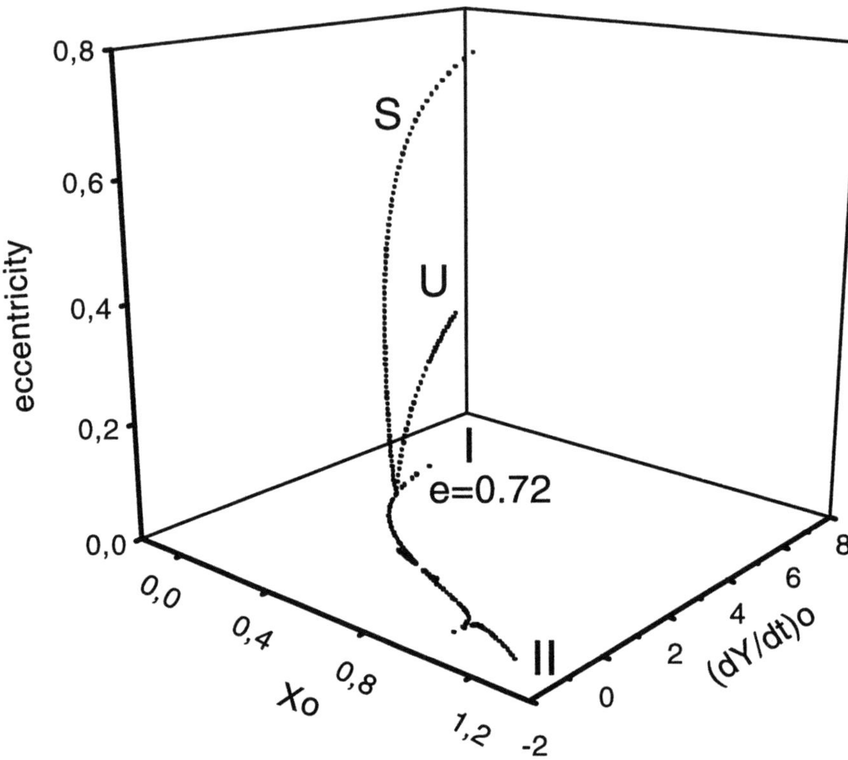

Fig. 3. The two families of periodic orbits of the elliptic restricted three body problem, in the rotating frame, in the x_0, $(dy/dt)_0$, e_j space.

3. The Averaged Hamiltonian at the 2:1 Resonance

The mapping model that we will construct is based on the averaged Hamiltonian of the elliptic restricted problem at the 2:1 resonance. A truncated form of this Hamiltonian has been given by Lemaitre and Henrard (1987) up to the eighth order in the eccentricity of the asteroid. However, as pointed out by the authors, the convergence of the perturbation series is poor and consequently the model, as it is, is not realistic. This can be seen if we compute the families of fixed points, and their stability, of the circular case, $e_j = 0$, using the 1st, 2nd, 3rd, ... 8th approximation, respectively, and comparing the results with each other and with the families of periodic orbits of the restricted circular three body problem. In particular, the even order approximations are far from the real model (we have stability instead of instability and vice versa). This is due to the fact that by the averaging method we force the system to behave as integrable, but for values of the eccentricity $e > 0.34$ the system behaves as strongly nonintegrable, as is clear from the Poincaré surfaces of section shown in Figure 2, where chaos appears. In fact, the perturbation series in the averaging method do not converge even for relatively small values of the eccentricity. Several methods have been developed to overcome this difficulty (Moons and Morbidelli (1993,1995), Morbidelli and Moons (1993), Ferraz-Mello (1987, 1988, 1989, 1994a,b), Ferraz-Mello and Sato (1989), Morbidelli and Giorgilli (1990a,b)), and averaged Hamiltonians were obtained that are valid for all eccentricities. However, these Hamiltonians are not expressed in a closed form, but are given numerically for a range of values of the initial conditions. In the present work we follow a different approach: we use the averaged Hamiltonian given by Lemaitre and Henrard (1987) and modify it by a small empirical term that introduces into the model the missing high eccentricity resonances and restores the correct position of the bifurcation point of the family of fixed points from the circular to the elliptic case and also gives the correct stability index for all the fixed points. This method has been applied successfully to the 3:1 resonance case (Hadjidemetriou (1993)).

The averaged Hamiltonian can be expressed in terms of the resonant action-angle variables at the 2:1 resonance , S, N σ, and ν given by

$$\begin{aligned} S &= \sqrt{(1-\mu)a}\,(1-\sqrt{1-e^2}), \quad \sigma = 2\lambda' - \lambda - \omega, \\ N &= \sqrt{(1-\mu)a}\,(2-\sqrt{1-e^2}), \quad \nu = -2\lambda' + \lambda + \omega'. \end{aligned} \tag{1}$$

We found that the averaged Hamiltonian that is closest to the actual model (i.e. its fixed points have the correct position and the correct stability) is obtained from the expansion given by Lemaitre and Henrard (1987) if we consider the first, second and seventh approximation of the circular case and add a small correction term. The corrected Hamiltonian has the form

$$\begin{aligned} H &= H_0(S,N) + \mu e H_1(\sigma) + \mu e^2 H_2(\sigma) + \mu e^7 H_7(\sigma) + \mu e^7 H_c(\sigma) \\ &\quad + \mu e_j H_j(S,N,\sigma,\nu), \end{aligned} \tag{2}$$

where

$$H_0 = -\frac{(1-\mu)^2}{2(N-S)^2} - B(N-S), \tag{3}$$

$$H_1 = A\cos\sigma, \quad H_2 = C + D'\cos 2\sigma, \qquad (D'\text{: corrected value}) \tag{4}$$

$$H_7 = P\cos\sigma + Q\cos 3\sigma + R\cos 5\sigma + T\cos 7\sigma, \tag{5}$$

$$H_c = T'\cos 8\sigma, \qquad \text{(correction term)} \tag{6}$$

and

$$H_j = J\cos\nu + \sqrt{\frac{2S}{N}}\,(F\cos(\sigma+\nu) + G\cos(\sigma-\nu)) + e_j H\cos 2\nu. \tag{7}$$

The eccentricity e that appears in the above expressions is a function of S and N, obtained from the equations (1). The coefficients are given by Lemaitre and

Henrard as: $A = 1.189$, $B = 2.00084$, $C = -0.3866$, $P = 1.598$, $Q = 6.964$, $R = -52.95$, $T = 51.95$, $J = -0.4273$, $F = 05739$, $G = 4.955$ and $H = -3.588$. The coefficient of the correction term is $T' = -20.0$. We have also corrected the coefficient D in the second order term H_2 of the expansion of the Hamiltonian, taking $D' = -0.30$ instead of $D = -1.691$. The value of the mass of Jupiter (in normalized units) is taken equal to $\mu = 0.00095387535$ and the value of the eccentricity of Jupiter is taken equal to $e_j = 0.048$.

The "corrected" Hamiltonian (2) has the same fixed points and the correct stability as the real model, for values of the eccentricity of the asteroid up to $e = 0.6$, the collision point with Jupiter, and up to values of the eccentricity of Jupiter equal to $e_j = 0.10$. However, the mapping model based on this Hamiltonian is acceptable for larger values of e, as we shall explain in the next section.

4. The Symplectic Mapping

The unperturbed motion of the asteroid is determined by the Hamiltonian H_0, given by (3), and the motion takes place on a 2-torus with constant radii S and N and constant frequencies $\partial H_0/\partial S$, $\partial H_0/\partial N$, respectively. On this torus we define a mapping by considering the points S_n, N_n at the times $t_n = nT$, $n = 1, 2, 3, \ldots$, where $T = 2\pi$ is the period of the motion of Jupiter. We define the mapping in this way (stroboscopic map at integral multiples of T), because we intend to study the elliptic restricted three body problem, which depends periodically on time, with period $T = 2\pi$.

The unperturbed mapping is given by the equations

$$\begin{aligned} \sigma_{n+1} &= \sigma_n + T\,\frac{\partial H_0}{\partial S_{n+1}} \qquad & S_{n+1} &= S_n, \\ \nu_{n+1} &= \nu_n + T\,\frac{\partial H_0}{\partial N_{n+1}} \qquad & N_{n+1} &= N_n, \end{aligned} \tag{8}$$

and is the analytic solution of the integrable part of the system.

The mapping (8) can be considered as derived from the generating function

$$W_0 = \sigma_n S_{n+1} + \nu_n N_{n+1} + T\,H_0(S_{n+1}, N_{n+1}), \tag{9}$$

through the equations

$$\begin{aligned} \sigma_{n+1} &= \partial W_0/\partial S_{n+1}, \qquad & S_n &= \partial W_0/\partial \sigma_n, \\ \nu_{n+1} &= \partial W_0/\partial N_{n+1}, \qquad & N_n &= \partial W_0/\partial \nu_n. \end{aligned} \tag{10}$$

We perturb now the generating function W_0 in such a way that the corresponding perturbed mapping has the same fixed points and the correct stability as the Poincaré map of the real system (here the elliptic restricted three body system Sun-Jupiter-asteroid). It can be proved (Hadjidemetriou 1991) that this can be achieved by taking the perturbed generating function

$$\begin{aligned} W &= \sigma_n S_{n+1} + \nu_n N_{n+1} + T\, H_0(S_{n+1}, N_{n+1}) && \text{(two-body problem)} \\ &+\mu T\, W_1(S_{n+1}, N_{n+1}, \sigma_n) && \text{(circular restricted problem)} \\ &+\mu e_j T\, W_2(S_{n+1}, N_{n+1}, \sigma_n, \nu_n), && \text{(elliptic restricted problem)} \end{aligned} \tag{11}$$

where

$$\begin{aligned} W_1(S, N, \sigma) &= eH_1(\sigma) + e^2 H_2(\sigma) + f(e^7 H_7(\sigma) + e^7 H_c(\sigma)), \\ W_2(S, N, \sigma, \nu) &= H_j(S, N, \sigma, \nu). \end{aligned} \tag{12}$$

This mapping is symplectic and has the same fixed points with the same stability index, as the averaged Hamiltonian H, and consequently it has the same fixed points as the Poincaré map of the real system (elliptic restricted problem). This is so because H is corrected in such a way that its fixed points correspond to the families of the periodic orbits of the real system. In addition, we have multiplied the high eccentricity terms H_7 and H_c in the generating function W by the coefficient $f = 0.02$, in order to reduce the chaotic regions of the mapping and make them comparable with the chaotic regions of the real system when $e_j = 0$.

The mapping obtained by the generating function (11) has the same topology as the Poincaré map of the real system. Thus, we consider it as a good model to study the long term behaviour of an asteroid close to the 2:1 resonance, because similar dynamical systems have the same generic properties and are expected to behave in the same way.

The equations of the complete mapping are obtained from the generating function W, given by (11), by making use of the transformation equations (10) (using W instead of W_0), and are given by

$$\begin{aligned} \sigma_{n+1} &= \sigma_n && +T\frac{\partial W_0}{\partial S_{n+1}} && +T\mu\frac{\partial W_1}{\partial S_{n+1}} && +T\mu e_j\frac{\partial W_2}{\partial S_{n+1}}, \\ S_n &= S_{n+1} && && +T\mu\frac{\partial W_1}{\partial \sigma_n} && +T\mu e_j\frac{\partial W_2}{\partial \sigma_n}, \\ \nu_{n+1} &= \nu_n && +T\frac{\partial W_0}{\partial N_{n+1}} && +T\mu\frac{\partial W_1}{\partial N_{n+1}} && +T\mu e_j\frac{\partial W_2}{\partial N_{n+1}}, \\ N_n &= N_{n+1} && && && +T\mu e_j\frac{\partial W_2}{\partial \nu_n}. \end{aligned} \tag{13}$$

These equations are in implicit form and at each step we have to solve (by the method of Newton-Raphson) the second and fourth equations (13) to obtain the values of S_{n+1} and N_{n+1} in terms of σ_n, ν_n, S_n, N_n and then proceed to the computation of σ_{n+1}, ν_{n+1} from the first and third equations (13). This however does not present any numerical difficulty.

The complete mapping (13) is in the four dimensional phase space σ, ν, S, N and can be considered as a model for the Poincaré map of the real system (the elliptic restricted three body problem). This latter system is a T-periodic time dependent dynamical system of two degrees of freedom and its Poincaré map is the stroboscopic map in the four dimensional phase space $\sigma,\ \nu,\ S,\ N$, at integral values of the period T of Jupiter. Thus, we see that the Poincaré map of the real system and the mapping (13) correspond to a system with the same number of degrees of freedom. On the contrary, by the averaging method we loose a degree of freedom, because a time dependent system is transformed to a time independent one. We can see this if we consider the Poincaré map of the averaged system: it is a map in a two dimensional phase space instead of the four dimensional phase space of the real system (because the averaged system is an *autonomous* system with two degrees of freedom). Thus, in comparing the real system with the averaged model and with the symplectic mapping model, we can say that in the averaged model the chaotic regions may be underestimated, whereas in the mapping model they may be overestimated. This in fact happened with our model, where more chaos appeared (in the circular case) than in the real model and we introduced the coefficient $f = 0.02$ to make the chaotic regions closer to reality.

From the mapping (13) we note that for $e_j = 0$ we have N =constant and consequently the first two equations are independent of the other two, with N as a parameter. The first two equations correspond to the circular restricted three body problem. To check the validity of our model, we computed several phase portraits of the mapping (13) for $e_j = 0$, that are shown in Figure 4. The axes are the variables $X = \sqrt{2S}\cos\sigma,\ Y = \sqrt{2S}\sin\sigma$. Comparing them with the Poincaré surfaces of section that are given in Figure 2, we note that the topology of the mapping model is close to that of the real system. (Note that $\sqrt{2S}$ is proportional to e).

Note that for $e_j = 0$ the two degrees of freedom $\sigma,\ S$ and $\nu,\ N$ are uncoupled. The coupling takes place through the term of order μe_j in the fourth equation (13), which is responsible for the change of the action N. This change however is very small, and we can assume that the asteroid stays for a long time in a plane $X\,Y$, corresponding to a fixed value of N (see Figure 4), and in addition there is a slow drift from one plane N=constant to the next, as the value of N varies slowly. This generates chaotic motion if we start close to a separatrix in the $X\,Y$ plane.

5. The Evolution of an Asteroid Close to the 2:1 Resonance

We shall use the mapping model (13) that we have developed, to study the motion of an asteroid in the region of phase space close to the 2:1 resonance. For the unperturbed motion, $\mu = 0$, the resonance is determined by the value of the semimajor axis a only, which in this case is equal to $a = 0.62976$ (in the normalized units we are using). From the definition of the action-angle variables $S,\ N,\ \sigma,\ \nu$, given by (1), we see that it is only the actions S and N that determine the value of a, and the resonance zone in the four dimensional phase space $S,\ N,\ \sigma,\ \nu$ is given by the relation $N = \sqrt{(1-\mu)a} + S$. Since a is fixed on the resonance, the value of S depends on the value of the eccentricity e only and is equal to zero for

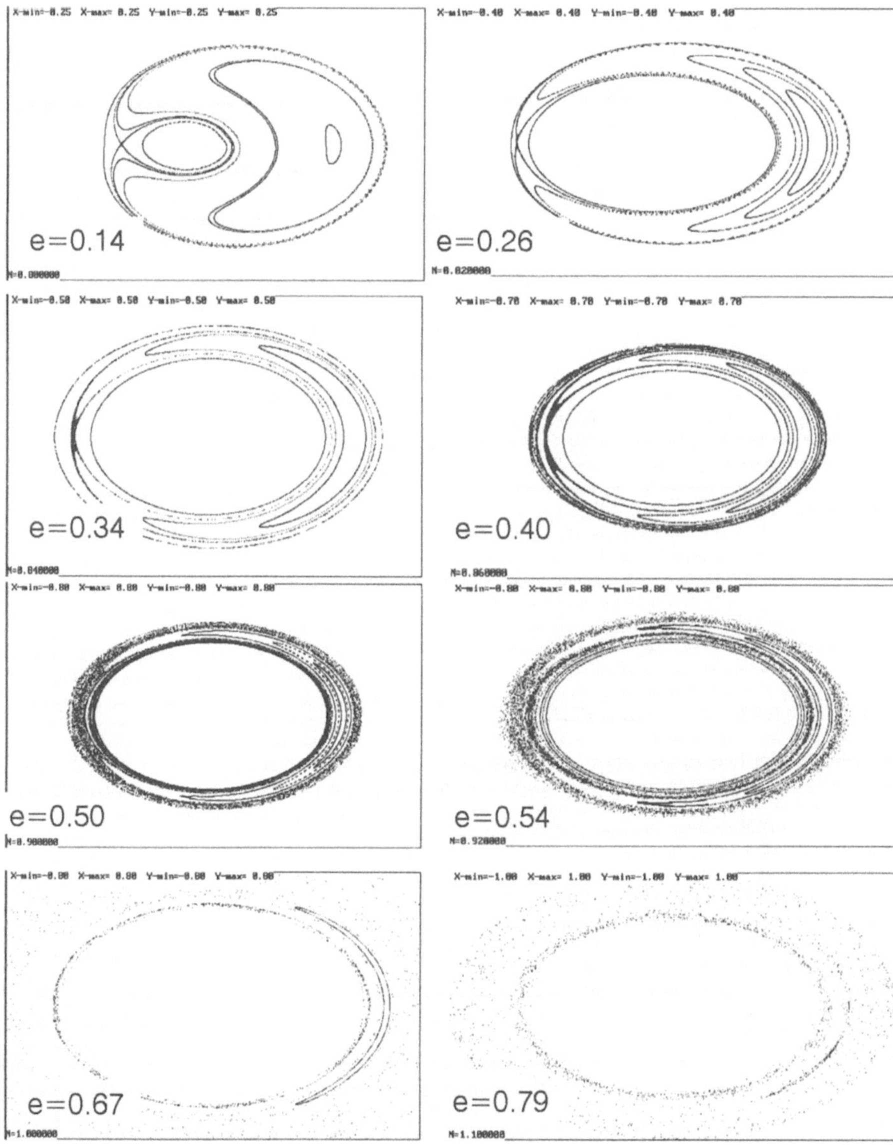

Fig. 4. The mapping (13) for $e_j = 0$, for several values of the parameter N. For comparison with Figure 2, we also give the value of the eccentricity of the resonant fixed point for each case. The axes are $\sqrt{S}\cos\sigma$ and $\sqrt{S}\sin\sigma$.

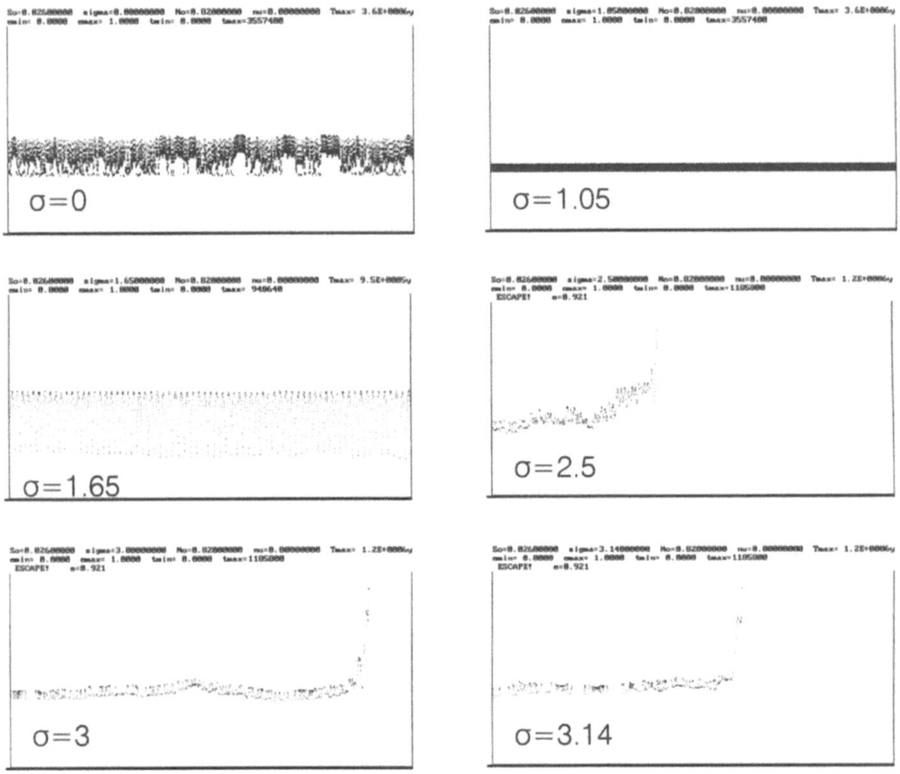

Fig. 5. The evolution of an asteroid, starting in the 2:1 resonance, with initial conditions $N = 0.82$, $S = 0.026$, $\nu = 0$, corresponding to $e = 0.25$, for different values of σ: (a) $\sigma = 0$, (b) $\sigma = 1.05$ rad, (c) $\sigma = 1.65$ rad, (d) $\sigma = 3$ rad.

$e = 0$. This means that the lowest value of N on the exact resonance is equal to $N = 0.7932$ and this value increases as the eccentricity of the resonance increases. Note that for $e = 0.59$ we have collision with Jupiter. The corresponding value of N is $N = 0.95$. The mapping model we have developed is valid up to this point. Beyond the collision with Jupiter the former unstable branch of periodic orbits/fixed points is stable in the real model (here the circular restricted three body problem), but unstable in the mapping model. This however does not have any practical effect, because the region of phase space of the real model well before the collision point with Jupiter is chaotic and the stable islands close to these stable points (after the collision) are very small, with no practical significance. This is clear in Figure 2.

As we mentioned above, the resonance is determined by the values of S and N only, but the initial position of an asteroid depends also on the angles σ and ν. We give in Figure 5 the evolution of an asteroid, starting very close to the resonance zone, taking $N = 0.82$ and $S = 0.026$. These values correspond to an initial eccentricity of the asteroid equal to $e = 0.25$. The evolution depends critically on the initial value of σ, as can be seen from Figure 5 (the initial value of ν does not play an important role). We note that we have bounded (ordered?) motion for initial values of σ in the range from $\sigma = 0$ to $\sigma = 1.7$ rad, with an amplitude that decreases from $\sigma = 0$ to $\sigma = 1$ rad, and then increases up to $\sigma = 1.7$ rad. From $\sigma = 1.7$ rad to $\sigma = 4.7$ rad the motion is chaotic: the eccentricity may remain small for several million years and then jumps to very large values. A typical evolution of this type is shown in Figure 5d, where we have a diffusion in the eccentricity after about one million years. In Figure 5 we present the results for values of σ up to $\sigma = \pi$, because the evolution for values in the range $\pi < \sigma < 2\pi$ are symmetric to those in the range $0 < \sigma < \pi$.

Similar results are obtained for different initial values of N. We have repeated the computations for the initial conditions: $N = 0.84$, $S = 0.047$, corresponding to $e = 0.34$, $N = 0.86$, $S = 0.067$, corresponding to $e = 0.40$ and $N = 0.88$, $S = 0.087$, corresponding to $e = 0.46$. As the initial eccentricity of the asteroid increases, the chaotic regions increase with a corresponding decrease of the ordered regions. The ordered regions are in the range $\sigma = 0.55$ to $\sigma = 1.55$ and $\sigma = 4.75$ to $\sigma = 5.7$ for $e = 0.34$, $\sigma = 0.70$ to $\sigma = 1.455$ and $\sigma = 4.80$ to $\sigma = 5.60$ for $e = 0.40$, and $\sigma = 0.855$ to $\sigma = 1.40$ and $\sigma = 4.90$ to $\sigma = 5.40$ for $e = 0.46$. (Note that the regions are symmetric with respect to $\sigma = \pi$).

From the above results we see that a large region of phase space at the 2:1 resonance zone is chaotic and the asteroid eventually diffuses to high values of the eccentricity and escapes from this region. The evolution however, inside the resonance zone, depends on the initial value of σ and the orbit stays in the resonance zone for a range of values of σ. This means that the 2:1 resonance zone is not completely depleted of asteroids, in the model of the elliptic restricted three body problem. Some additional mechanisms are needed to explain the gap in the 2:1 resonance, possibly the gravitational effect of Saturn. This has been studied by Morbidelli and Moons (1993), who considered the effects of secular resonances with Jupiter and Saturn inside the 2/1 mean motion resonance.

The above results are qualitatively similar to results obtained by Ferraz-Mello (1996a,b), who used a different symplectic mapping model.

Acknowledgements

This work was supported by the EU contracts
ERBCHRX-CT93-0330 and ERBCHRX-CT94-0445.

References

Ferraz-Mello, S. (1987): *Averaging the elliptic Asteroidal Problem near a First Order Resonance*, Astron. J. **94**, 208-212.

Ferraz-Mello, S. (1988): *The High-Eccentricity Libration of the Hildas*, Astron.J.**96**, 400-408.

Ferraz-Mello, S. (1989): *A Semi-Numerical Expansion of the Averaged Disturbing Function for some Very-High-Eccentricity Orbits*, Celes. Mech. **45**, 65-68.

Ferraz-Mello, S. (1994a): *Dynamics of the 2:1 Asteroidal resonance*, Astron. J. **108**, 2330-2337.

Ferraz-Mello, S. (1994b): *The convergence domain of the Laplacian expansion of the disturbing Function*, Cel. Mech. Dyn. Astron. **58**, 37-52.

Ferraz-Mello, S. (1996a): *On the Hecuba Gap*, preprint.

Ferraz-Mello, S. (1996b): *A Symplectic Mapping approach to the study of the Stochasticity of asteroidal Resonances*, Celest. Mech. (to appear).

Ferraz-Mello, S. and Sato, M. (1989): *The Very-High-Eccentricity Assymmetric Expansion of the Disturbing Function near Resonances of any Order*, Astron. Astrophys. **225**, 541-547.

Hadjidemetriou, J.D. (1991): *Mapping Models for Hamiltonian Systems with application to Resonant Asteroid Motion*, in *Predictability, Stability and Chaos in N-Body Dynamical Systems*, A.E. Roy (ed.), 157-175, Kluwer Publ.

Hadjidemetriou, J.D. (1993): *Asteroid Motion near the 3:1 Resonance*, Celest. Mech. **56**, 563-599.

Henrard, J. (1988): *Resonances in the Planar Elliptic Restricted Problem*, in *Long Term Dynamical Behaviour of Natural and Artificial N-Body Systems*, A.E. Roy (ed.), 405- 425, Kluwer Publ.

Henrard, J. and Lemaitre, A. (1987): *A Perturbative Treatment of the 2/1 Jovian Resonance*, Icarus **69**, 266- 279.

Henrard, J., Watanabe, N. and Moons, M. (1995): *A bridge between Secondary and Secular Resonances inside the Hecuba Gap*, Icarus **115**, 336-346.

Klafke, J.C., Ferraz-Mello, S., Michtchenko, T., (1992): *Very-high-eccentricity Librations at some higher order resonances*, I.A.U. Symposium 152 (to appear).

Lemaitre, A. and Henrard, J (1990): *On the Origin of Chaotic Behaviour in the 2/1 Kirkwood Gap*, Icarus **83**, 391-409.

Moons, M, and Morbidelli, A. (1993): *The Main Mean Motion Commensurabilities in the Planar Circular and Elliptic Problem*, Celest. Mech. Dyn. Astron. **57**, 99-108.

Moons, M. and Morbidelli, A. (1995): *Secular Resonances in Mean Motion Commensurabilities: the 4/1, 3/1, 5/2 and 7/3 cases*, Icarus **114**, 33-50.

Morbidelli, A. (1996): *On the Kirkwood Gap at the 2/1 Commensurability with Jupiter: numerical results*, preprint.

Morbidelli, A. and Giorgilli, A. (1990a): *On the Dynamics in the Asteroid Belt. Part I: General theory*, Celest. Mech. Dyn. Astron. **47**, 145-172.

Morbidelli, A. and Giorgilli, A. (1990b): *On the Dynamics in the Asteroid Belt. Part II: Detailed study of the main Resonances*, Celest. Mech. Dyn. Astron. **47**, 173-204.

Morbidelli, A. and Moons, M. (1993): *Secular Resonances in Mean Motion Commensurabilities: the 2/1 and 3/2 cases*, Icarus **102**, 316-332.

Šidlichovský M.: (1988), On the Origin of 5/2 Kirkwood Gap in *Proceedings of the IAU Colloq. 96, The Few Body Problem in Turku*, (ed. Valtonen M.), p. 117.

Šidlichovský M. (1992): *Mapping for the asteroidal resonances*, Astron. Astrophys. **259**, 341-348.

Wisdom, J. (1982): *The origin of the Kirkwood Gaps*, Astron. J. **87**, 577-593.

Wisdom, J.(1983): *Chaotic Behaviour and the Origin of the 3/1 Kirkwood Gap*, Icarus **56**, 51-74.

Wisdom, J. (1985): *A Perturbative Treatment of Motion Near the 3/1 Commensurability*, Icarus **63**, 272-289.

Wisdom, J. (1987): *Chaotic Dynamics of the Solar System*, Icarus **72**, 241-275.

SYMPLECTIC INTEGRATORS FOR HILL'S LUNAR PROBLEM

J. WALDVOGEL

Seminar für Angewandte Mathematik, Eidgenössische Technische Hochschule CH-8092 Zürich, Switzerland

Abstract. Hill's lunar problem is one of the simplest, yet realistic, gravitational three-body problems. It models, e.g., the motion of the Earth's moon or the motion of Saturn's coorbital satellites. It still has a great importance since it is being used as a reference problem in certain more accurate perturbation models of the moon's motion. In regularized coordinates Hill's lunar problem may be described by a polynomial Hamiltonian of degree 6.

Here we will use an additive splitting of this Hamiltonian in order to implement symplectic composition integrators for Hill's lunar problem. Experiments indicate that these integrators are very accurate for orbits consisting of many revolutions. E.g., in integrations of quasiperiodic or homoclinic orbits the Hamiltonian remains nearly constant until the orbit escapes. However, a comparison with high-order Taylor series methods shows that these methods are considerably faster and therefore quite competitive, although the Hamiltonian seems to grow linearly with time.

In an appendix we discuss new algorithms for evaluating the Stumpff functions since these functions are needed for constructing our symplectic integrators.

Key words: Hill's lunar problem – symplectic integrator – Stumpff functions

1. Introduction

Symplectic integrators are often seen as a major advance in the numerical treatment of Hamiltonian systems. These integrators have the property that the map corresponding to one integration step is a symplectic map, i.e. it shares important properties (e.g. conservation of areas) with the flow defined by the differential equation. A difficulty of symplectic integrators is the fact that symplecticity is lost when the step size is changed. Furthermore, many of the well known symplectic integrators are implicit and therefore rather slow.

For comprehensive articles on symplectic integrators the reader is referred to, e.g., review papers by Sanz-Serna [8] or Yoshida [17]. A comprehensive and modern account of the entire field of numerical treatment of ordinary differential equations is contained in the volumes by Hairer, Nørsett, and Wanner [1,2].

The purpose of this paper is to use a particular class of *explicit* symplectic integrators, namely the composition methods [5,17], and a particular problem of celestial mechanics, namely Hill's lunar problem [3], for demonstrating advantages and disadvantages of symplectic integration. At the same time, the construction of symplectic integrators in this example is explained.

In this study we use explicit methods in order to minimize the overhead of the method. A class of Hamiltonians permitting composition methods as examples of explicit methods are the *separable* Hamiltonians $H(q,p) = V(q) + T(p)$, where $q \in \mathbb{R}^n$ are the coordinates and p are the conjugated momenta. The concept of the composition methods, dating back as far as 1959 [13,7], is a generalization of separability and is applicable to differential equations of the form

$$x_1' = A(x) + B(x), \quad x \in \mathbb{R}^N, \quad ' \equiv \frac{d}{ds} \tag{1}$$

R. Dvorak and J. Henrard (eds.), The Dynamical Behaviour of our Planetary System, 291.

as long as the individual flows corresponding to

$$x_1' = A(x_1), \quad x_2' = B(x_2) \tag{2}$$

are known explicitly.

Hill's lunar problem, as one of the simplest gravitational problems, is chosen for several reasons:

i) It is a problem of practical relevance in lunar theories, coorbital motion etc.
ii) The complexity of the family of orbits leaves no doubt that the problem is non-integrable. The rigorous proof of this conjecture, however, is still missing.
iii) Hill's lunar problem has orbits with arbitrarily many revolutions about the origin [10]. The use of regularized coordinates permits a constant step size in the numerical integration of these orbits. Furthermore, the regularized Hamiltonian permits an additive splitting with explicit partial flows, thus allowing for the direct application of composition methods.

In the next section the regularized differential equations of Hill's problem as well as the partial flows will be summarized, and details about the individual symplectic maps will be given. Then, we describe the orbits considered, together with the performance of the symplectic integrators. Finally, the results are compared with a commercially available explicit integrator (ODE45 of MATLAB) and with high-order Taylor series integrators. These latter integrators turn out to be highly competitive for high-precision computations.

2. Hill's Lunar Problem Regularized

In this section we summarize part of the article [16] with the goal of collecting the relations defining the individual flows corresponding to Equ. (2). Hill's lunar equations describe the motion of the moon under the influence of the earth and an infinitely remote sun on a circular orbit. In a uniformly rotating geocentric frame of reference, where x is the tangential coordinate, the equations of motion are [3,10]

$$\begin{aligned} &\ddot{x} - 2\dot{y} - 3x + xr^{-3} = 0\,, \quad r = \sqrt{x^2+y^2} \\ &\ddot{y} + 2\dot{x} \qquad\; + yr^{-3} = 0\,, \end{aligned} \tag{3}$$

where dots denote derivatives with respect to time t. The equations of motion (3) imply the existence of the Jacobi integral

$$\frac{1}{2}\left(\dot{x}^2 + \dot{y}^2\right) - \frac{3}{2}\,x^2 - \frac{1}{r} = h\,, \tag{4}$$

where h is the Jacobi constant.

Following Levi-Civita's regularization procedure [4] for removing the collision singularity at $x = y = 0$, we construct a Hamiltonian equivalent to Equs (3). By using the coordinates and conjugated momenta

$$q_1 = x, \quad q_2 = y, \quad p_1 = \dot{q}_1 - q_2, \quad p_2 = \dot{q}_2 + q_1$$

Hill's lunar problem is represented by the Hamiltonian

$$H(q,p) = \tfrac{1}{2}\,(p_1^2 + p_2^2) + p_1 q_2 - p_2 q_1 - q_1^2 + \tfrac{1}{2}\,q_2^2 - \tfrac{1}{r}\,, \qquad r = \sqrt{q_1^2 + q_2^2}\,. \tag{5}$$

Levi-Civita's regularization consists of introducing a new independent variable s (instead of time t) and a new Hamiltonian K according to

$$dt = r\,ds, \qquad K = r(H - h)\,, \tag{6}$$

where the constant h is the fixed energy $H(q,p) = h$ of the orbit under consideration, and h agrees with the definition (4) of the Jacobi constant. Furthermore, new coordinates u_1, u_2 and new momenta v_1, v_2 are introduced according to the canonical transformation

$$q_1 + iq_2 = (u_1 + iu_2)^2, \quad p_1 + ip_2 = \frac{v_1 + iv_2}{2(u_1 - iu_2)}\,, \tag{7}$$

where complex notation has been used. In terms of the new coordinates and momenta the new Hamiltonian becomes

$$K(u,v) = K_1(u,v) + K_2(u) \tag{8}$$

where

$$K_1(u,v) = \tfrac{1}{8}\,(v_1^2 + v_2^2) - (u_1^2 + u_2^2)^2 \left[\tfrac{1}{2}\,(u_1 v_2 - u_2 v_1) + h\right] - 1 \tag{9}$$

$$K_2(u) = (u_1^2 + u_2^2)(-u_1^4 + 4u_1^2 u_2^2 - u_2^4)\,, \tag{10}$$

and the equations of motion (in vector notation) are

$$\frac{du}{ds} = \frac{\partial K}{\partial v}\,, \qquad \frac{dv}{ds} = -\frac{\partial K}{\partial u}\,, \qquad \frac{dt}{ds} = u_1^2 + u_2^2\,. \tag{11}$$

The purpose of the splitting (8) is to write the system (11) in the manner of Equ. (1). It turns out that both the Hamiltonian systems derived from K_1 and K_2 are integrable; therefore the splitting (8) leads to a family of explicit symplectic integrators.

The Hamiltonian system derived from $K_2(u)$ is trivially integrable; with the initial conditions

$$u(0) = u_0, \quad v(0) = v_0, \quad t(0) = t_0\,, \quad u_0, v_0 \in \mathbb{C} \tag{12}$$

we obtain

$$\begin{aligned} u(s) &= u_0 \\ v(s) &= v_0 - s\,\frac{\partial K_2}{\partial u}\,(u_0), \qquad \frac{\partial K_2}{\partial u} = \begin{pmatrix} 6u_1(-u_1^4 + 2u_1^2 u_2^2 + u_2^4) \\ 6u_2(u_1^4 + 2u_1^2 u_2^2 - u_2^4) \end{pmatrix} \\ t(s) &= t_0\,. \end{aligned} \tag{13}$$

The Hamiltonian system due to $K_1(u, v)$ is related to the regularized Kepler motion in rotating coordinates. To demonstrate its integrability we write the system in complex notation as

$$\begin{aligned} \frac{du}{ds} &= \frac{v}{4} - \frac{u\overline{u}}{2}\, iu \\ \frac{dv}{ds} &= u(2h + Im\,\overline{u}v) - \frac{u\overline{u}}{2}\, iv \\ \frac{dt}{ds} &= u\overline{u} \end{aligned} \tag{14}$$

and introduce new coordinates $U = e^{it/2}u$, $V = e^{it/2}v$ satisfying the initial conditions

$$U(0) = e^{it_0/2}u_0, \quad V(0) = e^{it_0/2}v_0\,. \tag{15}$$

The system (14) is transformed into

$$\frac{dU}{ds} = \frac{V}{4}, \quad \frac{dV}{ds} = (2h + U_1V_2 - U_2V_1)\,U, \quad \frac{dt}{ds} = U\overline{U}\,, \tag{16}$$

a system which has the angular momentum $Im(\overline{U}V) = U_1V_2 - U_2V_1$ as a first integral. Therefore the first two equations of (16) define a harmonic oscillator with frequency $\frac{\omega}{2}$, where

$$\omega := \sqrt{-2h - Im(\overline{u}_0 v_0)}\,, \tag{17}$$

and in view of the initial conditions (15) their solution is

$$\begin{aligned} U(s) &= e^{i\frac{t_0}{2}}\left[u_0 \cos\left(\frac{\omega}{2}\, s\right) + \frac{u_0}{2\omega} \sin\left(\frac{\omega}{2}\, s\right)\right] \\ V(s) &= e^{i\frac{t_0}{2}}\left[v_0 \cos\left(\frac{\omega}{2}\, s\right) - 2\omega\, u_0 \sin\left(\frac{\omega}{2}\, s\right)\right]. \end{aligned}$$

The final step is to determine $t(s)$ by integrating the third relation of (16).

We now collect the equations given in [16] defining the flow due to K_1. Since the frequency ω in (17) may be 0 or imaginary we write the map in terms of K. Stumpff's "c functions" [12,11]

$$\begin{aligned} c_0(z) &= \cos(\sqrt{z}) \qquad & c_1(z) &= \frac{\sin(\sqrt{z})}{\sqrt{z}} \\ c_2(z) &= \frac{1-\cos(\sqrt{z})}{z}\,, & c_3(z) &= \frac{\sqrt{z} - \sin(\sqrt{z})}{z^{3/2}}\,. \end{aligned} \tag{18}$$

Methods for accurately evaluating the Stumpff functions will be discussed in the appendix. Let $x_0 = (u_0, v_0, t_0)$ contain the initial point in phase space (complex notation for u, v) and the initial time. Then the time $t(s)$ and the position $(u(s), v(s))$ are given by

$$\begin{aligned} t(s) &= t_0 + u_0\overline{u}_0 \frac{s}{2} \left(1 + c_1(\omega^2 s^2)\right) + Re(\overline{u}_0 v_0) \frac{s^2}{2} c_2(\omega^2 s^2) \\ &\quad + v_0\overline{v}_0 \frac{s^3}{8} c_3(\omega^2 s^2) \\ u(s) &= e^{-\frac{i}{2}(t(s)-t_0)} \left[u_0\, c_0\left(\frac{\omega^2 s^2}{4}\right) + v_0 \frac{s}{4} c_1\left(\frac{\omega^2 s^2}{4}\right)\right] \\ v(s) &= e^{-\frac{i}{2}(t(s)-t_0)} \left[- u_0\, \omega^2 s\, c_1\left(\frac{\omega^2 s^2}{4}\right) + v_0\, c_0\left(\frac{\omega^2 s^2}{4}\right)\right] . \end{aligned} \tag{19}$$

3. Composition Methods

Consider the differential equation (1), and denote the individual flows induced by the vector fields A, B by $\exp_A(s)$, $\exp_B(s)$, respectively, i.e. the solution $x(s)$ of the initial value problem $x' = A(x)$, $x(0) = x_0$ is written as

$$x(s) = \exp_A(s)(x_0) .$$

Composition of flows is written as juxtaposition, e.g.

$$x(s_1 + s_2) = \exp_A(s_2) \;\; \exp_A(s_1)\,(x_0) .$$

A composition algorithm approximates the flow induced by the vector field $A + B$ by a repeated composition of the flows due to A and B. A popular and simple algorithm is the *leapfrog* method which uses 2 compositions and is of error order $p = 2$:

$$\exp_{A+B}(s) = \exp_A\left(\tfrac{s}{2}\right) \;\; \exp_B(s) \;\; \exp_A\left(\tfrac{s}{2}\right) + O(s^3) . \tag{20}$$

The general *symmetric* composition method of order p with $2m$ compositions is of the form

$$\begin{aligned} \exp_{A+B}(s) &= \exp_A(a_1 s) \;\; \exp_B(b_1 s) \;\; \exp_A(a_2 s) \ldots \exp_B(b_{1+[m/2]} s) \ldots \\ &\quad \ldots \exp_A(a_2 s) \;\; \exp_B(b_1 s) \;\; \exp_A(a_1 s) + O(s^{p+1}) \end{aligned} \tag{21}$$

if m is odd. If m is even the factor in the middle is $\exp_A(a_{1+[m/2]} s)$. Symmetric methods are preferred since they are automatically reversible (i.e. backwards integration of an orbit always ends at the starting point).

In the case of Hamiltonian systems the individual maps used in the composition (21) are all symplectic; therefore every composition method is a symplectic integrator.

The theory of composition methods [5] is based on Lie algebras and uses the Baker-Campbell-Hausdorff (BCH) formula describing the composition of flows,

$$\exp_A \ \exp_B = \exp_C, \tag{22}$$

where

$$C = A + B + \frac{1}{2}\,[A,B] + \frac{[A,[A,B]] + [B,[B,A]]}{12} + \ldots$$

and

$$[A,B] := A \cdot \nabla B - B \cdot \nabla A$$

is the commutator bracket of the vector fields A, B.

In order to find sets of coefficients $a_1, b_1, a_2, \ldots$ for a given error order p the BCH expansion of the composition must be equated with the Lie series expansion of the exact solution up to order p. This results in a system of algebraic equations for $a_1, b_1, a_2, \ldots$ in analogy to the determining equations for the classical Runge-Kutta coefficients. Therefore the resulting methods are referred to as (compositional) Runge-Kutta methods.

In the following we restrict ourselves to vector field splittings satisfying

$$[B,[B,[B,A]]] = 0 \tag{23}$$

since, in fact, this condition is met by the splitting (8). This relation considerably simplifies the conditions on the coefficients $a_1, b_1, a_2, \ldots$. Methods of this class are referred to as Runge-Kutta-Nyström (RKN) methods since the separable Hamiltonians $H(\boldsymbol{q},\boldsymbol{p}) = V(\boldsymbol{q}) + T(\boldsymbol{p})$ are in this class, and if $T(\boldsymbol{p}) = \frac{1}{2}\sum p_j^2$ the system is in the class considered by Nyström.

In the experiments of the next section we use a convenient Nyström method of order $p = 4$ with $2m = 8$ compositions, referred to as RKN-4 and defined by the coefficients [5]

$$a_1 = \frac{1}{2} - \sqrt{\frac{7}{72}}\,, \quad b_1 = 1, \quad a_2 = \sqrt{\frac{7}{72}} - \frac{1}{3}\,, \quad b_2 = -\frac{1}{2}\,, \quad a_3 = \frac{2}{3} \tag{24}$$

and a more accurate method of order $p = 6$ with $2m = 14$ compositions, referred to as RKN-6 and defined by

$$\begin{aligned}
a_1 &= 1.01308797891717472981, & b_1 &= 0.00016600692650009894,\\
a_2 &= 1.18742957373254270702, & b_2 &= -0.37962421426377360608,\\
a_3 &= -0.01833585209646059034, & b_3 &= 0.68913741185181063674,\\
a_4 &= 0.34399425728109261313, & b_4 &= 0.38064159097092574080.
\end{aligned} \tag{25}$$

MacLachlan [5] lists methods of orders up to $p = 8$.

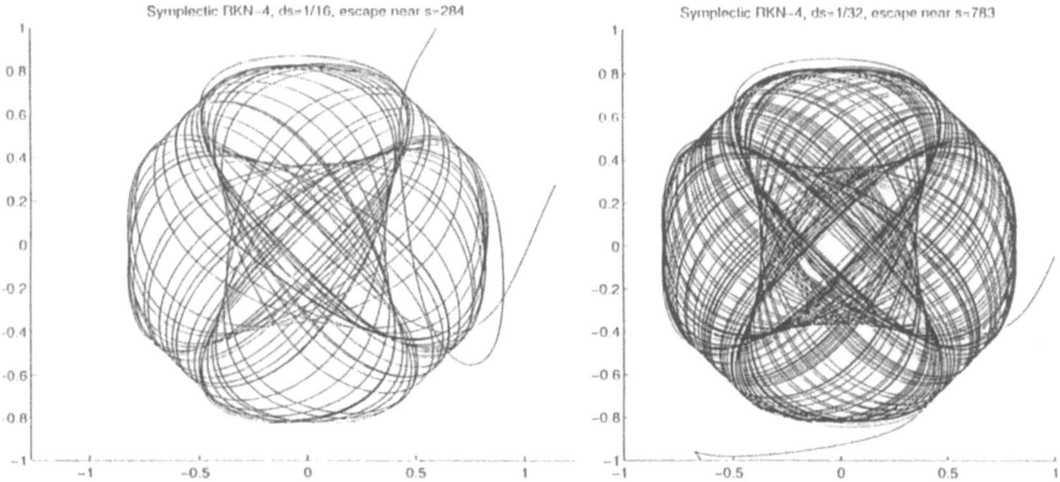

Fig. 1. Integrator RKN-4: the orbit escapes near $s = 284$ (left) when integrated with a step $\Delta s = 1/16$, and near $s = 783$ (right) when integrated with a step $\Delta s = 1/32$. The correct orbit escapes near $s = 424$, see Fig. 2, 4.

4. Results and Conclusions

In this section we report about a few experiments carried out with homoclinic orbits in Hill's lunar problem [10]. The family of orbits considered may be uniquely characterized by its Jacobi constant h and the property that

$$\lim_{t \to -\infty} x(t) =: c \,, \tag{26}$$

exists. These orbits are referred to as nonoscillating orbits [10], and we have $h = -0.375\, c^2$. For certain values of h the corresponding nonoscillating orbit chaotically revolves about the origin hundreds of times before escaping to infinity.

The comparisons of this section will be carried out using the nonoscillating orbit with the energy $h = -1.03895\,34169\,0923$. For convenience, the asymptotics of nonoscillating orbits [15] and a preliminary high-precision integration were used in order to generate the nearby point

$$\begin{aligned} u_0 &= 1.14311\,78537\,8775 + 0.27028\,78925\,4599 \cdot i \\ v_0 &= -2.73213\,07672\,5326 - 1.06280\,27746\,4126 \cdot i \end{aligned} \tag{27}$$

on the orbit to be considered.

For the symplectic integrations regularized variables were used. The vector $\boldsymbol{x}$ of dependent variables is defined as $\boldsymbol{x} = (Re\, u, Im\, u, Re\, v, Im\, v, t)$, and the maps $\exp_A(s)(\boldsymbol{x}_0)$ and $\exp_B(s)(\boldsymbol{x}_0)$ are given by Equs (19) and Equs (13), respectively.

Typical of many differential equations and of many numerical integration procedures is the sensitivity to changes in the initial data and to a badly chosen step size. This is very much so for the differential equation and integrators considered here.

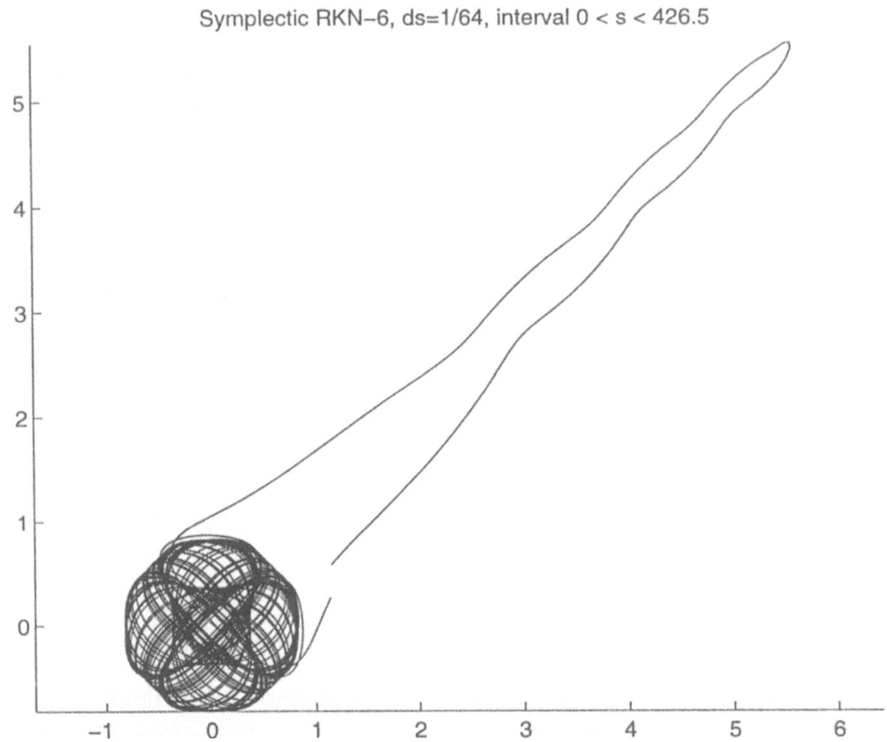

Fig. 2. Integrator RKN-6, escape near $s = 424$. The orbit is believed to be correct to drawing accuracy, except for the "pull back" of the escaping branch. This is an artefact of the regularized coordinates which would require the reduction of the step Δs inversely proportional to the distance from the origin.

In general, the deviation between neighbouring orbits grows exponentially in time. However, a symplectic integrator is expected to reproduce the correct qualitative behaviour of an orbit in spite of too large a step.

We consider Hill's lunar problem in the splitting (8) with the initial data (27). The orbits produced by the integrator RKN-4 (Equ. (24)) with steps $\Delta s = 1/16$ and $\Delta s = 1/32$ are shown in Fig. 1.

None of the orbits in Fig. 1 is correct after a long time; nevertheless they show the same qualitative behaviour. This is the most obvious advantage of symplectic integration. Unfortunately, it is rather difficult to predict a good step size. With $\Delta s = 1/256$ the true orbit up to the true escape time near $s = 424$ would be obtained at least in drawing accuracy. We point out, however, that such a *quantitatively* correct symplectic long-term integration is rather expensive if a low-order method is used.

Better accuracy and faster integration is obtained with a higher-order integrator, e.g. with RKN-6 (Equ. (25)). Here we claim that $\Delta s = 1/64$ suffices to integrate

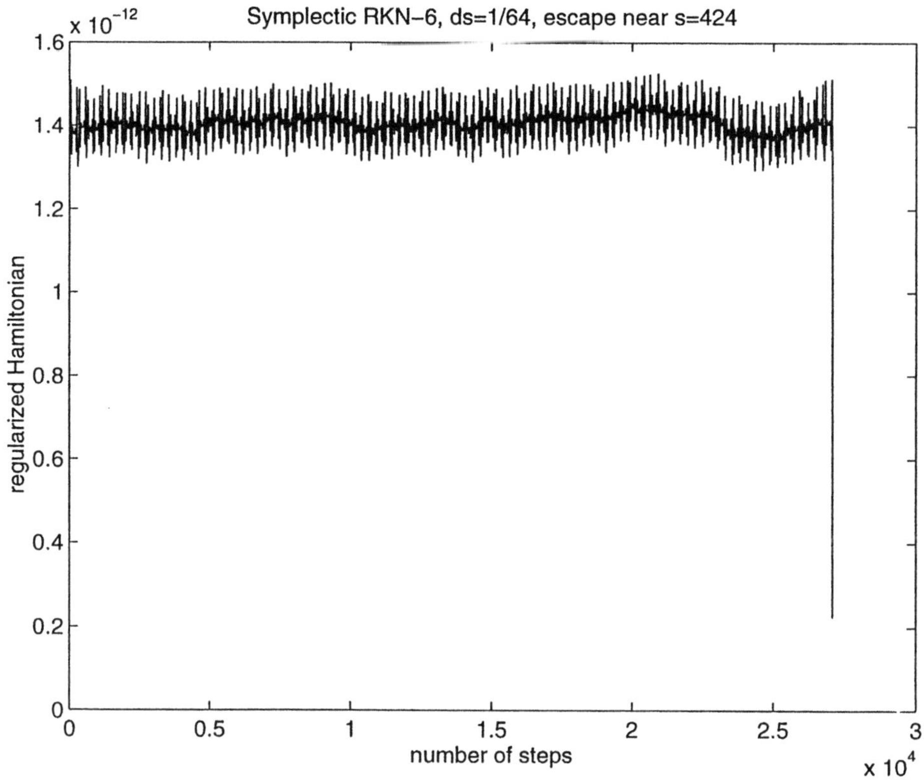

Fig. 3. The regularized Hamiltonian K for the integration of Fig. 2 (RKN-6). During the escape the situation changes dramatically due to the constant (instead of decreasing) step size.

the orbit reliably until it escapes near $s = 424$, see Fig. 2.

However, it is not possible to verify this claim on the basis of the information given so far. Also, the attempt of "computing" a long orbit may not be meaningful due to its inherent sensitivity to small changes of the initial data or small errors. In our case the correctness of an orbit may be established with high confidence by recomputing it with a smaller step or with a different algorithm, see Fig. 4.

The value of the Hamiltonian (a constant in theory) on the computed orbit provides a convenient rough estimate of the accuracy that is being achieved. Typically the regularized Hamiltonian K ($= 0$ on the orbit) oscillates chaotically but very stably at a very small scale and remains bounded away from 0, see Fig. 3. Only at the onset of the escaping phase of the motion the symplectic integration breaks down, and the behaviour of K changes dramatically. From the asymptotics of hyperbolic escape there follows that the ideal step Δs in regularized time should be inversely proportional to the distance from the origin.

To widen the spectrum of integrators the same orbit was recomputed by means of the commercially available integrator ODE45 of MATLAB (an explicit non-

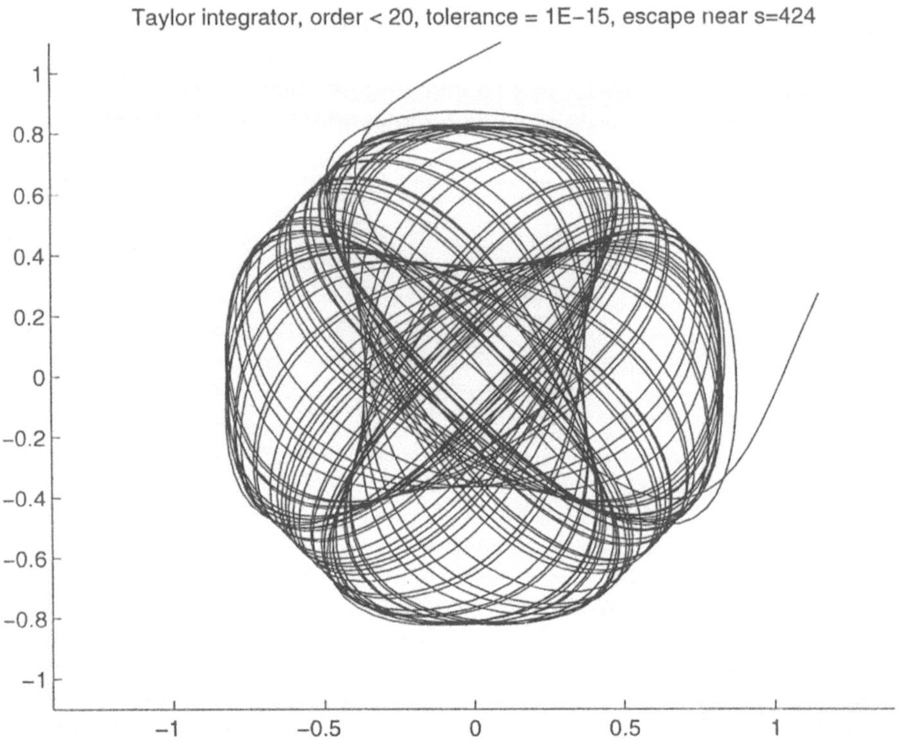

Fig. 4. Taylor integrator, $tol = 10^{-15}$, $10 \leq$ order ≤ 20, escape near $s = 424$. The orbit is correct at least to drawing accuracy.

symplectic Runge-Kutta method with step size control), using the smallest possible tolerance of $2.22 \cdot 10^{-14}$. The orbit of Fig. 2 was reproduced in every detail (except for the artefact). Therefore we believe this orbit to be correct to at least drawing accuracy. The execution time in MATLAB was roughly the same as for RKN-6 with $\Delta s = \frac{1}{64}$. It must be said, however, that time comparisons in MATLAB are to be handled with care.

On the other hand, the regularized Hamiltonian K shows an almost perfect linear growth from 0 to $K = -1.1 \cdot 10^{-12}$ at $s = 424$. This fact puts a warning sign next to explicit Runge-Kutta integrators for long-term use! Symplectic integrators do not seem to have this problem, see Fig. 3.

To conclude, we discuss the surprising performance of the Taylor series algorithm [1, 14] for this problem. A Taylor integrator, to be built for the individual differential equation under consideration, generates the Taylor series of the solution at the initial point and propagates the solution to the next point. The order and step size may be varied in a most flexible way. A simple strategy suggested by the author [14] is to sum the Taylor series up to a given error tolerance tol and keep the order between fixed bounds $p_{\min}$ and $p_{\max}$ by possibly halving or doubling the

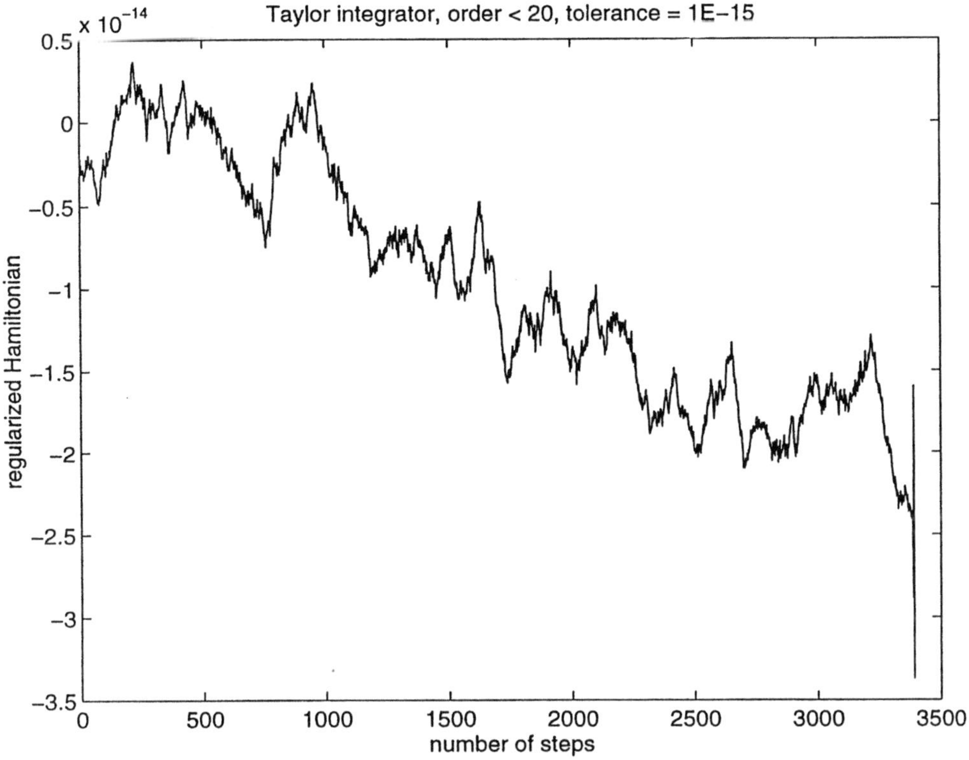

Fig. 5. The regularized Hamiltonian K for the integration of Fig. 4 (Taylor series).

step.

The orbit obtained with $tol = 10^{-15}$, $p_{\min} = 10$, $p_{\max} = 20$ is shown in Fig. 4. Every detail of Fig. 2 is reproduced in this integration that runs about twice as fast as ODE45. Furthermore, the deviation of the Hamiltonian K is reduced by a factor of 50 and stays at the remarkably low level of $2 \cdot 10^{-14}$, see Fig. 5. Due to the slight variation in the truncation order of the Taylor series K varies irregularly. It also shows the tendency of linear growth, however. Therefore symplectic integration seems to be a possible way of keeping the energy constant.

Appendix

The Stumpff Functions

The family of entire functions

$$c_n(z) := \sum_{k=0}^{\infty} \frac{(-z)^k}{(n+2k)!}, \quad n = 0, 1 \ldots \tag{28}$$

was introduced by K. Stumpff [12] in his theory of uniform treatment of Kepler motion. In Equs (18) the first 4 Stumpff functions are written in terms of trigonometric functions. Equivalent representations in terms of hyperbolic functions may be used [11], e.g.

$$c_0(z) = \cosh(\sqrt{-z}), \quad c_1(z) = \frac{\sinh(\sqrt{-z})}{\sqrt{-z}}, \quad c_2(z) = \frac{1-\cosh(\sqrt{-z})}{z}. \qquad (29)$$

Since $z = 0$ appears as a removable singularity, (18) or (29) may not be well suited for computing $c_n(z)$ if $|z| << 1$. The purpose of this section is to discuss methods (better than the Taylor series (28)) of evaluating $c_n(z)$ near the origin. In Section 2 the results are only needed for real values of z; however, they hold in the complex plane as well. Items i) through iv) below cover aspects of increasing complexity.

i) $c_0(z)$ and $c_1(z)$ may be accurately evaluated by means of (18) and (29), together with $c_n(0) = 1/n!$, $(n = 0, 1, \ldots)$, using the library functions since they usually provide constant relative accuracy, even near the origin.

ii) The identity

$$c_2(z) = \frac{1}{2}\left[c_1\left(\frac{z}{4}\right)\right]^2, \qquad (30)$$

equivalent to the duplication formula of the cosine, directly reduces $c_2(z)$ to $c_1(\frac{z}{4})$ in a numerically stable way.

iii) It is interesting to notice that (30) is the particular case $n = 1$ of the general quadrupling formula

$$2^n\, c_{n+1}(z) = c_1\left(\frac{z}{4}\right) c_n\left(\frac{z}{4}\right) + \sum_{k=1}^{[n/2]} \frac{c_{2k+1}(z/4)}{(n-2k)!}, \quad n = 0, 1, \ldots .$$

Unfortunately, for $n = 2$ this relation reads as

$$4c_3(z) = c_1\left(\frac{z}{4}\right) c_2\left(\frac{z}{4}\right) + c_3\left(\frac{z}{4}\right); \qquad (31)$$

therefore it does not allow the direct evaluation of $c_3(z)$. However, by repeated application of (31) the argument z may be reduced to arbitrarily small values. For sufficiently small arguments a finite section of the Taylor series (28) or an appropriate Padé approximant may be used. This method was suggested by S. Mikkola [6].

iv) An elegant algorithm for $c_3(z)$ is obtained via the triplication formula of the sine, which becomes

$$c_3(z) = \frac{1}{9}\left[c_3\left(\frac{z}{9}\right) + \frac{4}{3}\, c_1\left(\frac{z}{9}\right)^3\right]. \qquad (32)$$

This directly yields the infinite series

$$c_3(z) = \frac{4}{3} \sum_{k=1}^{\infty} 9^{-k}\, c_1(9^{-k} z)^3 . \tag{33}$$

In order to evaluate this series efficiently the number N of terms necessary to meet a given error tolerance ϵ in the domain $|z| < r$ has to be decided in advance. Let the successive terms of the series (33) be denoted by

$$a_k := 9^{-k}\, c_1(9^{-k} z)^3, \quad k = 1, 2, \ldots, N, \quad \boldsymbol{a} := (a_1, \ldots, a_N)^T . \tag{34}$$

Once a_N is known, $a_{N-1}, a_{N-2}, \ldots a_1$ can be stably computed recursively with about 5 multiplications each. Since the series (33) or

$$c_3(z) = \frac{4}{3} \sum_{k=1}^{N} a_k + R_N, \quad |R_N| < \epsilon \tag{35}$$

converges like a geometric series with convergence quotient $q = 9$ we follow the Neville-Aitken extrapolation procedure [9] (specialized to $q = 9$), and we consider the transformed sequence

$$\tilde{\boldsymbol{a}} = M\boldsymbol{a} \tag{36}$$

where M is the infinite matrix product

$$M = \ldots \left(I + \frac{I-J}{80}\right)\left(I + \frac{I-J}{8}\right)(I-J)^{-1} \quad \text{with } J = \begin{bmatrix} 0 & & & & \\ 1 & 0 & & & \\ & 1 & 0 & & \\ & & \ddots & \ddots & \\ & & & 1 & 0 \end{bmatrix}, \tag{37}$$

and $I \in \mathbb{R}^{N\times N}$ is the unit matrix. Equ. (35) then becomes

$$c_3(z) = \frac{4}{3}\, \tilde{a}_N + \tilde{R}_N . \tag{38}$$

The matrix M is handled by means of the function

$$g(x) := \prod_{k=1}^{\infty} \left(1 + \frac{1+x}{9^k - 1}\right) ; \tag{39}$$

then we have

$$M = g(-J)\,(I-J)^{-1} . \tag{40}$$

From the properties $g(-1) = 1$ and $g(9x) = (1+x)\,g(x)$ we obtain the Taylor series

$$g(x) = \gamma \sum_{k=0}^{\infty} \gamma_k\, x^k = \gamma\left(1 + \frac{x}{8} + \frac{x^2}{8\cdot 80} + \frac{x^3}{8\cdot 80\cdot 728} + \ldots\right) \tag{41}$$

with

$$\begin{aligned}\gamma_0 = 1,\ \gamma_k &= \prod_{\ell=1}^{k} (9^\ell - 1)^{-1}\,, \quad (k = 1, 2, \ldots) \\ \gamma &= \Big(\sum_{k=0}^{\infty} (-1)^k \gamma_k\Big)^{-1} = 1.14082\,27572\,64437\,28201\,66654\,27682\,.\end{aligned} \tag{42}$$

Now the following final result is obtained from (36), (40), (41), (42):

$$\tilde{a}_N = \gamma \sum_{k=0}^{N-1} \sigma_k\, a_{N-k} = \gamma \left(a_N + \frac{7}{8}\, a_{N-1} + \frac{561}{640}\, a_{N-2} + \ldots\right) \tag{43}$$

with

$$\sigma_k = \sum_{\ell=0}^{k} (-1)^\ell \gamma_\ell\,;$$

then $c_3(z)$ is given by (38). An analysis of the error $|\tilde{R}_N|$ shows that, e.g., in the disk $|z| \le 5$ (for $|z| > 5$ (18) or (29) are acceptable) Equ. (44) yields about $N(N + 2.5)/2$ correct decimal digits, hence for $N = 1, 2, 3, 4, 5, 6, 7, 8$ about $D_N = 1.75,\ 4.5,\ 8.25,\ 13,\ 18.75, 25.5, 33.25,\ 42$ correct digits are obtained, respectively.

Acknowledgements

The author is indebted to Dr. Willi Bucher of HTL Winterthur, Switzerland, for his helpful suggestions and for the comparison of various algorithms for the third Stumpff function.

References

[1] Hairer E., Nørsett S.P., Wanner G., 1994: Solving Ordinary Differential Equations I, Nonstiff Problems, Second Edition. Springer-Verlag, 480 pp.

[2] Hairer E., Wanner G., 1991: Solving Ordinary Differential Equations II, Stiff and Differential-Algebraic Problems. Springer-Verlag, 601 pp.

[3] Hill, G.W., 1886: On the part of the motion of the lunar perigee which is a fraction of the mean motions of the sun and moon. Acta Math. **8**, 1; also: Collected Mathematical works of G.W. Hill. Vol. 1. 1, p. 243. Cargenie Inst. of Wash., Washington, D.C., 1905.

[4] Levi-Civita, T., 1920: Sur la régularisation du problème des trois corps. Acta Math. **42**, 99-144.

[5] McLachlan, R.I., 1995: On the numerical integration of ordinary differential equations by symmetric composition methods. SIAM J. Sci. Comp. **16**, 151-168.

[6] Mikkola S., 1996: Personal communication.

[7] Ruth, R.D., 1983: A canonical integration technique. IEEE Trans. Nucl. Sci. NS-**30**, 2669-2671.

[8] Sanz-Serna, J.M., 1992: Symplectic integrators for Hamiltonian problems: An overview. Acta Numerica, 243-286.

[9] Schwarz H.R., 1989: Numerical Analysis. John Wiley, 517 pp.

[10] Spirig, F. and Waldvogel, J., 1991: Chaos in co-orbital motion. In: A.E. Roy (ed.), Predictability, Stability, and Chaos in N-Body Dynamical Systems. Plenum Press, 395-440.

[11] Stiefel E.L. and Scheifele G., 1971: Linear and Regular Celestial Mechanics. Springer-Verlag, 301 pp.

[12] Stumpff K., 1956: Himmelsmechanik, Vol. I. Deutscher Verlag der Wissenschaften.
[13] Trotter, H.F., 1959: On the product of semi-groups of operators. Proc. Am. Math. Soc. **10**, 545-551.
[14] Waldvogel J., 1984: Der Tayloralgorithmus. ZAMP **35**, 780-789.
[15] Waldvogel, J. and Spirig, F., 1988: Co-orbital satellites and Hill's lunar problem. In: A.E. Roy (ed.), Long-Term Dynamical Behavior of Natural and Artificial N-Body Systems. Kluwer, 223-243.
[16] Waldvogel, J. and Spirig, F., 1995: Chaotic motion in Hill's lunar problem. In: A.E. Roy and B.A. Steves (eds.). From Newton to Chaos: Modern Techniques for Understanding and Coping with Chaos in N-Body Dynamical Systems. Plenum Press, 217-230.
[17] Yoshida, H., 1993: Recent progress in the theory and application of symplectic integrators. Celest. Mech. **56**, 27-43.

ROTATION NUMBER AND GLOBAL STABILITY OF SYMPLECTIC INTEGRATOR

YI-SUI SUN AND JI-LIN ZHOU *
Department of Astronomy, Nanjing University, Nanjing 210093, China

Abstract. The global validity of symplectic integration methods or a mapping approach is discussed. The results show that in the regions of phase space where symplectic integration schemes and the Hamiltonian system possess the same topology, they are effective; but in the regions where the schemes possess other fixed points than those of the Hamiltonian system, their topologies are different from that of the actual system, thus the symplectic integration method or the mapping approach is not effective globally. We present a method to remove the main extra resonance islands in the symplectic integration schemes by changing the rotation numbers of the orbits. With this modification, the global stability of the symplectic integration method can be improved.

Key words: Symplectic integration methods – mapping approach – global stability

1. Introduction

Symplectic integration methods are effective for investigating the long-time evolution of dynamical systems, because symplectic integration algorithms preserve the symplectic structure; as a result the phase space volume is preserved [1,2,3]. Since a numerical integration method is a discretisation of continuous dynamical systems, it can be regarded as a mapping approach. Hadjidemetriou proposed a systematic method to construct a mapping model for a Hamiltonian system [4], the main idea is to preserve the basic features of the actual system in the mapping, i.e., to preserve the local topology of phase space not only qualitatively but also quantitatively. This can be achieved by preserving local fixed points and their stability indices. We find that the mapping constructed by Hadjidemetriou's method is exactly the same as the first order scheme of our symplectic integrator. So the first order scheme can preserve the local topology of phase space. Though higher order schemes are generally more accurate than the first order one, the fixed points they possess may be different from those of the first order one. If the differences are only of higher order of step size, and they have the same stability indices, higher order schemes should be better than the first order one, otherwise the situation is complicated. Moreover, for the mapping system there may exist $n-$periodic fixed points ($n \geq 2$), but for the actual Hamiltonian system, only equilibrium solutions can be found. So when the mapping system has $n-$periodic fixed points ($n \geq 2$), we cannot conclude that it has the same topology of phase space as the actual system. Thus the validity of the scheme becomes questionable. This relates to the problem of global validity for symplectic integration methods and the mapping approach. We study the problem with a simple dynamical model.

* Supported by the National Natural Science Foundation of China and a grant from the Ph.D. Foundation

2. Model and Schemes of Symplectic Integration

2.1. Hamiltonian system

We take the simple pendulum as a model, the corresponding Hamiltonian is

$$H = \frac{1}{2}p^2 + \cos q. \tag{1}$$

Obviously $p = p_0 = 0, q = q_0 = k\pi$ $(k = 0, \pm 1, ...)$ are equilibrium solutions of the Hamiltonian system (1) (we will restrict q to $q \in [0, 2\pi]$ later on). They are the periodic solutions with any period, the corresponding monodromy matrix is $M = e^{A\tau}$, where $A = \begin{pmatrix} 0 & \cos q_0 \\ 1 & 0 \end{pmatrix}$ is the matrix of variational equations. The stability index is defined as $K = trace M = \sum_{i=1}^{2} e^{\lambda_i \tau}$, where $\lambda_i (i = 1, 2)$ are the eigenvalues of A; we have $K = 2 + \cos q_0 \tau^2 + O(\tau^4) = 2 + (-1)^k \tau^2 + O(\tau^4), k = 0, 1$.

2.2. Symplectic integration schemes

The first, second and third order schemes of symplectic integration to be considered are

$$SI_1 : \begin{cases} p' = p + \tau \sin q, \\ q' = q + \tau p', \end{cases} \tag{2}$$

$$SI_2 : \begin{cases} p' = p + \tau \sin q + \frac{\tau^2}{2} p' \cos q, \\ q' = q + \tau p' - \frac{\tau^2}{2} \sin q, \end{cases} \tag{3}$$

and

$$SI_3 : \begin{cases} p' = p + \tau \sin q + \frac{\tau^2}{2} p' \cos q - \frac{\tau^3}{6}(p'^2 \sin q + \sin 2q), \\ q' = q + \tau p' - \frac{\tau^2}{2} \sin q - \frac{\tau^3}{3} p' \cos q, \end{cases} \tag{4}$$

respectively, where τ is the step size. For the above schemes, $n-$periodic fixed points exist for $n = 1, 2, ...$:

2.2.1. *1-periodic fixed points*

The 1-periodic fixed points of SI_1, SI_2 and SI_3 have the following general expressions, respectively:

$$\begin{cases} q_{01} = k\pi, & k = 0, 1 \\ p_{01} = \frac{2m\pi}{\tau}, & m \in \mathbf{Z} \end{cases} \tag{5}$$

$$\begin{cases} q_{02} = k\pi - \arctan(m\pi) + (-1)^k \tau^2 \frac{m\pi}{4(1+m^2\pi^2)^{3/2}} + O(\tau^4), & k = 0, 1 \\ p_{02} = \frac{2m\pi}{\tau} - (-1)^k \frac{\tau}{2} \frac{m\pi}{\sqrt{1+m^2\pi^2}} + O(\tau^3), & m \in \mathbf{Z} \end{cases} \tag{6}$$

and

$$\begin{cases} q_{03} = q^* + \tau^2 q^{**} + O(\tau^4), \\ p_{03} = \frac{2m\pi}{\tau} + (\frac{1}{2}\sin q^* + \frac{2}{3}m\pi\cos q^*)\tau + O(\tau^3), \end{cases} \tag{7}$$

where

$$\begin{cases} q^* = k\pi + \arctan \frac{m\pi}{\frac{2}{3}m^2\pi^2 - 1}, & k = 0,1 \\ q^{**} = (-1)^k \frac{m\pi[(\frac{1}{12}+\frac{4}{9}m^2\pi^2)(\frac{2}{3}m^2\pi^2-1)+\frac{2}{3}(\frac{2}{3}m^2\pi^2-1)^2+\frac{m^2\pi^2}{3}]}{[(\frac{2}{3}m^2\pi^2-1)^2+m^2\pi^2]^{3/2}}. & m \in \mathbf{Z} \end{cases} \tag{8}$$

When $m = 0$, $p_{01} = p_{02} = p_{03} = 0$, $q_{01} = q_{02} = q_{03} = k\pi, (k = 0, 1)$, coinciding with the equilibrium solution of Hamiltonian system (1). For $m \geq 1$ we have two classes of fixed points, corresponding to $k = 0$ and $k = 1$ respectively. Their stability can be judged from the stability index defined as $K = traceA$, where A is the matrix of the linearized mapping near the fixed points. After some manipulations, we have

$$\begin{aligned} K_1 &= 2 + \tau^2 \cos q_{01}, \\ K_2 &= 2 + \tau^2(\cos q_{02} - m\pi \sin q_{02}) + O(\tau^4), \\ K_3 &= 2 + \tau^2((1 - \tfrac{2m^2\pi^2}{3})\cos q_{03} - m\pi \sin q_{03}) + O(\tau^4) \end{aligned} \tag{9}$$

for SI_1, SI_2 and SI_3, respectively.

Thus for $m \geq 1$, (q_{01}, p_{01}) and (q_{02}, p_{02}) are stable when $k = 1$ and unstable when $k = 0$, while (q_{03}, p_{03}) is stable when $k = 0$ and unstable when $k = 1$. For $m = 0$, we have $K_1 = 2 + (-1)^k\tau^2$, $K_2 = 2 + (-1)^k\tau^2 + O(\tau^4)$, $K_3 = 2 + (-1)^k\tau^2 + O(\tau^4)$. In this case, for all the schemes SI_1, SI_2, SI_3 , the fixed points $(0, \pi)$ are stable while $(0, 0)$ is unstable. Moreover, one can see that the stability indices of $(0, \pi)$ and $(0, 0)$ for SI_1, SI_2 and SI_3 are, up to $O(\tau^4)$, the same as those of the Hamiltonian system (1). Thus in the neighbourhood of the fixed points $(0, 0)$ and $(0, \pi)$, the phase space of the Hamiltonian system and of the symplectic integration schemes possess the same topology. But for $m \geq 1$, the topologies between them are different, since as we know no such equilibrium solutions exist for the Hamiltonian system (1).

2.2.2. 2−periodic fixed points

For SI_1 , we have two sets of 2−periodic fixed points:

$$\begin{cases} q_{01}^{(2a)} = k\pi, & k = 0,1 \\ p_{01}^{(2a)} = \frac{m\pi}{\tau}, & m = 1,3,5,... \end{cases} \tag{10}$$

and

$$\begin{cases} q_{01}^{(2b)} = (k - \frac{m}{2})\pi - \frac{\tau^2}{4}\sin(k - \frac{m}{2})\pi + O(\tau^4), & k = 0,1 \\ p_{01}^{(2b)} = \frac{m\pi}{\tau} - \frac{\tau}{2}\sin(k - \frac{m}{2})\pi + O(\tau^3). & m = 1,3,5,... \end{cases} \tag{11}$$

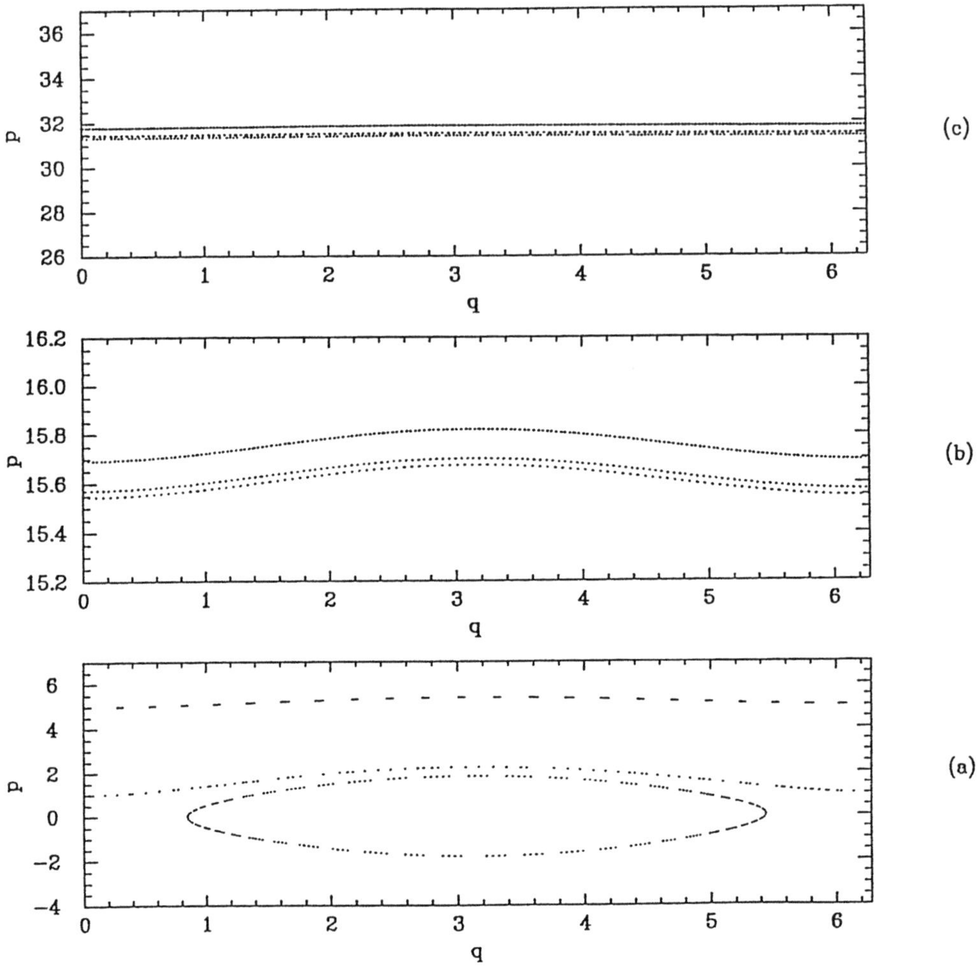

Fig. 1. Diagrams of Hamiltonian system (1) in the vicinity of (a) $p = 0$, (b) $p = 15.7$, (c) $p = 31.4$ with $\tau = 0.2$.

The stablility index is easily shown to be

$$K = 2+2\tau^2[\cos q+\cos(q+\tau p+\tau^2 \sin q)]+\tau^4 \cos q \cos(q+\tau p+\tau^2 \sin q)|_{(q_0,p_0)}, \quad (12)$$

where (q_0, p_0) stands for $(q_{01}^{(2a)}, p_{01}^{(2a)})$ or $(q_{01}^{(2b)}, p_{01}^{(2b)})$. Thus $K = 2 - \tau^4 + o(\tau^4)$ for $(q_{01}^{(2a)}, p_{01}^{(2a)})$ and $K = 2 + \tau^4 + o(\tau^4)$ for $(q_{01}^{2b}, p_{01}^{(2b)})$, which show the fixed points $(q_{01}^{(2a)}, p_{01}^{(2a)})$ are always stable, while $(q_{01}^{(2b)}, p_{01}^{(2b)})$ are always unstable.

For the 2−periodic fixed points of SI_2 and SI_3, they are rather laborious to find; we resort to numerical calculations.

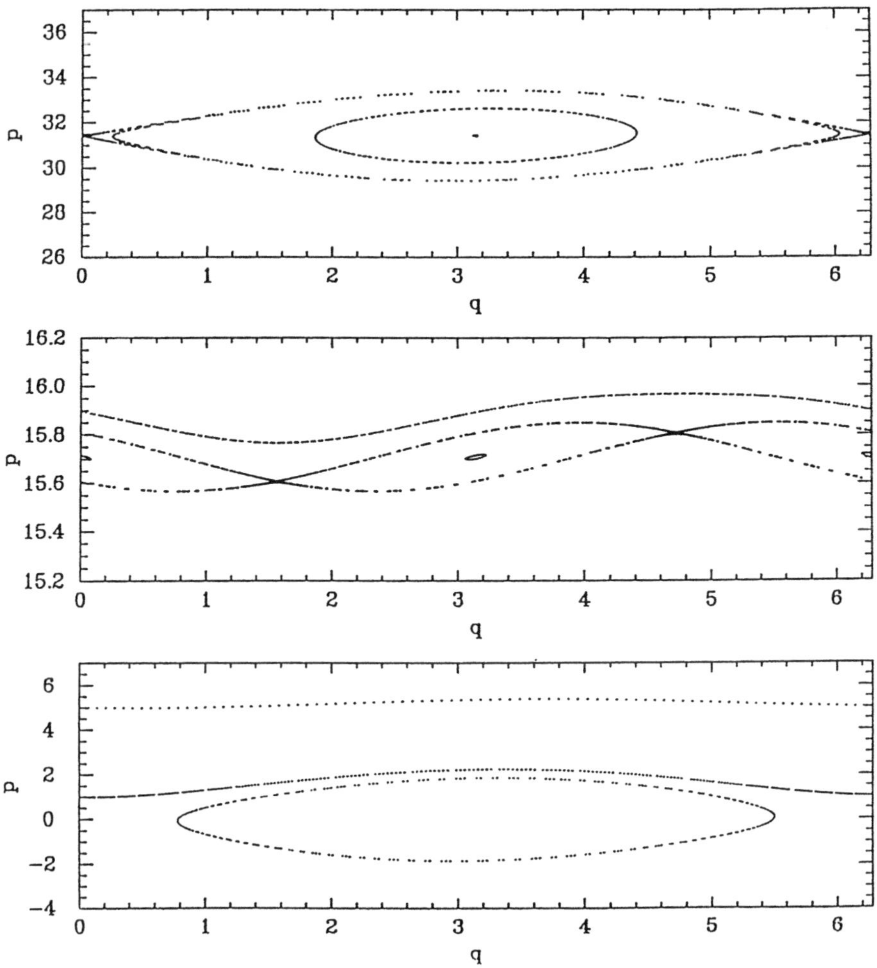

Fig. 2. Diagrams of first order schemes SI_1 in the vicinity of (a) $p = 0$, (b) $p = 15.7$, (c) $p = 31.4$ with $\tau = 0.2$.

3. Numerical Results

We study the structure of phase space near the fixed points for the Hamiltonian system (1) and the symplectic integration schemes (2),(3),(4), respectively. At first $\tau = 0.2$ is chosen. Fig.1 is the diagram of the Hamiltonian system in the neighborhood of $p = 0$, $p = \pi/\tau = 15.7$ and $p = 2\pi/\tau = 31.4$. This is a well-known picture. Fig.2, Fig.3, Fig.4 are the corresponding diagrams of SI_1, SI_2 and SI_3, respectively. We can see that, in the vicinity of $p = 0$, the Hamiltonian system and the SI_1, $SI2$ and $SI3$ have the same topology, it coincides with the theoretical analysis. But in the strip near $p = \pi/\tau$ and $p = 2\pi/\tau$, far from the q−axis, there exist 1−periodic and 2−periodic fixed points for symplectic integrators as

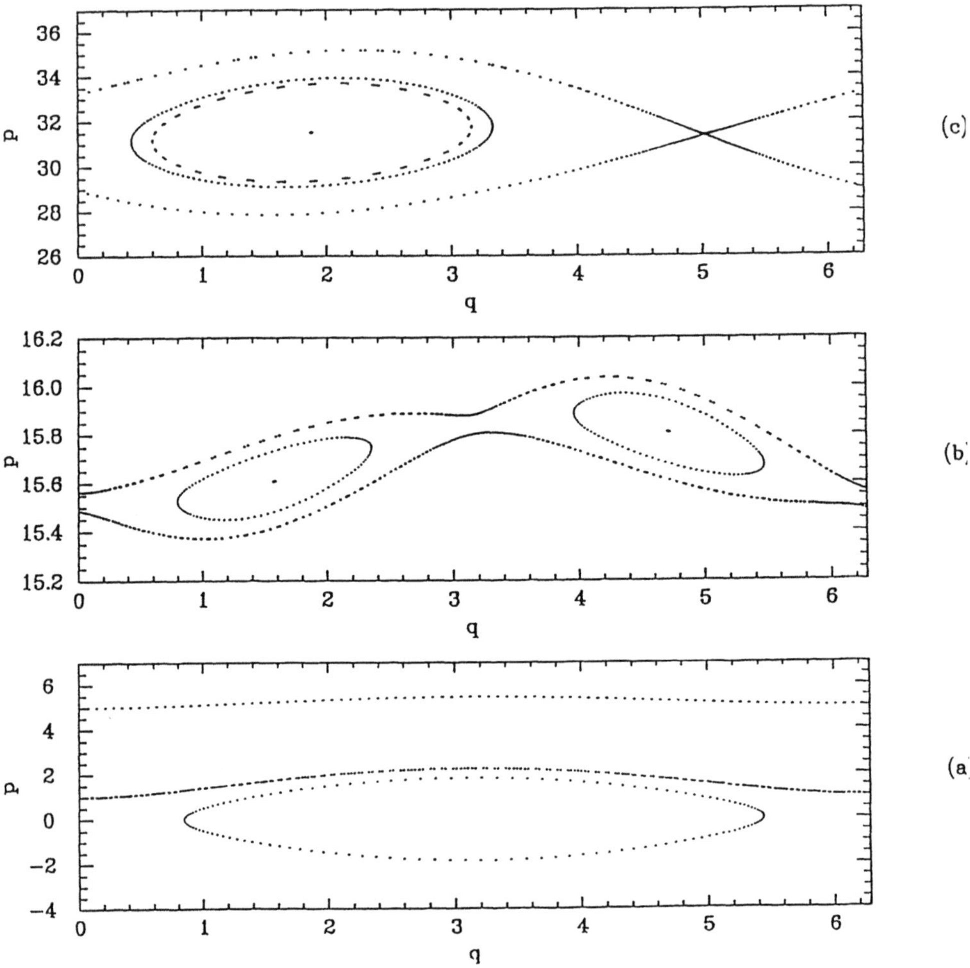

Fig. 3. Diagrams of second order schemes SI_2 in the vicinity of (a) $p = 0$, (b) $p = 15.7$, (c) $p = 31.4$ with $\tau = 0.2$.

indicated theoretically, while according to Fig.1, no equilibrium solutions exist for the Hamiltonian system, and the topologies between the Hamiltonian system and the symplectic integration schemes are different. Some n−periodic fixed points with $n > 2$ are also present (Fig.5 is the diagram of SI_1). When we use a smaller step size, e.g., $\tau = 0.05$, the topologies of phase space near $p = 15.7$ of SI_1, SI_2 and SI_3 become the same as that of the Hamiltonian system, because in this case there are no n−periodic fixed points of SI_1, SI_2 and SI_3 in the region (see Fig.6). But the n−periodic fixed points still exist, with larger p (Fig.7). According to the numerical results, we find that when the mappings possess some fixed points other than the ones of the actual system, we cannot conclude that higher order schemes

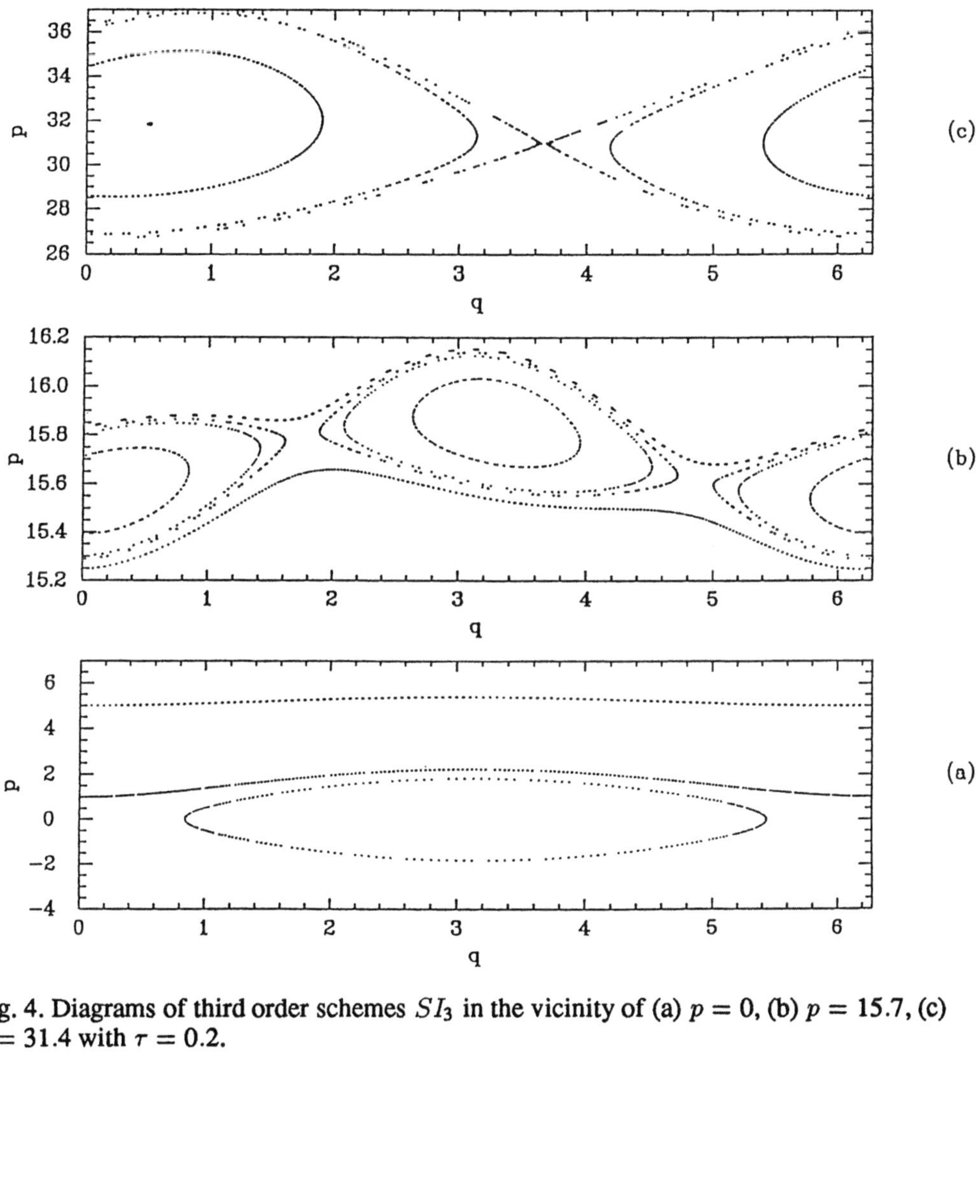

Fig. 4. Diagrams of third order schemes SI_3 in the vicinity of (a) $p = 0$, (b) $p = 15.7$, (c) $p = 31.4$ with $\tau = 0.2$.

must be better than lower order ones. However in the region where the symplectic integration schemes and the Hamiltonian system have the same topology, the higher order schemes will be better than lower order ones, as shown in Figs.1-4a, in which the invariant curves of the schemes near the fixed point $(0, \pi)$ are closer and closer to those of the Hamiltonian system with increasing the order of the schemes.

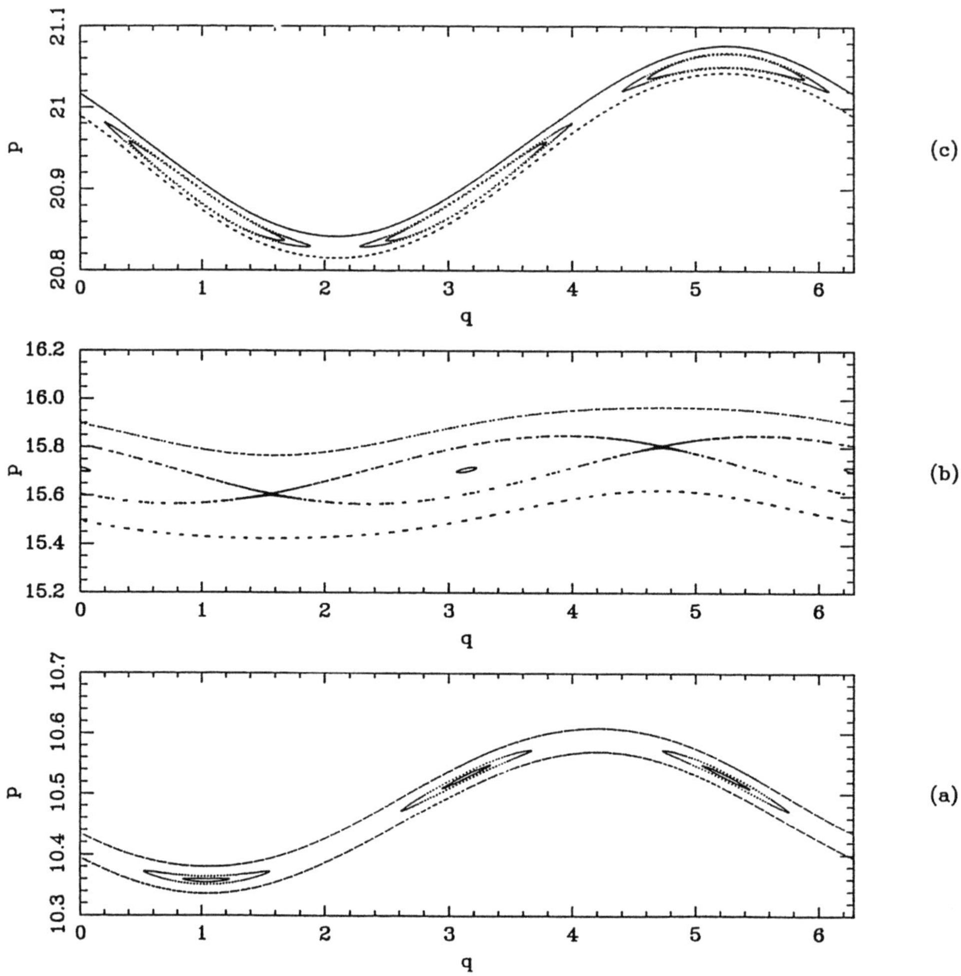

Fig. 5. 3-periodic islands of SI_1 with $\tau = 0.2$, with comparison of size to those of the 2-periodic islands.

4. Rotation Number

To remove some of the above extra resonance islands, we define the rotation number of an orbit as

$$\rho = \lim_{n\to\infty} (q_n - q_0)/(2n\pi), \tag{13}$$

where q_0 is the initial value of q, q_n is its value after n iterations.

We know the rotation number is a good indicator for the existence of resonance islands. Take SI_1 as an example. As it is well known, SI_1 corresponds to a Hamiltonian [6]

$$H_{map} = p^2/2 + 2\pi\delta(2\pi t/\tau)\cos q. \tag{14}$$

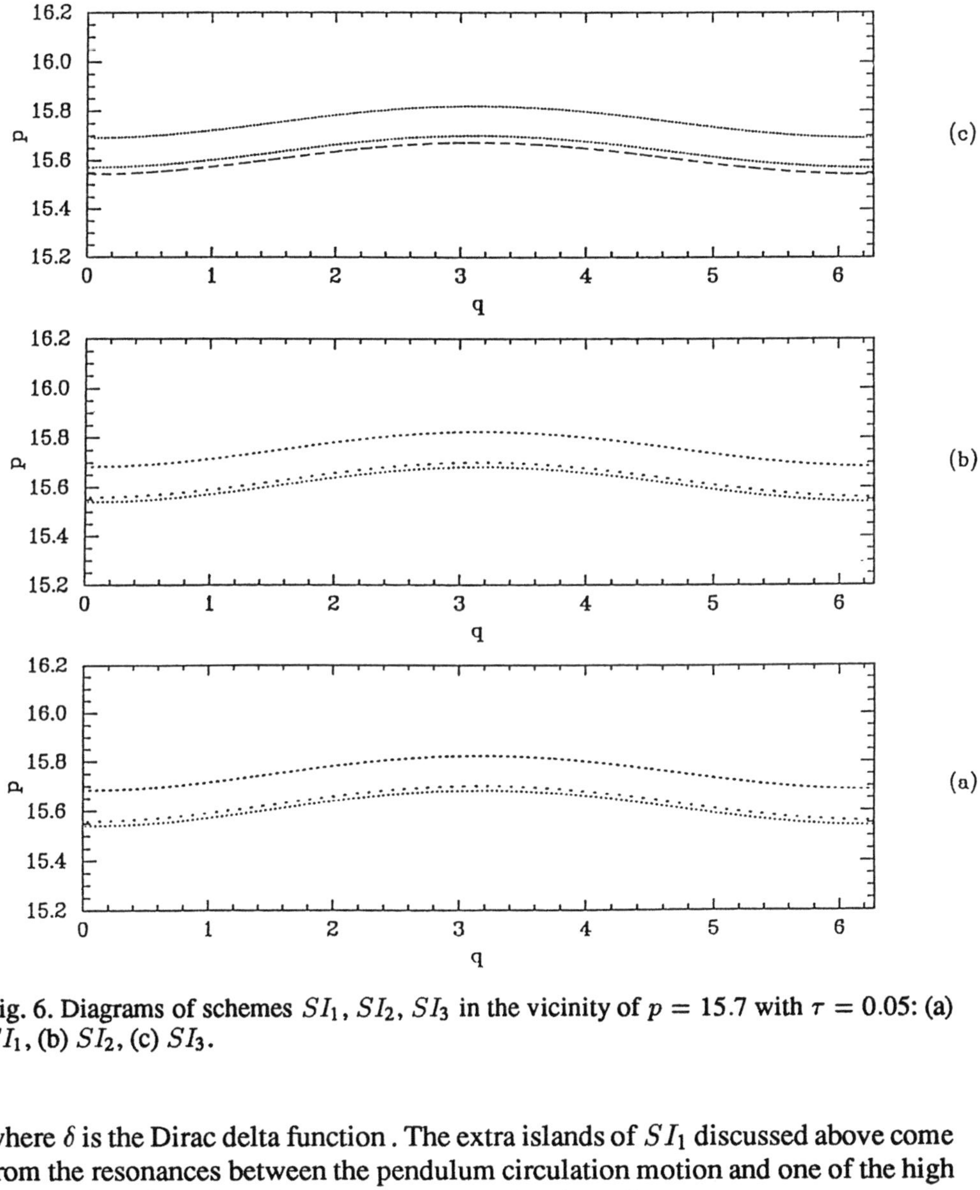

Fig. 6. Diagrams of schemes SI_1, SI_2, SI_3 in the vicinity of $p = 15.7$ with $\tau = 0.05$: (a) SI_1, (b) SI_2, (c) SI_3.

where δ is the Dirac delta function. The extra islands of SI_1 discussed above come from the resonances between the pendulum circulation motion and one of the high frequencies of the delta function, i.e., $q - m2\pi t/\tau \approx 0$.

According to the definition of the rotation number, $\rho = \omega/\omega_m$, where ω is the basic frequency of the pendulum motion (q, p), $\omega = \lim_{n\to\infty}(q_n - q_0)/(n\tau)$, and ω_m is the basic frequency of the high frequency oscillations of the delta function, $\omega_m = 2\pi/\tau$. Thus if $\rho = m/n$, m, n being integers, the corresponding orbit is in a resonance island.

We calculate the rotation numbers of orbits of SI_1 in the (q, p) plane. The initial q of the orbits are chosen as $q_0 = \pi$, which corresponds to the q-location of the

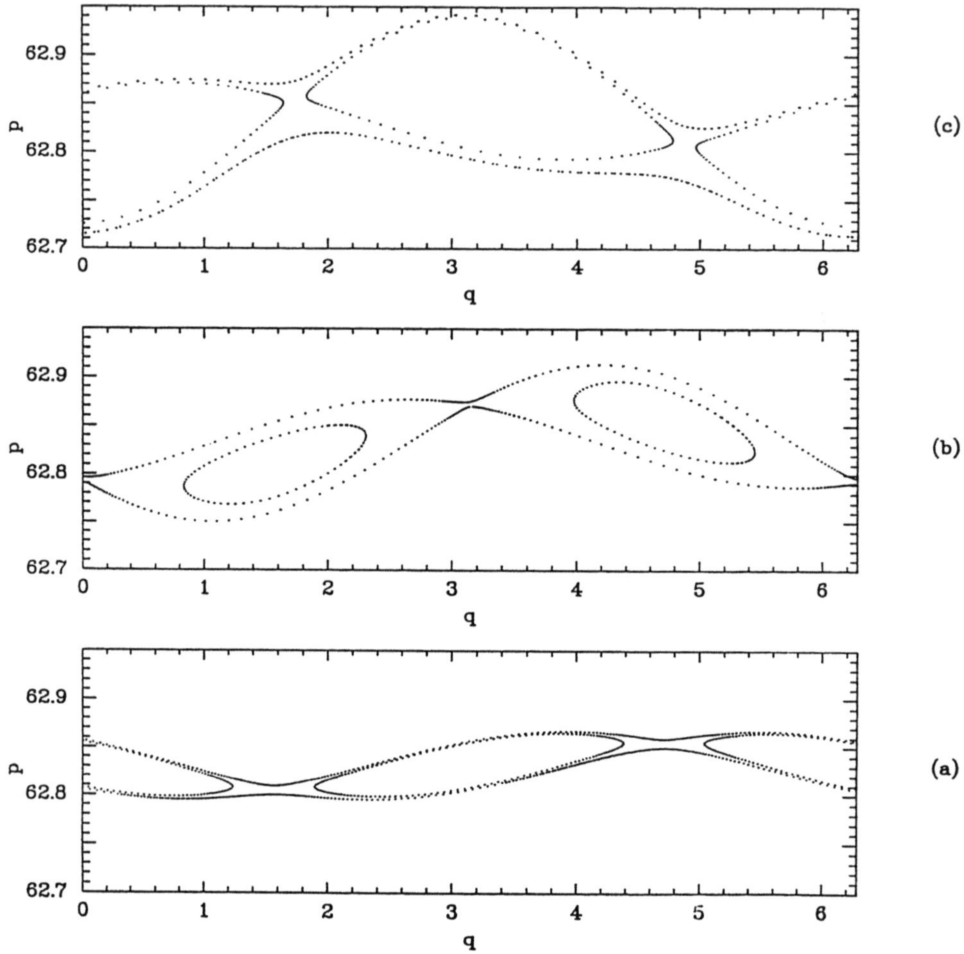

Fig. 7. Diagrams of schemes SI_1, SI_2, SI_3 in the vicinity of $p = 62.8$ with $\tau = 0.05$. (a) SI_1, (b) SI_2, (c) SI_3.

main fixed point $(\pi, 0)$, and the initial p are scanned from 0 to 40. Fig.8(a)(b) show the results. We obtain $\rho = 0$ in the libration region of the main resonance. When p_0 is near π/τ, $2\pi/\tau$, we have $\rho = 1/2$ and $\rho = 1$, respectively. Moreover, the widths of p-intervals where $\rho = 0, 1/2, 1$ coincide with those of the corresponding resonant regions of SI_1.

According to this, we can partly remove the extra resonance islands by changing the rotation number of the resonant orbits. This can be achieved by adding some extra terms in the mapping. As one can see from Fig.2, Fig.3, and Fig.4, the 1 : 1 resonance island is the largest (and thus the most unrealistic) resonance one in each case. If we add some new terms to the mapping so that ρ cannot reach 1, the

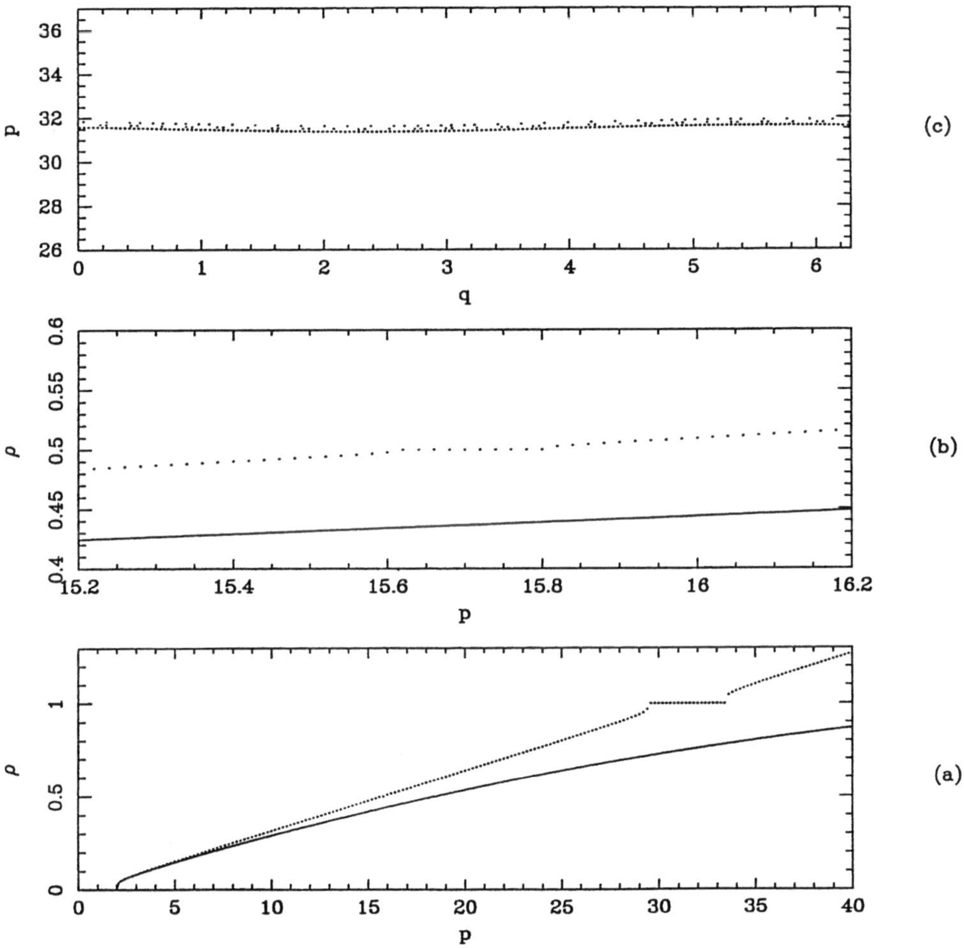

Fig. 8. (a) Variations of the rotation number with initial p_0 of orbits of SI_1 (dotted line) and SI_1' (solid line), the initial q_0 of the orbits are π; (b) is the enlargement of (a) in the corresponding region; (c) Diagrams of SI_1' in the vicinity of $p = 31.4$ with $\tau = 0.2$.

mapping may do not have the 1 : 1-resonance region and looks much better.

Take SI_1 for example. The simplest method of changing the rotation number of an orbit is adding some terms in the q' equation. We require the added terms to fulfill the following requirements:

(a) they are simple;

(b) they are much smaller than the $\tau p'$ term in the relevant region, so they change mainly the qualitative properties of the orbit, and the quantitative properties are not changed too much;

(c) the corrected mapping is measure-preserving.

Based on the above considerations, the generating function $F = qp' + \tau(\frac{1}{2}p'^2 + \cos q) - \frac{1}{3}\tau^4 p'^3$ is chosen, so that we have a corrected first order scheme of symplectic integration:

$$SI_1' : \begin{cases} p' = p + \tau \sin(q), \\ q' = q + \tau p' - \tau^4 p'^2. \end{cases} \tag{15}$$

Note that the added term is $O(\tau^4)$, so it is very small when p is small. When $p = 30$ (for $\tau = 0.2$), where the $1:1$ resonance island is located, the added term $(= 1.4)$ is still smaller than the second term $(= 6)$. For SI_1' we recalculated the three curves of Fig.2(c). The results are showed in Fig.8(c). This time all the three orbits circulate, and they are more realistic as compared to those of the original Hamiltonian system (Fig.1c).

The integrators SI_2 and SI_3 behave similarly. When we add the same term to SI_2, SI_3 as in SI_1, the $1:1$ resonance islands can also be removed. However, as compared to the last term in the q' equation of SI_2, SI_3, now the added term is no more small near $p = 2\pi/\tau$. Thus it may cause a loss of the numerical precision when we apply the method to higher order schemes of symplectic integration.

5. Conclusions and Discussions

According to the theoretical analyses and the numerical results, we can conclude that :

(1) when the symplectic integration schemes and the actual system possess the same topology of phase space, the symplectic integration method is effective, and higher order schemes are better than the lower order ones;

(2) when the symplectic schemes possess other fixed points than the ones of the actual system (including the n−periodic fixed points), the topologies of phase space are different from the actual system, in this case the symplectic integration methods are generally not effective globally, and we cannot conclude that higher order schemes are better than lower order ones.

Thus when we use symplectic integration methods (or a mapping approach), we have to examine whether or not the symplectic integration schemes have the same topology of phase space as the actual system. If the schemes have additional fixed points, in the vicinity of them the topology of phase space is distorted, and the symplectic integrators or the mapping approach cannot be applied. Because the fixed points depend strongly on the step size, we should choose it such that there are no other fixed points than those of the actual system in the interesting region, when we make use of a symplectic integration method or the mapping approach. Or alternatively, we can add an extra term in the symplectic integration schemes to remove part of the extra resonance islands.

It has already been noticed by Hadjidemetiou that when we construct an approximating system (eg. SI_1), spurious periodic solutions may appear[5]. But what he concerns most is the averaged Hamiltonian of the original system, whose fixed points may be in wrong positions, or sometimes some fixed points may be added or missed , as compared with the original system. In all these cases the topology of the mapping model is different from that of the real system. We are concerned with the construction of a mapping instead of obtaining of the averaged Hamiltonian. As we have shown in the paper spurious periodic solutions may also appear in this case.

Although symplectic integration methods can be of different types, e.g., implicit schemes (generating function methods, Runge-Kutta methods...) and explicit schemes (Ruth method, Neri method, Yoshida method...) (see [4]), the basic idea is to replace the original continuous system with a discrete one, or with a mapping. As we have stated in the introduction of the paper, the mapping generally has $n-$periodic fixed points, no matter which scheme. Moreover, since the mapping is symplectic, the $n-$periodic fixed points generally forms an $n-$islands chain, with alternating stable and unstable fixed points, just as we found . According to this we think, what we have found is a common phenomenon for all schemes of symplectic integration.

The extra resonance islands discussed above arise from the resonance between the pendulum circulation motion and one of the high frequencies of the delta function. Thus it is similar to the case in [6]. By our discussion, we find the existence of artifacts is a common phenomenon when we construct a mapping system for a Hamiltonian system. To remove these artifacts in the interesting region, one can choose an appropriate step size or add new terms to the mapping.

We have the pleasure to thank Drs. Saha and Waldvogel for their useful comments on the manuscript.

References

[1] Feng, K. and Qin, M.Z.: 1987, The symplectic method for the computation of Hamiltonian systems, *Lecture Note in Math*, **1297**, 1-37

[2] Channell, P.J. and Scorel, C.: 1990, Symplectic integration of Hamiltonian systems, *Nonlinearity*, **3**, 231-259

[3] Yoshida, H.: 1993, Recent progress in the theory and application of symplectic integration, *Celest. Mech. and Dyn. Astron.*, **56**, 27-43

[4] Hadjidemetriou, J.D.: 1991, Mapping models for Hamiltonian systems with application to resonant asteroid motion , in *Predictability, Stability and Chaos in N-body Dynamical System* , A.E.Roy (ed.), Plenum press, New York, 157-175

[5] Hadjidemetriou J. D.: 1995, Symplectic Mappings, *Proc. of IAU Sym.* **172**, Paris

[6] Wisdom J. and Holman M.:1992, Symplectic Maps for the N-Body Problem: Stability Analysis, *Astron. J.* **104**, 2022

ON A GLOBAL EXPANSION OF THE PLANAR DISTURBING FUNCTION

C. BEAUGÉ*

Instituto Astronômico e Geofísico, USP, Av. Miguel Stefano 4200, 04301-904 São Paulo, SP, Brazil

Abstract. Starting with a simple Taylor-based expansion of the inverse of the distance between two bodies, we are able to obtain a series expansion of the disturbing function of the three-body problem (planar elliptic case) which is valid for all points of the phase space outside the immediate vicinity of the collision points. In particular, the expansion is valid for very high values of the eccentricity of the perturbed body. Furthermore, in the case of a mean-motion resonant configuration, the above-mentioned expression is easily averaged with respect to the synodic period, yielding once again a global expansion of $\langle R \rangle$ valid for very high eccentricities. Comparisons between these results and the numerically computed exact function are presented for the 2/1 and 3/1 resonances.

Key words: disturbing function – three-body problem –mean-motion resonances

1. Taylor Series in True Angular Variables

1.1. Power Series

Let us suppose the planar elliptic restricted three-body problem, and adopt the following notation: m_1 the mass of the secondary body (i.e. perturber), r_1 its heliocentric distance, r the corresponding distance of the third body from the Sun, and S the angle between both radii. In terms of these, the complete disturbing function is given by

$$R = Gm_1 \left(\frac{1}{\Delta} - \frac{r}{r_1^2} \cos S \right) \tag{1}$$

with G the gravitational constant. Let us first concentrate on the direct part, which is the most problematical. Defining $\rho = r/r_1$ as the ratio between both radii, this function can be written as

$$\frac{r_1}{\Delta} = (1 + \rho^2 - 2\rho \cos S)^{-1/2} \tag{2}$$

where the angle S, in terms of the longitudes of perihelion (ϖ) and true anomalies (f) of both secondary masses, is given by $S = (f - f_1 + \varpi - \varpi_1)$.

Traditional expansions of this function follow one of two paths: *(i)* a Fourier series in angle S (i.e. Laplace, 1799) or *(ii)* a power series in ρ (i.e. Legendre, 1785). Both approaches are valid only as long as no orbit intersection occurs. A number of important systems, however, lie beyond this limit. The case of Pluto-Neptune is probably the most obvious, but we can also mention the problem of asteroidal and meteorite transportation to the inner solar system, the evolution

* Present Address: *Observatorio Astronómico, Universidad Nacional de Córdoba, Laprida 854, (5000) Córdoba, Argentina*

of Apollo-Amor-Atem bodies, high-eccentricity secular resonances, cometary dynamics, resonance capture through Poynting-Robertson drag, etc. In all these cases, no global analytical expansion exists to date, and numerical, semi-numerical approaches or local expansions (e.g. Ferraz-Mello and Sato, 1989, Morbidelli and Giorgilli, 1990, Yokoyama, 1994) must be used.

In this paper we present a new expansion which avoids these problems, and is valid for all values of ρ, be it lower or higher than unity. As a first step, we notice that the only pertinent "variable" in function $\frac{r_1}{\Delta}$ is

$$x = \rho^2 - 2\rho \cos S. \tag{3}$$

which marks the distance from the point in phase space to the singularity (i.e. collision point) located at $(\rho, \cos S) = (1, 1)$. This variable x is equal to -1 in the pole, and takes values larger than this for every point $(\rho, \cos S)$ outside. More important, written in terms of x, the original function acquires the form

$$\frac{r_1}{\Delta} = (1 + x)^{-1/2}, \tag{4}$$

which is much simpler than the original expression (see Laskar, 1991 and also Laskar and Robutel, 1995 for a similar application to planetary-type orbits). But the principal characteristic of equation (4) is that, by means of this definition, the number of variables explicit in the expansion has been reduced to one. Notice that the magnitude of ρ is not intrinsically significant, only the distance from the singularity is important.

The price payed for the simplification of the function is an increase in the range of the independent variable since, if $\rho \in (0, 2)$, now $x \in [-1, 8)$. Fortunately, this can be overcome applying a translation and scale-transformation, defining $t = 2(x - 7/2)/9$, whose range of values is now the orthodox $t \in [-1, 1)$. We can now find a power-series in t. We choose a Taylor series. After expanding $\Delta^{-1}(t)$, and passing to the original variable x, we obtain

$$\frac{r_1}{\Delta} = \sqrt{\frac{2}{9}} \sum_{n=0}^{\infty} a_n \left(\frac{2}{9}\right)^n \left(x - \frac{7}{2}\right)^n \tag{5}$$

which constitutes a power-series in $(x - 7/2)$ which is uniformly and absolutely convergent in the whole domain of x not including the singularity $x = -1$. The sequence $\{a_n\}$ contains the coefficients of the Taylor series of $(1 + t)^{-1/2}$.

Now, any numerical calculation of the series (5) will involve only a finite number of terms. We will make this fact explicit from the start and define N as the maximum value of n to be considered. In other words, let us truncate expansion (5) at some term, and work directly with the resulting polynomial. Applying Newton's binomial theorem to the right-hand side of (5), we can convert the (now) polynomial of $(x - 7/2)$ to one in x. Furthermore, we can invert the order of both sums and write

$$\frac{r_1}{\Delta} \simeq \sum_{j=0}^{N} c_j x^j$$

Notice that functions c_j are implicitly dependent on N, for the time being left indeterminate. Introducing equation (3) into expansion (6), we obtain, after some rather straightforward algebra,

$$\frac{r_1}{\Delta} \simeq \sum_{j=0}^{N} \sum_{k=0}^{j} c_j \binom{j}{k} (-2)^k \rho^{2j-k} \cos^k S \tag{7}$$

which constitutes the final form of our first development of the disturbing function.

We must remember, however, that (7) is not a series expansion, but a polynomial originating from the truncation (at N terms) of a Taylor series. This expansion was absolutely and uniformly convergent in all the domain of the independent variable so, in principle, expression (7) should be a very good aproximation of the exact function for every initial condition. The truncation error should also diminish as higher values of N are considered.

1.2. From S to True Anomalies

The next step in the Taylor series is to change from positive powers of the cosine of the relative angle S to trigonometric functions of multiples of the true anomalies and longitudes of perihelion. After a temporary switch from the cosines to their exponential counterparts, and after a reordering of terms of equal harmonics, we finally obtain

$$\frac{1}{\Delta} \simeq 2 \sum_{k=0}^{N} \sum_{i=0}^{N-k} A_{k,i}\; r^{(2i+k)} \left(\frac{1}{r_1}\right)^{(2i+k+1)} \cos k(f - f_1 + \varpi - \varpi_1) \tag{8}$$

where the final coefficients are now

$$A_{k,i} = (-1)^k \gamma_k \sum_{j=i}^{\min(2i,N-k)} c_{j+k} \binom{j+k}{2j-2i+k} \binom{2j-2i+k}{j-i} \tag{9}$$

Notice that $A_{k,i}$ are also implicitly dependent on N, both in the upper limit of the sum and through c_{j+k}. The temporary use of exponential functions has introduced a new parameter, given by

$$\gamma_k = \begin{cases} 1/2 & \text{if} \quad k = 0 \\ 1 & \text{if} \quad k > 0 \end{cases} . \tag{10}$$

However, we are still missing the indirect part of the disturbing function. >From equation (1), we can see that this contribution has exactly the same functional form as the terms of the expansion (8), so it is possible add this term directly in the definition of the coefficients used in this section. This can be done very simply modifying the coefficient $A_{k,i}$ with $k = 1, i = 0$ in the following way: $A_{1,0} \longrightarrow A_{1,0} - 1/2$.

With this simple change, we have finally obtained a truncated expansion of the complete elliptic planar disturbing function, as yet given in terms of the true anomalies and radii, that, treated as a series with undefined N, converges uniformly (although not absolutely) for all values of the ratio r/r_1 belonging to the open interval $(0, 2)$. However, this convergence is not absolute but *conditional*. In theory, expression (8) is a polynomial whose distance from the exact disturbing function should diminish as the number N is increased, leading to a convergence for every interval in the phase space not including the real singularities (i.e. $x = -1$). However, we must remember that we are dealing with a truncated conditionally convergent series. A rather unfortunate property of such an expansion is that the resulting coefficients $A_{k,i}$ grow exponentially to infinity as the parameter N is increased.

As a further note, we mention that all the substitutions performed in (7) to obtain expression (8) serve solely to "simulate" a Fourier-type formalism which will prove useful as the development of R proceeds.

2. Transformation from True to Mean Anomalies

The next step consists in the passage from the true anomalies to the mean angles. Since the radii also depend implicitly on f, and these appear in positive powers, the expansions we in fact need to construct are

$$\left(\frac{r}{a}\right)^m \cos kf = \sum_{s=0}^{\infty} B_{m,k,s} \cos sM \; ; \; \left(\frac{r}{a}\right)^m \sin kf = \sum_{s=0}^{\infty} \overline{B}_{m,k,s} \sin sM \quad (11)$$

$$\left(\frac{a_1}{r_1}\right)^m \cos kf_1 = \sum_{t=0}^{\infty} C_{m,k,t} \cos tM_1 \quad ; \quad \left(\frac{a_1}{r_1}\right)^m \sin kf_1 = \sum_{t=0}^{\infty} \overline{C}_{m,k,t} \sin tM_1$$

where both m and k are non-negative integers ($m \geq k$). The maximum value of the m-index is given by $\max(m) = 2N$, as can be seen from equation (8).

The calculations of the coefficients of series (11) and (12) will be done in two parts: first, a transformation to eccentric anomalies; secondly, a passage from these to the mean angles. In the case of the perturber there is no particular worry, since the eccentricity (e_1) of this body is rarely larger than 0.1 and most expansions cited in the literature will work perfectly well in this range. We therefore choose the classical expansion found in Brown and Shook (1933).

In the case of the orbit of the third body, things are much more complicated. Aside from the huge values of the m and k index, this time the eccentricity of the body is not restricted to low values but is expected to reach magnitudes close to parabolic escape ($e \geq 0.9$). As can be seen either by direct numerical computation of the previous expansion, or through the studies done by Brumberg and Fukushima (1994), the usual Brown-Shook expansion is practically invalid for high eccentricities, even for values of the index of the order of unity. In order to overcome this setback, Brumberg and Fukushima developed a different expansion based on a new intermediate angle, which they called the "elliptic anomaly". We prefer a different approach.

Making use of the fact that $m \geq k$, we begin by separating the original function through $(r/a)^m E^{ikf} = (r/a)^{(m-k)} \cdot (r/a)^k E^{ikf}$. Each of these products will now be treated by means of a finite expansion, also presented by Brown and Shook (1933). The advantages of this expansion with regards to the one by Brumberg and Fukushima are two. First, it is given in terms of the usual eccentric anomaly and not as function of the newly defined elliptic variable. Not only do we thus avoid possible problems in the posterior transformation to the mean anomaly, but we also maintain a certain coherence in the formalism of the theory. Second, the sums are finite, with the important consequence that the propagation of round-off error is thus diminished considerably.

Last of all, we must treat the transformation to the mean anomaly. For this step we will use the usual Fourier-Bessel expansions found in practically any book of celestial mechanics. Introducing the complete expressions into the expansions of the previous subsection, we finally obtain the passage from true to mean anomalies.

3. The Averaged Resonant Disturbing Function

Let us now suppose that the massless particle lies in the vicinity of a generic $(p+q)/p$ mean-motion resonance with the perturber. We can then define the usual canonical resonant angular variables as

$$\begin{aligned} q\sigma &= (p+q)\lambda_1 - p\lambda - q\varpi \\ q\sigma_1 &= (p+q)\lambda_1 - p\lambda - q\varpi_1 \\ q\sigma_2 &= \lambda - \lambda_1 \end{aligned} \tag{13}$$

where $\lambda_i = M_i + \varpi_i$ are the mean longitudes of the orbits. The last of variables (13) is the so-called synodic angle, whose derivative gives the frequency between two successive conjunctions. Introducing the inverse-transformations of (13) into the expansion, all periodic terms will now be given as functions of the resonant angles. Now, the averaging process we are considering is one over the synodic period of the system. After numerous algebraic calculations, we can finally write

$$\begin{aligned} \langle R \rangle \simeq & \sum_{k=0}^{N} \sum_{i=0}^{N-k} \sum_{u=0}^{\infty} \frac{1}{2\gamma_u} A_{k,i} \frac{a^m}{a_1^{m+1}} \\ & \Big[\mathcal{D}_{m,k,u} \cos\left((u|p| + k)\sigma - (u|p+q| + k)\sigma_1\right) + \end{aligned} \tag{14}$$

$$\overline{\mathcal{D}}_{m,k,u} \cos\left((u|p| - k)\sigma - (u|p+q| - k)\sigma_1\right)\Big]$$

where

$$\begin{aligned}\mathcal{D}_{m,k,u} &= (B_{m,k,u|p|} - \overline{B}_{m,k,u|p|})(C_{m+1,k,u|p+q|} - \overline{C}_{m+1,k,u|p+q|}) \\ \overline{\mathcal{D}}_{m,k,u} &= (B_{m,k,u|p|} + \overline{B}_{m,k,u|p|})(C_{m+1,k,u|p+q|} + \overline{C}_{m+1,k,u|p+q|})\end{aligned} \tag{15}$$

and where $m = 2i + k$.

Although expression (14) gives all the valuable information, it is possible to simplify the notation and, at the same time, unify all terms which contain the same multiple of the critical angle σ in the cosines. A further simplification, this time involving the neglect of certain terms, can also be performed. For this, we must note that the multiple of the σ_1 variable (call it $j \equiv u|p+q| \pm k$) is related, through the D'Alembert principle, to the lowest power of the perturber's eccentricity, as it appears in the corresponding coefficient. Since we are only expected to work with disturbing masses in quasi-circular orbits, it is not necessary to include absolute values of j above, say, $|j_{max}| = 3$. Even for values of $e_1 \sim 0.1$, the loss of precision due to this simplification is only of the order of 10^{-6}. The consequent economy of calculation is well worth it.

Introducing these two modifications in (14) we can furthermore obtain a much simplified expression for the disturbing function, now given by

$$\langle R \rangle \simeq \frac{1}{a_1} \sum_{u=0}^{\infty} \sum_{j=-3}^{3} \sum_{i=0}^{N-k} \frac{1}{2\gamma_u} A_{k,i} \alpha^m D_{m,u,j} \cos\left[uq\sigma - j(\sigma - \sigma_1)\right] \tag{16}$$

with

$$D_{m,u,j} = \begin{cases} \mathcal{D}_{m,k,u|p|} & \text{if} \quad u|p+q| < j \\ \mathcal{D}_{m,k,u|p|} + \overline{\mathcal{D}}_{m,k,u|p|} & \text{if} \quad u|p+q| = j \\ \overline{\mathcal{D}}_{m,k,u|p|} & \text{if} \quad u|p+q| > j \end{cases} \tag{17}$$

The k-index is now restricted to $k = |u|p+q| - j|$. One last simplification can be undertaken. Notice in equation (16) that the periodic terms are dependent only on the u and j index. All the remaining sums can be added to the amplitudes. Thus, we can finally write the disturbing function as

$$\langle R \rangle \simeq \sum_{u=0}^{\infty} \sum_{j=-3}^{3} G_{u,j} \cos\left[uq\sigma - j(\sigma - \sigma_1)\right] \tag{18}$$

where functions $G_{u,j}(a/a_1, e, e_1)$ are given by

$$G_{u,j} = \frac{1}{2a_1} \sum_{i=0}^{N-k} \frac{\alpha^{(2i+k)}}{\gamma_u} A_{k,i} D_{2i+k,u,j} \qquad (|k = u|p+q| - j|) \tag{19}$$

These coefficients are explicit functions of the semimajor axes and implicit in the eccentricities. More important, however, they are independent of all the angular variables, mainly both resonant arguments σ and σ_1. In general terms, coefficients (19) are analogous to the Fourier amplitudes of $\langle R \rangle$.

4. Comparison With the Exact Function

The time has now arrived to apply this expansion and calculate the value of $\langle R \rangle$ for given initial conditions and compare the results with those obtained by the numerical computation of the exact function. In all estimations, we shall consider α corresponding to exact resonance and $e_1 = 0$. This restriction to the circular problem has been adopted in order to separate the approximations introduced by considering only the low-order harmonics in j (see equation (18)) from those inherent to the expansion itself.

We will separate all calculations in two sections: first, we will show the results with a "low-precision" expansion; that is, choosing a value of N sufficiently low to allow direct determination of all terms through a numerical program using machine round-off error. Second, we will compare these results as higher values of N are considered. In both cases however, we will take m_1 and the gravitational constant equal to unity.

4.1. Low-Precision Results

We begin with the 2/1 resonance. Figure 1 shows, for six different values of the resonant angle σ, a comparison between the numerical calculation of the exact mean disturbing function (continuous lines) and our expansion (symbols) using $N = 160$. All graphs give the value of $\langle R \rangle$ as function of the eccentricity, from circular orbits to the parabolic limit. For further comparison, in broken lines we present the value of the disturbing function as predicted by the Laplace expansion truncated at second order in e. Let us recall that for this commensurability the limiting eccentricity for the convergence of the Laplace expansion, as given by Sundman's criterion, is $e_{Sund} \simeq 0.18$.

In the first two graphs, $\sigma = 0$ and $\sigma = \pi/5$, the present expansion reproduces extremely well the exact function, for all values of the eccentricity. The relative error never exceeds $\varepsilon \sim 10^{-4}$ and reaches a minimum of 10^{-5} for eccentricities of the order of $e = 0.7$. For these values of the critical angle, we are far away from the collision curve of the resonance, so all values attained by $\langle R \rangle$ are relatively low. Thus, a small number of terms of the series is sufficient to approximate the solution. For $\sigma = 2\pi/5$, in the third graph, we now approach the collision curve for very high eccentricities. Near the singularity the value of $\langle R \rangle$ obtained by our expansion begins to fall short of the exact function. Nevertheless, the overall agreement is still extremely good.

This underestimation is seen more clearly in the next graph, for $\sigma = 3\pi/5$, where the singularity is now completely present for elliptic orbits. Rigorously speaking the mean disturbing function is undefined at these points; nevertheless the steep increase of the function in the vicinity of the pole is real enough. Once again our expansion falls short in this range. This is expected, since an infinite number of terms (i.e. $N \to \infty$) should be necessary in order to approach the singularity with

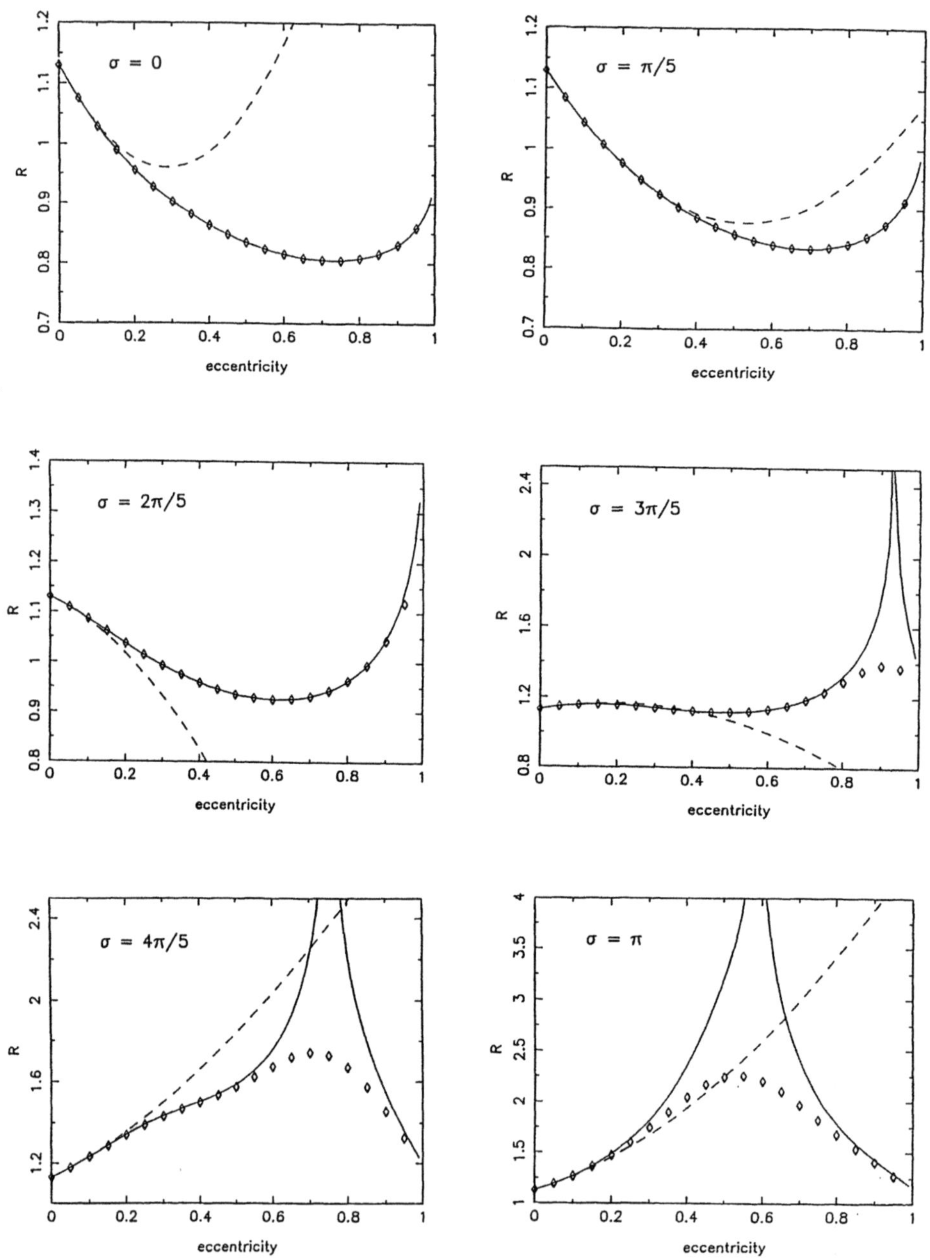

Fig. 1. 2/1 Resonance. Comparison between the exact mean disturbing function (continuous lines), the second-order Laplace expansion (broken lines) and the expansion presented in this paper (symbols). Graphs show $\langle R \rangle$ as function of the eccentricity, for six different values of the critical angle $\sigma = 2\lambda - \lambda_1 - \varpi$.

the original Taylor series (equation (7)). What is important, however, is that our results do not cease to be valid at this point. No false divisors or discontinuities appear in the convergence; the value of the exact function is simply underestimated. The behaviour of our expansion beyond the collision curve can be appreciated in the last two graphs, corresponding to $\sigma = 4\pi/5$ and $\sigma = \pi$. We can see that once the pole is passed, the analytical development simply returns to the exact function. In other words, far away from the singularities, the expansion continues to be as precise as in the first graphs.

The comparison between these results for the 2/1 resonance and the values predicted by the Laplace expansion is clear from Figure 1. For values of the eccentricity below e_{Sund}, the agreement of the Laplace series is very good for all values of σ. For values larger than this limit, the agreement is non-existent.

In view of these results, we can safely conclude that the present global expansion of the disturbing function works very well. It effectively approximates the behaviour of the exact function for any mean-motion resonance at any value of the eccentricity. The magnitude of the relative error is not orbitally-dependent, but solely a function of the magnitude of the exact function itself, just as has been predicted all through this paper. Nevertheless, we must still analyse the dependence of the relative error with the parameter N. This should be done with two aspects in mind: First, to visualize the actual convergence of the series. Second, the gradual improvement of the results near the singularities. Both these aspects of the same problem will be discussed presently.

4.2. High-Precision Results

All previous results have been obtained by choosing a value of N sufficiently low so as to allow direct calculation of all coefficients through a Fortran program using machine round-off precision. If more terms of the series are desired, we must adopt an algebraic manipulator in which the number of significant digits increases with N. With this in mind, we have chosen the *Mathematica 2.1* package (Wolfram, 1991) installed in an HP Apollo 9000/735 workstation. Now, even though this change in calculating tool effectively improves the precision of the results, it also increases tremendously the CPU time necessary for each coefficient.

We begin with the 3/1 resonance. A single initial condition was chosen, corresponding to the stable corotation center in the h-axis. Figure 2 shows the evolution of the relative error of the expansion as function of N. We can see, after the first few terms, the fall-off in error is practically exponential, reaching of the order of $\varepsilon \sim 10^{-13}$ at $N = 185$. In other words, for this point at least, the convergence of the series is relatively fast, and the final value of the expansion has a precision that rivals that obtained through numerical computation of the exact function itself.

Next, in the case of the 2/1 commensurability, we explore the region around the pole itself. We took the semimajor axis corresponding to exact resonance and a value of the eccentricity equal to $e = 0.7$. This latter value was chosen for

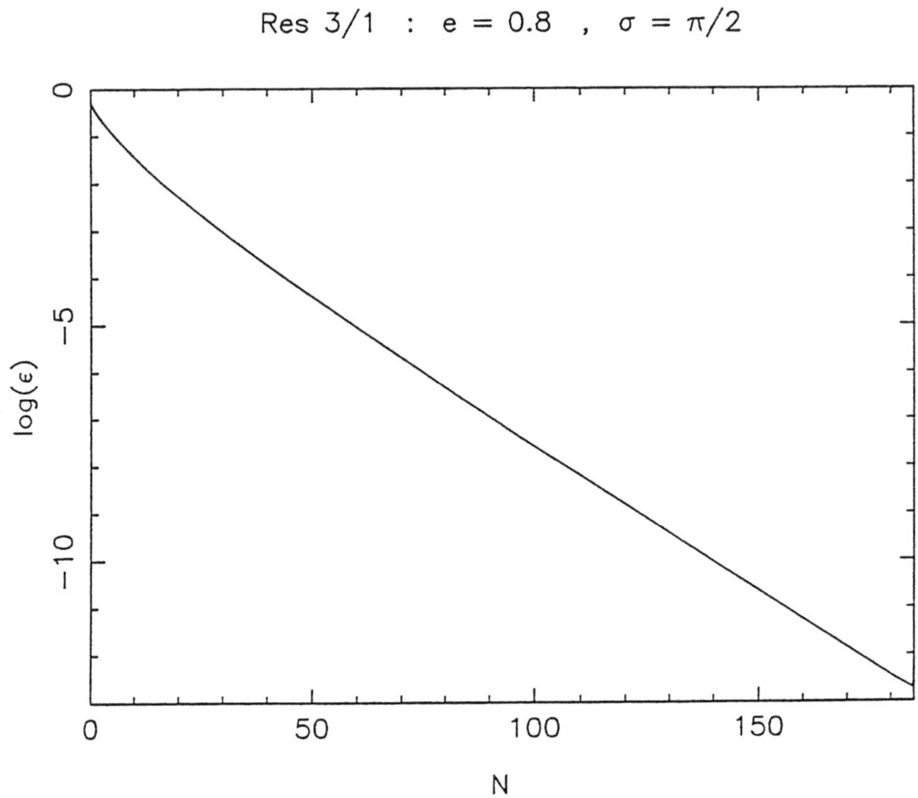

Fig. 2. 3/1 Resonance. For given initial conditions, the plot shows the decrease in relative error between our expansion and the exact $\langle R \rangle$ as a function of the number of terms N.

two reasons: first, it is very close to the apocentric collision point ($e_{apo} \simeq 0.58$), which means that the effect of the singularity should be very high. Second, the distance from e_{apo} is nevertheless sufficiently large so as to shown very clearly the existence of stable apocentric libration points beyond the collision curve (see Moons and Morbidelli, 1994). Thus, we particularly wished to see whether our expansion could in fact reproduce these new solutions.

Figure 3 shows the results of two applications of our series expansion, each for a different total number of terms ($N = 160$, and $N = 800$). Each graph presents the exact function, as function of σ, in broken lines. The singularity, which appears twice due to the symmetry of $\langle R \rangle$ with respect to $\cos\sigma$, is marked by the spikes. The stable apocentric libration center is seen as a local minimum of the function. Each analytical approximation is shown by continuous curves. In the top graph ($N = 160$) we can see that, although the agreement is very good outside the neighborhood of the pole, inside this region the underestimation is very acute. Moreover, the apocentric libration is not identified at all. In figure 3b ($N = 800$),

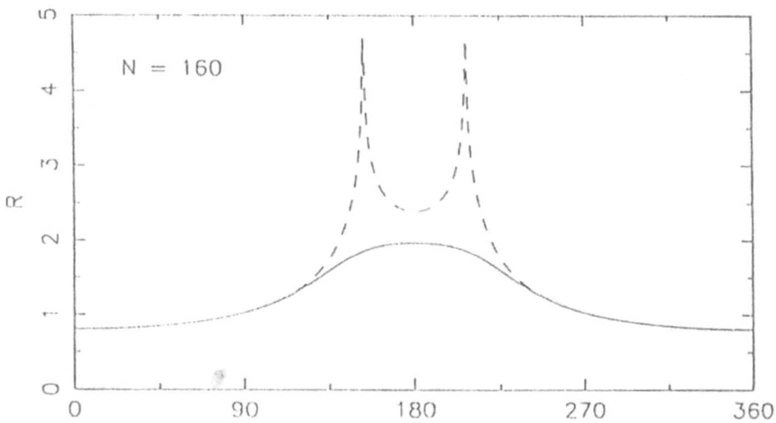

(a)

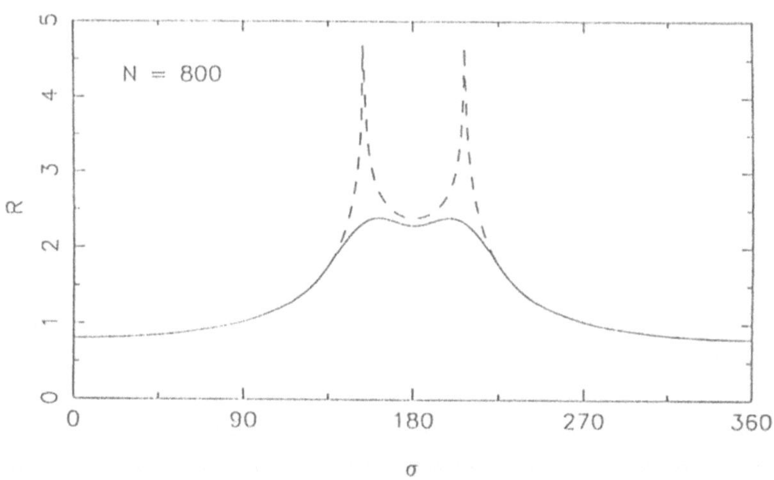

(b)

Fig. 3. 2/1 Resonance. Mean disturbing function in terms of the critical angle σ. Chosen eccentricity is $e = 0.7$. Exact function is shown in broken curves (spikes denote the singularity), while the analytical approximation is shown in continuous lines. (a). $N = 160$. (b). $N = 800$.

the agreement improves significantly. Not only is the "underestimation region" much reduced, but we also see the appearance of the apocentric libration as a marked local minimum at $\sigma = \pi$. Nevertheless, a higher number of terms should still be necessary to approximate the singularity region with good precision.

5. Conclusions

In the present work we have presented a newly derived series expansion of the planar elliptic disturbing function which is valid for all points of the phase space outside the collision points. Although in theory it is possible to determine the value of this function with vanishing error, this is extremely complicated in practice, due to the fact that the intermediate coefficients $A_{k,i}$ and $D_{k,u,j}$ increase drastically in magnitude with the total number of terms N. Thus, beyond a certain point, a gain in desired precision implies the use of algebraic manipulators with decreasing round-off error. This translates to a, once again, drastical increase in computer time. Thus, a certain compromise must be reached between expected precision and practicality.

Nevertheless, practically all motion of real bodies occurs far from the poles and in the region of relatively "low" values of $\langle R \rangle$. Thus, from the practical point of view, the quality of the representation near the singularities should not be too much of an issue, as long as we can guarrantee good precision outside them.

Acknowledgements

This work was supported by the São Paulo State Science Foundation (FAPESP). The author would like to express his gratitude to S. Ferraz-Mello for many valuable discussions throughout this work.

References

Brown, E.W. and Shook, C.A.: 1933, *Planetary Theory*, Cambridge University Press, UK.

Brumberg, E. and Fukushima, T.: 1994, *Cel. Mech. & Dynam. Astron.*, **60**, 60–80.

Ferraz-Mello, S. and Sato, M.: 1989, *Astron. Astrophys.*, **225**, 541–547.

Laplace, P.S.: 1799, *Mécanique Céleste*, English translation by N. Bowditch, Chelsea Pub. Comp. edition, N.Y., 1966.

Laskar, J.: 1991, Analytical framework in Poincaré variables for the motion of the solar system. In *Predictability, Stability in N-Body Dynamical Systems*, A. Roy (ed). Kluwer Publ., pp 93–114.

Laskar, J., and Robutel, P.: 1995, *Cel. Mech. & Dynam. Astron.*, **62**, 193–217.

Legendre, A.M.: 1785, *Mém. Math. et Phys. présentés à l'Ac. r. des sc. par divers savants*, **X**. bibitem Morbidelli, A., and Giorgilli, A.: 1990, *Celest. Mech.*, **47**, 145–172.

Moons, M. and Morbidelli, A. : 1993, *Cel. Mech. & Dynam. Astron.*, **55**, 99–108.

Wolfram: 1991, *Mathematica: A System for Doing Mathematics by Computer*, Addison-Wesley Pub. Co., Redbook City, CA.

Yokoyama, T.: 1994, *Cel. Mech. & Dynam. Astron.*, **60**, 387–400.

LINKS BETWEEN TIME SCALES USING BARYCENTRIC RELATIVISTIC EPHEMERIDES

X. MOISSON
Bureau des longitudes, URA 707
77, avenue Denfert-Rochereau
75014, Paris FRANCE

Abstract. Up to now, semi-analytical ephemerides of the solar system bodies were computed in the Schwarzschild approximation of the Einstein field equations and expressed in Barycentric Dynamical Time (TDB). The new resolutions of the IAU (IAU, 1991) specify the introduction of both barycentric and geocentric coordinate time scales, respectively TCB and TCG, as the independent argument of time of the ephemerides.
This paper is devoted to three main points :

- the integration of semi-analytical barycentric ephemerides constructed on the base of a metric expanded at the fourth order in $1/c$,
- the transformation of the solutions VSOP to the TCB time scale,
- the difference between Mercury, Venus, the Earth, the Moon and Mars planetocentric time scales and TCB.

For the second point, the use of J2000(TCB) as epoch of reference has been made. In doing so, we introduced a new reference frame, namely J2000(TCB), the precession quantities of which differ from the ones of the ecliptic reference frame J2000. This difference has been evaluated. As a first application of the third point, we derived a rough evaluation of the error made in planetary spacecraft ranging if we do not take into account the effect of these differences on the spin rate of the planet.

Key words: Time scales – General relativity theory – Ephemerides – Astronomical constants

1. Introduction

While the law of motion of the solar system bodies is expressed in the framework of the Newtonian gravitation theory, the time-like argument used in the differential equations and in their solutions is the uniform Newtonian time or a linear function of it. But it is a well known fact that the time is no longer uniform when we consider the special or general relativity theory (GRT). Moreover, concerning the relativistic equations of motion, the first-order post-Newtonian approximation to the GRT has been efficient for a long time. At this level of accuracy, the time scale used in the ephemerides is not significant.

But, in one hand, the present requirements in ephemeris precision imply to use the second-order approximation to GRT and on the other hand, we make a first-order approximation in using an ephemeris with TDB as time-like argument as if it was expressed in TCB.

According to the IAU resolutions, one should use in astronomical developments both Barycentric (BRS) and Geocentric (GRS) Reference Systems in their GRT sense, and the two time scales attached to them, TCB and TCG respectively.

These considerations imply to specify the spatial coordinate system used in representing the solutions of the law of motion, as it is done in the Newtonian theory but also to specify the time coordinate used, not only as an independent

R. Dvorak and J. Henrard (eds.), The Dynamical Behaviour of our Planetary System, 333.

variable for the equations, but mainly as one of the four coordinates used to represent the space-time which is a four dimension riemannian manifold in the GRT.

After a short description of the theoretical tools needed, we will present the barycentric equations of motion and the method used to integrate them. In a second part, we will discuss the substitution of TCB to TDB in the VSOP87 solutions constructed by (Bretagnon, Francou, 1988) and the modifications induced by the choice of J2000(TCB) as epoch of reference on the reference frame J2000(TDB). Finally, the differences between Planetocentric Time Scales and TCB are computed for Mercury, Venus, the Earth, Mars and the Moon as well as the effect they induce on the spin rate of these planets.

2. Theoretical Background

The space-time of GRT is the curved pseudo-Riemannian manifold described by the quasi-galilean metric $g_{\alpha\beta}$ [1]. The metric tensor is found by solving the Einstein's field equations with complementary boundary conditions.

For practical purposes in celestial mechanics, we consider the post-Newtonian expansion of both geocentric and barycentric metric:

$$g_{\alpha\beta} = \eta_{\alpha\beta} + h_{\alpha\beta}$$

where $\eta_{\alpha\beta}$ is the metric associated to the Minkovsky space-time :

$$\begin{cases} \eta_{00} &= 1 \\ \eta_{0i} &= 0 \\ \eta_{ij} &= -\delta_{ij} \end{cases} \tag{1}$$

and $h_{\alpha\beta}$ is expanded with respect to the linearized theory (Brumberg 1991):

$$\begin{cases} h_{00}(t,\mathbf{r}) &= c^{-2}h^{(2)}_{00} + c^{-4}h^{(4)}_{00} + c^{-5}h^{(5)}_{00} + \dots \\ h_{0i}(t,\mathbf{r}) &= c^{-3}h^{(3)}_{0i} + c^{-5}h^{(5)}_{0i} + c^{-6}h^{(6)}_{0i} + \dots \\ h_{ij}(t,\mathbf{r}) &= c^{-2}h^{(2)}_{ij} + c^{-4}h^{(4)}_{ij} + c^{-5}h^{(5)}_{ij} + \dots \end{cases} \tag{2}$$

Denoting x^k_A and $v^k_A = dx^k_A/dt = \dot{x}^k_A$ the BRS rectangular coordinates and velocity components of any body A of the solar system and writing $r^k_A = x^k - x^k_A$, $r_A = (r^k_A r^k_A)^{1/2}$, we have for the barycentric reference system in harmonic

[1] Here and below, Greek subscripts run from 0 to 3 and the Latin ones from 1 to 3. Moreover, the Einstein summation convention is applied

coordinates:

$$\left\{\begin{array}{lcl} h_{00}^{(2)} & = & -2\sum_A \frac{GM_A}{r_A} \\ h_{00}^{(4)} & = & -3\sum_A \frac{GM_A}{r_A}\dot{\mathbf{x}}_A^2 + 2\left(\sum_A \frac{GM_A}{r_A}\right)^2 + 2\sum_A \frac{GM_A}{r_A}\sum_{B\neq A}\frac{GM_B}{r_{AB}} \\ & & -\sum_A \frac{GM_A}{r_A}\left(\dot{\mathbf{x}}_A^2 - \mathbf{r}_A\ddot{\mathbf{x}}_A - \frac{1}{r_A^2}(\mathbf{r}_A\dot{\mathbf{x}}_A)^2\right) \\ h_{0i}^{(3)} & = & 4\sum_A \frac{GM_A}{r_A}(\dot{x}_A^i) \\ h_{ij}^{(2)} & = & 2\sum_A \frac{GM_A}{r_A}\eta_{ij} \end{array}\right. \tag{3}$$

This barycentric metric includes all the mutual relativistic perturbations of the planets in contrast with previous analytical works (Bretagnon, Francou, 1988) which only assumed the Schwarzschild relativistic perturbation of the Sun.

3. Barycentric Equations of Motion

Associated with the metric (2-3), we can construct the barycentric equations of motion following the Fock's method (Fock, 1955), on the form of perturbed motion:

$$\ddot{\mathbf{x}}_i = -\sum_{j\neq i}\frac{GM_j}{r_{ij}^3}\mathbf{r}_{ij} + \sum_{j\neq i}\frac{GM_j}{c^2}\left(A_{ij}\mathbf{r}_{ij} + B_{ij}\dot{\mathbf{r}}_{ij}\right) \tag{4}$$

for each body i, with :

$$\left\{\begin{array}{lcl} A_{ij} & = & \frac{\dot{\mathbf{x}}_i^2}{r_{ij}^3} - 2\frac{\dot{\mathbf{r}}_{ij}^2}{r_{ij}^3} + \frac{3}{2r_{ij}^5}(\mathbf{r}_{ij}\dot{\mathbf{x}}_j)^2 + G\left[5M_i + 4M_j\right]\frac{1}{r_{ij}^4} \\ & & +\sum_{k\neq i,j} GM_k\left[\frac{4}{r_{ij}^3 r_{ik}} + \frac{1}{r_{ij}^3 r_{jk}} + \frac{4}{r_{ij}r_{jk}^3} - \frac{7}{2}\frac{1}{r_{ik}r_{jk}^3} - \frac{1}{2}\frac{1}{r_{jk}^3 r_{ij}^3}(\mathbf{r}_{ij}\mathbf{r}_{ik})\right] \\ B_{ij} & = & \frac{1}{r_{ij}^3}\left(4\mathbf{r}_{ij}\dot{\mathbf{r}}_{ij} + \mathbf{r}_{ij}\dot{\mathbf{x}}_j\right) \end{array}\right. \tag{5}$$

To facilitate the integration of these equations, we introduce for a body the perturbative complex variable p and the real variable w defined by the relations:

$$\left\{\begin{array}{l} x^1 + ix^2 = a(1-p)e^{i\lambda} \\ x^3 = aw \end{array}\right. \tag{6}$$

where a is the semi-major axis, $\lambda = nt + \varepsilon$ the mean mean longitude, n being the mean motion, x^k the rectangular coordinates of the body and $i^2 = -1$.

Then, if we call X, Y and Z the components of the sum of the Newtonian and the relativistic perturbations, we can reformulate equations (4) for each body at hand on the form :

$$\left\{\begin{array}{l} \ddot{p} + 2in\dot{p} - \frac{3}{2}(p+\bar{p}) = n^2 P \\ \ddot{w} + n^2 w = n^2 W \end{array}\right. \tag{7}$$

with :

$$\begin{cases} P &= -\frac{3}{2}(p+\bar{p}) + (\frac{a^3}{r^3}-1)(1-p) - \frac{1}{n^2a}(X+iY)e^{-i\lambda} \\ W &= -(\frac{a^3}{r^3}-1)w + \frac{1}{n^2a}Z \end{cases} \tag{8}$$

and $r^2 = (x^1)^2 + (x^2)^2 + (x^3)^2$

Knowing P and W, the system (7) admits the solution:

$$\begin{aligned} p = \quad & Ae^{i\lambda} - 3\bar{A}e^{-i\lambda} + inB + 3inCt - 2C \\ & + \frac{3}{4}ine^{-i\lambda}\int(3P-\bar{P})e^{i\lambda}dt + \frac{1}{4}ine^{i\lambda}\int(3\bar{P}-P)e^{-i\lambda}dt \\ & - 2in\int Pdt - \frac{3}{2}n^2\int\int(P-\bar{P})dtdt \\ w = \quad & De^{i\lambda} + \bar{D}e^{-i\lambda} + \frac{1}{2}ine^{-i\lambda}\int We^{i\lambda}dt - \frac{1}{2}ine^{i\lambda}\int We^{-i\lambda}dt \end{aligned} \tag{9}$$

where A and D are complex integration constants, B and C real ones.

The Keplerian part of P is of second order in eccentricity and inclination and we use an iterative scheme to integrate equations (7). This means that we introduce a barycentric solution of the motion of the solar system bodies in the right-hand side of (8) and after computing p and w with (9), we derive a new solution with the help of (6).

As it has been previously emphasized, the analytical solution VSOP87 (Bretagnon, Francou, 1988) is expressed in terms of Fourier and Poisson's series of thousands years of Dynamical Barycentric Time (TDB=T) reckoned from J2000(TDB). For example, a rectangular coordinate of a body looks like :

$$x(T) = \sum_{\alpha=0}^{5} T^\alpha \sum_j \left[a_{\alpha j}\sin\left(\sum_{k=1}^{12} c_{\alpha jk}\lambda_k\right) + b_{\alpha j}\cos\left(\sum_{k=1}^{12} c_{\alpha jk}\lambda_k\right)\right] \tag{10}$$

where $c_{\alpha jk}$ belongs to $\mathbf{Z}$, the barycentric amplitudes $T^\alpha a_{\alpha j}$ and $T^\alpha b_{\alpha j}$ are expressed in "barycentric" astronomical unit and T is expressed in thousand of Julian years (TJY). The subscript k refers to the argument k of the table 1. For $k = 9, 10$ and 11, λ_k are Delaunay's arguments of the Moon and λ_{12} represents the mean longitude of the Moon ζ reckoned from equinox of date.

In the expressions (6 - 10) the mean mean longitude λ_k is equal to $n_k t + \varepsilon_k$ where the mean motion n_k and the phase ε_k at the epoch J2000(TDB), are expressed using the respective units : $radian$ and $radian/TJY$.

We present hereafter the links between the astronomical time scales which will be useful for the substitution of TCB in the ephemerides.

4. Links Between Time Scales

Mainly, the time scales TCB, TCG, TDB and TDT have their common origin at the epoch 1977 January 1, $0^h0^{mn}32^s.184$ TDB exactly, the unit of measurement of them being the SI second.

TABLE I
Initial phases and mean motions. TCB quantities are referred by a hat.

k	body	ε_k	n_k	$\hat{\varepsilon}_k$	$\hat{n}_k$
1	Mercury	4.402 608 842 40	26087.903 141 574 2	4.402 599 532 94	26087.902 736 815 2
2	Venus	3.176 146 696 89	10213.285 546 211 0	3.176 143 052 28	10213.285 387 749 9
3	Earth	1.753 470 459 53	6283.075 849 991 4	1.753 468 217 42	6283.075 752 508 2
4	Mars	6.203 476 112 91	3340.612 426 699 8	6.203 474 920 81	3340.612 374 869 5
5	Jupiter	0.599 546 697 39	529.690 965 094 6	0.599 546 308 37	529.690 956 876 3
6	Saturn	0.874 016 756 50	213.299 095 438 0	0.874 016 680 38	213.299 092 128 6
7	Uranus	5.481 293 871 59	74.781 598 567 3	5.481 293 844 90	74.781 597 407 0
8	Neptune	5.311 886 286 76	38.133 035 637 8	5.311 886 273 15	38.133 035 046 2
9	Delaunay's	5.198 466 741 03	77713.771 468 120 5	5.198 439 008 91	77713.770 262 376 0
10	arguments	1.627 905 233 37	84334.661 581 308 3	1.627 875 138 58	84334.660 272 839 3
11	of the Moon	2.355 555 898 27	83286.914 269 553 6	2.355 526 177 37	83286.912 977 340 6
12	Moon	3.810 344 546 97	83997.091 135 595 4	3.810 314 572 65	83997.089 832 363 9

TABLE II
Relationships between time scales. In the TDB-TDT relation, P is the periodic part. JD is the Julian date.

TT=TDT=TAI+32^s.184

TDB=TDT+P

TCG=TT+L$_G$(JD-2443144.5)$\times$86400

$$\text{dTCG/dTCB}=1+\tfrac{1}{2}c^{-2}h_{00}^{(2)}-\tfrac{1}{2}\tfrac{v^2}{c^2}-\tfrac{1}{8}\tfrac{v^4}{c^4}+\tfrac{3}{4}c^{-2}h_{00}^{(2)}\tfrac{v^2}{c^2}+\tfrac{1}{2}c^{-4}h_{00}^{(4)}-\tfrac{1}{8}(h_{00}^{(2)})^2+h_{0i}^{(3)}\tfrac{v^i}{c}$$

TCB=TDB+L$_B$(JD-2443144.5)$\times$86400

The TDB and TDT time scales have been introduced in 1976 by the IAU with the constraint that they differ only by periodic variations. The link between the two coordinate time TCG and TCB is found by matching the barycentric and the geocentric metric in the vicinity of the Earth using the tensor relation :

$$\hat{g}_{\alpha\beta}(w^\gamma)=\frac{\partial x^\mu}{\partial w^\alpha}\frac{\partial x^\nu}{\partial w^\beta}g_{\mu\nu}(x^\alpha) \tag{11}$$

where $g_{\alpha\beta}$ is the metric associated to the barycentric reference system (x^α) and $\hat{g}_{\alpha\beta}$ the metric associated to the geocentric one (w^α).

The mathematical expressions of those links are given at the fourth order in $1/c$ in table 2 in which $\mathbf{v}=d\mathbf{x}/dt$ is the barycentric coordinate velocity of the point where the comparison is performed. In this table, the value of the constant L_C is given below as the coefficient of the secular term in (23), L_G can be found in (Fukushima 1994) and we have $L_B=L_C+L_G$.

Once again, one should pay attention to integrate the differential relation between TCB and TCG given in table 2. Indeed, in this relation, the metric tensor components $g_{\alpha\beta}$ are expressed using barycentric coordinates $(x^\alpha) = (c\,TCB, x^1, x^2, x^3)$ where c is the speed of light. Since its right-hand member is computed using ephemeris expressed in TDB, this expression takes the form :

$$\int dTCG = \int f(TDB)\,dt$$

with t =TCB, which is obviously not consistent.

5. Substitution of TCB in the Ephemeris

5.1. Modification of VSOP87

We present here the modifications induced by the substitution of the new time scale TCB in an expression like (10). Let's write without loss of generality for our purpose the rectangular coordinates of a body under the simplified form :

$$\begin{aligned}\Phi_\alpha(T) &= (T-T_0)^\alpha \left[a_\alpha \sin\left(\textstyle\sum_k c_{\alpha k}\left[n_k(T-T_0)+\varepsilon_k\right)\right]\right. \\ &\qquad \left.+ b_\alpha \cos\left(\textstyle\sum_k c_{\alpha k}\left[n_k(T-T_0)+\varepsilon_k\right)\right]\right]\end{aligned} \tag{12}$$

where T represents TDB and the coefficients $(T-T_0)^\alpha a_\alpha$ and $(T-T_0)^\alpha b_\alpha$ are homogenous to lenghts.

Based on (Fukushima, 1986) and (Hellings, 1986), the changing from TDB to TCB in the equations of motion should be accompanied by the transformations :

$$\begin{aligned} t - t_0 &= \frac{1}{1-L_B}(T-T_0) + \Delta t \\ \hat{a}_\alpha &= \left(\frac{1}{1-L_B}\right)^{1-\alpha} a_\alpha \\ \hat{b}_\alpha &= \left(\frac{1}{1-L_B}\right)^{1-\alpha} b_\alpha \\ \hat{n}_k &= (1-L_B)n_k \\ (\widehat{GM}) &= \frac{1}{1-L_B} GM \end{aligned} \tag{13}$$

in which the quantities with a hat refer to the barycentric system of reference with the time scale TCB = t, t_0 =J2000(TCB), T_0=J2000(TDB) and Δt is the difference J2000(TDB)-J2000(TCB) expressed in SI seconds.

Concerning the mean elements of the planets, the difference in mean longitude due to Δt is absorbed in the phase at the epoch so that we have :

$$\hat{\varepsilon}_k = \varepsilon_k - \hat{n}_k \Delta t \tag{14}$$

The transformations (13) are equivalent to a conformal change of the metric g from the reference system $(c\,TDB, x^1, x^2, x^3)$ to the new one $(c\,TCB, \hat{x}^1, \hat{x}^2, \hat{x}^3)$ on the form :

$$\hat{g} = e^\gamma g \tag{15}$$

with $\gamma = 2\ ln\left(\frac{1}{1-L_B}\right)$. This fact insures that the angles are respected by the transformation.

Substituting (13) and (14) in (12) and writing $\hat{\Phi}_\alpha$ the argument (12) in TCB, we have :

$$\hat{\Phi}_0 = \left[\frac{1}{1-L_B}\right]\Phi_0 \tag{16}$$

and for $\alpha \geq 1$:

$$\hat{\Phi}_\alpha = \left[\frac{1}{1-L_B}\right]\Phi_\alpha + \alpha\Delta t\frac{\Phi_\alpha}{T-T_0} \tag{17}$$

Using (16) and (17), we transformed the solutions VSOP.

5.2. Influence of the Epoch of Reference on the Reference Frame

The recommendation 7 of the resolution A4 of the IAU (IAU, 1991) specifies that the new conventional celestial reference frame shall be as near as possible to the mean equator of J2000 and the origin of this principal plane be as near as possible to the dynamical equinox of J2000 so that the epoch J2000 in TDB should be kept as the reference of the time scale used in the ephemerides. In doing so, the continuity with the catalogue FK5 is respected.

For our purpose we make the use of J2000 (TCB) as the epoch of reference. Using the value $L_C = 1.480\,826\,858\,501 \times 10^{-8}$ (see eq. 23), correcting the one given in (Fukushima, 1994) (opposite sign for the post-Newtonian corrections), and the value of the potential on the geoid (Bursa, 1992), we have :

$$L_B = L_C + L_G = 1.5505197604 \times 10^{-8} \tag{18}$$

which leads to the value :

$$\Delta t = 11^s.25405695 \tag{19}$$

We give in table 3 the corrections to the precession quantities, the complete expression of which can be found in (Simon et al., 1994), induced by the difference Δt.

The amplitudes of corrections are smaller than the precision at which those quantities have been computed. Therefore, this choice does not affect significantly the determination of the reference frame J2000 (TCB) which could at this accuracy be identically mixed up with J2000 (TDB).

TABLE III
Corrections to precession quantities

Quantity	Value at J2000(TDB)	Correction
P_A	0	$1''.8 \times 10^{-5}$
ε_A	23°26′21″.412	$1''.7 \times 10^{-7}$
π_A	0	$1''.7 \times 10^{-7}$
Π_A	174°52′23″.433	$3''.1 \times 10^{-6}$
θ_A	0	$7''.1 \times 10^{-6}$
ζ_A	0	$8''.2 \times 10^{-6}$
z_A	0	$8''.2 \times 10^{-6}$
ω_A	23°26′21″.412	$\leq 10^{-18}$
ψ_A	0	$1''.8 \times 10^{-5}$
χ_A	0	$3''.8 \times 10^{-8}$

6. Applications

6.1. Difference between planetocentric and barycentric coordinate time scales

Let $u = TCP$ and $t = TCB$ be a planetocentric and barycentric coordinate time respectively. The proper time of an observer being the same in both barycentric and planetocentric reference system, we have :

$$\frac{du}{dt} = 1 + \frac{1}{2}c^{-2}h_{00}^{(2)} - \frac{1}{2}\frac{v^2}{c^2} - \frac{1}{8}\frac{v^4}{c^4} + \frac{3}{4}c^{-2}h_{00}^{(2)}\frac{v^2}{c^2} + \frac{1}{2}c^{-4}h_{00}^{(4)} - \frac{1}{8}(h_{00}^{(2)})^2 + c^{-3}h_{0i}^{(3)}\frac{v^i}{c} \tag{20}$$

where v^i are the components of the barycentric velocity of the observer.

Using the ephemeris in TCB, we compute and integrate this relation for Mercury, Venus, the Earth, Mars and the Moon.

The barycentric quantity $\widehat{GM}$ used in deriving these solutions are refered in table 4 and 5.

6.1.1. *Difference TCB-TCG*

We write as usual :

$$TCB - TCG = L_C t + \alpha t^2 + \cdots + P \tag{21}$$

where P is the sum of the periodic and the Poisson's parts of this difference. In the contributions of order c^{-4} to L_C, only the Sun's potential has been taken into

TABLE IV
VSOP's (GM) and barycentric $(\widehat{GM})$

Body	Mercury	Venus	Earth	Mars
GM	6.553709	96.63318	118.5685621	12.73979
$\widehat{GM}$	6.553710	96.63319	118.5685639	12.73979

TABLE V
VSOP's (GM) and barycentric $(\widehat{GM})$ (continued)

Body	Jupiter	Saturn	Uranus	Neptune	Moon
GM	37692.02	11283	1726.2	2043.9	1.4583956848
$\widehat{GM}$	37692.02	11283	1726.2	2043.9	1.4583957075

TABLE VI
Contributions to L_C induced by the mixed components h_{0i}. Unit is 10^{-20}

Body	Mercury	Venus	Mars	Jupiter	Saturn	Uranus	Neptune	Moon
δ	0.004	0.1	0.005	0.6	0.04	0.001	0.0004	1

account in the metric components. In doing so, we have neglected additive terms smaller than 10^{-19}.

Moreover, let us write :

$$\delta = \int c^{-3} h_{0i}^{(3)} \frac{v^i}{c}\, dt = 4c^{-4} \int \sum_{A \neq E} \frac{GM_A}{r_A} v_A^i\, v_E^i\, dt \tag{22}$$

where E denotes the Earth. The contribution of the planet A to the secular part of δ is reported in table 6. Those additive terms to L_C have been neglected too.

In the following expressions (25-27), t is expressed in thousand of Julian years and the last number is the period of the term in years. The amplitudes of the periodic terms are in microsecond.

We found :

$$\begin{aligned} TCB - TCG = {} & 1.480\,826\,858\,5 \times 10^{-8}\, t - 1.3 \times 10^{-17}\, t^2 \\ & +1656.674608120 \quad \sin(\ 6283.075752571\, t + 6.240052) \quad 1.0000 \\ & +22.417472345 \quad \sin(\ 5753.384795689\, t + 4.296975) \quad 1.0921 \\ & +13.839792612 \quad \sin(12566.151505142\, t + 6.196900) \quad .5000 \\ & +102.156727748\, t \quad \sin(\ 6283.075752571\, t + 4.249030) \quad 1.0000 \\ & +4.322988461\, t^2 \quad \sin(\ 6283.075752571\, t + 2.642891) \quad 1.0000 \\ & + \cdots \end{aligned} \tag{23}$$

In comparing this expression to (Brumberg et al., 1993), we notice three differences brought about :

- the contribution of the post-newtonian part (fourth order terms)

- the influence of Δt on the periodic and Poisson's parts
- the modification of the mean elements (13).

6.1.2. Difference between planetocentric (TCP) and barycentric coordinate times (TCB) at the planetocenter.

Moon

$$
\begin{array}{rlr}
TCB - TCP = 1.482\,539\,1330\,79 \times 10^{-8}\,t - 1.308 \times 10^{-17}\,t^2 & & \\
+1656.662173966 & \sin(\ 6283.075752571\,t + 6.240052) & 1.0000 \\
+126.419201864 & \sin(77713.770263153\,t + 2.056846) & .0808 \\
+13.839631785 & \sin(12566.151505142\,t + 6.196900) & .5000 \\
+102.155971231\,t & \sin(\ 6283.075752571\,t + 4.249030) & 1.0000 \\
+4.322946679\,t^2 & \sin(\ 6283.075752571\,t + 2.642892) & 1.0000 \\
& +\cdots &
\end{array}
\qquad (24)
$$

Mercury

$$
\begin{array}{rlr}
TCB - TCP = 3.825\,071\,637\,763 \times 10^{-8}\,t - 2.788 \times 10^{-17}\,t^2 & & \\
+12618.597594306 & \sin(\ 26087.902737076\,t + 3.050735) & .2408 \\
+1285.989592271 & \sin(\ 52175.805474152\,t + 6.101471) & .1204 \\
+196.411594751 & \sin(\ 78263.708211228\,t + 2.869021) & .0803 \\
+35.540757657 & \sin(104351.610948304\,t + 5.919756) & .0602 \\
+7.064149246 & \sin(130439.513685381\,t + 2.687307) & .0482 \\
+7.918032234 & \sin(\ 25558.211780194\,t + \ .668699) & .2458 \\
+4.601403875 & \sin(\quad 529.690956882\,t + \ .839286) & 11.8618 \\
+350.095887966\,t & \sin(\ 26087.902737076\,t + 1.515336) & .2408 \\
+71.357756883\,t & \sin(\ 52175.805474152\,t + 4.565944) & .1204 \\
+16.347851393\,t & \sin(\ 78263.708211228\,t + 1.333430) & .0803 \\
+3.944177674\,t & \sin(104351.610948304\,t + 4.384126) & .0602 \\
+5.025566728\,t^2 & \sin(\ 26087.902737076\,t + 6.225361) & .2408 \\
+2.013743027\,t^2 & \sin(\ 52175.805474152\,t + 3.011223) & .1204 \\
& +\cdots &
\end{array}
\qquad (25)
$$

Venus

$$
\begin{array}{rlr}
TCB - TCP = 2.047\,125\,757\,550 \times 10^{-8}\,t + 0.151 \times 10^{-17}\,t^2 & & \\
+571.059373337 & \sin(10213.285387852\,t + \ .879913) & .6152 \\
+15.578748651 & \sin(\ 9683.594430970\,t + 5.718814) & .6488 \\
+5.066410945 & \sin(\quad 529.690956882\,t + \ .375095) & 11.8618 \\
+3.348808904 & \sin(\ 9999.986295721\,t + 5.444595) & .6283 \\
+40.282679852\,t & \sin(10213.285387852\,t + 4.033572) & .6152 \\
& +\cdots &
\end{array}
\qquad (26)
$$

Mars

$$
\begin{array}{rlr}
\multicolumn{3}{l}{TCB - TCP = 0.971\,938\,157\,234 \times 10^{-8}\, t + 0.626 \times 10^{-17}\, t^2} \\
+11419.440440994 & \sin(\ 3340.612374903\, t + \ .338122) & 1.8808 \\
+532.340051144 & \sin(\ 6681.224749806\, t + \ .676243) & .9404 \\
+37.216327851 & \sin(10021.837124709\, t + 1.014363) & .6269 \\
+38.748142709 & \sin(\ 2810.921418021\, t + 2.454494) & 2.2352 \\
+11.638963032 & \sin(\ \ -3.523118294\, t + 1.960065) & 1783.3848 \\
+891.587590716\, t & \sin(\ 3340.612374903\, t + 5.174636) & 1.8808 \\
+83.124620189\, t & \sin(\ 6681.224749806\, t + 5.512671) & .9404 \\
+8.717062721\, t & \sin(10021.837124709\, t + 5.850819) & .6269 \\
+35.621379210\, t^2 & \sin(\ 3340.612374903\, t + 3.621946) & 1.8808 \\
+6.557148235\, t^2 & \sin(\ 6681.224749806\, t + 4.012377) & .9404 \\
 & +\cdots &
\end{array} \tag{27}
$$

Solutions are available at the nanosecond accuracy.

6.2. Effect of Time Scales on Angular Velocity

It has been emphasized by Hellings (Hellings, 1986) that the spin rate of a planet is constant (over short period of time) only if we consider it as a function of the planet's proper time.

In planetary spacecraft ranging, an electromagnetic signal is sent from Earth to a spacecraft where the signal is actively transponded to a planetary site (lander) located on the planet's surface.

Let us suppose that the planetary site is located on the edge of the planetary disk as seen by an Earth-based observer. Thus, if we forget about physical relevancy and use a TCB model, we introduce an error in the orientation of the planet whose amplitude could be roughly estimated as (Klioneer, Private communication):

$$
\varepsilon = Max\ |TCB - TCP|\ \frac{2\pi L}{T} \tag{28}
$$

where L is the radius of the body and T its period of rotation.

In computing ε, we only assumed the periodic part of the difference between coordinate time. The results are reported in the table 7 where the values of T and L are those given in (Bertotti, 1990). The effect is negligible for Venus and Mercury but comparable with current accuracy of observation for the Moon and Mars.

7. Conclusion

In this paper, we have presented a barycentric treatment of the equations of motion of the solar system bodies in the field of the general relativity theory based on the barycentric metric of (Brumberg 1991). In integrating these equations, we expect to obtain solutions with an accuracy ten to one hundred times better than the present VSOP's solutions.

TABLE VII
Effect of the non-constant spin rate

Body	T (*days*)	L (*km*)	ε (*cm*)
Mercury	58.650	2439.0	3.81
Venus	243.000	6052.0	0.10
Mars	1.026	3397.2	275.00
Moon	27.320	1738.0	0.77

The VSOP87 time-like argument has been converted to the Barycentric Coordinate Time including the choice of the epoch of reference J2000 (TCB) as reference epoch. The influence of this choice has been calculated and the conclusion is that this convention implies no significant difference on the precession quantities.

As a first result of this procedure, a coherent expression of the difference between TCB and TCP for the telluric planets has been computed, as the effect on the planetary spacecraft ranging induced by the variation of the spin rate of a planet as a function of the barycentric coordinate time.

References

Bertotti, B., Farinella, P.: 1990, *Physics of the Earth and the Solar System*, Kluwer academic publishers.
Bretagnon, P., Francou, G.: 1988, Planetary theories in rectangular and spherical variables. VSOP87 solutions,*Astro.&Astrophy.*, **202**, 309
Brumberg, V.-A., Kopejkin, S.-M.: 1990, Time scales in the solar system,*Celest. Mech*, **48**, 23
Brumberg, V.-A., Bretagnon, P., Francou, G.: 1991, Analytical algorithms of relativistic reduction of astronomical constant, in *Journées 1991 (ed. N. Capitaine)*, 141
Brumberg, V.-A.: 1991, *Essential Relativistic Celestial Mechanics*, Adam Hilger.
Brumberg, V.-A., Bretagnon, P., Francou, G.: 1993, Analytical relativistic transformations between time reference systems, *Astro.&Astrophy.*, **275**, 651
Bursa, M., Sima, Z., Kostelecky, J.: 1992, *Studia geoph. et geod.*, **36**, 101
Fairhead, L., Bretagnon, P.: 1990, An analytical formula for the time transformation TB-TT, *Astro.&Astrophy.*, **229**, 240
Fock, V. A.: 1955, The Theory of Space, Time and Gravitation, *Moscow: State technical Publications*, in Russian.
Fukushima, T., Fujimoto, M.-K., Kinoshita, H. and Aoki, S.: 1986, A System of Astronomical Constants in the Relativistic Framework, *Celest. Mech.*, **38**, 215
Fukushima, T.: 1994, Time ephemeris, *Astro.&Astrophy.*, **294**, 895
Guinot B., Seidelmann P.-K.: 1988, Time scales : their history, definition and interpretation, *Astro.&Astrophy.*, **194**, 304
Hellings, R. W.: 1986, Relativistic effects in astronomical timing measurements, *Astro. J.*, **91**, 650
Proceedings of the twenty-first general assembly, 1991 Kluwer academic publishers, Resolution A4, p. 52
Moyer, T.: 1981, Transformation from proper time on earth to coordinate time in solar system barycentric space-time frame of reference,*Celest. Mech.*, **23**, 33
Seidelmann P.-K., Fukusima, T.:1992, Why new time scales ?,*Astro.&Astrophy.*, **265**, 833
Simon J.-L., Bretagnon, P., Chapront, J., Chapront-Touzé, M., Francou, G., Laskar, J.: 1994, Numerical expressions for precession formulae and mean elements for the Moon and the planets, *Astro.&Astrophy.*, **282**, 663

ORBITAL STABILITY OF PLANETARY QUASI-SATELLITES

SEPPO MIKKOLA * AND KIMMO INNANEN †

*Turku University Observatory, Tuorla, 21500 Piikkiö, Finland.
†Department of Physics and Astronomy, York University,
4700 Keele Street, North York, Ontario, M3J 1P3, Canada.*

Abstract. An object having the same mean motion and mean longitude, but different eccentricity than a planet, appears to circle the planet like a retrograde moon. If the distance is large enough such that the object is not gravitationally bound to the planet, this type of motion is another kind of 1:1 resonant trajectory, in addition to the Trojan-type orbits. We have studied the stability of such a motion both numerically and analytically, and have found adiabatic invariants which govern this type of motion. As long as the orbits are almost coplanar, stability seems to persist in the restricted problem. This applies both to the circular and eccentric restricted three-body problems. Numerical computations in the actual Solar System suggest that at least the Earth and Venus could have such satellites at roughly the distance of 0.1 AU. In the case of Mars, the motions are unstable and lead to escape from such an orbit in a time-scale of few million years.

Key words: stability – satellites – asteroids

1. Introduction

Apart from satellites and rings orbiting relatively nearby to planets (*i.e.* within the so-called planetary spheres of influence) or Lagrange-point dynamics, there has not been much progress in finding other examples of stable, long-term motion in the presence of the Sun, one or more planets, and a test particle. During the course of our own recent numerical experiments with randomly placed asteroid-like particles between the planets in a self-consistent simulation of the Solar System (Mikkola and Innanen, 1995), we noted with interest the evident, but unexpected stability of one test particle. That somewhat serendipitous orbit triggered a much more careful investigation, much of which constitutes the main body of this paper. Actually these orbits are extensions of the well known (e.g. Szebehely 1967, Stumpff 1965, Hénon and Guyot 1970) retrograde periodic orbits in the circular restricted problem of three bodies. Here we do not restrict the results to periodic orbits and the stability study is extended to the actual Solar System. For the Earth, a region of long-term stability occurs at a distance of the order of 0.1 AU, or some 40 times the Earth-Moon separation. Whether or not there exist natural objects at this distance, interesting reasons may exist for man-made stations in such orbits. In fact a similar trajectory was planned by Kogan (1988) for the former Soviet Union's space-mission [1] to the neighborhood of the Martian moon Phobos (Mars, Phobos and the space probe being the primary, secondary and particle, respectively).

2. Theory of the Motion of the Quasi-Satellites

To understand the motion an, at least qualitatively correct, analytical model is helpful. We use the method of averaging to construct a such a theory. As a guide

[1] Neishtadt, A. 1996, private communication.

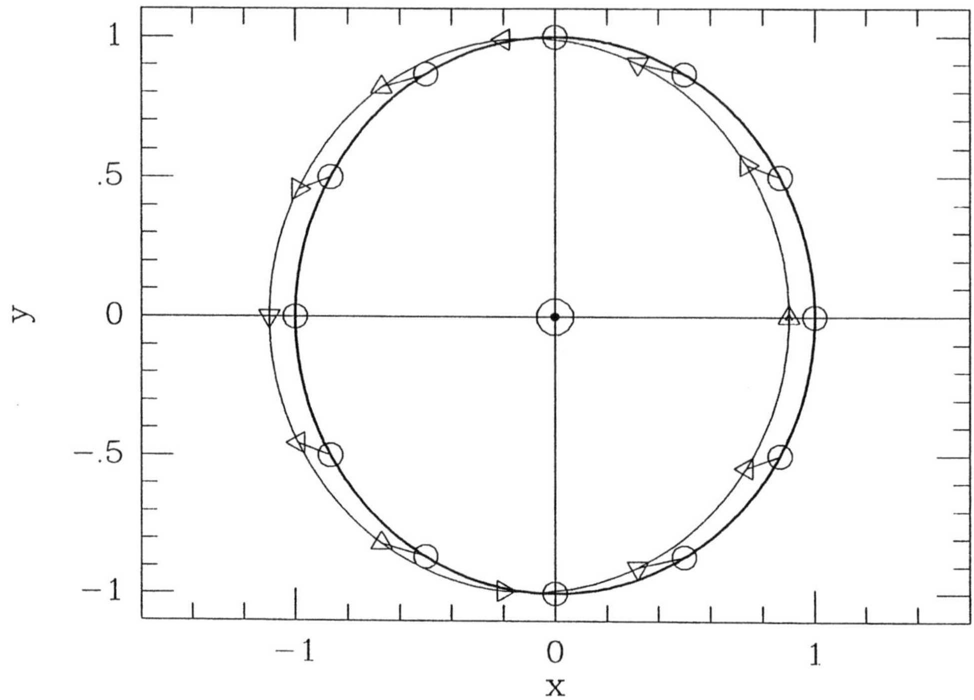

Fig. 1. The heliocentric motions of a planet (circular orbit) and an object having the same mean motion and mean longitude. Note that the particle always stays on the same side of the planet. The apparent position of the object, as viewed from the planet, is around the object's perihelion direction.

for the developments we use numerical computations. This is necessary for the confirmation of the validity of the approximations used, because the system is in fact non-integrable.

We use the standard notations for the variables of elliptic motion: a, e, i, ω , Ω, M, for the semi-major axis, eccentricity, inclination, argument of perihelion, longitude of the ascending node and the mean anomaly, respectively. Quantities related to the planet are declared by the subscript 1, e.g. M_1 is planet's mean anomaly, e_1 its eccentricity and m_1 the mass.

Instead of a, ω and M it is convenient to introduce the quantities $\epsilon = (a - a_1)/a_1$, $\tilde{\omega} = \omega + \Omega$, and $\theta = M + \tilde{\omega} - M_1 - \tilde{\omega}_1$, which thus are the relative deviation of the semi-major axis from that of the planet, the mean longitude of the perihelion and the difference of mean longitudes of the particle and the planet, respectively.

2.1. Quasi-Satellites in the Restricted Problem

To gain insight into the circumstances in question, we consider first the circular restricted problem and assume that the planet's mass is very small so that during a few periods the particle essentially moves in an unperturbed ellipse around the Sun. Suppose the planet has a circular orbit about the Sun. If a minor body has the *same orbit plane, same semi-major axis and the same mean longitude*, then for the differences of coordinates in the rotating system we have

$$dx_o = -ea_1 \cos(M) + .., \qquad dy_o = 2ea_1 \sin(M) + .., \tag{1}$$

to first order in eccentricity. We note that the apparent motion in the rotating coordinate system is along an ellipse with an axis ratio = 1 : 2 and it occurs in the retrograde direction circling the planet once in a planetary period. In this example the planet is at the center of the ellipse, which is not generally the case, but the center oscillates back and forth in the planet's neighborhood. In the non-rotating planetocentric system, however, the particle moves around the point $(-\frac{3}{2}ea,\ 0)$ (*i.e.* in the direction of the aphelion of the particle's heliocentric orbit) along a circle of radius $\frac{1}{2}ae$, completing this cycle twice in a planetary period.

The motions in heliocentric coordinates are illustrated in Fig. 1. A description of motion like this is a good approximation to the actual motion (for a few planetary periods) if the distance of the object from the planet is large enough (*i.e.* the eccentricity of particle's orbit is sufficiently large) and the planet's mass is sufficiently small. This is expected to be the case if planet's mass m_1 and object's eccentricity e satisfy the inequality $(m_1/M_\odot)^{1/3} << e$.

In the more general case in which the planet also has an eccentric orbit, the motions are more complicated. Let the coordinate system be chosen such that the planet's orbital plane is in the xy-plane and the x-axis is directed towards the perihelion of the planet. Then the mean longitude of the planet is equal to its mean anomaly M_1, while for the minor body this angle is $M + \tilde{\omega}$. By assumption these angles are near to each other. Thus it is convenient to use the angle $\theta = M + \tilde{\omega} - M_1$. In the development of the potential we take the elements $e,\ e_1,\ \theta$ and the (mutual) inclination i to be small quantities of the same order of magnitude and the expansions in terms of these are restricted to the leading order only. The relative variation of the semi-major axis is much smaller than these, a fact that can be used to simplify the analytic theory.

The heliocentric Hamiltonian may be written $H = -\mu/2a - U$, where $\mu = M_\odot + m_1$ and $M_\odot$ and m_1 are the masses of the Sun and planet respectively, a is the heliocentric semi-major axis of the particle, and

$$U = m_1(1/|\mathbf{r} - \mathbf{r}_1| - \mathbf{r} \cdot \mathbf{r}_1/r_1^3) \tag{2}$$

is the disturbing function due to the attraction of the planet.

If we think in terms of the Delaunay elements, and introduce the angle θ as a new canonical variable, then the Hamiltonian may be written $H = -\mu/2a - n_1\sqrt{\mu a} - U$, where $n_1 = \sqrt{\mu/a_1^3}$ is the planet's mean motion. After expanding in powers of ϵ (dropping the constant $-\frac{3}{2}\mu$), and averaging U over the fast variable M_1 one obtains the expression (Mikkola & Innanen 1992)

$$H_{sec} = -\frac{\mu}{a_1}\frac{3}{8}\epsilon^2 - R + .. \tag{3}$$

where $R =< U >$ is the secular perturbing function, which we calculate next. It is convenient to introduce the variables σ and ψ by the substitutions

$$e\cos(\tilde{\omega}) - e_1 = \sigma\cos(\psi), \qquad e\sin(\tilde{\omega}) = \sigma\sin(\psi) \tag{4}$$

which leads to the expression

$$|\mathbf{r} - \mathbf{r}_1|^2 = a_1^2(\frac{5}{2}\sigma^2 + \theta^2 - \frac{3}{2}\sigma^2\cos(2w) + 4\sigma\theta\sin(w) + 4s^2\sin(\psi - \Omega + w)^2) \tag{5}$$

where $w = M_1 - \psi$ and $s = \sin(i/2)$. After averaging over w and expanding in powers of θ, the secular perturbative function R takes the approximate form

$$R = \frac{m_1}{a_1}(\alpha/\rho + \beta\theta^2/\rho^3 + .. - 1 - \epsilon), \tag{6}$$

where

$$\rho^2 = \sigma^2 + (\xi - \eta\cos(2\psi - 2\Omega))s^2. \tag{7}$$

The values of the numerical factors α, β, ξ and η are given in the next subsection. The usual Lagrange equations for the adopted elements $\epsilon = (a-a_1)/a_1$, e, i, $\tilde{\omega}$, Ω and θ can be written in the form (with $V = \sqrt{1-e^2}$)

$$\begin{aligned}
\dot{\epsilon} &= \frac{2}{na^2}\frac{\partial R}{\partial \theta}\\
\dot{e} &= \frac{V^2-V}{na^2e}\frac{\partial R}{\partial \theta} - \frac{V}{na^2e}\frac{\partial R}{\partial \tilde{\omega}}\\
\dot{i} &= \frac{\cos(i)-1}{na^2V\sin(i)}\frac{\partial R}{\partial \theta} - \frac{1}{na^2V\sin(i)}\frac{\partial R}{\partial \Omega}\\
\dot{\tilde{\omega}} &= \frac{V}{na^2e}\frac{\partial R}{\partial e} + \frac{1-\cos(i)}{na^2V\sin(i)}\frac{\partial R}{\partial i}\\
\dot{\Omega} &= \frac{1}{na^2V\sin(i)}\frac{\partial R}{\partial i}\\
\dot{\theta} &= n - n_1 - \frac{2}{na^2}\frac{\partial R}{\partial \epsilon} + \frac{V-V^2}{na^2e}\frac{\partial R}{\partial e} + \frac{1-\cos(i)}{na^2V\sin(i)}\frac{\partial R}{\partial i}
\end{aligned} \tag{8}$$

Before attempting the solution of these equations we consider numerical results for a quasi-satellite of the Earth. In these calculations the Earth's present orbit was used and all the other planets were neglected. Thus the considered case is an elliptic restricted problem. Results are illustrated in Fig. 2 for 5000 years. We see that the quantities ρ and $\zeta = \sqrt{\sigma^2 + 4s^2}$ conserve their values fairly well and the oscillation of θ is very rapid compared with the variations of the other quantities. Because of this we can first solve for the motion of θ and ϵ; then a further averaging over the (θ, ϵ)-motion can be done.

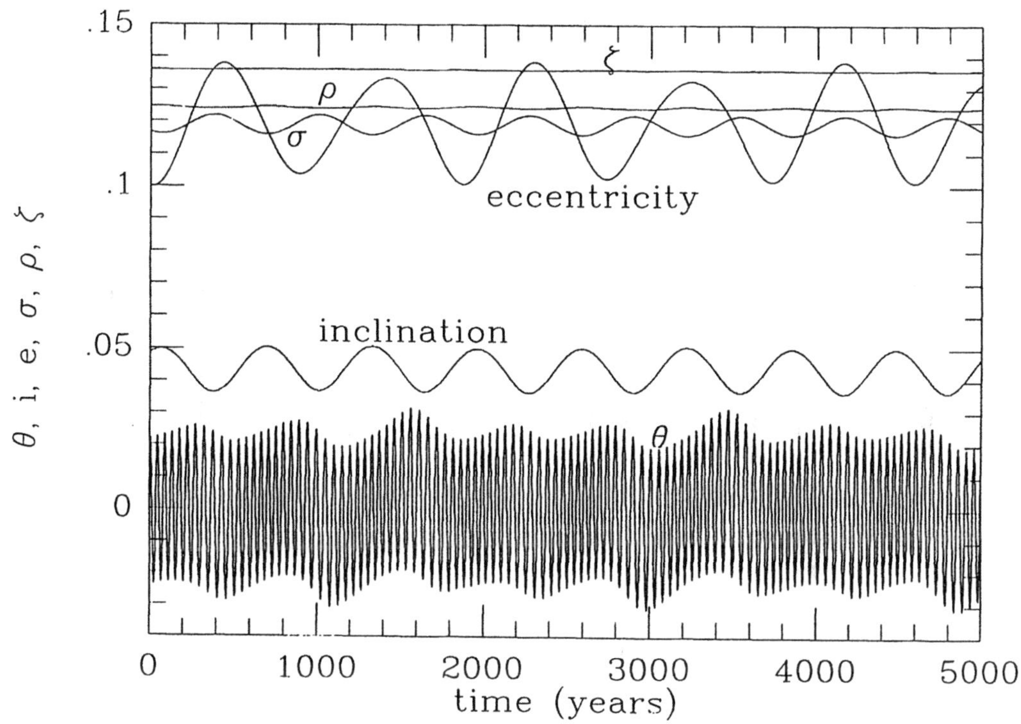

Fig. 2. An example of time evolution of θ, e, i, σ, ζ and ρ in the elliptic restricted problem (Earth in its present orbit + massless particle).

From the first and last equation we obtain to leading order

$$\dot{\epsilon} = \frac{4\beta m_1}{n_1 a_1^2 \rho^3}\theta, \qquad \dot{\theta} = n - n_1 + g = -\frac{3}{2}n_1\epsilon + g + .. \tag{9}$$

where g stands for the terms after $n - n_1$ in the expression for $\dot{\theta}$. Differentiating the above equations once more we get

$$\ddot{\theta} + \frac{6m_1\beta}{a_1^3\rho^3}\theta = \dot{g}. \tag{10}$$

Thus the pair θ and ϵ oscillate with the basic frequency

$$f_\theta = \sqrt{\frac{6m_1\beta}{a_1^3\rho^3}} \tag{11}$$

which is much faster than the fluctuations of g. An approximate solution may be written $\theta = \theta_0 \cos(f_\theta t) + \frac{\dot{\theta}_0}{f_\theta}\sin(f_\theta t) + \frac{\dot{g}}{f_\theta}$ and a similar expression is valid for ϵ, with $\dot{g}$ replaced by $\frac{4m_1\beta}{n_1(a_1\rho)^3}g$. Fig. 3 illustrates the motion of a and θ over a short time. Here the output interval for the data was a half period of the planet, so that the

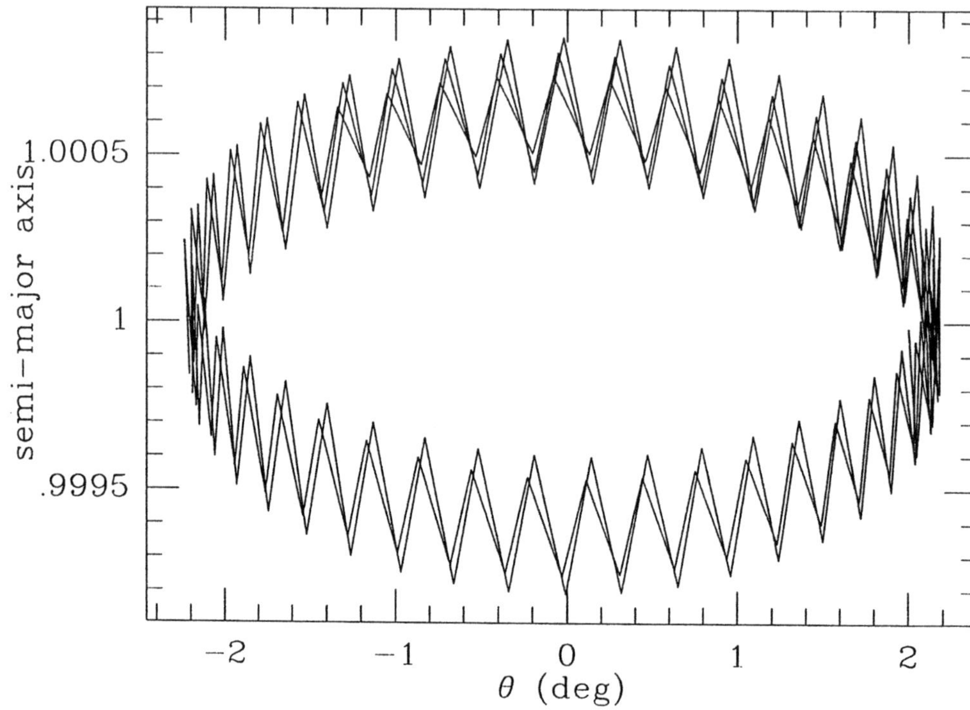

Fig. 3. The behavior of a versus θ. Values after each half planetary period are connected with line to illustrate the short period effect (period=planet's period). The theory averages out this effect resulting to harmonic oscillation for the considered variables.

rapid fluctuation seen here is the short period effect which is averaged out in the theory. To solve for the other variables, we use the secular Hamiltonian averaged further over the oscillation of those variables, *i.e.*

$$H_{sec} = -\frac{\mu}{a_1}\frac{3}{8} < \epsilon^2 > -\frac{m_1}{a_1}(\alpha/\rho + \beta < \theta^2 > /\rho^3 - 1 - \epsilon + ..), \tag{12}$$

which is equivalent to using the perturbing function

$$R = \frac{m_1}{a_1}(\alpha/\rho + \beta < \theta^2 > /\rho^3 - \epsilon + ..). \tag{13}$$

The above Hamiltonian is time independent and thus constant. From this we conclude that the quantity ρ is also a constant (approximate in practice). Similarly $\zeta^2 = \sigma^2 + 4s^2$ is invariant. This we discovered first numerically and an evaluation of its derivative with the above perturbing function indeed gives zero to the order of accuracy of the theory.

These two invariants allow elimination of σ by forming the difference $\zeta^2 - \rho^2 = (4 - \xi + \eta\cos(2\psi - 2\Omega)\ )s^2 =$ constant, which further gives for the inclination the approximation

$$i \approx < i > (1 - \frac{\eta}{8 - 2\xi} \cos(2\psi - 2\Omega)) + .. \approx < i > (1 - .094 \cos(2\psi - 2\Omega)). \quad (14)$$

We conclude that the inclination varies only by $< 10\%$ around its average value. Fig. 2 is obviously in good agreement with this result. With the secular perturbing function R, the Lagrange equations yield for $\psi = \mathrm{atan}(e \sin(\tilde{\omega})/(e \cos(\tilde{\omega}) - e_1))$ the leading order approximation

$$\dot{\psi} = \frac{2}{\sqrt{\mu a_1}} \frac{\partial R}{\partial(\rho^2)} = -\frac{\alpha m_1}{(\sqrt{\mu} a_1^{1.5} \rho^3)} (1 + 3 \frac{< \theta^2 >}{\rho^2}) + .. \quad (15)$$

The equation of motion for the longitude of the ascending node may be written

$$\dot{\Omega} = \frac{1}{2\sqrt{\mu a_1}} \frac{\partial R}{\partial(\rho^2)} (\xi - \eta \cos(2\psi - 2\Omega)) \quad (16)$$

(17)

which gives the secular rate of precession for Ω

$$\dot{\Omega} = \dot{\psi}(1 - \sqrt{(1 - \xi/4)^2 - (\eta/4)^2}) \approx 0.2935\dot{\psi} \quad (18)$$

in good agreement with our numerical integrations.

At this point the analytical problem has been essentially solved although the results are approximate and apply for small inclinations only. As long as this approximate treatment can be considered valid, the motion is stable and quasi-periodic. A more comprehensive study of the stability as a function of inclination and the amplitude of the θ-oscillation can be obtained by numerically evaluating the effective potential for the (θ, ϵ)-motions. As illustrated in Fig. 4, this gives an infinitely high effective potential well for the motion of the angle θ at $\theta = 2\sigma$ in the case of zero inclination, which suggests strong stability for this type of motion in the planar case. Consideration of inclined orbits shows that at $\psi = \Omega$ the effective potential becomes unstable if $s > 0.5\sigma$, which may be written approximately

$$i > \sigma = \sqrt{e_1^2 + e^2 - 2ee_1 \cos(\tilde{\omega})}. \quad (19)$$

If this condition is satisfied, then *instability is guaranteed*, while stability may be possible if the inclination is small and the oscillation of θ also remains sufficiently narrow.

2.2. SUMMARY OF ANALYTICAL RESULTS

Most of the results obtained above are valid in the limit of small inclination only. However, numerical experiments show that they still give a good qualitative description for all stable orbits.

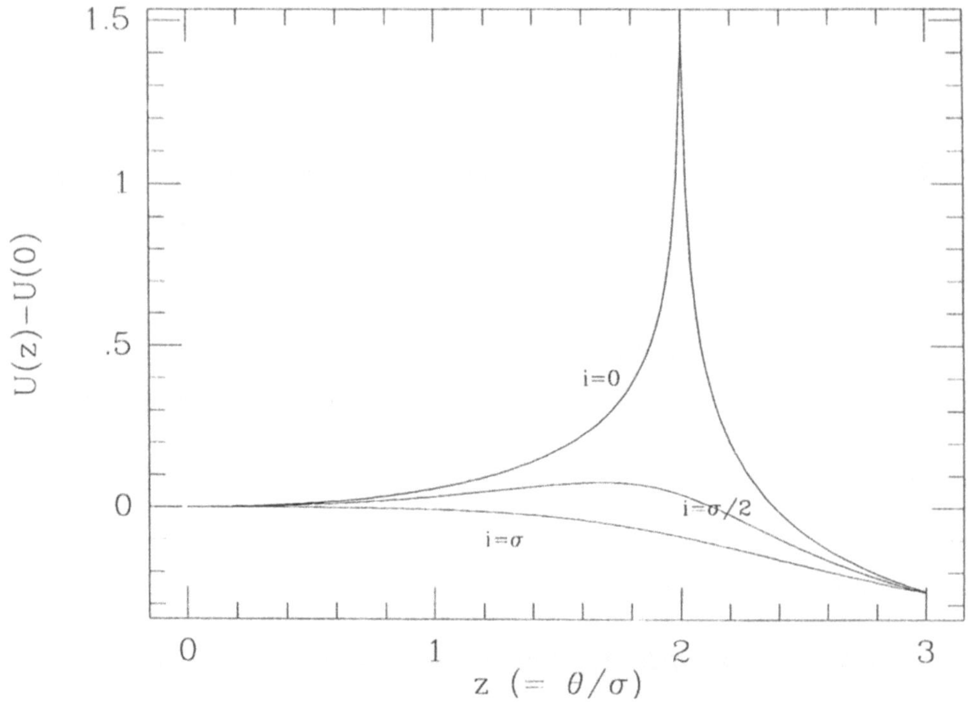

Fig. 4. The effective potential of θ for various values of inclination at $\psi = \Omega$. If $i = 0$ there is an infinite potential barrier at $\theta = 2\sigma$, while for $i > \sigma$ instability is guaranteed.

The most essential feature of these orbits is the (nearly) harmonic oscillation of the semi-major axis and the mean longitude difference (or the pair $(\epsilon,\ \theta)$ in our treatment). In this respect the quasi-satellite orbits resemble motion of the Trojan asteroids. As we demonstrated above (Fig.4), the $(\epsilon,\ \theta)$-oscillation is stable only for small enough inclination: $i < \sigma$ and for large amplitude motions the limit is even smaller. For a typical stable motion with $\sigma \leq 0.1$ this means that $i \lesssim 5^{o}$. The variations of the other orbital parameters are much slower than that of θ and seem not to have much direct effect on the stability as long as the external perturbations do not essentially modify the orbit.

The computational recipe for obtaining the approximate behavior of the orbit according to the above results is as follows: With the numerical constants and notations

$$\alpha \approx 0.68644, \quad \beta \approx 0.050158, \quad \xi \approx 1.12316, \quad \eta \approx 0.5386 \tag{20}$$

$$s = \sin(i/2), \quad \Psi = 2(\psi - \Omega), \quad \sigma^2 = e^2 + e_1^2 - 2ee_1\cos(\tilde{\omega}), \tag{21}$$

we can write the approximate constants of motion in the form

$$\zeta^2 = \sigma^2 + 4s^2, \quad \rho^2 = \sigma^2 + (\xi - \eta\cos(\Psi)\ )s^2 \tag{22}$$

and from the constancy of their difference one derives

$$s^2 = (\zeta^2 - \rho^2)/(4 - \xi + \eta\cos(\Psi)\,), \quad \Psi \approx \Psi_o + 1.4130\dot{\psi}t \tag{23}$$

where (and subsequently) the subscript $_o$ denotes initial values, t is the time and $\dot{\psi} = -\alpha m_1/(\sqrt{\mu}a_1^{1.5}\rho^3)(1 + 3\frac{<\theta^2>}{\rho^2})$, $\Omega \approx \Omega_o + 0.2935\dot{\psi}t$.

Further with $\psi \approx \psi_o + \dot{\psi}t$ one can calculate

$$\sigma^2 = \zeta^2 - 4s^2, \quad e\sin(\tilde{\omega}) = \sigma\sin(\psi), \quad e\cos(\tilde{\omega}) = e_1 + \sigma\cos(\psi) \tag{24}$$

from which e and $\tilde{\omega}$ can be obtained. From these equations it follows that the motion of $\tilde{\omega}$ is circulation if $e_1 < \sigma$, but libration about the mean value 0 if the eccentricity of the planet's orbit is large enough, *i.e.* if $e_1 > \sigma$. We note that the system has essentially three different frequencies

$$f_\theta = \sqrt{\frac{6m_1\beta}{a_1^3\rho^3}}, \quad f_\psi = -\alpha m_1/(\sqrt{\mu}a_1^{1.5}\rho^3), \quad f_\Psi = 1.431 f_\psi \tag{25}$$

related to the $(\epsilon,\ \theta)$ cycle, ψ (and $e,\ \tilde{\omega}$) variation and the inclination fluctuation, respectively. As long as f_θ is much larger than f_ψ (requiring e.g. that m_1 is small enough) the theory may be consider reasonable, otherwise the different periodicities are too close to each other for the averaging over θ-motion to be justified.

3. Numerical Results in the Solar System

Long term numerical simulations (up to 10 Myr) have confirmed the validity of the simple theory and suggest that such distant moons may exist in the real Solar System. We studied hypothetical quasi-satellites of Venus, Earth and Mars. All the major planet were included in our numerical model of the Solar System and the integration was carried out using the symplectic method of Wisdom and Holman (1991) with the small stepsize of only three days.

In Fig. 5 the evolution of the quantity σ is illustrated for the case of nearly equi-planar particle orbits in the neighborhood of the Earth (note that in this case $\sigma = \rho = \zeta$). Results for Venus were very similar, only the conservation of the values of σ was slightly better. However the quasi-satellites of Mars all finally escaped from the neighborhood of that planet, presumably by perturbations due to Jupiter. The slightly perturbed neighborhoods of the planets Earth and Venus look especially promising for real existence of quasi-satellites. An eccentricity value of ~ 0.1 seems most likely for a such a body.

A major result is that the quantity σ (and ζ and ρ for inclined orbits) is essentially constant. Moreover, numerical computations show that this is true even when the planet's eccentricity e_1 varies considerably, *i.e.* σ seems to be an adiabatic invariant in the planar case. The same applies in inclined orbits to the quantities ζ and ρ. During the orbit integrations the Liapunov exponents of the tests particle orbits

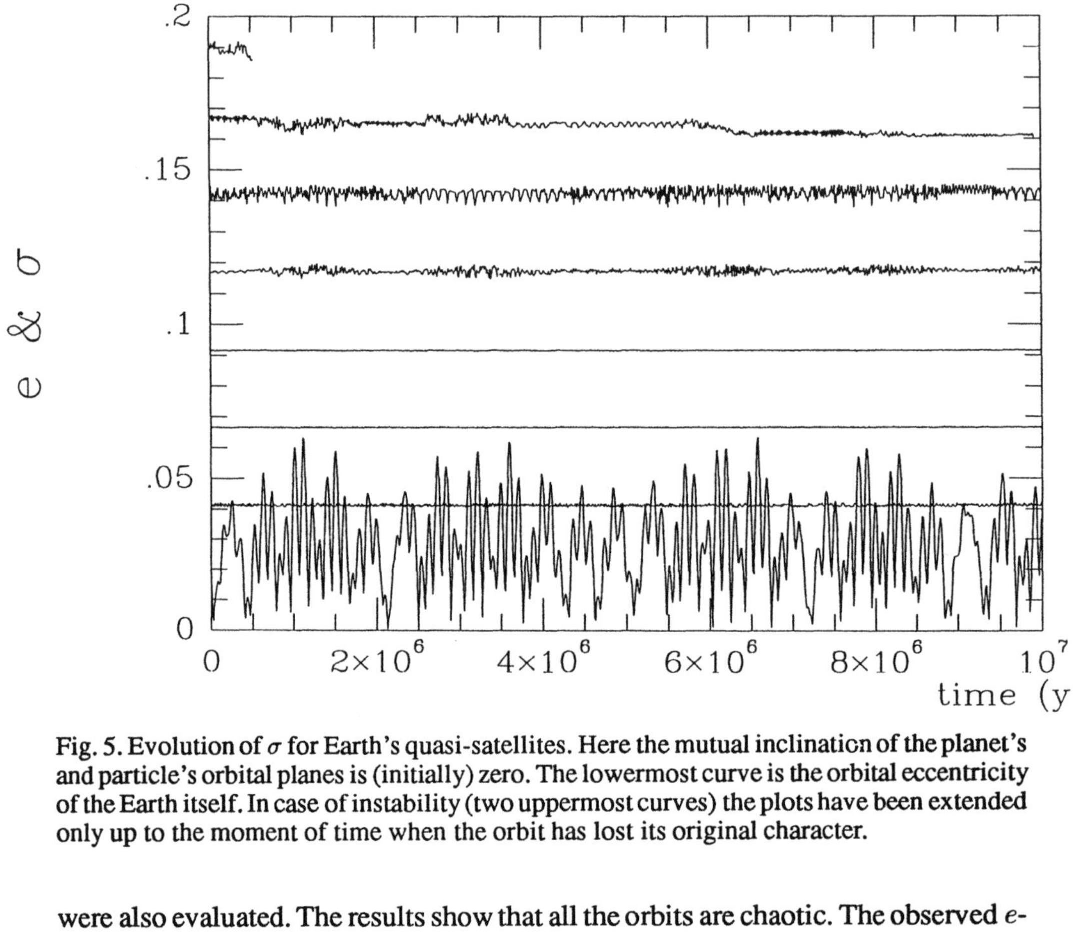

Fig. 5. Evolution of σ for Earth's quasi-satellites. Here the mutual inclination of the planet's and particle's orbital planes is (initially) zero. The lowermost curve is the orbital eccentricity of the Earth itself. In case of instability (two uppermost curves) the plots have been extended only up to the moment of time when the orbit has lost its original character.

were also evaluated. The results show that all the orbits are chaotic. The observed e-folding times varied from small values of less than 100 years up to 750,000 years. The longest Liapunov times belong to those particles with $\rho \sim 0.1$. Empirical formulae tying the Liapunov times to stability times have been published recently (Murrison, Lecar and Franklin 1994). These expressions suggest that particles with the largest Liapunov times could survive over the age of the Solar System. However, those particular orbits which lost their quasi-satellite character during the computation did so in much shorter times than what the Franklin-Lecar law suggested. Thus such relations are probably system dependent, and it remains to be investigated with longer integrations whether the orbits are really stable in the long run.

4. Conclusions

Self-consistent n-body integrations and analytical results have been used to establish the existence of long-term stability for quasi-satellites of a planet, itself in a nearly elliptic orbit about the Sun. Also adiabatic invariants governing the motion have been established. >From the planet, the object appears to be in retrograde motion. For the Earth or Venus, the most stable planetocentric (minimum) distance is about 0.1 AU, or roughly 40 times the Moon's geocentric distance. Our numerical results show that stability exists for nearly coplanar orbits for 10 Myr in a self-consistent model of the Solar System, in the neighborhood of Venus and Earth, while all our test particles in the vicinity of Mars lost their stability in a timescale of a few million years.

Our results are, to some extent, a generalization of a numerical investigation of the planar, restricted 3-body problem due to Hénon and Guyot (1970). In the latter work, no attempt appears to have been made to apply the results in the planetary context. No natural objects in this configuration are known in the Solar System, but this does not necessarily mean that such small bodies could not exist, for the search process requires a special observational strategy.

Acknowledgements

A portion of this work was written while KI was a special guest of Tuorla Observatory and Turku University in the spring of 1995, and he expresses his appreciation to SM, as the Acting Director of the Observatory. KI also expresses appreciation for partial support through an operating grant from the Natural Sciences and Engineering Research Council of Canada, and for assistance from the Finnish Academy in 1996.

References

Hénon, M. and Guyot, M.:1970, in *Periodic Orbits Stability and Resonances*, ed. G.E.O. Giacaglia, proceedings of a symposium in São Paulo, Brasil 1969, Reidel.

Kogan, A. Ya.: 1988, *Cosmic Research* (Trans. from Russian), **26**, No 6.

Murrison, M. A., Lecar, M. and Franklin, F. A.: 1994 *Astron. J.*, **108**, 2323–2329.

Mikkola, S. and Innanen, K.: 1992, *Astron. J.*, **104**, 1641–1649.

Mikkola, S. and Innanen, K.: 1995, *Mon. Not. R. Astr. Soc.* **277**, 497–501.

Wisdom, J. and Holman, M.: 1991 *Astron. J.* **102**, 1520.

Stumpff, K.: 1965, *'Himmelsmechanik'*, Band II, VEB Deutscher Verlag der Wissenschaften, Berlin.

Szebehely, V.: 1967 *'Theory of orbits'*, Academic Press, New York.

ALTERNATIVE DERIVATIONS OF PERTURBATIONS WITH USE OF A MIXED REFERENCE-PLANE SYSTEM: APPLICATION TO THE MOTION OF THE ORBIT PLANE OF HYPERION

P.J. MESSAGE
University of Liverpool,
L69 3BX, U.K.
sx20@liverpool.ac.uk

Abstract. The use of a mixed set of angle parameters in the theory of the motion of the orbit plane of Hyperion, and of the associated terms in the longitudes in the orbit plane, is approached by starting from a theory constructed by using a single reference plane, and then calculating the differences between the longitudes defined in the two systems. This is an alternative approach to that used in a previous paper, in which the theory was developed from the start using the mixed set. Comparison is then possible between results from the two approaches.

Key words: Hyperion – ephemerides – orbit plane

1. Introduction

In a presentation to I.A.U. Symposium No. 172, in Paris (Message 1996, referred to hereinafter as the "previous paper"), a development was given of the equations for the perturbations of the orbit of Saturn's satellite Hyperion, with especial reference to the perturbations of its orbit plane, and the effect of those perturbations on the motion in the orbit plane. The derivation used the set of parameters which it has become usual to use when developing theories of the motion of Hyperion, that is, to use the mean longitude, and the longitude of the apse, referred to the Ecliptic and Equinox (*i.e.* First Point of Aries), but to refer the orientation of the plane to the equator plane, or ring plane, of Saturn, to make use of the fact that the orbit planes of Hyperion and Titan (and in fact all the satellites except Iapetus) are inclined at quite small angles to that plane, and the equations for the rectangular-type orbital plane parameters may be treated as linear for any precision of theory so far required. The Lagrange equations were derived for this mixed set of parameters.

2. Orbital Parameters and Equations

In this paper, however, let us explore an alternative approach (the possibility of which was mentioned in the previous paper), which is to develop the theory with all longitudes referred to the same reference plane, and, subsequently, to calculate the differences between the mean longitudes, and also the apse longitudes, defined in the two systems. (Also as mentioned in the previous paper, the development of the theory using a single reference plane allows a consistent canonical set of orbital parameters to be constructed by taking the longitudes so defined as co-ordinates.) To gain the advantage, mentioned above, provided by the small inclinations, we will use the equator plane of Saturn as reference plane. Let us in fact measure

R. Dvorak and J. Henrard (eds.), The Dynamical Behaviour of our Planetary System, 357.

mean longitudes and apse longitudes from the equinox, along the ecliptic to the ascending node of Saturn's equator on the ecliptic, and then along Saturn's equator to the ascending node of the orbit on Saturn's equator, and then along the orbit. Let us denote by $\hat{\psi}$ a true longitude, by $\hat{\lambda}$ a mean longitude, and by $\hat{\varpi}$ an apse longitude, defined on this basis (*i.e.* using the ecliptic, equator plane of Saturn, and orbit plane). As in the previous paper, we will use the following notation:
i for the inclination of the orbit plane to Saturn's equator plane,
h for the longitude of the ascending node of the orbit, on Saturn's equator plane, measured from the ascending node of Saturn's equator on the Ecliptic,
I for the inclination of the orbit plane to the Ecliptic,
Ω for the longitude of the ascending node of the orbit, on the ecliptic, measured from the Equinox,
I_e for the inclination of the equator plane of Saturn to the Ecliptic,
Ω_e for the longitude of the ascending node of Saturn's equator on the Ecliptic, also measured from the Equinox.

We will use, as before, the rectangular-type parameters

$$p = \sin\, i\, \sin\, h = \sin\, I\, \sin\,(\Omega - \Omega_e)$$

$$q = \sin\, i\, \cos\, h = \sin\, I\, \cos\, I_e\, \cos\,(\Omega - \Omega_e) - \cos\, I\, \sin\, I_e$$

So let us now consider the equations for the perturbations, using the set of orbital parameters $(\hat{\lambda}, \hat{\varpi}, a, e, q, p)$. The Lagrange equations for these, with the disturbing function, R, expressed in terms of them, are then

$$\begin{aligned}
\frac{\mathrm{d}a}{\mathrm{d}t} &= \frac{2}{na}\frac{\partial R}{\partial \hat{\lambda}},\\
\frac{\mathrm{d}e}{\mathrm{d}t} &= \frac{Y}{na^2}\frac{\partial R}{\partial \hat{\lambda}} - \frac{X}{na^2}\frac{\partial R}{\partial \hat{\varpi}},\\
\frac{\mathrm{d}\hat{\lambda}}{\mathrm{d}t} &= n - \frac{2}{na}\frac{\partial R}{\partial a} + \frac{Y}{na^2}\frac{\partial R}{\partial e} - \frac{Z_e}{nab}\mathcal{P}_e,\\
\frac{\mathrm{d}\hat{\varpi}}{\mathrm{d}t} &= \frac{X}{na^2}\frac{\partial R}{\partial e} - \frac{Z_e}{nab}\mathcal{P}_e,\\
\frac{\mathrm{d}q}{\mathrm{d}t} &= -\frac{\cos\, i}{nab}\frac{\partial R}{\partial p} - \frac{Z_e}{nab}q\left\{\frac{\partial R}{\partial \hat{\lambda}} + \frac{\partial R}{\partial \hat{\varpi}}\right\},\\
\frac{\mathrm{d}p}{\mathrm{d}t} &= +\frac{\cos\, i}{nab}\frac{\partial R}{\partial q} - \frac{Z_e}{nab}p\left\{\frac{\partial R}{\partial \hat{\lambda}} + \frac{\partial R}{\partial \hat{\varpi}}\right\},
\end{aligned}$$

where

$$X = \frac{\sqrt{1-e^2}}{e},$$

$$
\begin{aligned}
Y &= X - \frac{1-e^2}{e}, \\
Z_e &= \frac{\cos i}{1+\cos\, i}, \\
\mathcal{P}_e &= p\,\frac{\partial R}{\partial p} - q\frac{\partial R}{\partial q},
\end{aligned}
$$

and

$$\cos\, i = \sqrt{1-(q^2+p^2)}.$$

3. The Perturbations due to Titan

As in the previous paper, we split the disturbing function, R_T, for the perturbation due to Titan, into two parts, to separate the part involving the orbit plane parameters from the rest. This time the split is

$$R_T = R_{0,2} + \delta R_2,$$

in which $R_{0,2}$ is that part of R_T independent of the orbit plane variables, q and p, when R_T is expressed in terms of
$\hat{\psi}_H - \hat{\psi}_T$,
r_H (the distance between Hyperion and Saturn),
r_T (the distance between Titan and Saturn),
and q and p,
and so δR_2 contains all of the terms involving the parameters, q and p, of the orbit planes. Then we find that to second order in q_H and p_H (the values of q and p for Hyperion),

$$\delta R_2 = G m_T r_H r_T \left\{ \frac{1}{\Delta_0^3} - \frac{1}{r_T^3} \right\} \mathcal{F},$$

in which
G is the constant of gravitation,
m_T is the mass of Titan,
Δ_0 is the distance between Hyperion and Titan, calculated as if the satellites' orbits were in the equator plane of Saturn (that is, omitting all terms in the q and p), and

$$
\begin{aligned}
\mathcal{F} = & -\frac{1}{4}\left\{ (q_H - q_T)^2 + (p_H - p_T)^2 \right\} \cos\,(\hat{\psi}_H - \hat{\psi}_T) \\
& + \frac{1}{4}\left\{ (q_H - q_T)^2 - (p_H - p_T)^2 \right\} \cos\,(\hat{\psi}_H + \hat{\psi}_T - 2\Omega_e)
\end{aligned}
$$

$$+\frac{1}{2}(q_H - q_T)(p_H - p_T)\sin(\hat{\psi}_H + \hat{\psi}_T - 2\Omega_e)$$
$$+\frac{1}{2}(p_H q_T - q_H p_T)\sin(\hat{\psi}_H - \hat{\psi}_T).$$

Substituting these terms into the Lagrange equations for the rates of change of the rectangular-type variables q_H and p_H, and recalling that these rates also have contributions from $R_{0,2}$ through the terms $\partial R/\partial\hat{\lambda} + \partial R/\partial\hat{\varpi}$ we find that $\mathrm{d}q_H/\mathrm{d}t$ has, to first order:

$$n_H m' \mathcal{K}\Big\{(p_H - p_T)\{\cos(\hat{\psi}_H - \hat{\psi}_T) + \cos(\hat{\psi}_H + \hat{\psi}_T - 2\Omega_e)\}$$
$$+(q_H - q_T)\{\sin(\hat{\psi}_H - \hat{\psi}_T) - \sin(\hat{\psi}_H + \hat{\psi}_T - 2\Omega_e)\}\Big\},$$

and that $\frac{\mathrm{d}p_H}{\mathrm{d}t}$ has, to first order:

$$n_H m' \mathcal{K}\Big\{(q_H - q_T)\{-\cos(\hat{\psi}_H - \hat{\psi}_T) + \cos(\hat{\psi}_H + \hat{\psi}_T - 2\Omega_e)\}$$
$$+(p_H - p_T)\{\sin(\hat{\psi}_H - \hat{\psi}_T) + \sin(\hat{\psi}_H + \hat{\psi}_T - 2\Omega_e)\}\Big\},$$

where

$$\mathcal{K} = \frac{1}{2} a_H r_H r_T \left\{\frac{1}{\Delta_0^3} - \frac{1}{r_T^3}\right\},$$

and m' is the mass of Titan in terms of that of Saturn.

Thus the contributions to the rates of change of q_H and p_H are, to this order, the same as in the previous paper, as we might expect. But now, to first order in the q and p, we find that $\mathrm{d}\hat{\lambda}_H/\mathrm{d}t$ and $\mathrm{d}\hat{\varpi}_H/\mathrm{d}t$ have no terms arising from derivatives of δR_2 with respect to q_H or p_H, whereas, in the previous paper, there were indeed terms in $\mathrm{d}\lambda_H/\mathrm{d}t$ and $\mathrm{d}\varpi_H/\mathrm{d}t$ arising from derivatives of δR. Of course there are also the terms

$$-\frac{2}{na}\frac{\partial\delta R_2}{\partial a} + \frac{Y}{na^2}\frac{\partial\delta R_2}{\partial e} \quad \text{in} \quad \frac{\mathrm{d}\hat{\lambda}}{\mathrm{d}t},$$

and

$$\frac{X}{na^2}\frac{\delta R_2}{\partial e} \quad \text{in} \quad \frac{\mathrm{d}\hat{\varpi}}{\mathrm{d}t},$$

but these provide no contribution of first order in q and p, as is seen from the form of δR_2 given above. (The corresponding terms in the approach of the previous paper do give such a contribution, of course, and we will return to this question later.) The equations for $\frac{\mathrm{d}q_H}{\mathrm{d}t}$ and $\frac{\mathrm{d}p_H}{\mathrm{d}t}$, to first order in the q and p then, are unchanged from the previous paper, and so also then will be the solution, to this order. The difference between the two approaches shows itself then, to this order, in the mean longitude and the apse longitude, and to these we now turn.

4. The Terms in the Longitude of the Satellite, and the Longitude of the Apse, due to the Use of a Mixed Reference System

We see that $\hat{\lambda}$ and $\hat{\varpi}$ contain no terms of first order in the q and p. We must now turn to the differences $\lambda - \hat{\lambda}$ and $\varpi - \hat{\varpi}$. Since $\lambda = \varpi + \ell$, $\hat{\lambda} = \hat{\varpi} + \ell$, $\psi = \varpi + f$, and $\hat{\psi} = \hat{\varpi} + f$, it is enough to consider $\varpi - \hat{\varpi}$, since $\lambda - \hat{\lambda}$ and $\psi - \hat{\psi}$ are each equal to it. From the definitions given above, $\hat{\varpi} = \Omega_e + h + g$, while $\varpi = \Omega + \omega$, so that $\varpi - \hat{\varpi} = \omega - g + \Omega - \Omega_e - h$. Now h and $\Omega - \Omega_e$ are two sides of a spherical triangle, which enclose the angle I_e, the opposite side being $\omega - g$, and the angle enclosed by $\Omega - \Omega_e$ and $\omega - g$ being $\pi - I$. Some use of the equations of spherical trigonometry gives (recalling the definitions of q and p given above)

$$\begin{aligned}\sin^2 I \sin(\varpi - \hat{\varpi}) = p\Big\{&\sin I_e \{1 - \cos I_e \cos i\} \\ &+ \{2\cos I_e \sin^2\frac{I_e}{2} + \sin^2 I_e \frac{\cos i}{(1+\cos i)}\}q \\ &- 2\sin I_e \sin^2\frac{I_e}{2}\frac{1}{(1+\cos i)}q^2\Big\}.\end{aligned}$$

Since i is small,

$$\sin^2 I \sin(\varpi - \hat{\varpi}) = p\{(1 - s\cos I_e)\sin I_e + q\sin^2\frac{I_e}{2}(1 + 3\cos I_e) + O(i^2)\}.$$

Since

$$\cos I = \cos I_e \cos i - q \sin I_e,$$

we have $I = I_e + O(i)$, and so

$$\varpi - \hat{\varpi} = p\tan\left(\frac{I_e}{2}\right) + O(i^2),$$

and thus also

$$\lambda - \hat{\lambda} = p\tan\left(\frac{I_e}{2}\right) + O(i^2).$$

These terms in λ and ϖ were given in the reduction of observational data by Taylor (1984).

5. Further Terms Derived by the Method of the Previous Paper

Now let us return to the terms in λ and ϖ, which involve the parameters of the orbit plane, in the equations derived in the previous paper for $\frac{d\lambda}{dt}$ and $\frac{d\varpi}{dt}$, each of which we found to contain the following terms, arising from derivatives of δR with respect to q and p,

$$n_H m' \mathcal{K} \tan\left(\frac{I_e}{2}\right)\left\{(q_H - q_T)\left\{-\cos(\psi_H - \psi_T) + \cos(\psi_H + \psi_T - 2\Omega_e)\right\}\right.$$

$$+(p_H - p_T)\sin(\psi_H + \psi_T - 2\Omega_e)$$

$$\left.-\left\{p_T + \left\{\cos I_e \sec^2\left(\frac{I_e}{2}\right) - 2\tan\left(\frac{I_e}{2}\right)\right\}p_H\right\}\sin(\psi_H - \psi_T)\right\}.$$

Using the notation of the previous paper, the mean value over $\lambda_H - \lambda_T$ of $\mathcal{K}\cos(\psi_H - \psi_T)$ is $\mathcal{A}_{0,0}$, and the mean value of $\mathcal{K}\cos(\psi_H + \psi_T - 2\Omega_e)$ is $\mathcal{C}_{0,0}$, so that $\frac{d\lambda_H}{dt}$ and $\frac{d\varpi_H}{dt}$ each contain the term

$$n_H m'(\mathcal{C}_{0,0} - \mathcal{A}_{0,0})\tan\left(\frac{I_e}{2}\right)(q_H - q_T).$$

Now, from the solution of the equations for $q_H - q_T$ and $p_H - p_T$, we see that the largest term in $q_H - q_T$ is $c\cos v$, where c is about 0.52°, and v is an argument, linear in the time, of period about 136 years. Thus the largest term in each of $\frac{d\lambda_H}{dt}$ and $\frac{d\varpi_H}{dt}$ from this source is

$$n_H m'(\mathcal{C}_{0,0} - \mathcal{A}_{0,0})\tan\left(\frac{I_e}{2}\right) c\cos v,$$

and so, since

$$\frac{dv}{dt} = n_H \chi_p \approx n_H m'(\mathcal{C}_{0,0} - \mathcal{A}_{0,0}),$$

the corresponding largest term in each of λ and ϖ is

$$tan\left(\frac{I_e}{2}\right) c\sin v.$$

Now $c\sin v$ is the largest term in p, and so this has confirmed, using the methods of the previous paper, the largest part of the term derived at the end of section 4 by the methods introduced at the beginning of the present paper.

There are also the terms

$$-\frac{2}{na}\frac{\partial\delta R}{\partial a} + \frac{Y}{na^2}\frac{\partial\delta R}{\partial e} \quad \text{in} \quad \frac{d\lambda}{dt},$$

and the terms

$$\frac{X}{na^2}\frac{\partial\delta R}{\partial e} \quad \text{in} \quad \frac{d\hat{\varpi}}{dt}.$$

Keeping only terms of first order in q and p, these give

$$n_H m' \left\{ -\frac{2}{na}\frac{\partial \mathcal{K}}{\partial a} + \frac{Y}{na^2}\frac{\partial \mathcal{K}}{\partial e} \right\} \tan\left(\frac{I_e}{2}\right)(p_H - p_T)\sin(\psi_H - \psi_T)$$

in $\frac{d\lambda}{dt}$, and

$$n_H m' \frac{X}{na^2}\frac{\partial \mathcal{K}}{\partial e} \tan\left(\frac{I_e}{2}\right)(p_H - p_T)\sin(\psi_H - \psi_T)$$

in $\frac{d\hat{\varpi}}{dt}$. Now suppose the largest term in

$$\left\{ -\frac{2}{na}\frac{\partial \mathcal{K}}{\partial a} + \frac{Y}{na^2}\frac{\partial \mathcal{K}}{\partial e} \right\} \sin(\psi_H - \psi_T)$$

to be $\mathcal{D}_{0,1}\sin\zeta$, and the largest term in

$$\frac{X}{na^2}\frac{\partial \mathcal{K}}{\partial e}\sin(\psi_H - \psi_T)$$

to be $\mathcal{E}_{0,1}\sin\zeta$.

Then $d\lambda/dt$ has, as its largest term from this source, $n_H m' \mathcal{D}_{0,1}\sin\zeta\; c\; \sin v$, and $d\varpi/dt$ has, as its corresponding largest term, $n_H m' \mathcal{E}_{0,1}\sin\zeta\; c\; \sin v$. Thus the largest term in λ from this source is within

$$\frac{1}{2} m' \mathcal{D}_{0,1} c \left\{ \frac{\sin(\zeta - v)}{\chi - \chi_p} - \frac{\sin(\zeta + v)}{\chi + \chi_p} \right\},$$

while the corresponding term in ϖ is within

$$\frac{1}{2} m' \mathcal{E}_{0,1} c \left\{ \frac{\sin(\zeta - v)}{\chi - \chi_p} - \frac{\sin(\zeta + v)}{\chi + \chi_p} \right\}.$$

The argument $\zeta - v$ has period about 21.8 years, while $\zeta + v$ has period about 16.5 years.

References

Message, P.J.: 1996, Foundations of a theory of the motion of the Orbit Plane of Hyperion, *Proceedings of I.A.U. Symposium 172*, ed. S.Ferraz-Mello., 127–136

Taylor, D.B.: 1984, A Comparison of the theory of the motion of Hyperion with observations made during 1967-1982, *Astron. Astrophys.* **141**, 151–158.

ASTRODYNAMIC STUDY OF THE EARTH ROTATION.

A possibility of large displacements of the poles with respect to the Earth crust.

C. MARCHAL
DES - ONERA
BP 72, 92322 Chatillon Cedex, France

Abstract. Earth rotation and the position of poles are essential elements of climate. Since centuries the motion of the poles with respect to the Earth's surface has remained very small, but this was not necessarily the case during geological times. The orogeny of a large mountain range, such as Himalaya, modifies the Earth's moments of inertia by a few millionths and this is sufficient for a large displacement of the stable equilibrium position of the poles; equilibrium position that is given by the main axis of largest inertia of the "non rotating Earth in isostatic equilibrium". The corresponding motion of the poles remains usually very slow, as slow as the corresponding orogeny, and the Earth's equatorial bulge doesn't forbid the transformation. This bulge is only slowly deformed until its installation in the new position of equilibrium. The stable equilibrium positions depend very much on the variations of the Earth's crust and the poles have certainly moved much more than the different lithospheric plates. Large orogenies and/or large subductions can move the poles by tens of degrees within a few million years.

Key words: Earth – rotation – polar motions

1. Introduction

Since about 1970 the theory of plate tectonics is widely accepted and most scientists think that the different lithospheric plates move with relative velocities of a few centimeters per year. The motor of these displacements seems to be the convection motions in the Earth mantle produced by the heat supplied essentially by the Earth's natural radioactivity.

This theory explains a wide variety of geological features and is confirmed by the measurements of space geodesy.

The relative velocity of the lithospheric plates, of course, at least implies variations of latitude of the same order of magnitude and thus displacements of the poles of similar amplitudes.

However there also remains the possibility of large general tiltings of the Earth's crust, or conversely large displacements of the poles, corresponding to periods of disequilibrium and rearrangements. The poles have certainly moved much more than the different lithospheric plates ...

2. Preliminary Notice (Numeration System)

In this paper with many very large and very small numbers we will use the convention of numeration by "figures and sizes" :

$$\text{Mass of Earth} = M = 5.9737 \times 10^{24}\ \text{kg} = \underbrace{5.9737}_{\text{figure}} \times \underbrace{\text{p24}}_{\text{size}}\ \text{kg},$$

R. Dvorak and J. Henrard (eds.), The Dynamical Behaviour of our Planetary System, 365.

Cavendish constant = G = 6.6726×10^{-11} m^3/s^2 kg = $\underbrace{6.6726}_{\text{figure}} \times \underbrace{\text{n11}}_{\text{size}}$ m^3/s^2 kg.

The letter p means "positive power of ten" and the letter n means "negative power of ten", while *the figure is always between one and ten* (we cannot write 0.59 p25 instead of 5.9 p24). *The size is thus always well defined,* it is of course the most important part of very large and very small numbers, it is even sometimes the only known part.

3. Main Elements of Earth

We have now a wide variety of informations about Earth, coming from many domains of Science.

Geodesy and space geodesy give the Earth shape and dimensions [Ref. 1].

Astronomy, astrodynamics and the motions of artificial satellites give the motions of the poles with respect to the stars and also with respect to the Earth's crust (the amplitude of the latter is less than 40 meters since one century, Ref. 2). They also give:

The Earth mass : M = { 5.9737 p24 $\pm$ 3 p20} kg,

The Earth main moments of inertia :

$$A = 0.329591 M R^2 \quad , \quad B = 0.329599 M R^2 \quad , \quad C = 0.330678 M R^2 \; , \tag{1}$$

with R = Earth equatorial radius =6 378 136 m.

The Earth main axes of inertia and its gravitational potential [Ref 8-10].

Notice that the angle between the main axis of largest inertia and the polar axis is very small, so small that most Earth models consider it as zero.

Geology, vulcanology, gravimetry, geomagnetism give many informations about the Earth's crust and our main source of information about inner Earth is seismology that reveals the existence of several discontinuities at widely separated levels [Ref. 3].

4. The Rotating and Non-Rotating Frames of Reference

Our $Oxyz$ rotating frame of reference will be the "International Terrestrial Reference Frame (ITRF)" that is very well related to the Earth crust (Fig.1 and Ref. 4).

We will also use a non-rotating frame of reference OXYZ with the same origin O. It is the geocentric equivalent of the "International Celestial Reference Frame" [Ref. 4].

We can define the Earth's axis and poles by the rotation vector $\boldsymbol{\omega}$ of Oxyz with respect to OXYZ.

The Earth's angular momentum **H** remains always extremely near to the direction of $\boldsymbol{\omega}$, their angle is usually a few arc-seconds and at most one or two minutes.

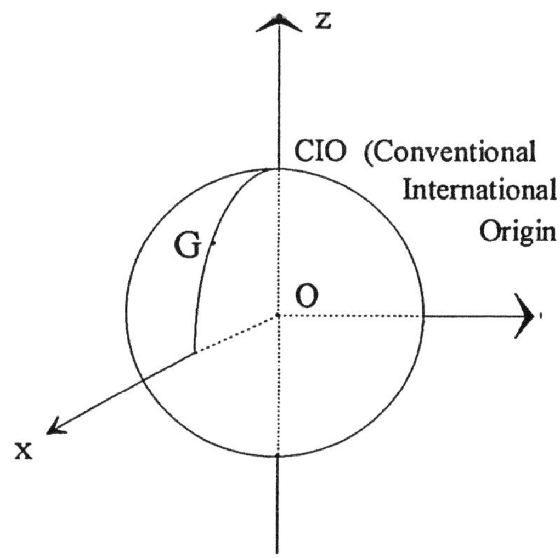

Fig. 1. The International Terrestrial Reference Frame $Oxyz$. O is the Earth center of mass, Oz passes through the Conventional International Origin, a few meters from the present North pole; it is defined by its position with respect to a few selected observatories. Oxz is along the International Reference Meridian (Greenwich). $Oxyz$ is a right and tri-rectangular trihedron.

Then, for the study of the large displacements of the poles with respect to the Earth's crust, it will be sufficient to study the evolution of **H** with respect to Oxyz.

5. The Four Earth Equilibria

Because of the presence of outer bodies, the Sun, the Moon, etc..., our planet cannot be in a state of perfect equilibrium. It permanently undergoes tides in its oceans and its crust. Furthermore its inner sources of energy draw along large masses at a very slow rate.

Fortunately these phenomena remain very small, at the scale of the Earth, and we can define the four following equilibria that are more or less near to the true Earth (figure 2).

A) The hydrostatic equilibrium of a non-rotating Earth.

B) The hydrostatic equilibrium of a rotating Earth.

C) The isostatic equilibrium of a non-rotating Earth.

D) The isostatic equilibrium of a rotating Earth.

The last equilibrium is very near to the true Earth but notice that the first one is already good; it is the basis of the PREM 1s model of Ref. 3.

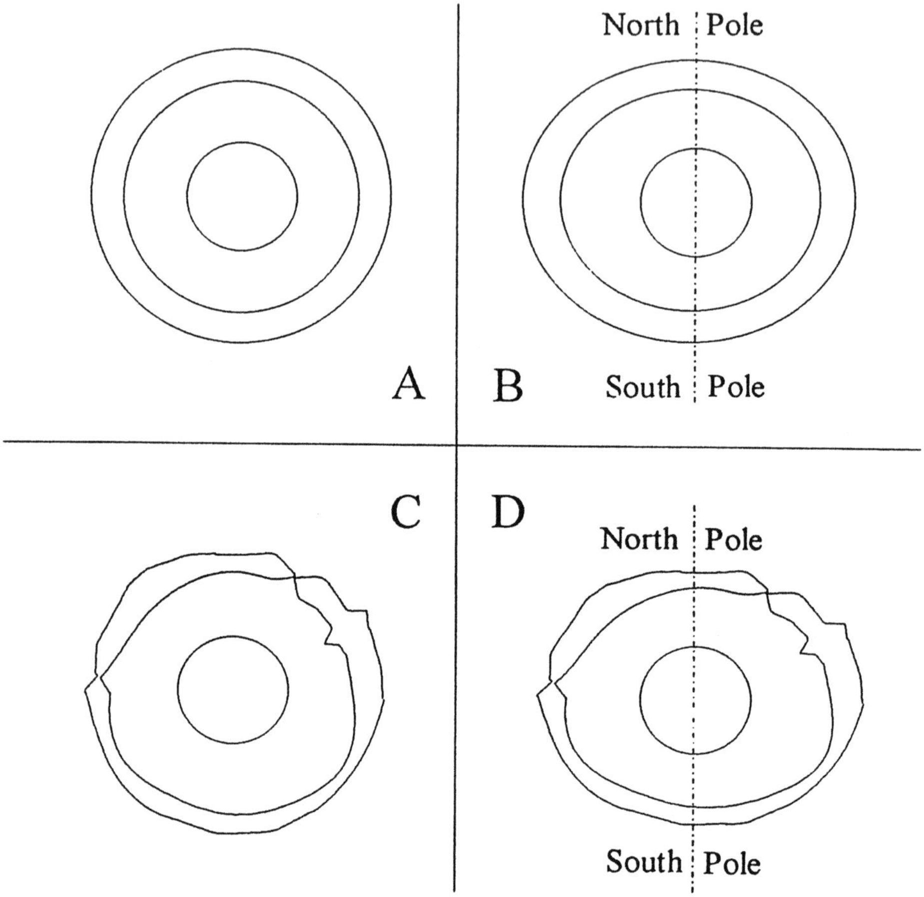

Fig. 2. The four equilibria of Earth. A) Hydrostatic equilibrium of a non-rotating Earth; the level surfaces are concentric spheres. B) Hydrostatic equilibrium of a rotating Earth; the level surfaces are concentric and coaxial oblate quasi-ellipsoids (slightly depressed at mid-latitudes). C) Isostatic equilibrium of a non-rotating Earth; the sea level is almost a sphere. D) Isostatic equilibrium of a rotating Earth; the sea level is almost an oblate ellipsoid. For B) and D), the Earth momentum H is 5.85989p33 kg m^2/s.

5.1. THE HYDROSTATIC EQUILIBRIUM OF A NON-ROTATING EARTH.

This first equilibrium (figure 2A) is simple and well known. It has spherical symmetry. The different layers are spherical and concentric with density increasing with depth.

5.2. THE HYDROSTATIC EQUILIBRIUM OF A ROTATING EARTH.

This second equilibrium (figure 2B) remains simple. It has spheroidal symmetry :

A) The axis of rotation is also an axis of symmetry of revolution.

B) The equatorial plane (normal to the axis of rotation through the center of mass) is a plane of symmetry.

C) On straight lines either parallel or perpendicular to the axis of rotation the density increases when we approach the center of mass.

This spheroidal symmetry is very general for hydrostatic equilibria (see Ref. 5, appendix 1 and Ref. 6).

The Earth has a rather slow rotation and its hydrostatic equilibrium has level surfaces that are concentric and coaxial oblate quasi-ellipsoids whose oblateness increases from the center to the surface. These quasi-ellipsoids are slightly depressed at mid-latitudes (the depression is only about 4.3 m at the Earth's surface at latitudes $\pm 45°$).

5.3. THE ISOSTATIC EQUILIBRIA.

With all its mountain ranges, its continents and its oceans the Earth is of course far from hydrostatic equilibrium and a much better approximation is that of isostatic equilibrium.

The idea of isostatic equilibrium comes from the large differences of viscosity and rigidity between the lithosphere and the asthenosphere. Vertical equilibrium of the lithosphere is achieved much more easily and faster than the horizontal one.

The upwards motion of Scandinavia, after the recent melting of the large Quaternary glacier, is about one meter per century and corresponds to re-equilibrium on the time scale of a few millenia only.

The theory of isostatic equilibrium requires the definition of a "level of compensation", the asthenosphere, a few hundred kilometers below the surface. Above this level only vertical equilibrium is required. We thus arrive at the usual pictures with a thick lithosphere under the continents, especially the mountain ranges, and a thinner one below the oceans (figures 2C and 2D).

Figure 2D corresponds to the isostatic equilibrium of the Earth with its true rotation. The main feature of the surface is its oblateness (the difference between the equatorial and polar radii is 21.4 km).

The mean sea level is not exactly a "geoid" i.e. the equipotential surface of level zero (because of sea currents and of variations of temperature and salinity) but these two surfaces remain within meters to each other and they are close to the reference ellipsoid.

Figure 2C corresponds to the isostatic equilibrium of a non-rotating Earth. Its mean sea level is of course almost spherical and we will see the interest of this equilibrium in section 7.

6. About the Periods of Disequilibrium

From time to time factors of disequilibrium appear in the Earth's crust.

For instance the Asia-India collision (fig.3) and the corresponding orogeny of the Himalaya range modify the Earth's inertia matrix by a few millionths (even if we assume the conservation of isostasy).

In front of the Indian plate the seabed (average level - 3800m) is pushed up in a large region : Hymalaya range, Tibet plateau, Hindu-Kush, Karakoram, Pamir, Kunlun, Chamdo, etc. . . about 10 millions km^2 with an average level of 2800 m.

Below this region the isostatic compensation reduces correspondingly the density, down to the "isostatic compensation level" near the bottom of continental plates.

By the definition of isostatic compensation these two opposite variations keep a constant pressure at the compensation level ; but they modify slightly the moments and products of inertia, the variations of which are not proportional to these of the pressure [Ref 5, pages 540-543].

Behind the Indian plate we have the opposite evolution : a thin oceanic crust takes the place of a thick continental crust. This second phenomenon reduces to almost nothing the variation of the average moment of inertia $(A + B + C)/3$, but it reinforces the variation of the product of inertia I_{YZ} that is modified by about 3n6 MR^2 during the orogeny.

This apparently limited transformation modifies the pole equilibrium positions by about 20° because it is of the order of the differences $C_C - A_C$ and $C_C - B_C$ presented in (17).

Variations of the Earth's inertia matrix of the same order of magnitude, or even a little larger, are obtained in the remarkable Ref.7. The authors consider even that the main source of variation of the Earth's inertia matrix is the motion of downgoing slabs during subductions.

Whatever the source of these variations they imply large polar displacements.

It may seem that the Earth's equatorial bulge would stabilize the rotation axis, however the Earth's viscosity and its tendency to the isostatic equilibrium allow the progressive deformation of the equatorial bulge until its installation in the new position of equilibrium.

On the other hand it is impossible to give a simple description of the period of disequilibrium. The Earth's elasticity, plasticity and viscosity as well as the more or less large transgressions and regressions of sea level that must be expected during the phases of variable rotation and isostatic re-equilibria lead to a huge complexity that even a full rheology study will not master.

Our only solid elements are the general theorems of Mechanics and the tendency towards the state of stable isostatic equilibrium, that is the state of minimum energy (for the given value of the Earth's angular momentum) because of the large losses of energy associated to non-pure rotations.

Hence we ignore the path followed on the way to the equilibrium but we can know that final state and we can even have an idea of the corresponding relaxation time that is about 100 000 years in the present state of the Earth [see Ref 5, pages 544-547].

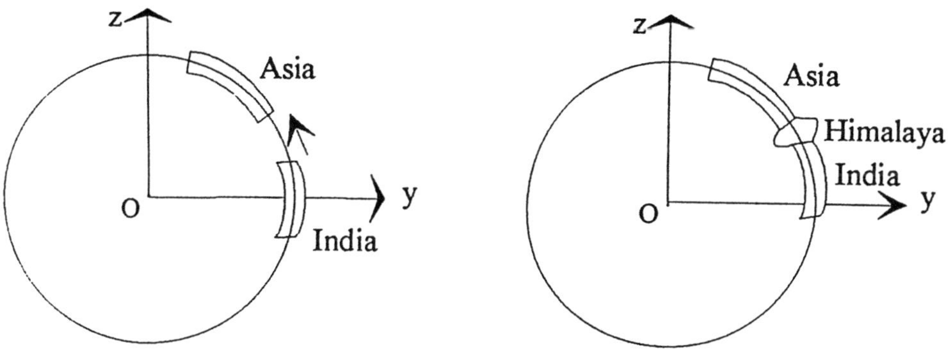

Fig. 3. The Asia-India collision and the orogeny of Himalaya

Usually the variations of the Earth's crust are slow and then the poles follow very nearly their position of stable equilibrium.

7. Application of the General Theorems of Mechanics

Let us first consider an isolated Earth, with then a constant angular momentum **H**.

We are interested into the relative Pole - Earth's crust motions, i.e. practically the evolution of **H** with respect to the rotating frame of reference of figure 1.

However, with the rotation vector ω of Oxyz with respect to OXYZ:

$$\left(\frac{d\mathbf{H}}{dt}\right)_{\mathrm{OXYZ}} = \left(\frac{d\mathbf{H}}{dt}\right)_{\mathrm{Oxyz}} + \omega \times \mathbf{H} = 0 \;, \tag{2}$$

that is:

$$\left(\frac{d\mathbf{H}}{dt}\right)_{\mathrm{Oxyz}} = \mathbf{H} \times \omega \;. \tag{3}$$

We thus need to know the relation between **H** and ω.

Then if the mass dm at the radius vector **r** has the velocity **V** with respect to OXYZ and the velocity **v** with respect to Oxyz we find:

$$\mathbf{V} = \mathbf{v} + \omega \times \mathbf{r} \;, \tag{4}$$

and, with M = mass of Earth:

$$\mathbf{H} = \int_M dm\, \mathbf{r} \times \mathbf{V} = \int_M dm\, \mathbf{r} \times (\omega + \mathbf{v}) = \mathcal{M}\omega + \int_M dm\, \mathbf{r} \times \mathbf{v} \;. \tag{5}$$

$\mathcal{M}$ is the Earth's inertia matrix with its classical expression:

$$\mathcal{M} = \begin{pmatrix} I_{y^2+z^2} & -I_{xy} & -I_{xz} \\ -I_{xy} & I_{x^2+z^2} & -I_{yz} \\ -I_{xz} & -I_{yz} & I_{x^2+y^2} \end{pmatrix} \quad ; \text{ with } \quad I_{xy} = \int_M dm\, xy \quad \text{etc.} \tag{6}$$

In the analysis of **v** we find:

A) The motions related to plate tectonics (velocity: a few centimeters per year).

B) The general motion of Earth mantel (velocity: a few meters per year).

C) The quasi-periodic motions related to the tides of the crust (amplitude up to one meter; periods from a few hours to a few years).

D) The motions of the oceans and the atmosphere: quasi-periodic tidal motions, permanent currents and winds, seasonal currents and winds, meteorology.

We can thus decompose **H** into three parts:

$$\left\{\begin{array}{ll} \mathbf{H} & = \mathbf{H}_\omega + \mathbf{H}_{qp} + \mathbf{H}_{er} \\ \mathbf{H}_\omega & = \int_M dm\, \mathbf{r} \times (\omega \times \mathbf{r}) = \mathcal{M}\omega \\ & \text{is the main part given by the Earth's rotation,} \\ & \text{its modulus is about 5.86p33 kg m}^2\text{s}^{-1}, \\ \mathbf{H}_{qp} & \text{is given by the tidal quasi-periodic component of } \mathbf{v}, \\ & \text{its modulus is about p27 kg m}^2\text{s}^{-1}, \\ \mathbf{H}_{er} & \text{is given by the remaining erratic part of } \mathbf{v}, \text{ essentially the} \\ & \text{oceanic currents; its modulus is smaller than 5p24 kg m}^2\text{s}^{-1}, \end{array}\right. \tag{7}$$

And thus finally:

$$\left(\frac{d\mathbf{H}}{dt}\right)_{\text{Oxyz}} = (\mathcal{M}\omega + \mathbf{H}_{qp} + \mathbf{H}_{er}) \times \omega\ . \tag{8}$$

The term $\mathbf{H}_{er} \times \omega$ is very small (an erratic polar displacement of less than 4 cm per day).

The term $\mathbf{H}_{pq} \times \omega$ is larger but with an average equal to zero. Its effect on one month is already very small and its effect in a few years is negligible.

The first term $(\mathcal{M}\omega) \times \omega$ is small for the present position of poles *because the present polar axis is practically a main axis of inertia* (the main axis of largest inertia). At 45° from that position this term would give an enormous eastward polar displacement of 65 km per day...

We must of course object that with such an axis the present Earth would not be in isostatic equilibrium and especially it would not keep its present equatorial bulge.

Hence we need to know the relation between the direction of ω and the inertia matrix $\mathcal{M}$ of an isostatic Earth.

A first approximation of this matrix is related with the four equilibria of figure 2.

The rotating and isostatic Earth corresponds to the figure 2D with the matrix $\mathcal{M}$ or $\mathcal{M}_D$ we are looking for. The inertia matrices $\mathcal{M}_A$ and $\mathcal{M}_C$ of the equilibria A and C are, by definition, independent of ω. Furthermore A and $\mathcal{M}_A$ have the spherical symmetry:

$$\mathcal{M}_A = \begin{pmatrix} I_A & 0 & 0 \\ 0 & I_A & 0 \\ 0 & 0 & I_A \end{pmatrix} , \tag{9}$$

where I_A is very slightly smaller than the average $(A + B + C)/3$ of (1) which is 0.329956 MR^2. $\mathcal{M}_C$ differs from $\mathcal{M}_A$ by about ten millionths.

The hydrostatic equilibrium B has the spheroidal symmetry, its inertia matrix $\mathcal{M}_B$ is function of four parameters only: the direction of its polar axis (two parameters) and its equatorial and polar moments of inertia A_B and C_B :

$$\mathcal{M}_B = \begin{pmatrix} A_B & 0 & 0 \\ 0 & A_B & 0 \\ 0 & 0 & A_B \end{pmatrix} + (C_B - A_B) \begin{pmatrix} \alpha^2 & \alpha\beta & \alpha\gamma \\ \alpha\beta & \beta^2 & \beta\gamma \\ \alpha\gamma & \beta\gamma & \gamma^2 \end{pmatrix} \tag{10}$$

where (α, β, γ) are the coordinates of the unit vector in the direction of ω.

The principle of small modifications allows us to write:

$$\mathcal{M}_D - \mathcal{M}_C = \mathcal{M}_B - \mathcal{M}_A + \text{"second order"} . \tag{11}$$

If we neglect the second order term (about n8 MR^2) and the difference $\mathcal{M} - \mathcal{M}_D$ between the true Earth and its isostatic approximation (about n7 MR^2) we can write:

$$\mathcal{M} \approx \mathcal{M}_D \approx \mathcal{M}_C + \mathcal{M}_B - \mathcal{M}_A , \tag{12}$$

and thus, since we always have $[(\mathcal{M}_B - \mathcal{M}_A)\omega] \times \omega = 0$,

$$\left(\frac{d\mathbf{H}}{dt}\right)_{\mathrm{Oxyz}} \approx (\mathcal{M}\omega) \times \omega \approx (\mathcal{M}_C\omega) \times \omega . \tag{13}$$

During the periods of tectonic quiescence on Earth the matrix $\mathcal{M}_C$ is constant or quasi-constant and the pole equilibrium positions correspond to $(d\mathbf{H}/dt)_{\mathrm{Oxyz}} = O$ that is to the eigenvectors of $\mathcal{M}_C$ and thus to its main axes of inertia.

The direction of stable equilibrium corresponds to the main axis of largest inertia of $\mathcal{M}_C$ and we will now verify that this axis is also the present polar axis and the present main axis of largest inertia of $\mathcal{M}$.

This situation is of course that of stable Euler-Poinsot rotations and has been recognize in Ref 11 as necessary for a stable Earth rotation.

Numerically the present value of $\mathcal{M}_C$ can be obtained from those of $\mathcal{M}$, $\mathcal{M}_A$, $\mathcal{M}_B$ with the present (α, β, γ) = (0,0,1): $\mathcal{M}_A$ is given in (9) and $\mathcal{M}_B$ in (10):

$$\mathcal{M}_B = \begin{pmatrix} A_B & 0 & 0 \\ 0 & A_B & 0 \\ 0 & 0 & C_B \end{pmatrix} , \tag{14}$$

where $(2A_B+C_B)/3 \approx I_A = MR^2(0.329956\pm \text{n}6)$, and C_B-A_B=1.070978n3MR^2 (Ref. 5, appendix 2).

$\mathcal{M}$ is given by the study of the Earth's potential (Ref.5, 8-10):

$$\mathcal{M} = MR^2 \begin{pmatrix} U - 3.64025\text{n}4 & 1.807\text{n}6 & 0 \\ 1.807\text{n}6 & U - 3.57727\text{n}4 & 0 \\ 0 & 0 & U + 7.21752\text{n}4 \end{pmatrix}, \quad (15)$$

with U less accurately known: $MR^2U = (A + B + C)/3$;$U = 0.329956 \pm \text{n}6$.

Finally, with $\mathcal{M}_C \approx \mathcal{M} + \mathcal{M}_A - \mathcal{M}_B$:

$$\mathcal{M}_C = MR^2 \begin{pmatrix} U_C - 7.032\text{n}6 & 1.807\text{n}6 & 0 \\ 1.807\text{n}6 & U_C - 7.34\text{n}7 & 0 \\ 0 & 0 & U_C + 7.766\text{n}6 \end{pmatrix} + \mathcal{O}(\text{n}7MR^2) \text{ ,(16}$$

with again U_C less accurately known: $U_C = 0.329956 \pm \text{n}6$.

Let us recall that as long as the Earth remains in a phase of tectonic quiescence the matrix $\mathcal{M}_C$ is constant or quasi-constant; it is also independent of the rotation vector ω which is not the case of $\mathcal{M}_B$, $\mathcal{M}_D$ and $\mathcal{M}$.

The corresponding main moments of inertia are then:

$$\begin{cases} A_C = (U_C - 7.514\text{n}6 \pm \text{n}7)MR^2 \\ B_C = (U_C - 2.52\text{n}7 \pm \text{n}7)MR^2 \\ C_C = (U_C + 7.766\text{n}6 \pm \text{n}7)MR^2 \,. \end{cases} \quad (17)$$

We can verify that C_C remains along the polar axis and remains the largest moment of inertia even if its differences with B_C and A_C are much smaller than $(C - B)$ and $(C - A)$ given in (1). The present poles are truly at their position of stable equilibrium.

The Earth is not isolated in space and we must consider the outer torques that give the tidal motions and also the precession and the nutation of the Earth's axis.

Fortunately the Earth rotation is fast and even if the motion of the Earth's axis with respect to the stars is large, the effect of outer torques on the relative Poles - Earth's crust motion is reduced by the lack of resonance to daily quasi-periodic variations of negligible amplitude [Ref.5, pages 538-540].

Thus even if the matrix$\mathcal{M}_C$ is much more "spherical" than the matrix $\mathcal{M}$ the term $(\mathcal{M}_C\omega) \times \omega$ remains in (13) and (8) the main term of $(d\mathbf{H}/dt)_{\text{Oxyz}}$.

That term is of course zero at the pole stable equilibrium position but, even if it doesn't reach 65 km/day as above, it would nevertheless give an eastward polar displacement of several hundreds of meters per day at 45° from the stable equilibrium position; with all the corresponding huge losses of energy.

Hence, at geological scale, the stability of the position of the poles with respect to the Earth crust is a strong one. The poles follow very nearly the possible slow variations of their positions of stable equilibrium, variations that correspond to these of the Earth crust itself.

8. Magnetic Poles and Geographic Poles

What information can we deduce from the motion of magnetic poles ?

The paleomagnetic data given in Ref 12 present the slow long-term velocities of magnetic poles with respect to several lithospheric plates (these of Africa, Eurasia, India and North America).

Since 200 million years these velocities never exceed 1.5 degree per million years ; while the velocity called “true polar wander", and corresponding to the long-term motion of magnetic poles with respect to the Earth mantle, remains always less than 0.7 degree per million years.

Unfortunately the relations between rotational and magnetic axes are complex. Their angle is presently 47° for planet Neptune and even 60° for planet Uranus.

For the Earth the magnetic axis has apparently many large short-period erratic variations and its minimum latitude was about 60° in the last 10 000 years.

The paleomagnetic studies generally assume that, over 20 000 years, the average direction of the magnetic axis is within 2° to 5° of that of the geographic axis ; which seems to be true in the last million years.

Taking account of all these precautions the reconstitution of past long-term polar motions shows indeed periods of quiescence with very slow relative motions of lithospheric plates and periods of disequilibrium with a general tilting of the Earth crust. These tiltings can reach 10° to 20° in 10 to 20 million years.

9. Conclusion

The study of the Earth's rotation and of the displacement of its poles over geological time is a complex but also very beautiful undertaking implying many domains of science from astronomy, mechanics and space dynamics to geology, geophysics, gravimetry, rheology and seismology.

The positions of poles at the Earth's surface are usually very near to their positions of stable equilibrium which are given by the main axis of largest inertia of the “non rotating Earth in isostatic equilibrium”. The corresponding relaxation time is usually about 100 000 years.

The stable equilibrium positions depend very much on the variations of the Earth's crust and the poles have certainly moved much more than the different lithospheric plates. Large orogenies and/or large subductions can move the poles by tens of degrees within a few million years.

We can hope that plate tectonics and paleogeography will be able to yield an accurate description of the successive polar shifts.

References

[1] Bursa, M.: 1992, Parameters of common relevance of Astronomy, Geodesy and Geodynamics, *Geodetic bulletin*,**66**, (The Geodesist's Handbook - Tscherning C.C. ed.)

[2] Bureau des Longitudes: 1984, *Encyclopédie Scientifique de l'Univers - La Terre, les eaux, l'atmosphère*, Gauthier Villars.

[3] Dziewonsky, A.M. and Anderson, D.L.: 1981, Preliminary reference Earth model, *Physics of the Earth and planetary interiors*,**25**, 297–356.
[4] Arias, E.F., Charlot, P., Feissel, M. and Lestrade, J.F.: 1995, The extragalactic reference system of the International Earth Rotation Service, I.C.R.S., *Astron. and Astrophys.*, **303**, 604.
[5] Marchal, C.; 1996, Earth's polar displacements of large amplitude: a possible mechanism, *Bulletin du Muséum d'Histoire Naturelle*, - Paris - 4è sfie - Section C - Tome 18 - N°2-3.
[6] Marchal, C.:1968, Figures d'équilibre séculairement stables des masses fluides hétérogènes en rotation,*Bulletin astronomique*, - Série 3 - Tome III - Fascicule 3, 341-360.
[7] Ricard, Y., Spada, G. and Sabadini, R.: 1993, Polar wandering of a dynamic Earth, *Geophysical Journal*, **113**, 284–298.
[8] Balmino, G., Reigber, Ch. and Moynot, B.: 1978, Le modèle de potentiel gravitationnel terrestre GRIM2, *Annales de Géophysique*, **34**, 55–78.
[9] Lerch, F.J., Klosko, S.M., Laubscher, R.E. and Wagner, C.A: 1979 ,Gravity model improvement using GEOS-3 (GEM-9 and 10), *Journal of Geophys. Res.*, **84**, 3897–3915).
[10] Reigber, CH., Schwintzer, P., Barth, W., Massmann, F.H., Raimondo, J.C., Bode, A., Li, H., Balmino, G., Biancale, R., Moynot, B., Lemoine, J.M., Marty, J.C., Barlier, F. and Boudon, Y.: 1993, GRIM4-C1-C2p : Combination solutions of the global Earth gravity field, *Surveys in Geophysics*, **14**, 381–393.
[11] Goldreich, P. and Toomre, A.: 1969, Some remarks on polar wandering, *Journal of Geophys. Res.*,**74**, 2555–2567.
[12] Besse, J. and Courtillot, V.:1991, Revised and synthetic apparent polar wander paths of the African, Eurasian, North American and Indian plates, and true polar wander since 200 Ma, *Journal of Geophys. Res.*, **96**, 4029–4050.

ORBITAL EVOLUTION OF ASTEROIDS IN THE HECUBA GAP

S.FERRAZ-MELLO and T.A.MICHTCHENKO
Instituto Astronômico e Geofísico,
Universidade de São Paulo,
Caixa Postal 9638, São Paulo, SP, Brasil.
Email sylvio@vax.iagusp.usp.br

Abstract. Numerical integrations of the motion of an asteroid in the Hecuba gap under the perturbations of Jupiter and Saturn. Most of the runs led to an escape from the gap. Examples are shown of the importance of resonances of f_{ϖ} and g_6 in the increase of the libration amplitude and subsequent escape.

Key words: Asteroid – Hecuba gap

1. Introduction

In previous papers (Ferraz-Mello, 1994a,b; Ferraz-Mello *et al.*, 1995), we have collected several evidences pointing that the global stochasticity observed inside the 2/1 asteroidal resonance is enough to deplete this resonance in a timescale in the range $10^7 - 10^9$ years and can, thus, explain the origin of the Hecuba gap. It was also shown that the only notorious difference between the motion of an asteroid in the Hecuba gap and those forming the Hilda group is the slower diffusion timescale of the 3/2 resonance. Our calculations have indicated, in the Hilda group, a characteristic time scale generally larger than the age of the solar system, which is consistent with the existence of about 60 observed asteroids in this group, notwithstanding all similarities between the dynamics of the 2/1 and 3/2 resonances.

This theory contradicted some previous ideas in several respects. First of all, it is well known that the inner chaotic region of the 2/1 resonance, in the planar elliptic restricted three-body problem, is shielded from the outside by an extended bunch of regular motions (see the Poincaré maps in Ferraz-Mello 1994a,b). It is also known that, when the long-period perturbations of Jupiter's orbit are taken into account, these regular barriers are not destroyed (Morbidelli and Moons, 1993). Thus, the possibility of a chaotic diffusion fast enough to drive these orbits off this region in a physically significant time was discarded. On the other hand, notwithstanding the fact that all Lyapunov times computed in the resonant region point to a global stochasticity of the same kind observed when secondary resonances overlap (Chirikov regime), some expectation that Nekhoroshev's inequality could prevail in the resonance region, limiting the diffusion, existed. But, the best estimation obtained for the superior limit of the mass of the disturbing body, allowing the Nekhoroshev theorem to be applied to the restricted three-body problem, is more than 20 orders of magnitude smaller than Jupiter's mass, ruling this question (Niederman, 1995).

R. Dvorak and J. Henrard (eds.), The Dynamical Behaviour of our Planetary System, 377.

The most serious criticisms against our conclusions came from the almost absence of precise simulations actually showing the diffusion. However, they now exist and show similar results. Up to very recently, sufficiently extended numerical integrations of fictitious asteroids in the 2/1 resonance showing a significant increase in eccentricity were almost not available. They were some of the 5 numerical integrations reported by Wisdom (1987) and one by Scholl (see fig. 7 in Ferraz-Mello, 1994b). This situation has improved. Workstations of high performance could be dedicated largely to this problem and many runs showing macroscopic instabilities able to drive the asteroid off the 2/1 resonance were published (Henrard *et al.*, 1995; Morbidelli, 1996). However, in what concerns the asteroids of Hilda group, many billions of years are necessary and no such long runs are so far possible.

In the meantime, some other results were obtained. For instance, using a symplectic mapping, we have shown that planar averaged models can depict a strong chaotic behaviour when the short-period perturbations whose arguments are $(2\lambda_{\rm Jup} - 5\lambda_{\rm Sat})$ and $(\lambda_{\rm Jup} - 2\lambda_{\rm Sat})$ are added to the long-period perturbations of Jupiter's orbit. The mapping also confirmed the huge difference between the timescale of the long-term orbital diffusion of the 3/2 and 2/1 resonances.

The stochasticity observed in spatial models was explained by Henrard *et al.* (1995) by means of an extension of the previously used averaged models so as to include the secular resonances involving the nodes of the asteroid and Jupiter's orbits. That paper confirmed the importance of the third dimension in the diffusive orbital processes in the Hecuba gap and showed paths, through moderate inclinations, leading from the inner chaotic region ($e \sim 0.1$) to high eccentricities, in a few tens of Myrs.

This paper summarizes almost one hundred simulations of asteroids in the 2/1 resonance, most of which showing significant increase in eccentricity in a time scale of the order $10^7 - 10^8$ years. All simulations were done using a three-dimensional Sun–Jupiter–Saturn–asteroid model.

2. The Simulations

The integrations reported here aimed at studying the modal transitions during the orbital evolution of an asteroid in the Hecuba gap. Everhart's integrator RA15 (Everhart, 1985) was chosen for its local precision even in high eccentricities. Low-pass filters were used to eliminate short-period oscillations. These filters ranged from 100 years, just filtering out the short-period perturbations of the asteroid orbit, to 1000 years, in which case also the libration is filtered out. The errors of the whole procedure in a timespan of 200 Myrs, checked with a two-body unperturbed model are of order 10^{-10} what is far less than the errors due to the fact that the model does not include the inner planets. Generally, the runs were followed up to the escape of the fictitious asteroid from the gap, which happened in a timescale between 15 and 150 Myr. In several cases, an escape was not observed in the run time (200

TABLE I
Results and initial orbital elements of the asteroid

Run	Semi-axis	Eccentricity	Result	T_{esc}
37B	3.276 AU	0.12	$e \Rightarrow 1$	80.5 Myr
13B	3.276	0.32	$e \Rightarrow 1$	15.3 Myr
14B	3.276	0.28	$e \Rightarrow 1$	28.8 Myr
15B	3.276	0.24	$e \Rightarrow 1$	9.8 Myr
16B	3.276	0.20	$e \Rightarrow 1$	61.0 Myr
18B	3.280	0.32	$e \Rightarrow 1$	13.7 Myr
19B	3.280	0.26	$e < 0.37; i < 6°$	> 200 Myr
25B	3.284	0.28	$e \Rightarrow 1$	136.7 Myr
23B	3.284	0.26	$e \Rightarrow 1$	62.3 Myr
26B	3.284	0.20	$e \Rightarrow 0.999$	24.3 Myr
41B	3.288	0.12	$e < 0.23; i < 8°$	> 200 Myr
29B	3.288	0.02	$e \Rightarrow 1$	45.5 Myr
27B	3.288	0.20	$e \Rightarrow 1$	50.0 Myr
47B	3.288	0.23	$e \Rightarrow 1$	57.2 Myr
44B	3.292	0.26	$e \Rightarrow 1$	57.2 Myr
30B	3.296	0.20	$e < 0.32; i < 24°$	> 200 Myr
40B	3.300	0.12	$e \Rightarrow 1$	78.9 Myr
48B	3.300	0.16	$e < 0.23; i < 7°$	> 200 Myr
32B	3.300	0.26	$e \Rightarrow 1$	112.3 Myr
49B	3.305	0.16	$e \Rightarrow 0.999$	50.5 Myr
50B	3.310	0.16	$e \Rightarrow 0.999$	196.7 Myr
45B	3.312	0.02	$e \Rightarrow 1$	5.8 Myr
42B	3.312	0.12	$e \Rightarrow 1$	25.7 Myr
33B	3.312	0.20	$e \Rightarrow 1$	118.7 Myr
34B	3.312	0.26	$e \Rightarrow 0.999$	20.2 Myr
51B	3.315	0.16	$e \Rightarrow 1$	77.2 Myr
35B	3.320	0.26	$e \Rightarrow 1$	12.7 Myr
43B	3.324	0.12	$e \Rightarrow 1$	173.1 Myr
31B	3.328	0.20	$e < 0.24; i < 4°$	> 200 Myr
46B	3.334	0.20	$e \Rightarrow 0.39; a \Rightarrow 4$AU	67.3 Myr
38B	3.346	0.20	$e \Rightarrow 1$	24.3 Myr

Myr). These integrations are valid only as explorers of the gap; the escape times in these runs are of the same order of those previously reported by Wisdom, Scholl and Henrard *et al.*, but much smaller than those recently obtained by Morbidelli (1996). The experiments done have shown that the results are extremely sensitive to the initial value of the angular elements of the asteroid and Jupiter. In particular, the diffusion speed is affected by the initial frequency of the angle $2\lambda_{Jup} - 5\lambda_{Sat}$ (Michtchenko & Ferraz-Mello, 1997). In order to verify the effects of using different integration techniques, the fictitious asteroid for which Morbidelli found a lifetime 469 Myrs was integrated with our programmes and survived the standard 200 Myr.

TABLE II
Adopted initial orbital elements

Element	Jupiter	Saturn	Asteroid
Semi-major axis (AU)	5.2031024	9.5219707	(see Table 2)
Eccentricity	0.048273	0.0542255	(see Table 2)
Longitude of Perihelion	180°	257.95053°	180°
Mean Anomaly	180°	175.88022°	0°
Inclination	1.3°	2.48242°	1°
Longitude of Node	265°	278.20601°	0°

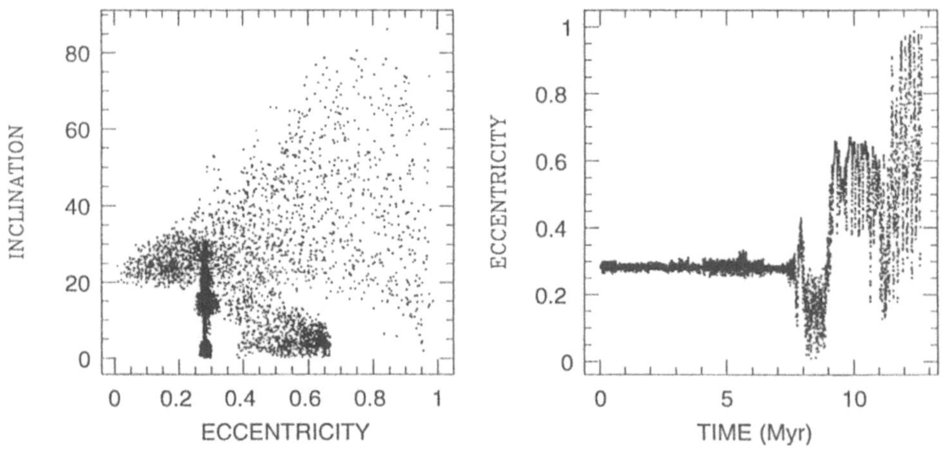

Fig. 1. Evolution of orbital parameters in run 35B

Attention has been also paid to the fact that the results of these simulations are spanned over some thousand of Lyapunov times making them sensitive to small changes in the integration parameters.

Some of the results obtained are shown in Table 1. The initial orbital elements corresponding to these runs are shown in Table 2. The values adopted for Jupiter and Saturn are close to the osculating elements of these planets in the epoch JD 2448800.5 (June 27th, 1992), except the initial mean anomaly of Jupiter taken as 180° (instead of the actual 151°). The whole system was rotated in such a way that Jupiter's longitude of the perihelion at the epoch is 180°. The initial conditions are such that $\sigma_0 = (2\lambda_{\rm Jup} - \lambda - \varpi)_0 = 0$ and $(\varpi - \varpi_{\rm Jup})_0 = 0$.

Figures 1 and 2 show two typical evolution patterns. In both cases $e_0 = 0.26$ and $i_0 = 1°$. Also, in both cases the initial libration amplitude is large. The 'fountain' pattern of fig. 1 is typical and, with small differences, visible in all runs starting with low inclination and eccentricities in the range $0.2 - 0.3$. The solution spent a long time in the vertical 'jet' before spreading out. Generally, the orbit remains in the

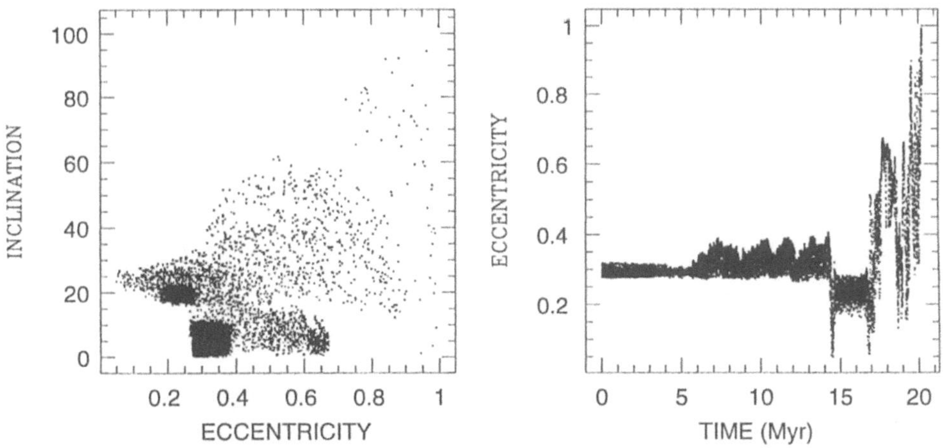

Fig. 2. Evolution of orbital parameters in run 34B

jet up to reach the Kozai resonance (when $i \sim 30°$). The fast and large fluctuations in e and i start and remain up to the final transition to high eccentricity. In figure 2, the 'fountain' has a different shape, but the evolution is similar. It is worth noting that after escaping from the jet, the solution remained stuck for almost 3 Myr in the dark spot at $i \sim 20°$. In the runs leading to a later escape, the remarkable point is that the solution may remain in the 'jet' for more than 100 Myr before evolving to high inclinations and eccentricities.

3. Chaotic Diffusion

In order to understand how the solutions escape from the 2/1 resonance, we have studied the chaotic diffusion of some solutions in the phase space through the analysis of their power spectra and the time evolution of independent frequencies. Here, we concentrate our attention in the middle-eccentricity region of phase space with initial small amplitudes of libration and small inclinations. This region appears stable in low-order analytical models (Henrard *et al*, 1995), but it was conjectured that very slow chaotic diffusion could be forced by secondary and secular resonances of higher order. The results of simulations done by Morbidelli (1996) pointed out that some solutions with initial elements in this region can survive even 1-Gyr runs, notwithstanding the chaotic behaviour.

We have applied the techniques of Fourier and wavelet spectral analyses to several solutions inside this region. To access the character of asteroidal motion, the Fourier spectral analysis of 5-Myr runs was done. The results show broad-band spectral components, mainly in the semi-major axis power spectrum, a feature due to the time variation of the fundamental frequencies, which is characteristic of chaotic behavior (Michtchenko & Ferraz-Mello 1995). Some previous studies done in the frame of Sun–Jupiter–asteroid model, have shown that the spectra of

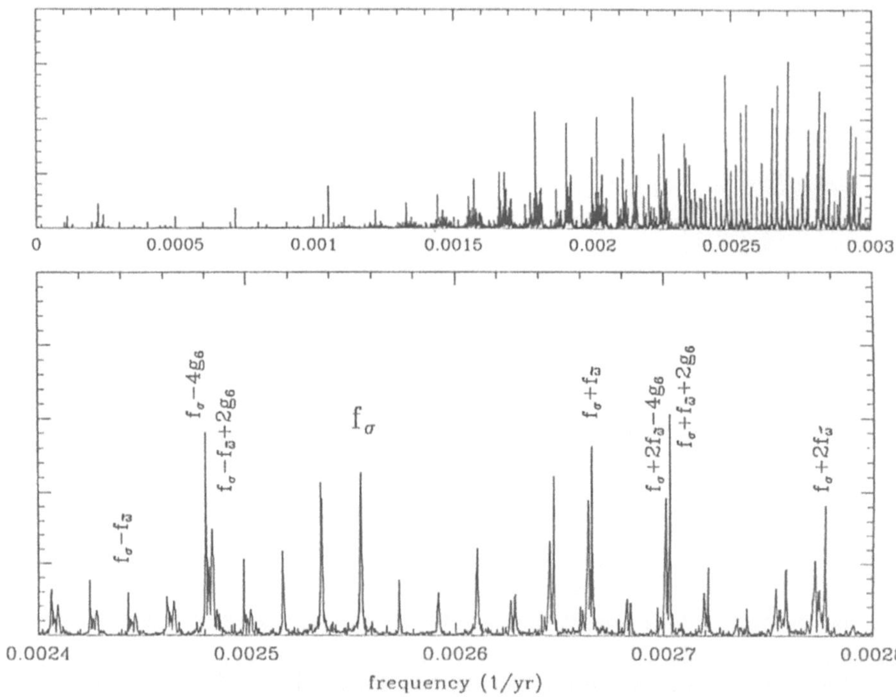

Fig. 3. *Top:* Semi-major axis power spectrum of the solution of initial elements $a_0 = 3.283$AU, $e_0 = 0.3$, $i = 1°$. *Bottom:* Expanded view around the region of the frequency of libration, f_σ, showing the various spectral lines corresponding to linear combinations $f_\sigma \pm m\, f_\varpi \pm n\, g_6$ (The frequency unit is rev/year).

asteroidal motions with small libration amplitude in the middle-eccentricity region have a complex structure with many spectral lines of comparable amplitude formed by the linear combination of the independent frequencies (It is worth emphasizing that this is a notorious difference between the *Hecuba* gap and *Hilda* group). The action of Saturn introduces many additional spectral lines of comparable amplitudes, mainly associated with the proper frequency of Saturn's perihelion, g_6, which create multiplets of higher order with other asteroidal modes. In the expanded view shown in Fig. 3, many groups of lines are seen at frequencies which are beats of g_6 with $f_\sigma \pm m f_\varpi$ ($m = 0, 1, 2, ...$). Because of the resonance $f_\varpi / g_6 \approx 6$ which occurs in this example, many of these lines tend to group in a series of equidistant sets. This occurs always when f_ϖ / g_6 is close to an integer; besides, when this almost commensurability occurs, the differences between the many lines of each set are small and may become of the order of the proper frequency of Jupiter's perihelion, g_5 (which is almost 7 times smaller than g_6), giving rise to new higher-order secondary resonances involving g_5. The experiments done clearly show that

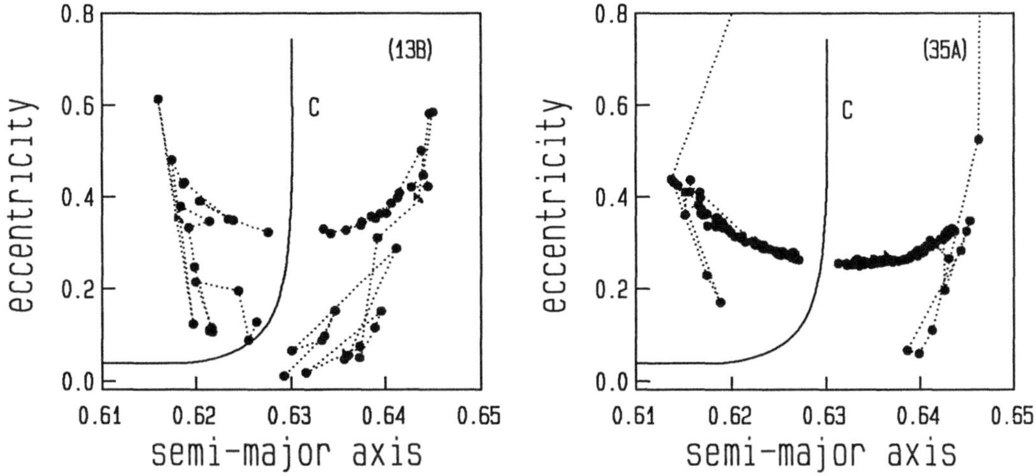

Fig. 4. Intersection of the solutions 13B and 35A with the surface defined by $\sigma = \Delta\varpi = \Delta\Omega = 0$; $\Delta\dot{\varpi} < 0$ and $\Delta\dot{\Omega} < 0$ (solution 35A is similar to 35B).

these secondary resonances always occur when a faster diffusion is verified.

When the inclination of the orbit varies only slightly, the chaotic diffusion may be studied also in the (a,e)–plane of the 2/1 resonance through the plot of the intersections of the solution with the surface defined by $\sigma \simeq \Delta\varpi \simeq \Delta\Omega \simeq 0$, for which $\Delta\dot{\varpi} < 0$ and $\Delta\dot{\Omega} < 0$. This representation gives us information about the directions of chaotic diffusion and the size of the overlapping zone. Due to the fact that, for small inclination, the regions of the main secular and secondary resonances in the (a,e)–plane are well known, the paths of asteroidal escape may be achieved. Two examples with fast diffusion are shown in fig.4. In these examples, the orbits move away from the pericentric branch (locus of stable libration centers) along the lines $f_\varpi/g_6 = 5$ and $f_\varpi/g_6 = 6$, respectively, up to reach, after some time, large amplitude of libration; then, they reach the regions of the phase space where the stronger secular resonance ν_{16} provokes the increase of the asteroidal inclination, the solution is driven to the domain of influence of the ν_5 and Kozai resonances and to huge increases in eccentricity and escape from the resonance.

Acknowledgements

This investigation was supported by the Research Foundation of São Paulo State (FAPESP).

References

Everhart, E.: 1985, "An efficient integrator that uses Gauss-Radau spacings", In *IAU Colloquium 83*, edited by A.Carusi and G.B.Valsecchi (Reidel, Dordrecht), 185.

Ferraz-Mello, S.: 1994a, "Kirkwood gaps and groups", In Proc. IAU Symposium 160, edited by A.Milani *et al.* (Kluwer, Dordrecht), pp. 175-188.

Ferraz-Mello, S.: 1994b, "Dynamics of the 2:1 asteroidal resonance". *Astron. J.* **108**, 2330-2337.

Ferraz-Mello, S.: 1996 "A symplectic mapping approach to the study of the stochasticity of asteroidal resonances" *Cel. Mech. Dyn. Astr.* (submitted).

Ferraz-Mello, S., Dvorak. R. & Michtchenko, T.A.: 1995, "Chaos and the depletion of asteroids in resonant orbits", In *From Newton to Chaos*, edited by A.E.Roy (Plenum Press, New York), pp. 157-169.

Henrard, J., Watanabe, N. and Moons, M.: 1995, "A bridge between secondary and secular resonances inside the Hecuba gap" *Icarus* **115**, 336-346.

Michtchenko, T.A. and Ferraz-Mello, S.: 1995, "Comparative study of the asteroidal motion in the 3/2 and 2/1 resonances with Jupiter. I. Planar model." *Astron. Astrophys.* **303**, 945-963.

Michtchenko, T.A. and Ferraz-Mello, S.: 1996, "Comparative study of the asteroidal motion in the 3/2 and 2/1 resonances with Jupiter. II. Three-dimensional model." *Astron. Astrophys.*, 1021-1035.

Michtchenko, T.A. and Ferraz-Mello, S.: 1997, "Escape of orbits from the Hecuba gap" *Planet. Sp. Sci.* (submitted).

Morbidelli, A.: 1996, "On the Kirkwood gap at the 2/1 commensurability with Jupiter" *Astron.J.* (submitted).

Morbidelli, A. and Moons, M.: 1993, "Secular resonances in mean-motion commensurabilities. The 2/1 and 3/2 cases". *Icarus* **102**, 316-332.

Niederman, L.: 1995, "Stability over exponentially long times in the planetary problem" In *From Newton to Chaos*, edited by A.E.Roy (Plenum Press, New York), pp. 109-118.

Wisdom, J.: 1987, "Chaotic dynamics in the Solar System". *Icarus* **72**, 241-275.

A STUDY OF THE LOCAL LYAPUNOV NUMBERS FOR ORBITS IN THE OUTER SOLAR SYSTEM

E. LOHINGER

Institut für Astronomie, Universität Wien, Türkenschanzstraße 17, A-1180 Vienna, Austria.

Abstract. We present the distributions of the local Lyapunov characteristic numbers (LLCNs) of Pluto's orbit and of some fictitious neighbouring orbits in the outer solar system, where the Sun (whose mass was increased by the masses of the inner planets) and the giant planets (Jupiter to Neptune) represent the perturbing bodies. According to a former study – where the extension of a "stable region" near Pluto's orbit has been determined – we compared the distributions of the LLCNs of orbits inside this region with those outside. Furthermore, the study of the LLCNs of each orbit was carried out for different time intervals.

Key words: Lyapunov numbers – Pluto – chaotic motion

1. Introduction

Former studies of the outer solar system indicated a regular motion for the giant planets and a chaotic behaviour for Pluto's orbit – since its largest Lyapunov exponent (LCE) was found to be positive (cf. Sussman & Wisdom, 1988 and 1992).

The extension of the study of LCEs, i.e. the computation of the LCEs of Pluto's orbit and some neighbouring orbits led to a "stable" zone[1] in the vicinity of Pluto's orbit and an adjoined strong chaotic zone (cf. Dvorak & Lohinger, 1996). In this former study the computations were carried out for three models – the circular and the elliptic restricted three body problem and the outer solar system – for a time interval of 1 million years.

The present numerical investigation about the motion of Pluto and 9 fictitious neighbouring bodies in the outer solar system was carried out for a time interval of about 20 million years, where we have found again a "stable" zone by means of the distributions of the local Lyapunov characteristic numbers (LLCNs).

The first study of the LLCNs was carried out in 1992, where the distributions were analyzed by calculating the moments of a distribution – the mean, *r.m.s*, asymmetry and flatness (cf. Froeschlé et al., 1993; Lohinger et al., 1993). In the following years the analysis of the LLCNs was continued by many scientists (e.g. Voglis & Contopoulos, 1994; Contopoulos et al., 1995; Smith & Contopoulos, 1995) whereby a new expression – the *stretching parameter* (Contopoulos) – was introduced.

In the study of Voglis & Contopoulos (1994) it was shown, among others, that the distributions were invariant for orbits of the standard map.

A similar study about the invariability of the distributions is presented in this paper.

[1] In this context "stable" means that the LCE of an orbit tends to zero within the considered integration time

R. Dvorak and J. Henrard (eds.), The Dynamical Behaviour of our Planetary System, 385.

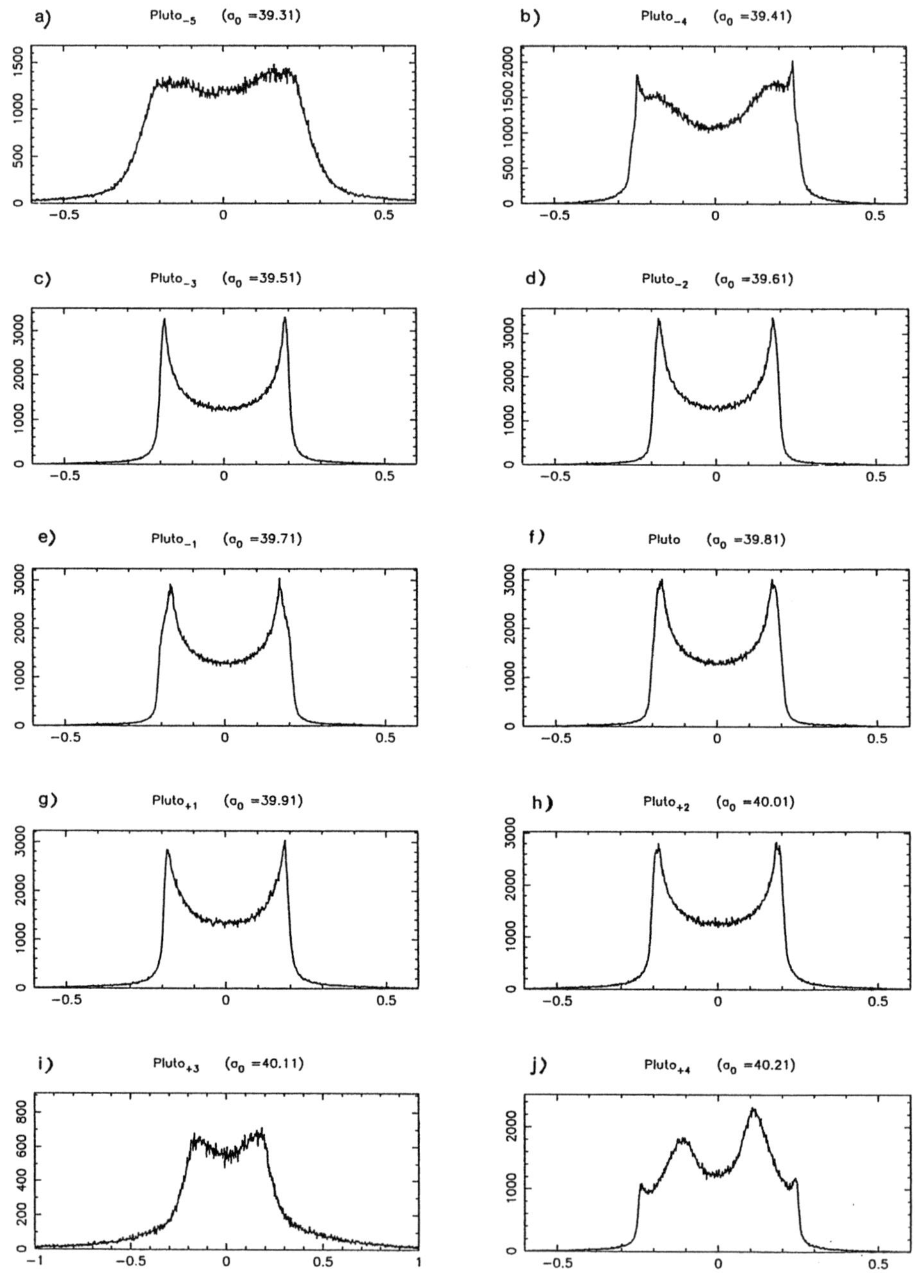

Fig. 1. The distributions of the LLCN[1] of the different orbits computed over the whole integration time of 20 million years. The shape of the distributions of the orbits in the "stable" zone (c)-(h) is quite uniform (bimodal an symmetric around zero), while in case of the chaotic orbits (a)-(b) and (i)-(j) the shape varies significantly.

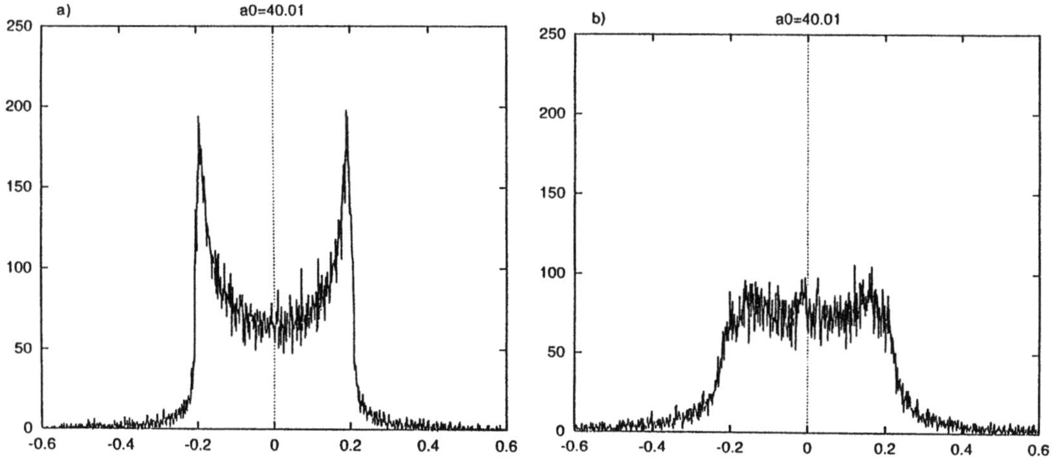

Fig. 2. (a) The typical shape of a distribution of a subinterval of 1 million years of orbits in the "stable" zone. (b) The strongest variation of a distribution of the outer orbits in the "stable" zone.

2. The Dynamical Model and the Computations of the LLCNs

For the computations we considered the Sun – whose mass was increased by the masses of the inner planets – and the giant planets Jupiter to Neptune as massive bodies and studied the dynamical behaviour of Pluto and some orbits in its vicinity, i.e. the neighbouring orbits resulted from a variation of Pluto's initial semi-major axis:

$$a_j = a_{Pluto} - 0.1 \cdot j \quad [AU] \qquad \text{with } j = 1, \ldots, 5$$
$$a_k = a_{Pluto} + 0.1 \cdot k \quad [AU] \qquad \text{with } k = 1, \ldots, 4$$

the other orbital elements were the same as for Pluto's orbit.

The numerical calculations of the dynamical system were carried out by means of the Burlisch-Stoer integration method. The program, which was provided by R. Gonczi (of the OCA Nice), employs the Gram-Schmidt procedure to replace the evolved tangent vectors by new orthonormalized ones, therefore, the local Lyapunov characteristic numbers are defined as:

$$LLCN^i = \frac{1}{\tau} \cdot \ln ||(\mathbf{v}_k^i)_\perp||$$

i.e., the logarithm of the length of the component of the evolved vector $\mathbf{v}_k^i$ (i depends on the dimension of the phase space and k defines the integration time when we renormalize) which is orthogonal to all previously orthonormalized vectors divided by the renormalizations time — in our case $\tau = 30$ years.

As mentioned above all orbits were analyzed for a time of about 20 million years.

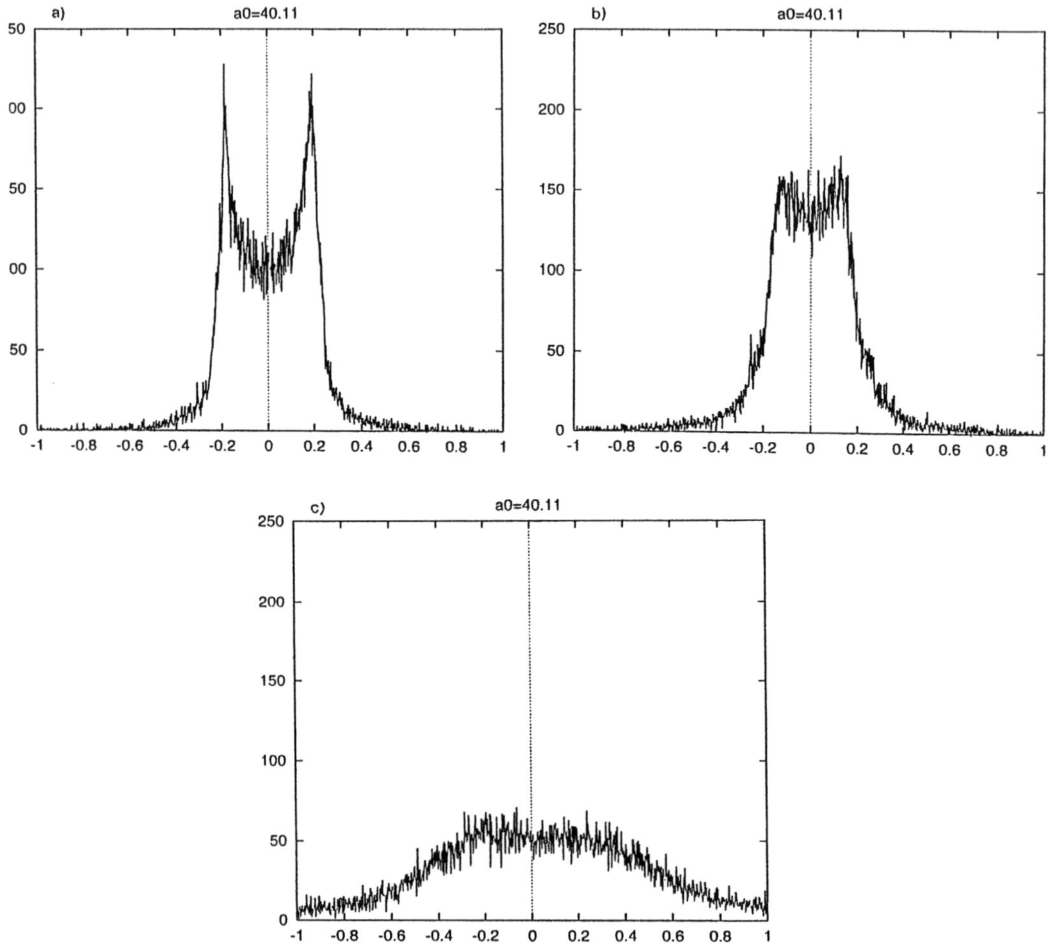

Fig. 3. The variation of the distributions in 3 subintervals of the orbit of Pluto_{+3} ($a_0 = 40.11$).

3. Results

The distributions of the LLCN^1 (which yields to the largest – the most important – LCE) for the different orbits computed over the whole integration time of 20 million years are shown in fig.(1). Here, the x-axis is defined by $[min(\text{LLCN}^1), max(\text{LLCN}^1]$ and is divided into 1000 bins. This study defines a "stable" motion for the orbits in the range $39.51 \leq a_0 \leq 40.01$ (from Pluto_{-3} to Pluto_{+2}), which is characterized by a bimodal distribution which is symmetric around zero so that the LCE tends to zero within the considered time interval. Furthermore, the shape of the distributions within this zone is quite uniform. In contrast thereto we have found significant changes for the distributions in case of

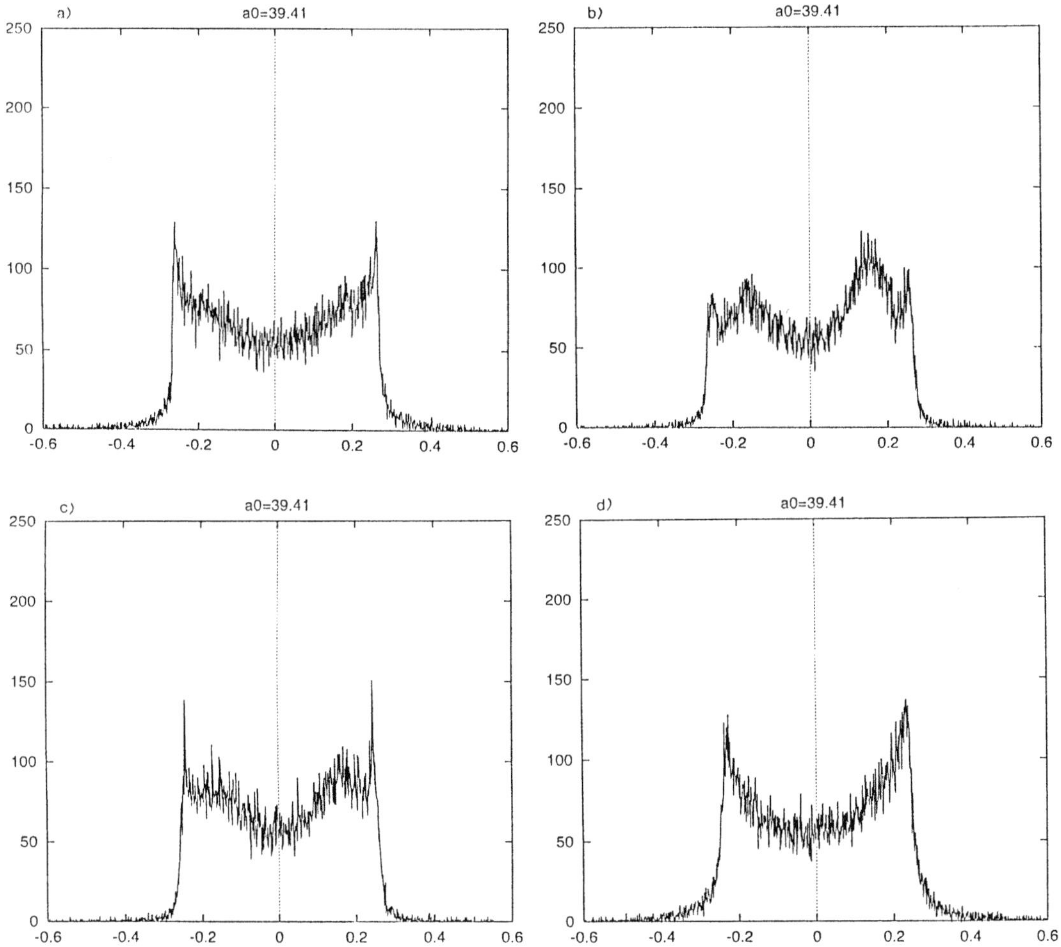

Fig. 4. The variation of the distributions in 4 subintervals of the orbit of Pluto$_{-4}$ ($a_0 = 39.41$).

the chaotic motion ($a_0 \leq 30.41$ and $a_0 \geq 40.11$).

For the study of the invariability of the distributions of the different orbits we divided the whole time interval of 20 million years into subintervals of 1 million years and computed the distributions of the LLCN[1] for each of them. Furthermore, we calculated for each orbit from the 20 distributions a mean distribution as standard of comparison. Fortunately the shapes of the mean distribution are very similar to those shown in fig.(1), therefore, we were of opinion that the distributions are invariant, as it was already found by Voglis & Contopoulos (1994) for orbits of the standard map. But a detailed analysis showed that we have to distinguish between 3 different behaviours:

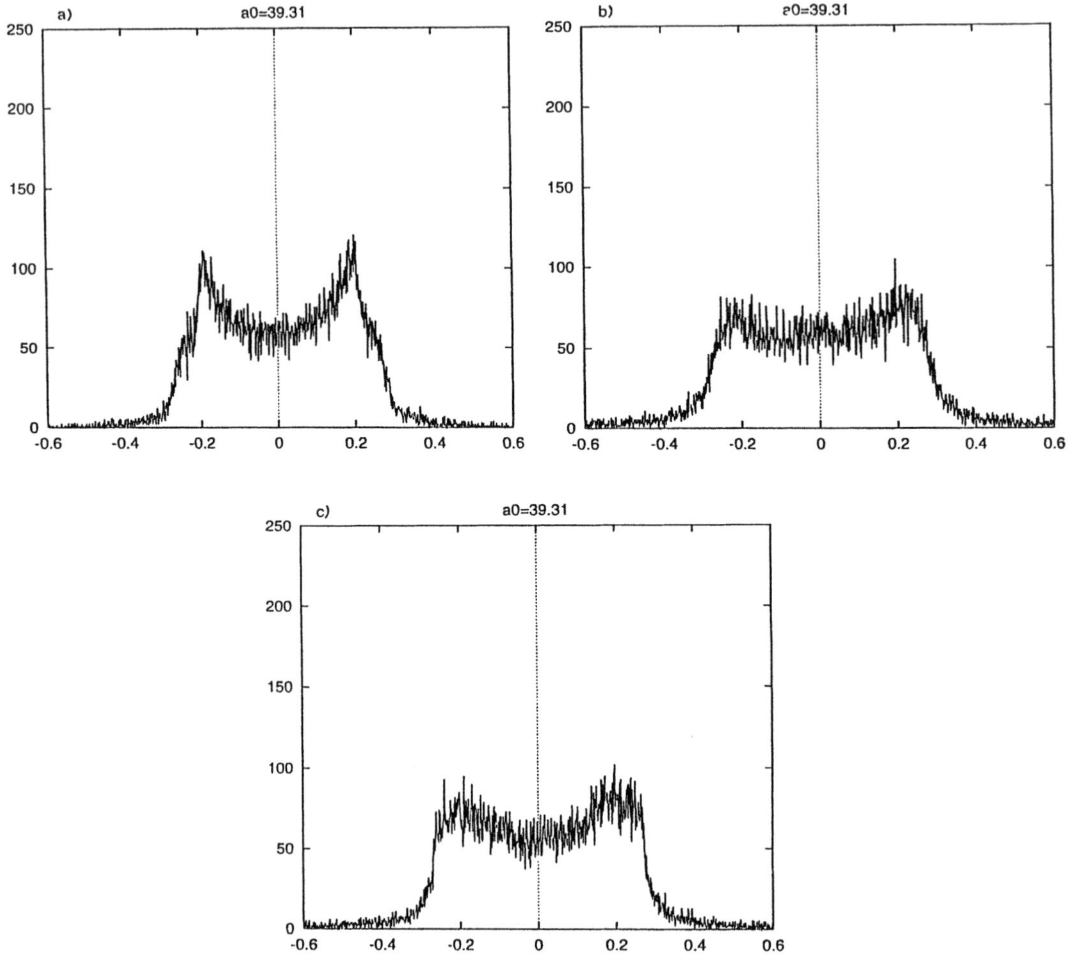

Fig. 5. The variation of the distributions in 3 subintervals of the orbit of Pluto_{-5} ($a_0 = 39.31$).

- For the orbits in the range $39.61 \leq a_0 \leq 39.81$ (Pluto_{-2}, Pluto_{-1} and Pluto) the distributions remain quite constant – the average variation with respect to the corresponding mean distribution is always about 7 %. Therefore, the shapes are quite uniform like the one shown in fig.2(a).
- For the orbits of Pluto_{+1} ($a_0 = 39.91$), Pluto_{+2} ($a_0 = 40.01$) and Pluto_{-3} ($a_0 = 39.51$) – which represent the outer orbits of the "stable" zone – we have found mostly a quite constant shape of the distributions again with an average variation of about 7 %. But in one or two subintervals of each orbit the distribution changes significantly – as it is shown in fig.2(b), so that the percentage of the average variation increases to about 20 %.

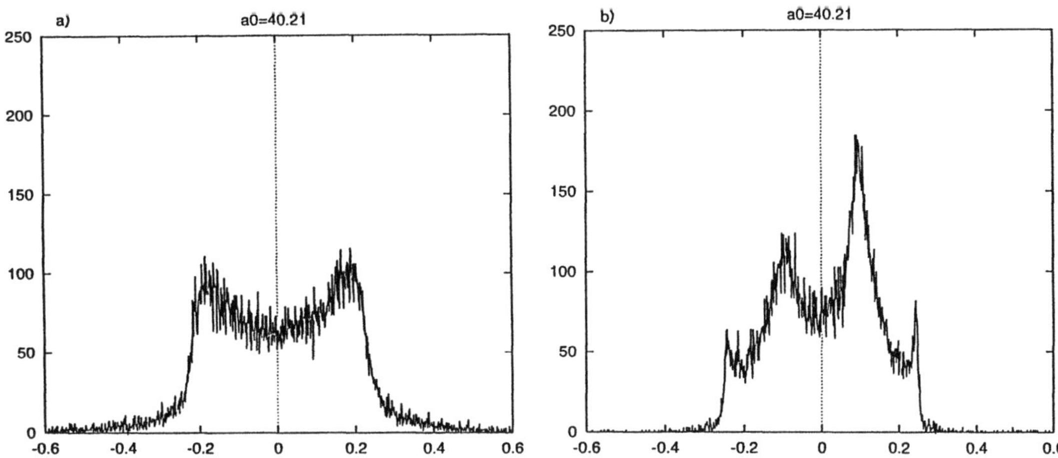

Fig. 6. (a) One of the distributions in the first subintervals of the orbit of Pluto$_{+4}$. (b) The typical distribution of this orbit, which was found in all subintervals from interval 7 on.

– For the chaotic orbits – i.e. Pluto$_{+3}$ and Pluto$_{+4}$ $(a_0 \geq 40.11)$, Pluto$_{-4}$ and Pluto$_{-5}$ $(a_0 \leq 39.41)$ – where we could not determine a uniform shape for the distributions of the whole time interval (see fig.1(a-b) and (i-j)) – strong variations from one subinterval to the other have been found, especially in the case of Pluto$_{+3}$ (see fig.3(a-c)), Pluto$_{-4}$ (see fig.4(a-d)) and Pluto$_{-5}$ (see fig.5(a-c)). The distributions of the LLCN[1] of the orbit of Pluto$_{+4}$ show such strong variations only in the first 6 subintervals while from the interval [6-7][2] on we have found a quite uniform distribution of the kind shown in fig.6(b).

4. Conclusion

As already shown in former studies (cf. Dvorak et al., 1995; Dvorak & Lohinger, 1996) we found again a "stable" zone in the vicinity of Pluto's orbit for the time interval of 20 million years. This zone is still of the same size as it was determined by means of the LCEs for a time interval of 1 million years (cf. Dvorak & Lohinger, 1996).

Of course we know that Pluto's orbit was found to be chaotic with a Lyapunov time of about 20 million years, but note that the chosen time interval for which the computations were carried out is small compared to that which enables the determination of a positive largest LCE for Pluto's orbit (cf. Sussman & Wisdom, 1988).

But maybe the sporadically stronger variations of the distributions of the outer orbits in the "stable" zone indicate already the transition to chaotic orbits if we increase the computation time. This means that the defined "stable" zone will get

[2] [6-7] defines the subinterval from 6 million years to 7 million years integration time.

smaller and smaller when increasing the time interval and will finally disappear. Concerning the invariability of the distributions we could find similar results for the "stable" zone as Voglis & Contopoulos in 1994. The strong variations in case of the chaotic orbits should vanish when we increase the integration time and by an enlargement of the subintervals – since it takes a longer time to get the final (invariant) distribution in case of a chaotic orbit (cf. Froeschlé, 1995; Gonczi, 1995).

Acknowledgements

The author wants to thank Dr. R. Dvorak for the discussions and the critical reading of this paper.
The research described in this paper was supported by the Austrian *Fonds zur Förderung der wissenschaftlichen Forschung* (project P 9921 – AST).

References

Contopoulos, G., Voglis, N., Efthymiopoulos, C., Grousouzakou, E.: 1995, "Invariant Spectra of Dynamical Systems", preprint.

Dvorak, R., Lohinger, E., Maindl, T.I.: 1995, "On the chaoticity of Pluto's orbit", *Perturbation theory and chaos in nonlinear dynamics with emphasis to celestial mechanics* (Hagel et al., eds.), Universidade da Madeira and University of Vienna (ISBN 3-901 617-00-0), p. 20-29.

Dvorak, R., Lohinger, E.: 1996, "Pluto's Lyapunov numbers in different dynamical models", Proceedings of the IAU Symposium No. **172**, in press.

Froeschlé, C., Froeschlé, Ch., Lohinger, E.: 1993, "Generalized Lyapunov Characteristic Indicators and corresponding Kolmogorov like Entropy of the Standard Mapping", *Celest. Mech.*, **56**, p. 307-314.

Froeschlé, C.: 1995, private communication

Gonczi, R.: 1995, private communication

Lohinger, E., Froeschlé, C., Dvorak, R.: 1993, "Generalized Lyapunov Exponents Indicators in Hamiltonian Dynamics: an Application to a Double Star System",*Celest. Mech.*, **56**, p. 315-322.

Smith, H., Contopoulos, G.: 1995, "Spectra and Lyapunov Numbers in Pulsating Systems.", preprint.

Sussman, G.J., Wisdom, J.: 1988, "Numerical Evidence that the Motion of Pluto is chaotic", *Science*, **241**, p. 433-437.

Sussman, G.J., Wisdom, J.: 1992, "Chaotic Evolution of the Solar System", *Science*, **257**, p. 56-62.

Voglis, N., Contopoulos, G.: 1994, "Invariant spectra of orbits in dynamical system", *J. Phys. A*, **27**, p. 4899-4909.

ON THE INFLUENCE OF SUPERNOVA SHOCKFRONTS ON THE STABILITY OF THE SOLAR SYSTEM

W. SCHAFFENBERGER, A. HANSLMEIER
Inst.f. Astronomie, Univ.-Platz 5, A-8010 Graz, Austria

Abstract. We studied the influence of a supernova shock wave on the orbits of celestial bodies of our solar system. The influence of the shock waves for planets is negligible. We found considerable effects only for smaller objects like comets and hypothetical objects in the Kuiper belt.

Key words: Supernova shockfront – solar system – stability

1. Introduction

In this paper we study the influence of a hypothetical supernova shock wave on the dynamics of the solar system. The stability of the orbits of planets, asteroids, comets and hypothetical objects in the Kuiper belt (Dyson 1994, Fernandez 1994) is tested by integrating the equations of motion. To describe the shock wave the Sedov-Taylor similarity solution (Sedov 1959, Landau & Lifshitz 1959) for strong shock waves was used. The main result was that such shock waves have a considerable influence on smaller objects (comets, objects in the Kuiper belt).

Supernovae are major sources of interstellar shock waves (Draine & McKee 1993). Let us consider a supernova with a kinetic energy of the ejecta of $E = 10^{44}J$ and a mass of the ejecta of one solar unit. Then the velocity of the ejected gas is in the range of $10000 km/s$. A few days after the explosion, the ejecta are cooled by adiabatic expansion and the speed of sound in the ejecta is $< 10 km/s$. The ejected gas acts on the ambient material like a piston with a velocity greater than the sound speed. Two shockwaves are formed: a shock propagating into the interstellar medium and a second shock propagating back into the ejecta.
There are three phases during the evolution of a supernova remnant (Spitzer 1978):

During the first phase, the speed of the outer shock is almost constant. The inner shock wave decelerates, the supernova remnant develops a structure which can be approximated by the Sedov-Taylor similarity solution (Ostriker & McKee 1988). During this phase the radiadive losses are small. The shock is nonradiative.

Finally, the radiative losses become important and the shock is now radiative. The shock wave becomes almost isothermal and a dense shell forms (Cioffi et all. 1988). Most of the supernova remnants have not reached this stage.

2. Supernova Shock Waves

In this section we describe the Sedov-Taylor similarity solution which we used for the modelling the shock wave. This solution of the hydrodynamic equations describes the propagation of a strong, spherically symmetric shockwave in an ideal gas with constant adiabatic index. The pressure behind the shockfront is much

R. Dvorak and J. Henrard (eds.), The Dynamical Behaviour of our Planetary System, 393.

greater than the pressure of the unshocked gas. So we can neglect the pressure of the unshocked gas.

There are only two parameters which determine the flow, the density of the unshocked gas ρ and the energy of the explosion E. With these two parameters and the variables r and t we can build only one dimensionless parameter :

$$\xi = r(\frac{\rho_1}{Et^2})^{\frac{1}{5}} \tag{1}$$

Therefore, we have a similarity flow. The flows at two different times are connected by scaling. In particular the radius of the shockfront corresponds to a special value of ξ:

$$r_0 = \xi_0(\frac{Et^2}{\rho_1})^{\frac{1}{5}} \tag{2}$$

This is the law for the propagating of the shockfront. Differentiating this equation with respect to t we get the velocity of the shock wave relative to the unshocked gas:

$$u_1 = \frac{dr_0}{dt} = \frac{2}{5}\frac{r_0}{t} \tag{3}$$

The density ρ_2, the pressure p_2 and the velocity of the gas relative to the shockfront u_2 behind the shock are:

$$\rho_2 = \frac{\kappa+1}{\kappa-1}\rho_1 \quad u_2 = \frac{\kappa-1}{\kappa+1}u_1 \quad p_2 = \frac{2\rho_1 u_1^2}{\kappa+1} \tag{4}$$

Where κ is the adiabatic index. One can derive these equations using the conservation laws of mass, momentum and energy. The velocity of the flow behind the shock relative to the center of the explosion is:

$$v_2 = u_1 - u_2 = \frac{2}{\kappa+1}u_1 \tag{5}$$

Now, we calculate the flow in the whole region behind the shock wave. The hydrodynamic equations for an adiabatic and spherically symmetric flow are:

$$\begin{aligned} \frac{\partial v}{\partial t} + v\frac{\partial v}{\partial r} &= -\frac{1}{\rho}\frac{\partial p}{\partial r} \\ \frac{\partial \rho}{\partial t} + \frac{\partial(\rho v)}{\partial r} + \frac{2\rho v}{r} &= 0 \\ (\frac{\partial}{\partial t} + v\frac{\partial}{\partial r})\ln\frac{p}{\rho^\kappa} &= 0 \end{aligned} \tag{6}$$

To solve these equations, we introduce the new dimensionless variables p', v' and ρ':

$$v = \frac{4}{5(\kappa+1)}\frac{r}{t}v' \quad \rho = \rho_1\frac{\kappa+1}{\kappa-1}\rho' \quad p = \frac{8\rho_1}{25(\kappa+1}\frac{r^2}{t^2}p' \tag{7}$$

They are functions of the dimensionless variable ξ only. Introducing these variables into the hydrodynamic equations we get three ordinary differential equations. The boundary conditions are:

$$\rho'(\xi_0) = p'(\xi_0) = v'(\xi_0) = 1 \tag{8}$$

There is an integral of this system corresponding to the conservation of the total energy of the gas behind the shockwave:

$$\frac{p'}{\rho'} = \frac{v'^2(\kappa + 1 - 2v')}{2\kappa v' - \kappa - 1} \tag{9}$$

The integral fullfills the boundary conditions. With this integral, the equations can be integrated analytically:

$$\begin{gathered}
(\frac{\xi_0}{\xi})^5 = v'^2[\frac{5(\kappa+1) - 2(3\kappa-1)v'}{7-\kappa}]^{\nu_1}[\frac{2\kappa v' - \kappa - 1}{\kappa - 1}]^{\nu_2} \\
\rho' = [\frac{2\kappa v' - \kappa - 1}{\kappa - 1}]^{\nu_3}[\frac{5(\kappa+1) - 2(3\kappa-1)v'}{7-\kappa}]^{\nu_4}[\frac{\kappa + 1 - 2v'}{\kappa - 1}]^{\nu_5} \\
\nu_1 = \frac{13\kappa^2 - 7\kappa + 12}{(3\kappa - 1)(2\kappa + 1)} \quad \nu_2 = -\frac{5(\kappa+1)}{2\kappa+1} \quad \nu_3 = \frac{3}{2\kappa + 1} \\
\nu_4 = \frac{13\kappa^2 - 7\kappa + 12}{(3\kappa - 1)(2\kappa + 1)(2 - \kappa)} \quad \nu_5 = \frac{2}{\kappa - 2}
\end{gathered} \tag{10}$$

And ξ_0 satisfies the following equation:

$$\xi_0^5 \frac{32\pi}{25(\kappa^2 - 1)} \int_0^1 (\rho' v' + p')(\frac{\xi}{\xi_0}^4 d(\frac{\xi}{\xi_0}) = 1 \tag{11}$$

For monatomic gases : $\xi_0 = 1.15$ The above formulae give the complete solution for the flow behind the shockfront.
From these formulae one can derive the behavior of the solution for small values of the distance from the center of the explosion:

$$\frac{v}{v_2} \propto \frac{r}{r_0} \quad \frac{\rho}{\rho_2} \propto (\frac{r}{r_0})^{\frac{3}{\kappa - 1}} \tag{12}$$

Figure 1 shows the velocity, the density and the pressure as a function of the distance from the center of the explosion.

3. Modification of the N-Body Problem

Now, we have to introduce the influence of the shock wave into the dynamics of the solar system. The differential equations for the N-body problem are :

$$\ddot{\mathbf{x}}_i = \sum_{j=1}^{N} \frac{G m_j (\mathbf{x}_i - \mathbf{x}_j)}{\mid \mathbf{x}_i - \mathbf{x}_j \mid^3} \tag{13}$$

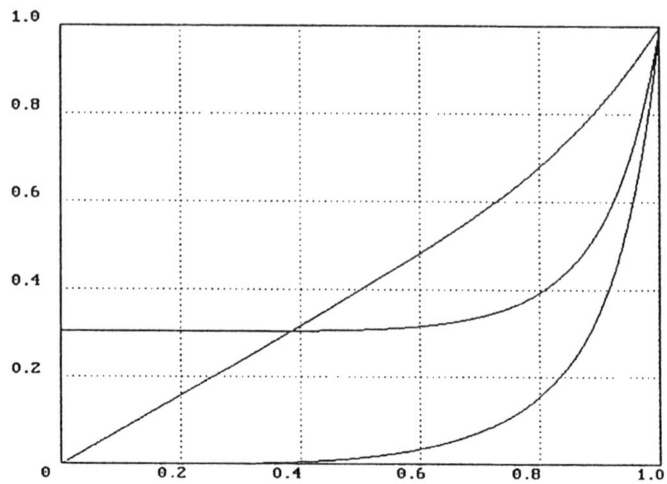

Fig. 1. The velocity, the density and the pressure as a function of the distance from the center of the explosion for a monatomic gas

where G is the gravitational constant, $\mathbf{x}_i$ is the position of the i-th body and m_i is the mass of the i-th body. The additional force of the shock wave is:

$$F = \frac{1}{2} c_w \rho_2 v^2 A \tag{14}$$

A is the sectional area of the body, and v is the velocity of the fluid relative to the body. c_w is a dimensionless factor. It depends on the figure of the body and the speed of the flow. Because this factor can not be calculated we set $\frac{1}{2}c_w = 1$ because the order of this dimensionless factor is of the order of unity for a supersonic flow. We assume that the celestial bodies are spheres and therefore: $A = \pi r^2$ where r is the radius of the body. The modified equations, including the shock wave are:

$$\ddot{\mathbf{x}}_i = \sum_{j=1}^{N} \frac{G m_j (\mathbf{x}_i - \mathbf{x}_j)}{\mid \mathbf{x}_i - \mathbf{x}_j \mid^3} + \frac{\rho_2 (\mathbf{v} - \dot{\mathbf{x}}_i)^2 \pi r_i^2}{m_i} \frac{\mathbf{v} - \dot{\mathbf{x}}_i}{\mid \mathbf{v} - \dot{\mathbf{x}}_i \mid} \tag{15}$$

$\mathbf{v}$ is the velocity of the flow relative to the center of the explosion. If we approximate the supernova blast wave by a strong spherically symmetric shockfront, we can use the formulae of the previous section. Therefore, the density ρ and the velocity $\mathbf{v}$ in the perturbed equations are known as functions of the time.

We consider a supernova at a distance of R_0. The time intervall for the propagation of the shockwave to the radius R_0 is:

$$t_0 = \sqrt{\frac{\rho_1 R_0^5}{E \xi_0^5}} \tag{16}$$

This time intervall is also the characteristic time scale for the change of the flow at this distance. To obtain the behavior of the flow, we substitute in the asymptotic formulae (12)

$$\frac{R_0}{r_0} = \frac{R_0}{\xi_0}\left(\frac{\rho_1}{Et^2}\right)^{\frac{1}{5}} \tag{17}$$

The equations (12) can be written as:

$$v \propto \frac{R_0}{t}$$

$$\rho \propto \rho_1^{\frac{5\kappa-2}{5(\kappa-1)}} E^{-\frac{3}{5(\kappa-1)}} R_0^{\frac{3}{\kappa-1}} t^{-\frac{6}{5(\kappa-1)}} \tag{18}$$

If we use the dimensionless ratio $\frac{t}{t_0}$, the latter can be written in the form:

$$v \propto \sqrt{\frac{E}{\rho_1 R_0^2}\frac{t_0}{t}}$$

$$\rho \propto \rho_1 \left(\frac{t}{t_0}\right)^{-\frac{6}{5(\kappa-1)}} \tag{19}$$

These formulae are valid for $t \gg t_0$. For $\kappa = 5/3$ the density decreases with $t^{-1.8}$. The time dependence of the velocity is asymptotically independent from the adiabatic index.

To cause an appreciable effect, the supernova must be verry close to our solar system. Let us consider a supernova at a distance of $1pc$. The energy of the supernova is $10^{44}J$ and the density of the interstellar medium is $10^{-20}kg/m^3$. According to equation (16) the propagation time of the shock is $t_0 = 37.6a$. This is not the correct propagation time, because at the beginning the shock propagates with almost constant velocity. The real propagation time is greater, but this time intervall is the time scale for the asymptotic time dependence of the flow. When the shock front reaches the solar system, the velocity of the flow is much greater then the speed of the celestial bodies. This velocity is:

$$v_2 = \frac{4\xi_0}{5(\kappa+1)}\left(\frac{E}{\rho_1 t^3}\right)^{\frac{1}{5}} \tag{20}$$

With the above values for the parameters of the flow one gets $v_2 = 7800km/s$. Therefore, one can neglect the speed of the body and the force of the flow is:

$$F = c_w \rho_2 v^2 A \tag{21}$$

The additional acceleration due to the shock wave is greater for small bodies, because the sectional area grows with the square of the radius but the mass grows with the cube of the radius. Therefore, the acceleration is inverse proportional to the radius of the body. The effect of the shock wave should be greater for comets and asteroids than for planets. Consider a planet like the earth with a radius of $r = 6000km$ an a mass of $m = 6 * 10^{24}kg$. Then the force of the flow acting on this planet is $F \approx 2 * 10^{8}N$. The gravitational force of the sun in a distance of $1AU$ is $F_g = 3.55 * 10^{22}N$. And the ratio of these forces is $\frac{F}{F_g} \approx 10^{-14}$. For a comet with a radius of $r = 5km$ and a density of $\rho = 600kg/m^3$ the force due to the shock wave is $F \approx 200N$. The gravitational force of the sun in a distance of $30AU$ is $f_g = 2 * 10^{9}N$. The ratio of these forces is $\frac{F}{F_g} = 10^{-7}$. This values are valid only when the shock wave arrives at the solar system. Using the equations (18) we obtain for the force:

$$F \propto A\rho_1^{\frac{5\kappa-2}{5(\kappa-1)}} E^{-\frac{3}{5(\kappa-1)}} R_0^{\frac{2\kappa+1}{\kappa-1}} t^{-\frac{2(5\kappa-2)}{5(\kappa-1)}} \tag{22}$$

With the ratio t/t_0 this can be written as:

$$F \propto A\frac{E}{R_0^3}\left(\frac{t}{t_0}\right)^{-\frac{2(5\kappa-2)}{5(\kappa-1)}} \tag{23}$$

For $\kappa = 5/3$ the force decreases proportional to $t^{-3.8}$. The influence of the shock wave on the dynamics of the solar system would not only be determined by the value of the force but also by the duration of its action. Therefore, a meassure of the effect of the shock is the integral of the force over the time. Because the force is only important if the velocity of the flow is much greater than the velocity of the celestial bodies one can neglect the speed of the body. For a rough estimation we use the asymptotic formulae for the flow and integrate up to unity. Then the integral of the force is:

$$\int_{t_0}^{\infty} F dt \propto A\sqrt{\frac{\rho_1 E}{R_0}} \tag{24}$$

It is inverse proportional to the square of the distance of the supernova. Therefore, the dependence of the influence of the shock wave on the distance should not be important. The force is smaller for greater distances but its duration of action is greater.

Integrating the differential equations (13) and (15) we used the classical fourth order Runge-Kutta method with constant step size. We integrated the perturbed and the unperturbed equations alternately and compared the two solutions. We calculated the difference of the positions, the semimajor axes and the inclinations of the orbits. In order to test the validity of our calculations, we checked the conservation of the total angular momentum. We considered a hypothetical supernova at a distance of 1pc and for a second test a supernova at a distance of 5pc. Including the planets, seven asteroids and the two comets Halley and Encke, we integrated the equations (13) and (15) for 4000 years with a step size of 0.05d. Including only the outer planets and 15 hypothetical cometlike objects modelling the Kuiper belt,

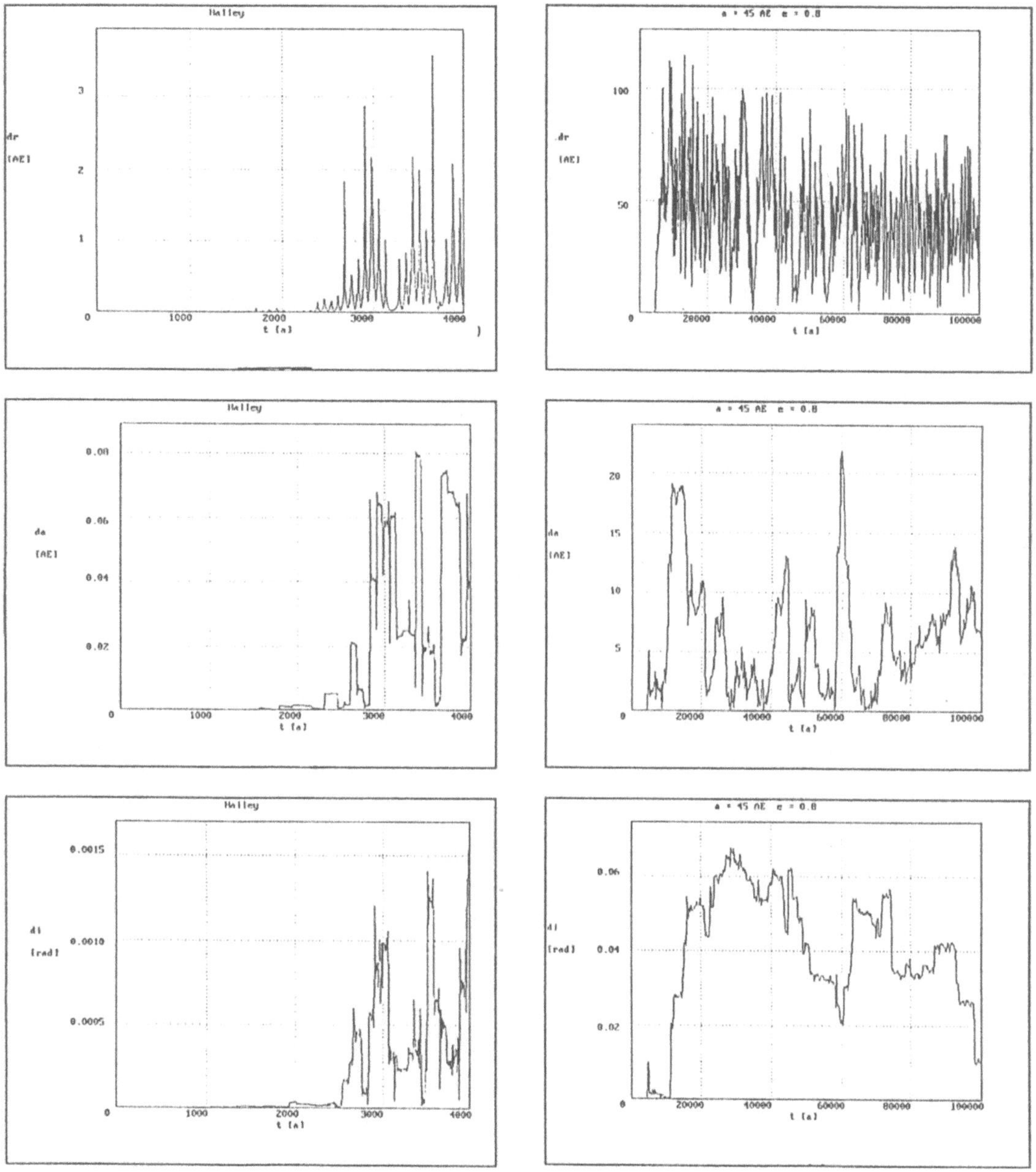

Fig. 2. Influence of a SN shockwave on orbital elements of Halley (a – to the left) and on orbital elements of a hypothetical Kuiper (b – to the right)

we integrated the equations for 100000 years with a step size of 0.25d and 0.5d. The errors were found to be smaller then 10^{-12}. For the hypothetical objects and the two comets we assume a diameter of 10km and a density of $600kg/m^3$. The semimajor axes of the hypothetical objects are between 30AU and 50AU and the excenticities of their orbits are between 0.2 and 0.8.

4. Results

4.1. PLANETS AND ASTEROIDS

The results for a supernova at a distance of 5pc were similar to the results for a supernova at a distance of 1pc. Therefore we discuss the results for the second case only. The effects for the planets are very small. The differences of the positions were in the order of $10^{-8}AU$ for the inner planets and in the order of $10^{-7}AU$ for the outer planets. The changes of the semimajor axes are $10^{-12}AU$. The changes of the position are 10^4 times larger, because the shockwave changes the period of revolution of the body. Therefore, the perturbed and the unperturbed position, moving on almost the same orbit, diverge. For the asteroids, the effects are about 10 times larger. We included in our calculations only the larger asteroids. The influence on the small asteroids should be greater.

4.2. COMETS AND HYPOTHETICAL OBJECTS

The comets and hypothetical objects show much greater effects than the planets and asteroids. For example figure 2a show the results for Halley. The changes of the position due to the shock wave are about 3AU. Remarkable are the sharp peaks and the increase of the deviations after 2500 years. There are significant changes in the orbital parameters too. The results for Encke are similar to that for Halley.

For the hypothetical objects we found the largest effects. After some 10000 years the changes of the position are of the same order as the orbit itself. The canges of the semimajor axes are about 1AU and the changes of the inclinations are about 3 degrees. The effects are greater for objects with very excentric orbits. Figure 2b show the result for an object in an orbit with a semimajor axis of 45AU and an excentricity of 0.8. The changes for the semimajor axis are about 20AU!

5. Conclusions

The influence of a hypothetical Supernova blast wave is extremly small for the planets. However, the effects for comets and the hypothetical objects are considerable. Therefore it is possible, that the influence of the supernova changes the orbits of some objects in the Kuiper belt so that they arrive the interrior of the solar system. The distance of the supernova is irrelevant while the distance of the supernova is not too great. If the distance of the supernova is too large, the shockfront becomes radiative and the Sedov-Taylor similarity solution is no longer valid.

To obtain a great influence, the supernova must be close to the solar system. Such an event is very rarely. With one supernova during a time intervall of 30 years in our galaxy, one can estimate one supernova during a time intervall of 5 billion years within a sphere of 10pc. During the main sequence phase of a star like the sun, only one supernova appears in a distance closer than 10pc.

References

Cioffi, McKee, Bertschinger: 1988, *Ap. J.*, **334**, 252–65
Draine & McKee: 1993, *Ann. Rev. Astron. Astrophys.*, **31**, 373–432
Dyson: 1994 *Sky Telesc.*, **87**, 27–30
Fernandez: 1994, *IAU Symposium No. 160 Asteroids, Comets, Meteors 1993*, 223–40
Landau & Lifshitz: 1959, *Fluid Mechanics*, Oxford Pergamon
Sedov: 1959, *Similarity and dimensional methods in mechanics*, Academic Press New York
Spitze: 1978, *Physical Processes in the interstellar Medium*, John Wiley & Sons

HENRI POINCARÉ:

A DECISIVE CONTRIBUTION TO SPECIAL RELATIVITY

C. MARCHAL
DES, ONERA,
BP 72, 92322 CHATILLON Cedex, FRANCE

Abstract. Maxwell's electromagnetism equations and the old notions of Newtonian absolute time and Euclidian absolute space were contradictory with the impossibility of detection of the Earth's absolute motion.

Henri Poincaré and Hendrik Antoon Lorentz exchanged many friendly letters on these subjects over more than ten years. They improved their analysis step by step and this situation led Henri Poincaré to consider that the notions of absolute time and space as well as the corresponding "ether" are artificial and do not really exist. The modifications of inertial reference frames do not follow the Galilean rules but those of the Lorentz transformation that can be deduced from Poincaré's relativity principle presented in 1904 at the world scientific congress of Saint Louis (Missouri).

Poincaré's later works of June and July 1905 have led to the full developmentof Special Relativity with its links to electromagnetism and the first mention of gravitational waves.

Unfortunately Henri Poincaré did not survive much to his gigantic pioneering work, he got cancer in 1909 and died in 1912.

Key words: Special relativity – Poincaré

1. Introduction

The theory of relativity is the result of a very long maturation of the knowledge and reflexion of mankind confronted with the properties of matter, energy, space and time.

Let us begin with the state of this confrontation in the second half of the nineteenth century.

The main elements were then the following :

I) *Galilean Relativity.*

For centuries it was believed that force was proportional to velocity : you push on an object and it moves, you stop pushing and it stops... It requires difficult observations and reflexion on rubbings to understand that in the absence of force the motion remains rectilinear and uniform (Galileo) and that force is proportional to acceleration (Newton).

The real motive of Galileo was the understanding of Earth's orbital motion. He needed what we now call the Galilean relativity : the Earth does not lose its atmosphere and its oceans along its orbit!. . . "A mechanical experiment gives the same results in a fixed laboratory and in a laboratory with a rectilinear and uniform motion" i.e.in practical terms, you can drink your coffee as usual as long as your airplane is flying in a rectilinear and uniform motion and is not buffeted by the wind [1].

[1] For this principle and several other fundamental philosophical reflexions Galileo is considered by the scientists as one of the founding fathers of modern science while he is known among the public

R. Dvorak and J. Henrard (eds.), The Dynamical Behaviour of our Planetary System, 403.

II) *The Earth's Motion.*

Copernicus and Galileo had no physical proof of Earth's motion, so Copernicus presented his work as an hypothesis even if Galileo was more adventurous. Fortunately by the middle of the nineteenth century this motion was firmly established on its three classical proofs : A) The aberration of stars (Bradley 1727), B) The parallax of stars (Bessel 1840) and C) The Foucault pendulum (1851).

III) *Absolute or "Newtonian" Time.*

"Tempus absolutum, verum et mathematicum. . ."

"True mathematical absolute time is, by its very nature, independent of all other variables. It flows uniformly and shall be called duration".

Apparent, relative and common time is the more or less accurate and subjective measure of the duration by the motions of celestial bodies as is usually done, instead of the true time, with the hours, days, months, years. (Newton Philosophiae naturalis Principia mathematica, 2 nd edition, 1713).

In Newton's day and even two centuries later, no clock was able to reveal the small differences related to the relativistic effects. It was therefore very natural to assume the existence of "absolute time", this essential parameter of so many physical laws, and Newton's definition of this time appeared then essentially as a warning : "Let us be careful ; Earth's rotation is perhaps slightly irregular".

IV) *Absolute Euclidian Space and the Notion of Force.*

The law of inertia :

acceleration = force / mass

is valid only in the "Galilean reference frames" or "inertial reference frames" that have no rotation and the relative motions of which are rectilinear and uniform.

In the second half of the nineteenth century the non-euclidian geometries of Lobatchevsky, Bolyai and Riemann were considered as mathematical curiosities without much interest and everybody considered the physical space as Euclidian.

The tremendous success of Newton's theory of universal attraction, with the accurate description of planetary motions and the discovery of planet Neptune (1846) after the long computations of Leverrier and Adams, emphasizes the validity of all above notions.

By 1850 all laws of Mechanics satisfied the Galilean relativity, being conserved by the usual Galilean transformations of reference frames, for instance by the classical expression:

$$\begin{cases} x_1 = x - Vt & : \text{ constant velocity V of the second set of axes with respect to the fir} \\ y_1 = y & \\ z_1 = z & \\ t_1 = t & : \text{ absolute time .} \end{cases}$$

essentially because of his trial of 1633.

But notice the irony and the historical chance: it is because he was under house arrest in his country home of Arcetri near Florence that he found the time necessary for philosophical reflexions. Without his condemnation he would have probably remained the busy professor and the burning, violent and sometimes unfair polemist he had been all his life.

V) *Maxwell's Electromagnetism Equations (1864).*

Maxwell's equations represent a major progress in the knowledge of matter, perhaps a progress of the order of magnitude of Newton's law of universal attraction. But they are also the source of the trouble : they are not conserved in Galilean transformations of reference frames.

Let us consider their simplest expression in vacuum. The electric field **E** and the magnetic induction **B** are related by the four following equations :

$$\left\{\begin{array}{llll} \mathrm{div}\,\mathbf{E} = 0 & ; & \mathrm{curl}\,\mathbf{E} = -\partial\mathbf{B}/\partial t & ; \\ \mathrm{div}\,\mathbf{B} = 0 & ; & \mathrm{curl}\,\mathbf{B} = \mu_0\epsilon_0\partial\mathbf{E}/\partial t & ; \end{array}\right. \tag{2}$$

with

μ_0 = magnetic permeability of vacuum = $4\pi \times 10^{-7}$ henry per meter
ϵ_0 = permittivity of vacuum = 8.854188×10^{-12} farad per meter

The simplest solutions are the plane waves, for instance those propagating in the Ox direction :

$$\left\{\begin{array}{l} u = x - ct \quad , \quad \text{with} \quad c = \left[\sqrt{\mu_0\epsilon_0}\right]^{-1} = 299792458 \text{ m/s} , \\ \mathbf{E} = (0, E_y(u), 0) \quad ; \quad \mathbf{B} = (0, 0, B_z(u)) \quad ; \quad E_y \equiv cB_z \ . \end{array}\right. \tag{3}$$

Thus, with respect to the suitable Maxwell's reference frame $Oxyzt$ in which (2) is valid, the plane waves are moving with the velocity c, the velocity of electromagnetic waves. This velocity was also recognized as the velocity of light after the Hertz experiments on the similarities of electromagnetism and light.

Unfortunately the Galilean transformation (1) does not conserve the velocity c and thus we must choose between the two following possibilities.

Either :

A) Maxwell's equations are valid with respect to a particular frame of reference $Oxyzt$ and are only approximative for slowly moving frames of reference (as those of our usual experiments on Earth).

Or :

B) Maxwell's equations are valid for all inertial frames of reference and the relativity can be extended from Mechanics to electromagnetism and optics. But there is a price to pay : the notions of absolute time and absolute space must be forsaken since they are contradictory with the invariance of the velocity of light.

The Newtonian absolute time seemed so obvious and so natural that the hypothesis A was immediately adopted. The hypothetical reference frame $Oxyzt$ takes a concrete form with "ether", a medium assumed to be very light and subtle that was considered as playing for light and electromagnetism the role of air for sound. Of course this ether offers a vivid picture of the "absolute space".

The next step was obviously the research of the properties of ether and also the determination of the "absolute" Earth's motion by suitable optic or electromagnetic experiments.

The Fizeau experiment (measurement of the velocity of light in a stream of water ; 1851) and the Airy experiment (measurement of the angle of aberration in a telescope full of water ; 1871) seem to show a "partial ether drag" by transparent media.

Using all kinds of ideas and equipments many experimenters (Trouton and Noble, Lodge, Kennedy and Thorndyke etc...) tried to study the ether and to determine the absolute Earth's motion, but without success.

The most famous experimenters were Michelson and Morley. Their experiment (1887) was unable to detect an anisotropy of the velocity of light in spite of an accuracy that would theoretically enable the detection of an Earth's motion by a few km/s only.

It is very fortunate that the Earth's orbital motion was firmly established in the minds of the scientists of this century. Two centuries earlier the simplest explanation would have been : "The Earth is not moving...".

Whilst these experiments were carried out, the theorists obtained some interesting results.

Lorentz and Fitzgerald noticed that a suitable contraction by the "wind of ether" could explain the apparent isotropy given by the Michelson and Morley experiment.

In 1887 Voigt obtained a transformation of coordinates that preserves Maxwell's plane and spherical waves.

In 1895 Lorentz noticed that the first-order of Voigt transformation preserves the first-order of Maxwell's equations.

Larmor gave the second-order a little later.

In his great memorandum of May 1904, "Magnetic phenomena in a system moving with any velocity smaller than that of light" (Ref.1), Lorentz gave an extension of the Voigt transformation which preserves Maxwell's equations in vacuum.

The greatest progress was achieved by the mathematician, physicist and philosopher Henri Poincaré who was a friend of Lorentz. They exchanged many scientific letters dating from 1895 and improved their analysis step by step.

Poincaré's successive main ideas and improvements are the following :

A) In the book "La Science et l'Hypothèse" (1902) in pages 111, 245, 246 (Ref. 2).

"*There is no absolute space* and we only conceived relative motions".

"*There is no absolute time,* the equality of two durations can be defined only by convention and we have no intuition of the simultaneity of events occuring in different places".

"Is the hypothesis of ether true ? This is only a question for metaphysicians. This hypothesis is convenient for the explanation of phenomena but it can be sacrificed. Usually this hypothesis is conserved for clarity, *but this reason is the only one.*"

B) The world scientific congress of Saint-Louis (Missouri, September 1904, published in November 1904).

Henri Poincaré was asked to present a general lecture on : "The present state and the future of mathematical Physics".

He boldly added to the five classical principles of Physics the "Principle of Relativity":

"The laws of physical phenomena must be the same for a fixed observer and for an observer in rectilinear and uniform motion *so that we have no possibility of perceiving whether or not we are dragged in such a motion*"[2] (Ref.11).

This principle was of course essentially based on the negative results of all ether experiments of this time and most of the lecture was devoted to the defense of the new principle.

The conclusion of the lecture :

"Thus the Relativity Principle was these last years valiantly defended, but the energy of defense proves how serious was the attack... *We will perhaps be forced to build a complete new Mechanics in which the inertia increasing with velocity, the velocity of light would become an unsurpassable limit*".

C) The note to the French Academy of Science (June 5, 1905, published June 9, 1905, Ref. 3).

Poincaré states the Principle of Relativity once again and analyses the "change of variables" presented by Lorentz in his memorandum. He simplifies its presentation and gives its present name : "The essential point, established by Lorentz, is that Maxwell's equations are preserved by a transformation to which I will give the name of Lorentz" (later, in 1914, Lorentz will correct this statement : "My demonstration was not general, only that of Poincaré and later those of Einstein and Minkowsky were complete".)

Poincaré notices that the Lorentz and associated transformations are the elements of a mathematical group (today the Poincaré group) and this allows him to give the value of the coefficient L used by Lorentz in his transformation : this coefficient must be identical to one.

Finally he notices that this relativity theory implies the existence of "gravitational waves" moving at the velocity of light, but his later research in this direction will remain unsuccessful.

We must remark that the second time t' appearing in the Lorentz transformation is *as physical as the first* because of the inexistence of ether and absolute time and because of the perfect symmetry of the transformation. Poincaré had already given a physical meaning to this time t' by the synchronisation between distant clocks with light signals, and with the invariance of the velocity of light (Ref. 4).

It is essential to notice that the Lorentz transformation is a direct consequence of the relativity principle, without need for an invariant velocity of light.

[2] Surprisingly this very first expression of the Relativity Principle at its true level is not mentioned in the very interesting and generally well documented ref 12. I don't find it either in ref 13 in spite of its presence in ref 14 and also in the famous Encyklopädie der Mathematischen Wissenschaften (ref 15).

Let us look for that transformation along the axes Ox and $O'x'$ moving along each other with a relative velocity V.

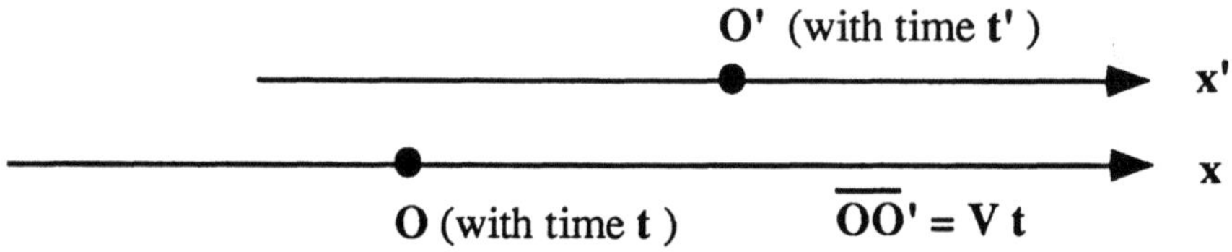

In order to achieve a perfect symmetry between the two frames of reference we will reverse the$O'x'$ axis

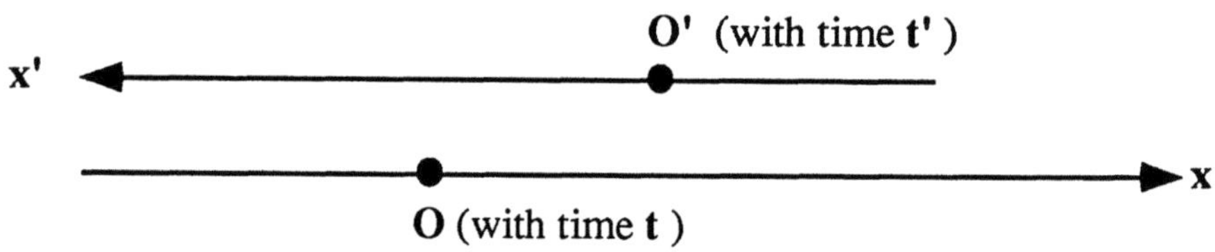

The homogeneity will lead to a linear transformation, and if we choose $t = t' = 0$ at O and O' when they cross each other the transformations $(x, t) \longrightarrow (x', t')$and $(x', t') \longrightarrow (x, t)$ are given by the following, with eight suitable constants A to D' :

$$\begin{cases} x' = Ax + Bt & ; \quad x = A'x' + B't' ; \\ t' = Cx + Dt & ; \quad t = C' + D't' . \end{cases} \tag{4}$$

The relativity principle and the symmetry lead to :

$$A = A' \quad ; \quad B = B' \quad ; \quad C = C' \quad ; \quad D = D' . \tag{5}$$

Furthermore at O' we have $x' = 0$ and $x = Vt$, hence from $x' = Ax + Bt$ we deduce $AV + B = 0$ and from $x = Ax' + Bt'$ and $t = Cx' + Dt'$ we obtain :$B = VD$ and then $A = -D$.

Finally the coherence gives :

$$\begin{cases} x = Ax' + Bt' = A(Ax + Bt) + B(Cx + Dt) = (D^2 + VCD)x , \\ t = Cx' + Dt' = C(Ax + Bt) + D(Cx + Dt) = (D^2 + VCD)t , \end{cases} \tag{6}$$

hence $D^2 + VCD = 1$, i.e.: $C = (1 - D^2)/DV$.

The transformation $(x, t) \longrightarrow (x', t')$ then becomes :

$$\begin{cases} x' = -Dx + DVt , \\ t' = [(1 - D^2)/DV]x + Dt . \end{cases} \tag{7}$$

The only remaining unknown, D, is a function of V that can be determined by the comparison of several velocities.

Let us again reverse $O'x'$ and have all three axes Ox, $O'x'$, $O''x''$ in the same direction:

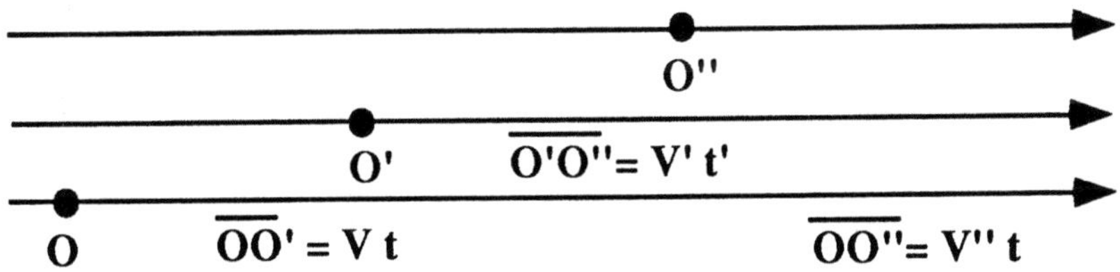

The relation (7) becomes, with opposite sign of x' :

$$x' = Dx - DVt \quad ; \quad t' = [(1 - D^2)/DV]x + Dt \ , \tag{8}$$

and similarly, with D' for V' and D'' for V'' :

$$x'' = D'x' - D'V't' \quad ; \quad t'' = [(1 - D'^2)/D'V']x' + D't' \ , \tag{9}$$

$$x'' = D''x - D''V''t \quad ; \quad t'' = [(1 - D''^2/D''V'']x + D''t \ , \tag{10}$$

The equations (8) and (9) give also :

$$\begin{cases} x'' &= D'(Dx - DVt) - D'V' \{[(1 - D^2)/DV]x + Dt\} \\ t'' &= [(1 - D'^2)/D'V'](Dx - DVt) + D' \{[(1 - D^2)/DV]x + Dt\} \ . \end{cases} \tag{11}$$

That is :

$$\begin{cases} x'' &= \left\{ D'D + \dfrac{D'V'}{DV}(D^2 - 1) \right\} x - DD'(V + V')t \\ t'' &= \left\{ \dfrac{D - DD'^2}{D'V'} + \dfrac{D' - D^2 D'}{DV} \right\} x + \left\{ \dfrac{DV}{D'V'}(D'^2 - 1) + D'D \right\} t \ . \end{cases} \tag{12}$$

The comparison between (10) and (12) gives :

$$D'' = D'D + [D'V'(D^2 - 1)/DV] \tag{13}$$

$$D''V'' = DD'(V + V') \tag{14}$$

$$(1 - D''^2)/D''V'' = [(D - DD'^2)/D'V'] + [(D' - D^2D')/DV] \tag{15}$$

$$D'' = [DV(D'^2 - 1)/D'V'] + D'D \tag{16}$$

Hence from (13) and (16) :

$$D'V'(D^2 - 1)/DV = DV(D'^2 - 1)/D'V' \tag{17}$$

and we can define K by :

$$K = D^2V^2/(D^2 - 1) = D'^2V'^2/(D'^2 - 1) \tag{18}$$

The quantity K has the same value for two arbitrary velocities (and their corresponding D) it is then a constant for all velocities. On the other hand the case $V = 0$ gives $x = x'$; $t = t'$ in (8) and then $D = +1$, hence we must choose the positive solution of (18) :

$$D = 1/\sqrt{1 - (V^2/K)} \tag{19}$$

We thus obtain, with (8), the transformation$(x,t) \longrightarrow (x',t')$ and Poincaré has no difficulty in developing it to $(x,y,z,t) \longrightarrow (x',y',z',t')$:

$$\begin{cases} x' &= (x - Vt)/\sqrt{1 - (V^2/K)} \\ y' &= y \\ z' &= z \\ t' &= [t - (Vx/K)]\,/\sqrt{1 - (V^2/K)}\,. \end{cases} \tag{20}$$

It remains to determine the constant K that gives the Galilean transformation (1) if$K = +\infty$ and the usual Lorentz transformation if $K = c^2$.

The constant K cannot be negative (it would become possible to go back in time) and its square root appears as the limit and unsurpassable velocity. This is confirmed by the square root $\sqrt{1 - (V^2/K)}$ and also by the composition of velocities deduced from (14) and (13) :

$$V" = (V + V')/[1 + (VV'/K)]\,, \tag{21}$$

that is, with $k = \sqrt{K}$

$$\frac{k - V''}{k + V''} = \frac{k - V}{k + V} \times \frac{k - V'}{k + V'}\,. \tag{22}$$

Hence $\mid V \mid < \sqrt{K}$ and $\mid V' \mid < \sqrt{K}$ imply always $\mid V'' \mid < \sqrt{K}$.

Very naturally Poincaré and Lorentz have chosen $K = c^2$ that agrees with the invariance of the velocity of light and with the conservation of Maxwell equations in all inertial reference frames. However notice that, if necessary, it is possible to assume that K is very slightly larger than c^2. The photons would then have a very small but non zero mass and their velocity, the velocity of light, would be almost $\sqrt{K}$ and would be an increasing function of their energy.

D) The final fundamental work of Poincaré on relativity is his paper "Sur la dynamique de l'électron" in which he develops and demonstrates the ideas of his note to the Academy (Ref. 5, July 1905, published in January 1906).

The expression of the transformation of the electromagnetic field is particularly impressive : the electromagnetism appears as the association between electrostatic and relativity.

The Lorentz - Poincaré theory then led to the relative character of physical space and physical time, it is in agreement with the Relativity Principle, with Maxwell's equations not only in vacuum but in all cases, with the ether drag experiments (Fizeau, Airy, Michelson, etc. . .) and with the classical results of electromagnetism as discovered by the pioneers: Coulomb, Ampère, Volta, Laplace, Gauss, Oersted, Faraday... The theory of Special Relativity was then complete.

Meanwhile Einstein prepared and published his first and most famous paper on Relativity "Zur Elektrodynamik der bewegten Körper" (Ref. 6). This paper was presented without reference and is considered for that reason by some people as a compilation of former works (Ref. 7)

The basic idea of Einstein is the invariance of the velocity of light (which forbids the slightest mass for the photon).

Einstein was led to the Relativity Principle. He obtained all the same results as Poincaré. He mentioned that the Lorentz and associated transformations are the elements of a mathematical group but made no use of this property.

Was Einstein aware of Poincaré's work ? This is a difficult question.

On the one hand he wrote a letter to Carl Seelig in 1955 :

"There is no doubt, that the special theory of relativity, if we regard its development in retrospect, was ripe for discovery in 1905. Lorentz has already observed that for the analysis of Maxwell's equations the transformations which later were known by his name are essential, and Poincaré had even penetrated deeper into these connections. Concerning myself, I knew only Lorentz' important work of 1895 - "La théorie électromagnétique de Maxwell" and "Versuch einer Theorie der elektrischen und optischen Erscheinungen in bewegten Körpern" but not Lorentz' later work, nor the consecutive investigations by Poincaré. In this sense my work of 1905 was independent." (Ref. 8, page 11).

On the other hand :

The 1905 Einstein's paper presents all the same results as Poincaré and even the group property of the elements of the Lorentz and associated transformations.

This notion of a mathematical group was then very new and practically ignored among physicists. Einstein makes no use of it. Einstein had of course no opportunity to read the July 1905 Poincaré's paper while writing his own paper but the June 1905 "Note to the Academy" arrived in Bern, in time, by June 12 or 13, and it was a part of Einstein's job to read it. According to his friends Maurice Solovine and Carl Seelig, Einstein read Poincaré's book "La Science et l'Hypothèse" (no absolute time, no absolute space, no ether. . .) during the years 1902 - 1904. This book was discussed at their lecture circle "Olympia Academy" over several weeks (Ref. 8 pages 129 and 139, and Ref. 9).

Nevertheless, even if the Principle of Relativity could be called Poincaré's Principle and even if Einstein was not the first, we owe him not only the General Relativity of 1916 but also a magnificent popularization of the Special Relativity. This is fortunate because the health of Poincaré was bad and he didn't survive much to his gigantic work[3].

[3] Poincaré became ill in 1909 and died in 1912 at the age of 58.

The bad health of Poincaré and the absence of reference in Einstein's 1905 paper are of course not the only reasons for which Poincaré is so little known and Einstein so famous.

If a great physicist such as Paul Langevin (who discussed Physics with Poincaré during the weeks of their trip to Saint-Louis in 1904) had supported Poincaré the evidence would have been recognized immediately.

If Poincaré had had the possibility of publishing in a famous physics publication, such as the Einstein paper in the "Annalen der Physik", he would have had a large audience. But he only found the "Rendiconti del Circolo Matematico di Palermo" for his July 1905 paper... a minor journal of Mathematics which was not known among physicists.

It seems incredible that Poincaré had almost no possibility of publishing in a journal of Physics, but the physicists of this time refused to consider that this prodigious mathematician was also one of them. Even today some physicists still believe that the physical character of the variables x' and t' of the Lorentz transformation was not sufficiently emphasized by the author of the Principle of Relativity !

Furthermore, and last but not least, most professors of the French Universities were politically from the left during this time of hard political confrontation (Dreyfus affair, separation of Church and State). They refused to support Henri Poincaré who was identified with his cousin Raymond Poincaré, leader of the right and future President of the French Republic. This is perhaps the true reason for the silence of Paul Langevin.

Poincaré was not a self-publicist. He had given to Lorentz more than his share of credit, which was fair-mindedly refused by Lorentz. He had called Fuchsian functions, functions of Professor Fuchs, some functions for which he had completed two thirds of the work...

Fortunately the friendship of Lorentz saved him. In 1921, after the triumph of the eclipse of 1919, the Nobel committee met and initially decided : "We must give the Nobel prize to Einstein for Relativity". But Lorentz, Nobel prize 1902 for Physics, protested : "This is unfair !" and he published the notice on Poincaré's life he had written in 1914 (Ref. 10)... "I have not established the Relativity Principle as rigorously and universally true. On the contrary Poincaré has obtained a perfect invariance (of Maxwell's equations) and has expressed the Postulat of Relativity (that is the Relativity Principle), expression that he was the first to use..." The embarrassed Nobel committee decided to think the subject over carefully and a few months later finally gave the Nobel prize to Einstein, not for the Relativity...but for the photoelectric effect !

Thus, in spite of his modesty and his shyness, Henri Poincaré must be considered not only as an excellent philosopher of science[4] and one of the greatest mathematicians. He is also a major physicist (electromagnetism and radio, op-

[4] For instance : "The definition that are good for theoreticians are not necessarily the best for children and students" (Ref. 16). If we had not forgotten that Poincaré's wise statement we would not have taught "modern mathematics" to millions of unfortunate pupils...

tics, fluorescence, kinetic theory of gas, quanta, etc...), the father of the Relativity Principle and the founder of the Special Relativity.

References

[1] Lorentz H.A.: 1904, Electromagnetic phenomena in a system moving with any velocity less than that of light, *Proc. Royal Acad. Amsterdam*, **6**, 809.

[2] Poincaré H.: 1902, *La Science et l'Hypothèse*, Paris - Flammarion.

[3] Poincaré H.: 1905, June 5, Sur la dynamique de l'électron, *Comptes rendus de l'Académie des Sciences de Paris*, **140**, 1504–1508.

[4] Poincaré H.: 1898, La mesure du temps, *Revue de métaphysique et de morale*, **6**, 371–384.

[5] Poincaré H.:1906, Sur la dynamique de l'électron, *Rendiconti del Circolo Matematico di Palermo*, **21**, 129–175, received July 23, 1905, published in January 1906.

[6] Einstein A.: 1905, Zür Elektrodynamik der bewegten Körper, *Annalen der Physik*, **17**, 891–921, received June 30, 1905, published September 26, 1905.

[7] Leveugle J.: 1994, *Poincaré et la Relativité*, La Jaune et la Rouge, Avril, p.31–51.

[8] Miller A.I.: 1981, *Albert Einstein's Special Theory of Relativity*, Addison - Wesley Publishing Compagny Inc., Reading Mass.

[9] Solovine M.: 1956, *Lettres à Maurice Solovine*, Paris Gauthier - Villars ed. VIII.

[10] Lorentz H.A.: 1921, Deux mémoires de Henri Poincaré dans la Physique mathématique, *Acta Matematica*, **38**, 298.

[11] Poincaré H.: 1904, L'état actuel et l'avenir de la physique mathématique, *Bulletin des Sciences Mathématiques*, **28**,(2e série (reorganized 39-1)), 306–904.

[12] Tonnelat M.A.: 1971, *Histoire du Principe de Relativité*, Paris, Flammarion.

[13] Ginzburg V.L.: 1979, *On the theory of relativity*, Nauka, Moscow.

[14] Bol'shaia Sovetskaïa Entsiklopedia: 1974, Great Soviet Encyclopedia - A translation of the third edition, Volume 18, Macmillan, Inc New-York, Collier Macmillan Publishers London : *see Relativity, Theory of*, 653.

[15] Pauli W.; Kottler, F.:*Encyklopädie der Mathematischen Wissenschaften*, Leipzig - Verlag und Druck Von BG Teubner, see *Relativitätstheorie V-2 - p. 545-546 (1904-1922)* and *Gravitation und Relativitätstheorie VI-2-2 - p. 171 (1922-1934)*.

[16] Poincaré H.: 1908, *Science et Méthode*, Paris - Flammarion, p. 123.

PERIODIC ORBITS IN THE SITNIKOV PROBLEM

JOSEF KALLRATH
BASF - AG
ZX/ZC - C13, D-67056 Ludwigshafen, GERMANY
e-mail: kallrath@zx.basf-ag.de

RUDOLF DVORAK
Institut für Astronomie der Universität Wien
Türkenschanzstr. 17, A-1180 Wien, AUSTRIA
e-mail: dvorak@astro.ast.univie.ac.at

and

JOHANNES SCHLÖDER
IWR der Universität Heidelberg
Im Neuenheimerfeld 368, D-69120 Heidelberg, GERMANY
e-mail: j.schloeder@iwr.uni-heidelberg.de

Abstract. Periodic solutions of differential equation systems play an important rôle in the analysis of dynamical systems. Extensive research in the restricted three body problem has been devoted to the task of determining periodic solutions. This paper presents a method to determine the period and a parameterization of periodic solutions for the Sitnikov configuration. Periodic orbits are determined as solutions of a formally overdetermined system of equations. Both, period and initial values, are computed by relaxation of these equations and applying a boundary value problem approach [Bock (1981,1987)] designed for least squares parameter estimation in systems of differential equations. We determine periodic orbits in the $n : m$ resonances (n is the number of periods of the third massless body, m is the number of periods of the primaries revolution) for $n < 10$, $m < 10$, and $|n - m| \leq 4$. Tables and plots show the location of the periodic orbits in the phase space and how it depends on the eccentricity of the primaries' orbit.

Key words: periodic solutions – Sitnikov problem

1. Introduction

The knowledge of periodic solutions of differential equation systems is essential for the analysis of dynamical systems and especially important in celestial mechanics. Extensive research in the restricted three body problem has been devoted to the problem of finding periodic orbits (=PO): see for instance Broucke (1969), Shelus and Kumar (1970), Dvorak and Kribbel (1989), or Hadjidemetriou (1984). Our paper deals with the computation of PO in the Sitnikov problem which is defined as follows: a massless body m_3 moves along the z-axis through the barycenter of the primaries which are assumed to be of equal masses. Due to the symmetry conditions the motion of m_3 is confined to the z-axis perpendicular to the orbital plane of m_1 and m_2.

In order to investigate the phase space of the Sitnikov problem and to find periodic solutions we use a general algorithm described in Appendix A. For a review and literature list on the numerical computation of periodic solutions in general we refer the reader to Reithmeier (1991). Our approach is similar to the one found in Deuflhard (1984). We can compute both, periods and periodic solutions of

R. Dvorak and J. Henrard (eds.), The Dynamical Behaviour of our Planetary System, 415.

dynamical autononomous and nonautonomous systems. The method is based on a boundary value approach to solve nonlinear least squares problems in differential equation systems invented by Bock (1981,1987); a brief review of this method is also found in Kallrath *et al.* (1993). In contrast to Deuflhard's program, we use the software package *PARFIT* developed by Bock and co-workers [*cf.* Bock (1987)].

2. The Sitnikov Problem

The Sitnikov problem defined above can also be regarded as a special case of the three-dimensional elliptic restricted three-body problem in which the massless body oscillates through the barycenter of the primaries. The primaries have equal masses and move on orbits with eccentricities $e \in [0, 1)$.

For describing the motion of the primaries in the elliptic problem we follow the units and nomenclature found in Broucke (1969), *i.e.*, we use the eccentricity e and the normalized mass μ (note that in the Sitnikov problem we have $\mu = 0.5$) of the smaller body as the parameters of the problem and the semi-major axis, $a \equiv 1$, of the relative orbit. With respect to the barycentric inertial frame of reference, the coordinates of the primaries are given by

$$\begin{aligned} \xi_1 &= -\Delta \cos v = -(\cos E - e) \quad , \quad \Delta = 1 - e\cos E = \frac{p}{1 + e\cos v} \qquad (1) \\ \eta_1 &= -\Delta \sin v = -\sqrt{1-e^2}\sin E \\ \xi_2 &= +\Delta \cos v = (\cos E - e) \\ \eta_2 &= +\Delta \sin v = \sqrt{1-e^2}\sin E \end{aligned}$$

where Δ is the distance between both primaries, E is the eccentric anomaly, v is the true anomaly, and the semilatus rectum $(1 - e^2)$ has been called p. The eccentric anomaly E is related to the time t through Kepler 's equation

$$t + \chi = E - e \sin E \quad . \qquad (2)$$

The phase constant $\chi = 0$ corresponds to a minimum elongation (periapsis) of the two primaries at $t = 0$, whereas $\chi = \pi$ corresponds to maximum elongation (apoapsis) which we chose as initial condition of the primaries.

The massless body moves along the z-axis through the barycenter of the primaries. Therefore, any net force upon the massless body in x- and y-direction vanishes. Thus we only need to consider the equation in z-axis. In inertial barycentric coordinates with the time t as the independent variable the equations of motion reads as

$$\ddot{z} = -\left(\frac{m_1}{r_1^3} + \frac{m_2}{r_2^3}\right) \cdot z$$

with

$$r_1 = \sqrt{\xi_1^2 + \eta_1^2 + z^2} \quad , \quad r_2 = \sqrt{\xi_2^2 + \eta_2^2 + z^2} \quad . \qquad (3)$$

Sincc $m_1 = m_2$ we observe

$$\xi_1^2 + \eta_1^2 = \xi_2^2 + \eta_2^2 \quad , \quad r_1 = r_2 \tag{4}$$

and thus define for convenience the distance, R,

$$R := \sqrt{\xi_1^2 + \eta_1^2} = \sqrt{\xi_2^2 + \eta_2^2} \tag{5}$$

of the primaries from the z-axis measured in the orbital plane of the primaries, and the distance, r, of m_3 from the primaries

$$r := \sqrt{R^2 + z^2} = \sqrt{R^2 + x_1^2} \quad . \tag{6}$$

By putting $(x_1, x_2) = (z, \dot{z})$ we get the first order system

$$\begin{aligned} \dot{x}_1 &= x_2 \\ \dot{x}_2 &= -\frac{1}{r^3} \cdot x_1 \quad . \end{aligned} \tag{7}$$

Changing the independent variable from time, t, to the true anomaly, v, leads to

$$\dot{x} = \frac{\mathrm{d}x}{\mathrm{d}t} = \frac{\mathrm{d}x}{\mathrm{d}v} \cdot \frac{\mathrm{d}v}{\mathrm{d}t} = x' \cdot \frac{\mathrm{d}v}{\mathrm{d}t} = x' \cdot \dot{v} = x' \cdot \frac{\sqrt{p}}{\Delta^2} \tag{8}$$

and thus to

$$\begin{aligned} x_1' &= \frac{\Delta^2}{\sqrt{p}} \cdot x_2 \\ x_2' &= -\frac{\Delta^2}{\sqrt{p}} \cdot \frac{1}{r^3} \cdot x_1 \quad . \end{aligned} \tag{9}$$

Note that independent on whether we use the time or the true anomaly we are studying a dynamical system with one degree of freedom dynamical. In both cases the right-hand side depends periodically on time.

3. Phase Space Structure

3.1. The circular case

When the two primary bodies move on a circle the problem is mathematically equivalent to a special case of the so–called two fixed center problem which was already integrated by means of quadratures by Euler (1764). MacMillan (1913) found solutions of the circular Sitnikov problem in terms of elliptic integrals in Legendre's normal form [see for instance Stumpff (1963)]. All the bounded motions in the circular case are periodic orbits or quasi periodic ones on a 2 dimensional torus. Fig. 1 shows for some initial conditions the Poincaré surfaces of section (=SOS) which are defined as usual as the crossing point of the orbit in phase space

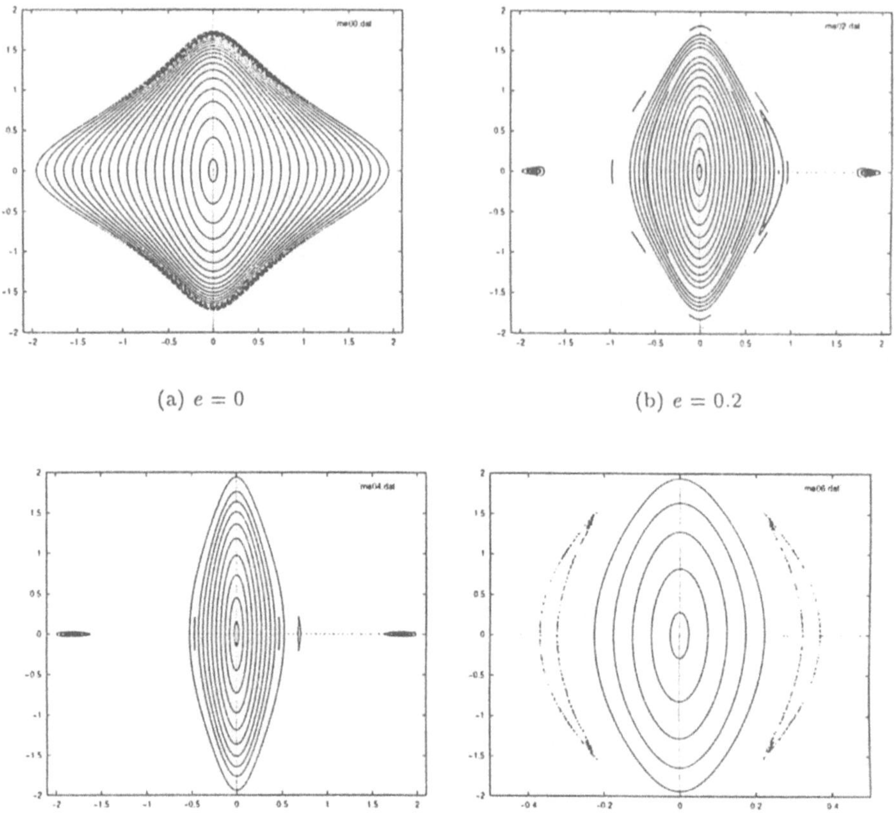

(a) $e = 0$ (b) $e = 0.2$

(c) $e = 0.4$ (d) $e = 0.6$

Fig. 1. Phase space structure in the circular case (upper left); $e = 0.2$ (upper right), $e = 0.4$, (lower left) and $e = 0.6$ (lower right). The surface of section is defined as z versus $\dot{z}$, the primaries are in pericenter position. Note how the Poincaré surfaces of section change with the eccentricity of the primaries 'orbit. In Fig. 1d ($e = 0.6$) the scale is different to be able to distinguish the islands outside the invariant curves.

with a special chosen plane. (In this case we plotted the distance and the velocity of the third body for multiples of the primaries' orbit to be able to compare it with the elliptic problem.) The integrable case of the Sitnikov Problem can be used as a starting point in perturbation analysis [see for instance Wodnar (1991), Hagel (1992, 1993)] of the elliptic Sitnikov problem with small e.

In the circular problem it is possible to find periodic orbits, *i.e.*, initial conditions, for any period T. The $n : m$ resonance solutions are $T = 2\pi \cdot m$-periodic solutions with an additional symmetry property. They can be composed by n identical T^0-periodic solutions with

$$T^0 = \frac{1}{n} \cdot 2\pi \cdot m \quad . \tag{10}$$

The graph $z(t)$ in the time interval $[0, T^0]$ looks the same as the graph in $[k \cdot T^0, (k+1) \cdot T^0]$ or more general

$$z(t) = z(\tau) \quad , \quad \dot{z}(t) = \dot{z}(\tau) \quad , \quad \tau = t - T^0 \cdot int\left(\frac{t}{T^0}\right) \quad . \tag{11}$$

Thus, each $n : m$ resonance solution with period $T = 2\pi \cdot m$ is also a T^0-periodic solution. This property is later used in a homotopy method initializing and tracking the desired $n : m$ resonance from $e = 0$ to $e = 1$. In the circular case, for a given $n : m$ resonance z_0 is computed such that

$$z(0) = z(T^0) = z_0 \quad , \quad \dot{z}(0) = \dot{z}(T^0) = 0 \quad . \tag{12}$$

This procedure provides numerical advantages since it is sufficient to consider a smaller time interval.

3.2. The Elliptic Case

In a fundamental paper, Sitnikov (1960), for the first time, demonstrated the existence of so-called oscillatory motion in the three body problem for m_3 small compared to the primaries. Later on much work has been devoted to that problem, setting the mass of the third body $m_3 = 0$. Moser (1973) constructed a mapping and proved the existence of motions which are widely cited as typical for a chaotic system: according to the initial conditions one can find orbits which cross the orbital plane of the primaries consecutive only after any sequence of integer numbers of primary periods (there exist a lower limit N_0 of this number depending on the eccentricity of the primaries' orbit). Liu and Sun (1990) used a mapping method to investigate the phase space structure for high energies (large velocities of m_3) and small eccentricities of the primaries orbit. Alfaro and Chiralt (1993) found that the fixed point at the center ($z = \dot{z} = 0$) may become unstable for certain values of e.

Numerical investigation were undertaken by Dvorak (1993) using the technique of Poincaré surfaces of section to visualize the phase space structures depending on e. Dvorak *et al.* (1993) gave a review on most of the available material dealing with Sitnikov's problem.

Fig. 1 shows how the Poincaré surface of section changes with the eccentricity of the primaries' orbit. In the elliptic case we used that point in time when the primaries are in their pericenter position to plot the distance z and the velocity $\dot{z}$ of the third body. The region of regular motion is more and more (with increasing eccentricity) confined to an area around the center fixed point. Larger and smaller islands appear and disappear. They are connected to periodic orbits. The determination of the location of those orbits is exactly the topic of this study.

In the elliptic case the mutual distance of the primaries and therefore the distance from the barycenter periodically varies with time as described by equation (1). Consequently in general the distances of the primaries to the third body are different for two consecutive crossings of m_3 through the orbital plane of the primaries. It is evident that then different forces will act on m_3 and therefore energy is pumped into or extracted from its motion.

This matter is well visible in Fig. 2: always during such a crossing of the plane the third body jumps from one orbit (closed curve in the $z-\dot{z}$ plane corresponding to an orbit in the circular case) to another one (Fig. 2a). It is also visible on Fig. 2b where the position relative to the orbital plane of the primaries is plotted versus the time (in radians as units of the phase of the primaries). The character of the 8π-periodic curve changes during such a barycentric passage. In fact the change of energy during the crossing of the primaries' orbit depends on the difference of the direction of motion of the primaries with respect to their barycenter. When the primaries approach their pericenter positions before the crossing of the massless body and still approach after the crossing the primaries gravitational forces act such that the deceleration is larger after the crossing than the acceleration before the crossing. As a consequence the third body loses velocity and the orbit "jumps" to a low energy orbit (corresponding to a different orbit in the case with $e = 0$) which causes z_{max} to be smaller than before the crossing (point A in Fig. 2). When, – vice-versa – before and after the crossing the primaries are moving away from their pericenter, the acceleration before the crossing will be larger than the deceleration after the crossing and the new orbit will lead to a larger z_{max} and also a larger period of m_3 (point B in Fig. 2).

Note that in the 7 : 4 resonance the position of m_3 is exactly the same after 4 primary periods (8π, point C in Fig. 2). Also the variety of symmetries of that orbit is remarkable.

4. The Rôle of the Periodic Orbits

Periodic orbits are often called the backbone of the phase space of dynamical systems. The fact that a stable PO is surrounded by a n-dimensional torus with quasiperiodic motions on its surface underlines the importance of POs. We calculated – using a boundary value problem approach described in Appendix A – the POs in the elliptic problem starting from the periodic orbits in the circular case. POs in the elliptic case are so-called resonance orbits and we concentrated on resonant POs $m : n$ with $m < 10$, $n < 10$ and $|n - m| \leq 4$ (see Table). Note that in the Sitnikov problem we can have periodicity only for resonant motion between the primaries and the body m_3. We have to distinguish in what concerns the starting position of the primaries: peri- or apocenter [$\chi = 0$ and $\chi = 180^o$, formula (2)] but in this paper we concentrate only on pericenter starting positions.

In Fig. 3 we plotted the locations of some selected POs in the SOS and visualized also some orbits in the vicinity of the POs: which may be either small islands around a fixed point in the SOS (=PO in phase space) or points close to a separatrix. The abscissa shows z while the ordinate shows $\dot{z}$.

TABLE I

The table lists $n : m$ resonances where m gives the number of revolutions of the primaries, and n counts the number of cycles of m_3. Integrations start with primaries at periapsis, $\dot{z}_0 = 0$ and z_0 as listed. The period of an $n : m$ resonance solution is $T = 2\pi \cdot m$. Although we have calculated many more POs, for reasons of space we show only resonances up to order 4 *i.e.* $|n - m| \leq 4$.

Resonance	z_0 for				
	$e = 0.0$	$e = 0.2$	$e = 0.4$	$e = 0.6$	$e = 0.8$
1:1	1.043698	0.872719	0.691427	0.496293	0.278653
1:2	1.848460	1.836787	1.816155	1.786429	1.747073
1:3	2.495394	2.924811	3.024252	3.048240	3.040044
1:4	3.064713	3.071739	3.071158	3.062020	3.042885
1:5	3.584274	3.137770	3.090335	3.067454	3.043343
2:1	0.449870	0.384097	0.316117	0.241345	0.150484
2:3	1.476417	1.077935	0.782990	0.530543	0.285864
2:5	2.184436	2.901572	3.022219	3.048070	3.040039
3:2	0.684979	0.582443	0.471090	0.347991	0.204884
3:4	1.340935	1.048001	0.778327	0.530061	0.285847
3:5	1.605885	1.082559	0.783083	0.530549	0.285864
3:7	2.075629	2.901207	3.022216	3.048070	3.040039
4:3	0.783862	0.663455	0.530493	0.382939	0.215804
4:5	1.270330	1.014006	0.763980	0.526762	0.285559
4:7	1.667973	1.082631	0.783083	0.530549	0.285864
5:2	0.245698	0.212414	0.182814	0.149810	0.103043
5:3	0.598686	0.509953	0.413897	0.305129	0.173718
5:4	0.839599	0.707161	0.558066	0.393830	0.217167
5:6	1.226913	0.997069	0.773163	0.526713	0.285558
5:7	1.395983	1.067891	0.778196	0.530060	0.285847
5:8	1.554551	1.082047	0.783076	0.530549	0.285864
5:9	1.704871	1.082632	0.783083	0.530549	0.285864
6:5	0.875536	0.734193	0.572832	0.397759	0.217382
6:7	1.197489	0.982746	0.790489	0.534720	0.285553
7:3	0.315207	0.270387	0.227101	0.179633	0.118175
7:4	0.559058	0.476573	0.388453	0.289970	0.170865
7:5	0.742504	0.630211	0.508833	0.376279	0.215609
7:6	0.900672	0.752193	0.580559	0.398868	0.217406
7:8	1.176222	0.973322	0.758222	0.526398	0.285553
7:9	1.300847	1.019860	0.764013	0.526762	0.285559
8:5	0.631957	0.537764	0.434143	0.314328	0.174133
8:7	0.919253	0.764799	0.569429	0.474038	0.217410
8:9	1.160130	0.965837	0.757207	0.526371	0.285553
9:5	0.536183	0.457236	0.373472	0.280310	0.167252
9:7	0.815115	0.688139	0.545564	0.385902	0.215837
9:8	0.933553	0.773889	0.586932	0.399342	0.217410

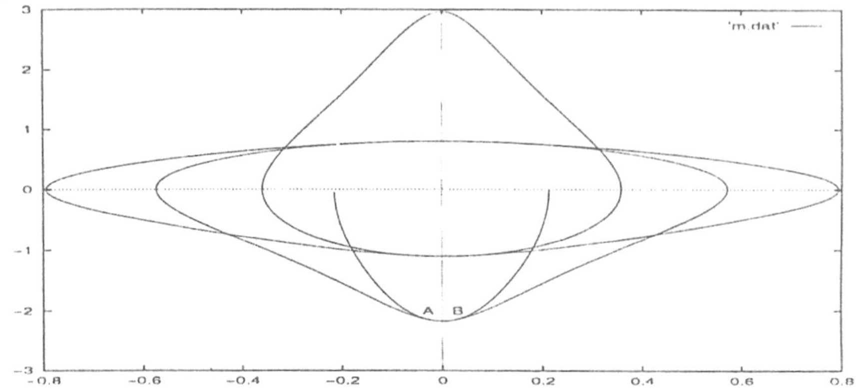

(a) Projection of the orbit on the z versus $\dot{z}$ plane

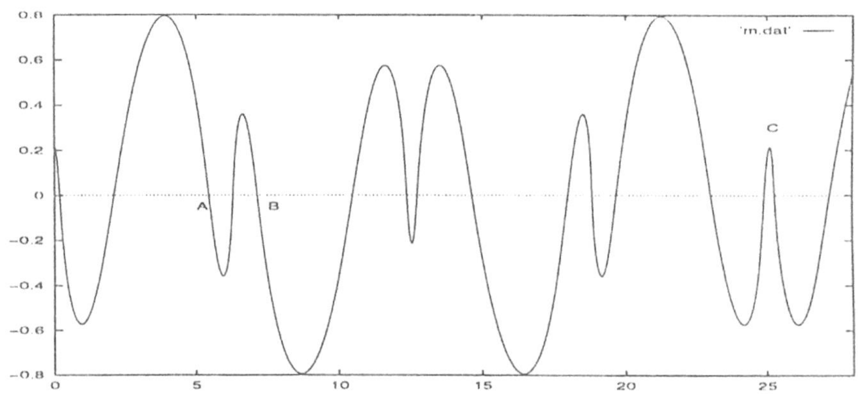

Fig. 2. A periodic orbit in the elliptic Sitnikov problem, upper picture projection of the orbit on the z versus $\dot{z}$ plane; lower picture distance from the barycenter versus the time. The orbit is the $7:4$ resonance with $e = 0.732$ for pericenter initial conditions of the primaries and $z_0 = 0.214725$ and $\dot{z}_0 = 0$. Point A: m_3 looses energy, Point B: m_3 gains energy, point C: after the time 4π m_3 finishes the 7:4 resonant periodic orbit

5. Characteristic Diagram: $z_0\,(e;\ n:m)$

We calculated periodic orbits for various initial conditions $x_{10} = z_0$, $x_{20} = 0$ and eccentricities $e \in [0, 1)$ using variable numbers of equidistant multiple shooting nodes between 81 and 401. The primaries were started in periapsis position ($\chi = 0$). For this experiment we used inertial barycentric coordinates, the computations were always started with $e = 0$. To select an $n : m$ resonance, at first, we computed $T^0 = 2\pi \cdot m/n$-periodic solutions. Due to the symmetry property (11) of $n : m$ resonances in the circular case it is always possible to find such solutions. The

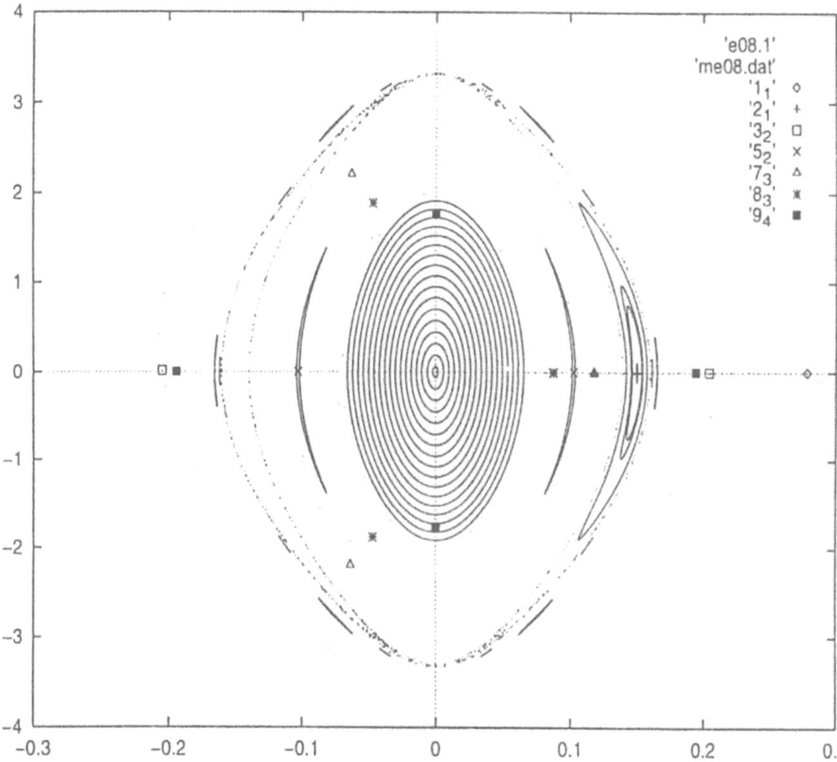

Fig. 3. SOS and position of selected PO for $e = 0.8$. Displayed are the $1:1, 2:1, 3:2, 5:2, 7:3, 8:3, 7:4$, and $9:4$ resonances for pericenter initial conditions of the primaries.

result of this computation, *i.e.*, the whole trajectory is used as input data initializing the states at the multiple shooting nodes for the next eccentricity $e^H = e + \Delta e$, $\Delta e = 0.001$, where e^H is the next homotopy target eccentricity. For the computation of solutions for higher eccentricities even smaller step-sizes, say $\Delta e = 0.0001$ are sometimes neccessary. Note that the homotopy includes e as a free parameter which then is forced to fulfill the equation

$$e - e^H = 0 \quad . \tag{13}$$

In this manner we set up a homotopy with adaptive step size Δe controlled by the accuracy with which the periodicity constraints (16) are fulfilled. The violation of (16) is less than 10^{-7}. By monitoring changes in sign of $z(t)$ we also ensure that we 'follow' the correct resonance. The respective results are summarized in *characteristic curves* in the $e - z_0$ diagram which consist of all initial conditions representing periodic solutions in the Sitnikov problem. In Fig. 4 such characteristics $z_0(e; n : m)$ are plotted for several $n : m$ resonances. Our calculations reach up to high eccentricities larger than $e = 0.95$. But so far, we have not yet

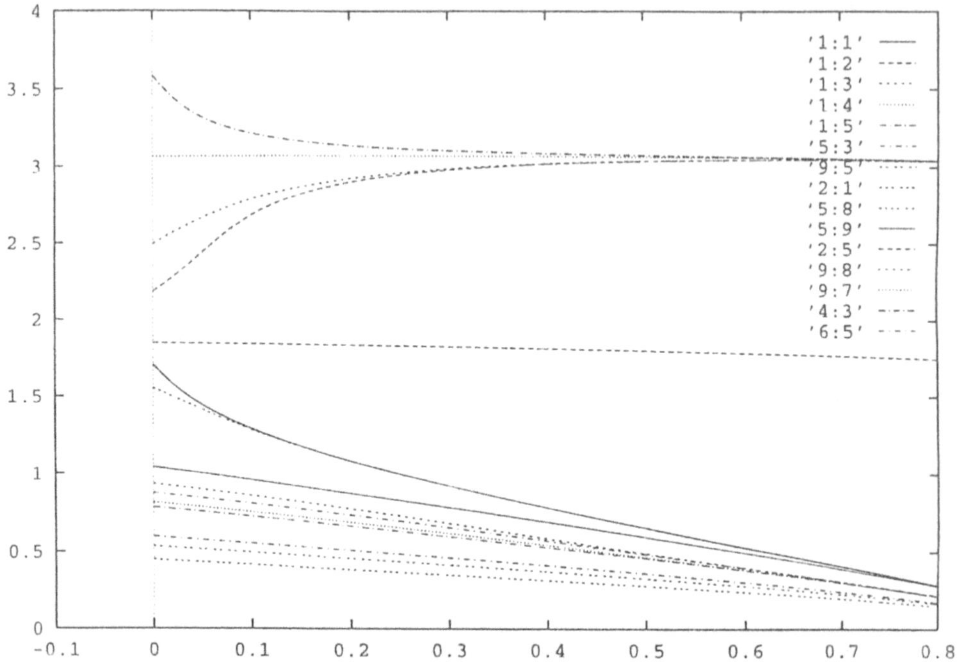

Fig. 4. Characteristic diagram with characteristic curves for some selected resonances *i.e.* z_0 as a function of eccentricity e. Note that curves never intersect. For higher eccentricities curves show an intersting asymptotic behaviour establishing two or three limit regions.

been able to determine POs for all resonances in the high eccentricity region. Therefore, we postpone a detailed discussion of the properties of curves in the characteric diagram, and in particular the asymptotic behaviour when approaching higher eccentricieties or even $e = 1$, to a forthcoming paper.

6. Conclusion and Remarks

In this paper we derived periodic orbits in the Sitnikov problem. We computed the PO and list initial conditions for resonant POs for all resonances $n : m$ up to $n < 10$, $m < 10$, and $|n - m| \leq 4$. From circular primaries' orbit on up to high eccentricities ($e = 0.9999$ in some cases) we succeeded to find the location of the PO in the surface of section representing the phase space. To be able to have detailed knowledge of the phase space structure with respect to the location of the POs we have to distinguish apocentric and pericentric initial conditions of the primaries for uneven n values in the $n : m$ resonances. Consequently two diagrams (two plots) characterize the complete phase space structure.

Our work and this paper will be continued by studying the stability of the resonant POs and investigating the limit behavior of the curves in the characteristic diagram when approaching $e = 1$.

Appendix

A. Periodic Solutions in Differential Equation Systems

A.1. Definition of the Problem

Consider a dynamical system

$$\mathbf{x} = f(t, \mathbf{x}) \quad , \quad \mathbf{x} \in \Re^n \tag{14}$$

with unknown initial values

$$\mathbf{x}_0 = \mathbf{x}(t = 0) \quad . \tag{15}$$

The solution of (14) and (15) is denoted by $\mathbf{x}(t; \mathbf{x}_0)$. The objective is to determine solutions of (14) with a positive period $T \in \Re^+$ such that

$$\mathbf{x}(T, \mathbf{x}_0) = \mathbf{x}_0 \quad . \tag{16}$$

If $f(t, \mathbf{x})$ depends explicitly on time and if this dependence is periodic then resonant orbits are the only periodic solution. This is the situation we meet in the elliptic Sitnikov problem.

Formally, on the one hand (16) describes n conditions. On the other hand, in autonomous systems, (14) and (16) provide $n + 1$ degrees of freedom: n initial coordinates and the period T. If (14) is derived from a physical system described by a conservative Hamiltonian the orbits are restricted to a smooth manifold specified by the energy. Therefore, an energy constraint

$$g_E : \Re^n \to \Re \quad , \quad g_E(\mathbf{x}_0; T) = 0 \tag{17}$$

could be added in order to determine T implicitly.

However, the remaining n conditions (16) do not uniquely determine the n initial values $\mathbf{x}_0$ since every point of a periodic orbit may serve as an initial value. Therefore, we introduce the following definition: A pair $(T, \mathbf{x}_0)$ is called a **periodic path** of (14) if the initial values $\mathbf{x}_0$ generate a solution $\mathbf{x}(t; \mathbf{x}_0)$ fulfilling the periodicity constraint (16). The set of points

$$X_T = \{\, \mathbf{x}(t; \mathbf{x}_0) \mid \dot{\mathbf{x}} = f(t, \mathbf{x}) \, , \, \mathbf{x}(T, \mathbf{x}_0) = \mathbf{x}_0 \, , \, \mathbf{x} \in \Re^n \, , \, 0 \le t \le T \,\} \tag{18}$$

generated by $(T, \mathbf{x}_0)$ is called a **periodic orbit**. If two pairs $(T, \mathbf{x}_{01})$ and $(T, \mathbf{x}_{02})$ generate the same periodic orbit X_T then there exists a shift parameter $t_\lambda \in \Re$ such that

$$\mathbf{x}(t; \mathbf{x}_{02}) = \mathbf{x}(t + t_\lambda; \mathbf{x}_{01}) \quad , \quad \forall t \quad . \tag{19}$$

The shift parameter t_λ provides an additional degree of freedom. Therefore, it is possible to add a constraint

$$g_S : \Re^n \to \Re \quad , \quad g_S(\mathbf{x}_0) = 0 \tag{20}$$

to (16). The **shift constraint** (20) may be used to shift the initial values to a convenient location, *e.g.*, to the x_1-axis by requiring $x_2 = 0$. Some care is necessary when specifying a shift constraint since it might happen that for all points of a periodic orbit (20) is violated.

In the Sitinikov problem we are interested in periodic solutions with initial points above the orbital plane of the primaries and starting with zero velocity.

Let us summarize: in the autonomous (non-autonomous) case we have a system of $n+2$ (1) nonlinear equations [(16), (17), (20)] and $n+1$ (n) unknown quantities: the initial values $\mathbf{x}_0$, and the parameter T. Under appropriate assumptions [existence and uniqueness of (14), consistency of (17) and (20), existence of periodic orbits] this system of nonlinear equations yields periodic solutions. In non-autonomous ODEs the period T is no longer a free parameter, *e.g.* in the case of the eccentric Sitnikov problem only resonance solution are possible and the initial conditions $\mathbf{x}_0$ are the only degrees of freedom. The shift constraint (20) may force one degree of freedom (usually the initial velocity) to zero. No energy constraint is applied. In contrast to the eccentric case, the circular Sitnikov problem has periodic solutions for any period fixed in advance. In that case, the energy constraint might be used, or the period could be fixed in advance.

A.2. Transformation to a Standard Problem

In order to get a fixed integration interval, e.g., $[0, 1]$, the following transformation is applied:

$$t = T \cdot \tau \quad , \quad \frac{\mathrm{d}}{\mathrm{d}t} = \frac{1}{T} \cdot \frac{\mathrm{d}}{\mathrm{d}\tau} \tag{21}$$

which leads to the problem of solving

$$\mathbf{x}' = \frac{\mathrm{d}\mathbf{x}}{\mathrm{d}\tau} = T \cdot f(T \cdot \tau, \mathbf{x}) \quad , \quad \mathbf{x} \in \Re^n \quad , \quad \tau \in [0, 1] \tag{22}$$

under the periodicity constraint

$$\mathbf{x}_1 = \mathbf{x}(\tau = 1; T, \mathbf{x}_0) = \mathbf{x}_0 \quad . \tag{23}$$

Now, T appears as an unknown parameter. Note that this transformation is not an equivalence transformation if $T = 0$ would be allowed. It has a fixed integration range as desired but unfortunately, $T = 0$ provides an unwanted trivial periodic solution $\mathbf{x}(\tau; T, \mathbf{x}_0) = \mathbf{x}_0$.

Also for non-autonomous system (in that case the period T is fixed in advance) we use this transformation.

A.3. Numerical Computation of Periodic Solutions of Differential Equation Systems

In our approach periodic solutions of ODEs are numerically computed as solutions of the relaxation of a system of boundary conditions for the differential equations model (14). We solve the resulting problem with a boundary value problem approach invented by Bock (1981,1987). This method is further described by Kallrath *et al.* (1993) and Schulz *et al.* (1997). It combines a multiple shooting technique with a generalized Gauss-Newton algorithm. The software package `PARFIT` is an implementation of this method. `PARFIT` is called by our program `PERIOD` which setups the least squares conditions, initializes a variable number of multiple shooting nodes, and controls several homotopy methods.

It was already noted that in the non-autonomous case there are $n+1$ equations to be fulfilled and on the other hand there are n degrees of freedom available. For that reason we use the following procedure: one set of equations, the periodicity conditions, are only considered in a relaxed form, i.e. they are coded as n least squares conditions

$$w_i = x_i(\tau = 1; T, \mathrm{x}_0) - x_{0i} \quad , \quad i = 1, \ldots, n \quad . \tag{24}$$

This set is companied by the shift constraint

$$g_S : \Re^n \rightarrow \Re \quad , \quad w_{n+1} = g_S(\mathrm{x}_0, T) = 0 \quad , \tag{25}$$

which we have coded as an equality condition. If one would just be interested in computing periodic solutions the shift constraint (25) is neither necessary nor recommended. However, since we are also interested in the characteristic diagram we need to find initial points which start exactly with zero velocity. Furthermore, periodic orbits resp. $n:m$ resonance solutions have $2n$ zero velocity points within the interval $[0, 2\pi m]$. Thus, the right zero velocity point is defined by the symmetric peaks of the circular periodic orbit solutions. By tracking the characteristic curves from $e=0$ towards $e=1$ we ensure that we keep the 'correct' zero velocity point.

The sum of the squared equations of condition or the sum of least squares

$$F(T, \mathrm{x}_0) = \sum_{i=1}^{n} w_i^2(T, \mathrm{x}_0) \tag{26}$$

identifies whether a periodic solution exists or not. If a solution $(T > 0)$ is found with $F(T, \mathrm{x}_0) = 0$ then (T, x_0) represents a periodic solution. Otherwise, if it is not possible to find a solution (T, x_0) with $F(T, \mathrm{x}_0) = 0$, this might indicate that no periodic solutions exist.

Acknowledgements

The authors greatfully acknowledge helpful comments of H.G. Bock and that he provided the package `PARFIT` to us.

References

Alfaro, J. and Chiralt, C.: 1993, 'Invariant rotational curves in sitnikov's problem', *Celestial Mechanics and Dynamical Astronomy*, **55**, 351–367.

Bock, H. G.: 1981, 'Numerical treatment of inverse problems in chemical reaction kinetics'. In *Modelling of Chemical Reaction Systems* (Ebert *et al.* eds.), Series in Chemical Physics, pages 102–125, Springer, Heidelberg.

Bock, H. G.: 1987, 'Randwertproblemmethoden zur Parameteridentifizierung in Systemen nichtlinearer Differentialgleichungen', Preprint 142, Universität Heidelberg, SFB 123, Institut für Angewandte Mathematik, 69120 Heidelberg.

Broucke, R.: 1969, 'Periodic orbits in the elliptic restricted three body problem', JPL Technical Report 32-1360, Jet Propulsion Laboratory, Pasadena, California.

Deuflhard, P.: 1984, 'Computation of Periodic Solutions of Nonlinear ODEs', *BIT*, **24**, 456–466.

Dvorak, R.: 1993, 'New Numerical Results to the Sitnikov Problem'. In *Qualitative and Quantitative Behaviour of Planetary Systems* (Dvorak and Henrard eds.), Dordrecht, Kluwer Academic Publishers.

Dvorak, R. and Kribbel, J.: 1992, 'A catalogue of periodic orbits in the elliptic restricted -body problem'. In *Chaos, Resonance and Collective Dynamical Phenomena in the Solar System* (Ferraz-Mello ed.), number 23 in IAU Symposium, pages 171–174.

Dvorak, R., Vrabec, F., and Wodnar, K.: 1993, 'The Sitnikov Problem: A Short Review'. In *Proceedings of Primo Convegno Nazionale di Meccania Celeste*, pages 16–22, Università di L'Aquila.

Euler, L.: 1764a, 'De motu corporis ad duo centra virium fixa attracti', *Nov. Comm. Petrop*, **10**, 207.

Euler, L.: 1764b, 'De motu corporis ad duo centra virium fixa attracti', *Nov. Comm. Petrop*, **11**, 152.

Hadjidemetriou, J.: 1984, 'Periodic orbits', *Celestial Mechanics and Dynamical Astronomy*, **34**, 379–393.

Hagel, J.: 1992, 'A New Analytic Approach to the Sitnikov Problem', *Celestial Mechanics*, **53**, 267–292.

Hagel, J.: 1993, 'A Computer Aided Analysis of the Sitnikov Problem'. In *Qualitative and Quantitative Behaviour of Planetary Systems* (Dvorak and Henrard eds.), Dordrecht, Kluwer Academic Publishers.

Kallrath, J., Schlöder, J. P., and Bock, H. G.: 1993, 'Least squares parameter estimation chaotic differential equations', *Celestial Mechanics and Dynamical Astronomy*, **56**, 353–371.

Liu, J. and Sun, Y.-S.: 1990, 'On the Sitnikov Problem', *Celestial Mechanics*, **49**, 285–302.

MacMillan, W. D.: 1913, 'An integrable case in the restricted problem of three bodies', *Astron. J.*, **27**, 11–13.

Moser, J.: 1973, *Random and Stable Motion in Dynamical Systems*, Princeton University Press.

Reithmeier, E.: 1991, *Periodic Solutions of Nonlinear Dynamical Systems*, Lecture Notes in Mathematics, Springer, Berlin.

Schulz, V., Bock, H. G., and Steinbach, M.: 1997, 'Exploiting Invariants in the Numerical Solution of Multipoint Boundary Value Problems for DAE', *Siam J. Sci. Comp.*, **submitted**.

Shelus, P. and Kumar, S.: 1970, 'Some periodic orbits in the elliptic problem of three bodies', *Astron. J.*, **75**, 4.

Sitnikov, K.: 1960, 'Existence of Oscillatory Motions for the Three-Body Problem', *Dokl. Akad. Nauk*, **133**, 303–306.

Stumpff, K.: 1965, *Himmelsmechanik II*, VEB, Berlin.

Wodnar, K.: 1991, 'New Formulations of the Sitnikov Problem'. In *Predictability, Stability and Chaos in N-Body Dynamical Systems* (Roy ed.), New York, Plenum Press.

MIX
Papier aus verantwortungsvollen Quellen
Paper from responsible sources
FSC® C105338

If you have any concerns about our products,
you can contact us on
ProductSafety@springernature.com

In case Publisher is established outside the EU,
the EU authorized representative is:
Springer Nature Customer Service Center GmbH
Europaplatz 3, 69115 Heidelberg, Germany

Printed by Libri Plureos GmbH
in Hamburg, Germany